D1683462

Energetic Materials
Edited by U. Teipel

Also of Interest

Kubota, N.

Propellants and Explosives
Thermochemical Aspects of Combustion

2002, ISBN 3-527-30210-7

Hattwig, M., Steen, H. (Eds.)

Handbook of Explosion Prevention and Protection

2003, ISBN 3-527-30718-4

Meyer, R., Köhler, J., Homburg, A.

Explosives

Fifth, Completely Revised Edition

2002, ISBN 3-527-30267-0

Energetic Materials

Particle Processing and Characterization

Edited by Ulrich Teipel

WILEY-VCH Verlag GmbH & Co. KGaA

Prof. Dr.-Ing. Ulrich Teipel
Fraunhofer Institut
Chemische Technologie
Joseph-von-Fraunhofer-Str. 7
76327 Pfinztal (Berghausen)
Germany

1st Edition 2005
1st Reprint 2005

All books published by Wiley-VCH are carefully produced. Nevertheless, authors, editor, and publisher do not warrant the information contained in these books, including this book, to be free of errors. Readers are advised to keep in mind that statements, data, illustrations, procedural details or other items may inadvertently be inaccurate.

Library of Congress Card No.: Applied for

British Library Cataloguing-in-Publication Data
A catalogue record for this book is available from the British Library.

Bibliographic information published by Die Deutsche Bibliothek
Die Deutsche Bibliothek lists this publication in the Deutsche Nationalbibliografie; detailed bibliographic data is available in the Internet at <http://dnb.ddb.de>

© 2005 WILEY-VCH Verlag GmbH & Co. KGaA, Weinheim

Printed on acid-free paper.

All rights reserved (including those of translation in other languages). No part of this book may be reproduced in any form – by photoprinting, microfilm, or any other means – nor transmitted or translated into machine language without written permission from the publishers. Registered names, trademarks, etc. used in this book, even when not specifically marked as such, are not to be considered unprotected by law.

Composition Typomedia GmbH, Ostfildern
Printing Strauss GmbH, Mörlenbach
Bookbinding Litges & Dopf Buchbinderei GmbH, Heppenheim

Printed in the Federal Republic of Germany.

ISBN 3-527-30240-9

Table of Contents

Energetic Materials. Edited by Ulrich Teipel
Copyright © 2005 WILEY-VCH Verlag GmbH & Co. KGaA, Weinheim
ISBN: 3-527-30240-9

Preface

Nowadays, propellants, explosives and pyrotechnics are composed mainly of particulate energetic materials. These are gaining in importance as a way to optimize the performance, burning behavior, stability, detonation properties, processing characteristics and, above all, the low sensitivity of these systems. By varying the characteristic profile of these materials, product design provides for particulate components specially optimized for the application in question. Well-known solid formation processes are often used to create particles of energetic materials, such as crystallization or precipitation, comminution or atomization. Although there is a certain amount of information available about these processes for particle syntheses or processing, there is currently a lack of detailed comprehension in some points, which is needed to be able to completely control the particle formation process or to make reliable predictions about the profile required by the user for the particles so created. Vital questions and tasks remaining for particle technology of energetic materials are the comprehensive characterization of particulate components, the experimental determination of the kinetics of particle formation processes, the influence of the manufacturing process on important particle characteristics, such as particle size distribution, morphology or polymorphy, the simulation and modeling of particle formation processes and particulate materials, as well as the creation of low-defect particles with regard to the sensitivity of propellants and explosives. This book is targeted at those working in industry, government or R&D, and involved in the fascinating field of energetic or other special materials. It will hopefully contribute to summarizing our current level of knowledge.

This volume begins with an introduction to the topic, with a focus on novel energetic materials. One of the main subjects, namely production, is described in chapters two to four, beginning with a look at processes used to reduce the size of the materials, followed by a detailed treatment of crystallization. After covering certain basics, the possibilities of designing hexogen, octogen, CL 20, NTO, ammonium nitrate and ammonium dinitramide in particular using crystallization are examined. In addition, the possibilities currently offered by simulation and the potential of crystallization with compressed gas are looked at. Alongside the mixing process of disperse systems important for particle processing, a whole chapter deals with the product design of particulate materials using microencapsulation and particle coating. This is followed by the increasingly important topic of nano-

Energetic Materials. Edited by Ulrich Teipel
Copyright © 2005 WILEY-VCH Verlag GmbH & Co. KGaA, Weinheim
ISBN: 3-527-30240-9

particles. The remaining chapters deal primarily with the second main topic of this book, the characterization of particle characteristics. They present the possibilities and limitation of particle size analysis, microstructures, polymorphy and morphology as well as the analysis of chemical and thermal properties and wettability. The rheological behavior of dispersions composed of particulate energetic materials and the relevant binder materials as well as that of solids are treated separately in Chapter 12. The whole is rounded off with a look at the performance of energetic materials, including the influence of particle size on reactions, that of crystal defects on the sensitivity of formulations and the diagnostics of shock wave and combustion processes.

The authors have incorporated in this work their excellent scientific expertise, knowledge and experience in particle technology as related to energetic materials. It was vital for this publication to win renowned colleagues as expert co-authors, and as its editor, I would like to thank all the authors for their willingness to work on this book. My gratitude is also extended to those who worked in different ways "in the background" for the individual authors and the editor. In particular I wish to sincerely thank Ulrich Förter-Barth and Hartmut Kröber as well as Irma Mikonsaari for their continuous and varied support in the preparation and carrying out of the project, as well as the reviewing, processing and correcting of the manuscripts. I also wish to thank the staff at Wiley-VCH for their cooperation from the start of the project until completion of the book.

Pfinztal, September 2004 *Ulrich Teipel*

List of Contributors

Editor

Prof. Dr.-Ing. Ulrich Teipel
Fraunhofer Institut für Chemische Technologie (ICT)
Department of Energetic Materials
Particle Technology
Joseph-von-Fraunhofer-Strasse 7
76327 Pfinztal, Germany

Authors

Prof. Dr. A. Yu. Babushkin
Krasnoyarsk State Technical University
26, Kirensky St.
Krasnoyarsk 660074, Russia
(Chapter 7)

Dr. Yuri A. Biryukov
Tomsk State University
Innovate-Technological Scientific-Educational Center
36, Lenin
Tomsk 634050, Russia
(Chapter 7)

Dr. Lionel Borne
French-German Research Institute of Saint-Louis (ISL)
5 Rue Du General Cassagnou
P.O. Box 34
68301 Saint – Louis, France
(Chapter 9, 13)

M. Sc. Julie K. Bremser
Material Characterization Laboratory
Los Alamos National Laboratory
P.O. Box 1663, MS G770
Los Alamos, NM 87545, USA
(Chapter 8)

Dr. G. A. Chiganova
Krasnoyarsk State Technical University
26, Kirensky St.
Krasnoyarsk 660074, Russia
(Chapter 10)

Dr. Helmut Ciezki
DLR – Deutsches Zentrum für Luft- und Raumfahrt, Raumfahrtantriebe
Lampoldshausen
74239 Hardthausen, Germany
(Chapter 13)

Dr. Norbert Eisenreich
Fraunhofer Institut für Chemische Technologie (ICT)
Department of Energetic Systems
Joseph-von-Fraunhofer-Strasse 7
76327 Pfinztal, Germany
(Chapter 13)

Dr. Jerry W. Forbes
University of California
Lawrence Livermore National Laboratory
P.O. Box 808, L – 282
Livermore, CA 94551, USA
(Chapter 13)

Dipl.-Ing. Ulrich Förter-Barth
Fraunhofer Institut für Chemische Technologie (ICT)
Department of Energetic Materials
Joseph-von-Fraunhofer-Strasse 7
76327 Pfinztal, Germany
(Chapter 12)

Dr. Alexander E. Gash
University of California
Lawrence Livermore National Laboratory
P.O. Box 808, L – 282
Livermore, CA 94551, USA
(Chapter 7)

Dipl.-Ing. Thomas Heintz
Fraunhofer Institut für Chemische Technologie (ICT)
Department of Energetic Materials
Joseph-von-Fraunhofer-Strasse 7
76327 Pfinztal, Germany
(Chapter 5)

Dr. Michael Herrmann
Fraunhofer Institut für Chemische Technologie (ICT)
Department of Energetic Materials
Joseph-von-Fraunhofer-Strasse 7
76327 Pfinztal, Germany
(Chapter 9)

Dr. D. Mark Hoffman
University of California
Lawrence Livermore National Laboratory
P.O. Box 808, L – 282
Livermore, CA 94551, USA
(Chapter 12)

Ing. Aat C. Hordijk
TNO Prins Maurits Laboratory
Research Group Pyrotechnics and Energetic Materials
Lange Kleiweg 137, P.O. Box 45
2280 AA Rijswijk, The Netherlands
(Chapter 6, 12)

Dr.-Ing. Christof Hübner
Fraunhofer Institut für Chemische Technologie (ICT)
Department of Polymer Engineering
Joseph-von-Fraunhofer-Strasse 7
76327 Pfinztal, Germany
(Chapter 12)

Dr. Manfred Kaiser
WIWEB Wehrwissenschaftliches Institut für Werk-, Explosiv- und Betriebsstoffe
Außenstelle Swisttal-Heimerzheim
Großes Cent
53913 Swisttal, Germany
(Chapter 10)

Dr. John Kendrick
ICI Technology
Wilton Research Center
P.O. Box 90
Wilton, Middlesbrough
Cleveland TS 90 8JE, UK
(Chapter 3)

Prof. Dr. Kwang-Joo Kim
Korea Research Institute of Chemical Technology
Chemical Engineering Divison
P.O. Box 107, Yuseong
Taejon 305–600, Korea
(Chapter 3)

Dr. Horst Krause
Fraunhofer Institut für Chemische Technologie (ICT)
Department of Energetic Materials
Joseph-von-Fraunhofer-Strasse 7
76327 Pfinztal, Germany
(Chapter 1)

Dipl.-Ing. Hartmut Kröber
Fraunhofer Institut für Chemische Technologie (ICT)
Department of Energetic Materials
Joseph-von-Fraunhofer-Strasse 7
76327 Pfinztal, Germany
(Chapter 3, 4)

Dr. Ronald S. Lee
University of California
Lawrence Livermore National Laboratory
P.O. Box 808, L – 282
Livermore, CA 94551, USA
(Chapter 13)

Dr. Stefan Löbbecke
Fraunhofer Institut für Chemische Technologie (ICT)
Department of Energetic Materials
Joseph-von-Fraunhofer-Strasse 7
76327 Pfinztal, Germany
(Chapter 10)

Dr. Alexey I. Lyamkin
Krasnoyarsk State Technical University
26, Kirensky St.
Krasnoyarsk 660074, Russia
(Chapter 7)

Dipl.-Ing. Irma Mikonsaari
Fraunhofer Institut für Chemische Technologie (ICT)
Department of Energetic Materials
Joseph-von-Fraunhofer-Strasse 7
76327 Pfinztal, Germany
(Chapter 2, 11)

Dr. Rudolf Nastke
Fraunhofer Institut für Angewandte Polymerforschung (IAP)
Geiselbergstraße 69
14476 Golm, Germany
(Chapter 5)

M.Sc. Kirk E. Newman
Energetic Materials and Technology Department
Naval Surface Warfare Center, Indian Head Divison
Bldg 457, Manley Road
Yorktown, VA 23691–0160, USA
(Chapter 12)

Dr. Michael Niehaus
Orica Germany GmbH
Kaiserstr.
53840 Troisdorf, Germany
(Chapter 5)

Prof. Dr. Ernesto Reverchon
Dipartimento di Ingegneria Chimica e Alimentare
Universita di Salerno
Via Ponte Don Melillo
84084 Fisciano (SA), Italy
(Chapter 4)

Prof. Dr. Eberhard Schmidt
Bergische Universität Wuppertal
Sicherheitstechnik/Umweltschutz
Rainer-Gruenter-Strasse 21
42119 Wuppertal, Germany
(Chapter 5)

Dr. Ferenc Simon
Research Institute of Chemical and Process Engineering
University of Kaposvár/Campus Veszprém
Egyetem u.2, P.O. Box 125
8200 Veszprém, Hungary
(Chapter 3)

Dr. Randall C. Simpson
University of California
Lawrence Livermore National Laboratory
P.O. Box 808, L – 282
Livermore, CA 94551, USA
(Charter 7)

M.Sc. Cary B. Skidmore
High Explosives Science and Technology
Los Alamos National Laboratory
P.O. Box 1663, MS C936
Los Alamos, NM 87545, USA
(Chapter 9)

Prof. Dr.-Ing. Ulrich Teipel
Fraunhofer Institut für Chemische Technologie (ICT)
Department of Energetic Materials
Particle Technology
Joseph-von-Fraunhofer-Strasse 7
76327 Pfinztal, Germany
(Chapter 2, 3, 4, 5, 8, 11, 12)

M.Sc. Fred Tepper
Argonide Corporation
291 Power Court
Sanford, Florida 32771, USA
(Chapter 7)

Dr. Joop ter Horst
Delft University of Technology
Laboratory for Process Equipment
Leeghwaterstraat 44
2628 CA Delft, The Netherlands
(Chapter 3)

Dr. Simon Torry
Future Systems Technology, QinetiQ
Fort Halstead
Sevenoaks
Kent, TN 14 7BP, UK
(Chapter 11)

Prof. Dr. Victor Valtsifer
Institute of Technical Chemistry
Russian Academy of Science
13, Lenin
Perm 614600, Russia
(Chapter 12)

Dr. Antoine E.D.M. van der Heijden
TNO Prins Maurits Laboratory
Research Group Pyrotechnics and Energetic Materials
Lange Kleiweg 137, P.O. Box 45
2280 AA Rijswijk, The Netherlands
(Chapter 3, 6)

Prof. Dr. Alexander Vorozhtsov
Tomsk State University
54, Belinsky
Tomsk 634050, Russia
(Chapter 7)

Prof. Vladimir E. Zarko
Institute of Chemical Kinetics and Combustion
Russian Academy of Science, Siberian Branch
Nowosibirsk 630090, Russia
(Chapter 7)

1
New Energetic Materials

Horst H. Krause

1.1
Introduction

For many years there was little discussion of or activity to develop new energetic materials for military applications. However, since the end of the Cold War there have been significant new activities in such materials. Particularly in the last 10 years, a number of new synthesized energetic materials have been reported and generated much discussion. Some of the most interesting newly developed materials include the following:

- **TNAZ** (1,3,3-**tri**nitro**az**etidine);
- **HNIW** (**h**exa**n**itrohexaza**i**so**w**urtzetane or **CL-20**);
- **ONC** (**o**cta**n**itro**c**ubane);
- **FOX-7** (1,1-diamino-2,2-dinitroethene);
- **ADN** (**a**mmonium **din**itramide). A number of questions can be raised with respect to this new generation of explosive materials:
- Do the new substances offer significant advantages compared with currently existing materials?
- What range of applications might be expected for the new materials?
- Have the new substances been sufficiently characterized and developed?
- Does processing or manufacture of these new substances pose particular compatibility or safety problems?
- Will the chemical stability and aging behavior of these new substances result in formulations with adequate service life?

1.1.1
Applications of Energetic Materials

To assess the potential of the newly developed materials, their energetic characteristics must be compared with those of contemporary materials. The values of some key characteristic properties of energetic materials, such as density, formation

Energetic Materials. Edited by Ulrich Teipel
Copyright © 2005 WILEY-VCH Verlag GmbH & Co. KGaA, Weinheim
ISBN: 3-527-30240-9

Table 1.1. Properties of existing and new energetic materials.

Abbreviation	Name	Applications[a]	Density (g/cm^3)	Oxygen balance (%)	Formation energy (kJ/mol)
Existing energetic materials					
TNT	2,4,6-Trinitrotoluene	HX	1.65	−74.0	−45.4
RDX	Cyclo-1,3,5-trimethylene-2,4,6-trinitramine	HX; RP; GP	1.81	−21.6	92.6
HMX (β)	Cyclotetramethyl-enetetranitramine	HX; RP; GP	1.96	−21.6	104.8
PETN	Pentaerythrol tetranitrate	HX	1.76	−10.1	−502.8
NTO	3-Nitro-1,2,4-triazol-5-one	HX	1.92	−24.6	−96.7
NG	Nitroglycerine	RP; GP	1.59	3.5	−351.5
NC	Nitrocellulose (13% N)	RP; GP	1.66	−31.8	−669.8
AN	Ammonium nitrate	HX; RP	1.72	20.0	−354.6
AP	Ammonium perchlorate	RP; HX	1.95	34.0	−283.1
New energetic materials					
TNAZ	1,3,3-Trinitroazetidine	HX; RP; GP	1.84	−16.7	26.1
CL-20 (HNIW)	2,4,6,8,10,12-(Hexanitro-hexaaza)tetracyclododecane	HX; RP; GP	2.04	−11.0	460.0
FOX-7	1,1-Diamino-2,2-dinitroethene	HX; GP; RP	1.89	−21.6	−118.9
ONC	Octanitrocubane	HX	1.98	0.0	465.3
ADN	Ammonium dinitramide	RP; HX; GP	1.81	25.8	−125.3

a HX, explosive; RP, solid rocket propellant; GP, gunpowder.

energy and oxygen balance, are listed in Table 1.1, providing a comparison of these new substances with currently existing energetic materials. A wide variety of materials are currently used in the energetic materials sector, some of which required decades of research and development. Table 1.1 lists the most important contemporary energetic materials and their primary application(s) in one or more of three key areas: explosives (HX), gunpowder (GP) and solid rocket propellants (RP).

Nitrocellulose remains the leading major ingredient of gunpowder formulations. One distinguishes between single- and double-base formulations, i. e. between pure nitrocellulose and a combination of nitrocellulose and a high-energy plasticizer. Triple-base formulations are also possible, where the third component is a solid energetic component such as nitroguanidine (NIGU).

Most solid rocket propellants fall into one of two classes: double-base (NC/NG) and so-called composite propellants. Composites consist of a fuel and oxidizer (e.g. aluminum and ammonium perchlorate) bound together in a polymeric matrix. Only a limited range of oxidizer candidates can be employed in solid propellants. It is therefore crucial to consider the oxidizer properties (including a positive oxygen balance) in the propellant development process. The property data shown in Table 1.1 indicate that ADN is probably the only viable oxidizer alternative to AP for solid rocket propellants; it therefore holds considerable promise for future development efforts.

The most important property for a good explosive is the material's density. CL-20 (or HNIW) has the highest density (2.04 g/cm^3) of the organic substances listed. It also exhibits the highest formation energy. This results from the stored energy, which is due to the highly stretched bonds in the ring system of this so-called 'cage' molecule.

Octanitrocubane belongs to the same group of cage compounds as CL-20. Octanitrocubane was synthesized only recently. The measured density of the ONC molecule, 1.979 g/cm^3 is unfortunately much lower than that expected based on computer-based simulation predictions.

1.2 Application Requirements

Of the materials listed in Table 1.1, the primary candidates that offer potential to increase the performance of future energetic materials are ADN, TNAZ and CL-20. The energy content of FOX-7 is comparable to that of RDX; however, it is significantly less sensitive and therefore a promising material for further development. Except for NC-based formulations, polymeric binders are used as the matrix of energetic materials. Newly developed energetic binders offer the possibility of improving the performance of composite systems. The optimal combination of solid material and binder is critical to the development of improved performance systems across the range of applications.

Based purely on chemical structure, CL-20, TNAZ and ADN are promising candidates as formulation ingredients for any of the three application areas (HX, GP or RP), because of their performance advantages. However, performance is not the only criterion that determines a material's suitability for practical application; other important considerations include the following:

- availability (and price);
- thermal and mechanical sensitivity [insensitive munitions (IM) characteristics];
- processability;
- compatibility;
- chemical and thermal stability;
- temperature-dependent mechanical behavior;
- burn rate behavior (for solid rocket propellants and propellant powders).

Each of the application areas has its own specific requirements with respect to the properties listed above.

1.2.1 Explosives

Material density clearly plays the most important role in developing a high-performance explosive. For example, the density is directly related to detonation velocity and Gurney energy of the formulation. This relationship is described by the Kamlet-Jacobs [1.1] equations:

$$D = A \cdot \left[N \cdot M^{0.5} \cdot (-\Delta H_d^\circ)^{0.5} \right]^{0.5} \cdot (1 + B \cdot \rho_0) \quad (1.1)$$

$$P_{CJ} = K \cdot \rho_0^{\,2} \cdot \left[N \cdot M^{0.5} \cdot \left(-\Delta H_d^\circ \right)^{0.5} \right] \quad (1.2)$$

where

D = detonation velocity (mm/μs);
A = 1.01;
B = 1.3;
K = 15.85;
N = mol gas per g explosive;
M = mean molecular weight of the gas (g);
ΔH_d^o = heat of detonation (cal/g);
ρ_0 = density of the explosive (g/cm^3);
P_{CJ} = detonation pressure (kbar).

In addition to the goal of high density, other primary considerations in developing explosives include processability and the ability to attain insensitive munitions (IM) characteristics. IM properties actually depend on the complete system rather than single ingredients. However, one can often make good predictions of the IM properties of the system based on single-ingredient properties such as thermal or friction/impact sensitivity. Finally, in developing a formulation one must search for the optimal balance of chemical architectures, because generally as the performance increases, so does the material's sensitivity. However, there are certain chemical combinations that yield relatively insensitive compounds despite their relatively high energetic content.

Again, it is important to distinguish between the sensitivity of individual ingredients and that of the final, tailored formulation. For instance, pure CL-20 is relatively sensitive to both impact and friction. However, as a component in a PBX formulation, the sensitivity of the resulting formulation is only slightly greater than that of a comparable formulation based on HMX. The fact that the CL-20-based formulation exhibits higher performance makes it an interesting candidate for certain applications.

In fact, the potential to develop energetic materials with IM properties is not limited to new materials. The sensitivity of well-established energetic materials can be reduced through various material improvements, such as better crystal quality, reducing crystal or molecular defects, eliminating voids, chemical impurities or the existence of multiple phases. Properties that are advantageous for IM systems include the following:

- high decomposition temperature;
- low impact and friction sensitivity;
- no phase transitions when the substance is subjected to rapid volume expansion or contraction;

- no autocatalytic decomposition;
- spherical crystal morphology;
- good adhesion of the binder matrix;
- no voids brought about by solvent or gas bubbles;
- high chemical purity;
- phase purity. Performance characteristics and IM properties of various materials are given in Tables 1.2 and 1.3.

Table 1.2. Performance characteristics of explosive components and example formulations[a].

Substance	ΔH_f (kcal/kg)	ϱ (g/cm^3)	D_{calc} (m/s)	P_{CJ} (GPa)	ΔE at $V/V_0 = 6.5$ (kJ/cm^3)	V_{gas} at 1 bar (cm^3/g)
TNT	−70.5	1.654	6881	19.53	−5.53	738
RDX	72.0	1.816	8977	35.17	−8.91	903
HMX	60.5	1.910	9320	39.63	−9.57	886
PETN	−407.4	1.778	8564	31.39	−8.43	852
TATB	−129.38	1.937	8114	31.15	−6.94	737
HNS	41.53	1.745	7241	23.40	−6.30	709
NTO	−237.8	1.930	8558	31.12	−6.63	768
TNAZ	45.29	1.840	9006	36.37	−9.39	877
CL-20	220.0	2.044	10065	48.23	−11.22	827
FOX-7	−85.77	1.885	9044	36.05	−8.60	873
ADN	−288.5	1.812	8074	23.72	−4.91	987
LX-14:						
95% HMX/5% estane	10.07	1.853	8838	35.11	−8.67	880
LX-19:						
95% CL-20/5% estane	161.6	1.972	9453	42.46	−10.07	827
Composition B:						
60% RDX/40% TNT	9.55	1.726	7936	27.07	−7.23	840
60% RDX/40% TNAZ	55.4	1.801	8827	34.16	−8.81	894
Octol:						
75 HMX/25 TNT	27.76	1.839	8604	33.54	−8.41	850
75 HMX/25 TNAZ	56.73	1.892	9237	38.69	−9.52	883

a ΔH_f = heat of formation; ρ_0 = density (TMD = total maximum density);
D_{calc} = calculated detonation velocity; P_{CJ} = calculated detonation pressure;
ΔE at $V/V_0 = 6.5$ = calculated Gurney energy at an expansion ratio of 6.5;
V_{gas} = gas volume at 1 bar of 1 g of explosive

Table 1.3. Insensitive munitions properties of existing and new energetic materials.

Substance	Friction sensitivity (N)	Impact sensitivity (N m)	Deflagration point (°C)
TNT	353	15	300
RDX	120	7.4	230
HMX	120	7.4	287
CL-20	54	4	228
TNAZ	324	6	>240
TATB	353	50	>325
FOX-7	216	15–40	>240

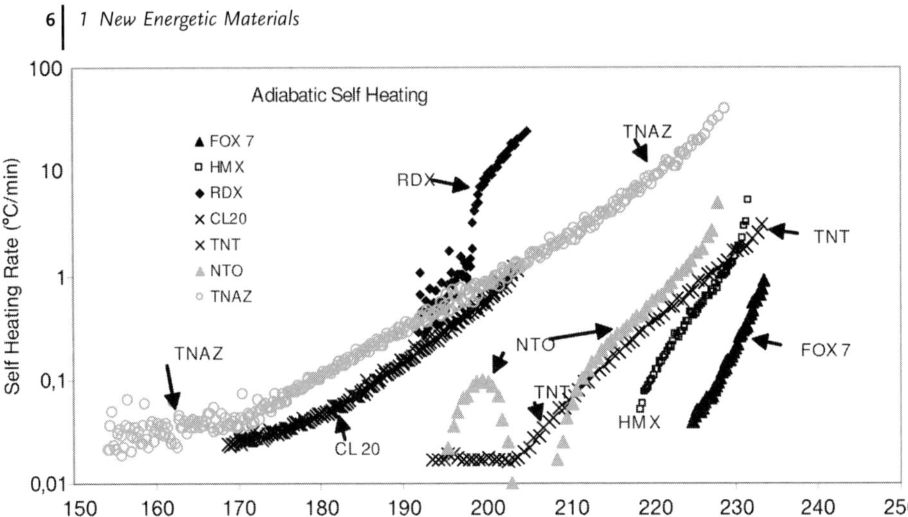

Figure 1.1. Adiabatic self-heating rate of various energetic materials.

The onset of decomposition temperature is only one criterion that impacts IM properties. Just as important is the material's heat quantity release characteristics, which cause material self-heating and accelerated decomposition. The **a**diabatic **r**ate **c**alorimeter (ARC) can be used to measure self-heating temperatures and rates, allowing these properties to be compared for different materials. Fig. 1.1 shows values of the self-heating rate for various pure compounds. The most sensitive substances tend to exhibit a low decomposition temperature and a rapid increase in self-heating rate over the temperature range examined.

It is important to distinguish between the sensitivity of raw materials and that of the finished product. A goal of modern explosive processing methods is to reduce considerably the sensitivity characteristics of the formulation compared with those that of the individual components.

In addition to active charges such as Composition B or octol, which use castable TNT more or less as a binder, **p**lastic **b**onded e**x**plosives (PBX) also consist of a polymeric binder that serves as a matrix for energetic fillers. Cleverly tailored formulations can be developed for specific applications by incorporating specially selected additives. NTO (3-nitro-1,2,4-triazol-5-one) has proven to be an important component of **i**nsensitive **h**igh **e**xplosive (IHE) formulations. Incorporating NTO significantly reduces a formulation's sensitivity, with only a relatively small decrease in performance characteristics. Nevertheless, the majority of active charge formulations currently used are still based on well-known energetic components such as TNT, PETN, RDX and HMX. Common examples include the following:

- Composition B (60% RDX/40% TNT) and
- octol (75% HMX/25% TNT);
- C4 (91% RDX/9% polyisobutylene) as explosive plasticizer;

- A5 (99% RDX/1% stearic acid);
- LX-14 (95% HMX/5% estane).

Nitramine/wax and nitramine/estane formulations with high filler content are often employed as hollow charges.

For explosives, materials with high densities are attractive, because the detonation velocity and maximum detonation pressure are highly dependent on density. CL-20 has the highest density and, as a result, has the highest detonation velocity [1.2]. Due to the high energy content of this cage-like molecule, the detonation pressure and Gurney energy are also very high, with values more than double those of TNT.

Most properties of TNAZ fall between those of RDX and HMX. Of interest, however, is its low melting temperature of 101 °C, which is near that of TNT (80.8 °C). This indicates that TNAZ could be used instead of TNT in melt or casting processes, while potentially providing improved performance.

ADN is an oxidizer. A wide variety of explosive charges [1.3] exist, particularly for underwater use, including aluminum as fuel and AP as oxidizer. In general, AP reduces the detonation velocity of an energetic system because it is not an ideal explosive material. Using ADN in place of AP could result in improved properties in such formulations.

FOX-7 is extremely interesting as an insensitive explosive in the performance class of RDX. The complex synthesis required for this material, however, means that it is expensive and only small quantities are commercially available. Nevertheless, FOX-7 certainly has huge potential as a component of future formulations with improved IM properties. The emphasis in recent years on IM properties, even at the expense of performance increases, has resulted in increased attention on good IM candidate materials such as FOX-7.

1.2.2
Solid Rocket Propellants

The most important application properties for solid rocket propellant formulations are the specific impulse (I_{sp}) and burn rate characteristics:

$$I_{SP} = k_1 \cdot (T_c N)^{0.5} = k_2 \cdot (T_c / M)^{0.5} \tag{1.3}$$

where T_c = flame temperature in the combustion chamber; N = number of moles per unit weight; M = mean molecular weight of the combustion gases; and k_1, k_2 = constants. The specific impulse is high

- when the heat of reaction is high;
- when the products have a high flame temperature;
- when the mean molecular weight of the reaction products is small.

Other important parameters besides specific impulse include the burn rate and pressure exponent. In the ideal case, the burn rate is independent of pressure ($n = 0$;

i.e. plateau burning). Generally, however, the propellant's burn rate increases with increasing energy input into the system. If the pressure exponent is too high ($n > 0.7$), practical applications may not be feasible, even for an otherwise promising energetic system. The incorporation of specialized burn rate modifiers, which typically have a catalytic influence on the burn rate process, can alter the burn rate and pressure exponent sufficiently to enable practical applications to be realized:

$$r = a \cdot p^n \tag{1.4}$$

where r = burn rate; a = constant; p = pressure; and n = pressure exponent. This equation is not generally applicable and is relevant only over a small pressure range. Performance data for solid rocket propellants are given in Table 1.4.

Specific impulse is commonly compared at a chamber pressure of 7 MPa, which results from the American standard of 1000 psi, which equates to 6894 MPa.

Differences in the various types of solid propellants can be summarized as follows. Double-base (DB) solid propellants consist primarily of nitroglycerine and nitrocellulose mixed with processing and burn rate additives. The use of such propellants is widespread. Their specific impulse ranges from 2100 to 2300 N s/kg. DB propellants have a relatively low or moderate burn rate of r = 10–25 mm/s at 7 MPa with a pressure exponent n = 0–0.3. DB propellants are low signature, but they are relatively sensitive to detonation and pose difficulties under low-temperature conditions.

Composite propellants based on ammonium perchlorate, aluminum and HTPB comprise the second group of commonly employed propellants. They have significantly higher performance than DB propellants, with specific impulses in the range 2500–2600 N s/kg for aluminum-based formulations. So-called reduced smoke formulations without aluminum have a specific impulse of 2400–2500 N s/kg, with a burn rate r of 6–40 mm/s at 7 MPa and a pressure exponent of n = 0.3–0.5.

Table 1.4. Performance data for solid rocket propellants (values at 7 MPa).

Type	I_{SP} (N s/kg)	r (mm/s)	n	Signature	Status
Composite, AP/Al/HTPB/nitramine	2500–2600	6–40	0.3–0.5	High	Existing
Composite (reduced smoke), AP/HTPB/nitramine	2400–2500	6–40	0.3–0.5	Reduced	Existing
DB RP, NC/NG	2100–2300	10–25	0–0.3	Low	Existing
AN RP, GAP/AN/nitramine/plasticizer	2200–2350	5–10	0.4–0.6	Low	R&D tested
Nitramine RP, GAP/nitramine/AP/plasticizer	2300–2450	10–30	0.4–0.6	Low	R&D tested
ADN RP, GAP/ADN/nitramine/plasticizer	2400–2600			Low	R&D new
ADN/Al RP, GAP/Al/ADN/nitramine/plasticizer	2500–2700			Low	R&D new

Disadvantages of composite formulations include the primary and secondary signatures emitted and their relatively high friction and impact sensitivity. However, on the positive side, their detonation sensitivity is relatively low and HTPB-based formulations generally exhibit good mechanical properties. Another disadvantage of formulations containing ammonium perchlorate is the undesirable emission of hydrochloric acid as an exhaust product. High-performance propellants usually contain aluminum fuel (owing to its high heat of combustion) and AP as the oxidizer. Hot microparticles of aluminum oxide (Al_2O_3) are released as an exhaust product, as is HCl. This combination disperses and absorbs radiation over all relevant wavelengths, making detection by an adversary relatively simple. In addition, the rocket's own external radar- or laser-based guidance system is adversely affected by the strong exhaust signature. Moreover, the emission of large amounts of environmentally damaging hydrochloric acid is very undesirable from an environmental perspective [1.4].

Development of a smokeless, low-signature solid propellant with performance equivalent to currently employed AP/Al composite formulations (~2600 N s/kg) is a major goal of the energetic materials industry.

One development approach is to use nitramines such as RDX or HMX with a corresponding reduction in the AP content. CL-20 offers an attractive option as an alternative high-energy nitramine oxidizer. In fact, the possibility exists that CL-20-based formulations may allow the development of materials that contain little or no AP. A second possibility is to replace AP with the alternative oxidizer ADN. ADN is also chlorine free and propellants based on this oxidizer emit environmentally acceptable exhaust gases [1.5].

To achieve performance increases beyond existing AP/Al composite formulations requires the combustion of substances such as Al or ALH_3 with ADN as an oxidizer. Such formulations would still emit a significant signature as a result of Al_2O_3 in their exhaust gases. They would, however, exhibit performance advantages beyond anything previously developed in the area of composite solid rocket propellants.

In summary, ADN is the material of choice to be developed as an alternative oxidizer to AP. However, CL-20 combined with an energetic binder could allow the development of smokeless solid propellants with relatively high performance. Experimental studies still need to be conducted to determine whether the burn rates and pressure exponents of formulations based on such new components will be in the acceptable range for practical applications.

1.2.3 Propellant Powder

For propellant powders, the specific energy (force, E_S) is often used to compare different formulations:

$$E_s = N_G \cdot R_0 \cdot T_{Ex} \tag{1.5}$$

where N_G = number of moles of the gas; T_{Ex} = flame temperature; and R_0 = gas constant. Additional parameters shown in Table 1.5 include the combustion tem-

Table 1.5. Performance data of propellant powders (at charge density 0.1, calculated ICT code).

Powder type and formulation	E_s (J/g)	Q_{ex} (J/g)	T_{Ex} (K)	*MM* (g/mol)
Conventional powders				
Single-base A5020, 92% NC (13.2% N)	1011	3759	2916	23.98
Double-base JA-2, 59.5% NC/15% NG/25% DGDN	1141	4622	3397	24.76
Triple-base M30, 28% NC/22.5% NG/47.5% NIGU	1073	3980	2996	23.20
HTPB formulations				
30% HTPB/70% RDX	874	3702	2046	17.16
30% HTPB/70% HMX	867	3668	2034	17.19
30% HTPB/70% CL-20	930	4008	2286	17.99
30% HTPB/70% TNAZ	926	4020	2184	17.57
30% HTPB/70% ADN	915	3329	1935	17.58
Semi-nitramine formulations				
30% NC/30% NG/40% RDX	1248	5700	3921	26.11
30% NC/30% NG/40% HMX	1246	5681	3914	26.12
30% NC/30% NG/40% CL-20	1224	5972	4042	27.46
30% NC/30% NG/40% TNAZ	1239	5984	4016	26.95
GAP formulations				
30% GAP/70% RDX	1190	4116	2838	19.83
30% GAP/70% HMX	1181	4082	2816	19.83
30% GAP/70% CL-20	1280	4409	3332	21.64
30% GAP/70% TNAZ	1272	4420	3204	20.94
25% GAP/75% ADN	1294	5454	3604	23.15

perature and the mean molecular mass (*MM*) of the reaction gases. Studies conducted in the USA often consider the heat of explosion when comparing different powder formulations.

The lower the combustion temperature, the lower is the erosion within the combustion barrel or nozzle caused by the burning powder. For combustion temperatures above 3500 K, erosion within the combustion chamber can become critical. At combustion temperatures below 3000 K, erosion is typically negligible. Powders in which the mean molecular mass of the exhaust gases is low exhibit the best combustion efficiency.

The specific energy value of 1141 J/g exhibited by JA-2 is a yardstick for conventional high-performance propellant powders. JA-2, which consists of 59.5% NC, 15% NG and 25% DGDN, has close to the optimum energy achievable among compositions consisting purely of NC and nitrate acid ester plasticizers. However, various strategies can be employed to attain propellant powders with still higher performance characteristics. For instance, the so-called semi-nitramine powders use NC as the binder matrix and incorporate nitramine solid fillers. Such formulations exhibit higher performance than JA-2. This high energy content leads to a relatively high combustion temperature (~4000 K). The combustion temperature can, however, be suppressed to an acceptable level by the addition of appropriate plasticizers. In addition to the conventional nitramines RDX and HMX, CL-20 and FOX-7 can also be employed as the nitramine filler in semi-nitramine formulations.

Formulations with the inert binder HTPB and energetic filler up to solids concentrations of 70% exhibit relatively low performance, with a maximum specific energy <1000 J/g. Further increases in the solids content results in systems that are extremely detonation sensitive. An alternative approach is to use an energetic binder as the binder matrix. The use of GAP with nitramine fillers up to 70% solids results in formulations with specific energies well above conventional state of the art high-energy JA-2. Despite their high energy content, GAP-based propellants burn at reasonable combustion temperatures ranging from 2800 to 3300 K. In particular, GAP/TNAZ and GAP/CL-20 formulations exhibit significantly higher energy content at practical combustion temperatures. This indicates the enormous energy increases potentially available through the use of such formulations.

In addition to GAP, there are a number of other energetic binder candidates which, combined with other new energetic materials, have excellent potential for development as future high-energy propellant powders. However, as is the case for solid rocket propellants, the burn rate behavior of such formulations is very difficult to predict and must be determined experimentally through further research and development.

1.3 New Energetic Materials

New materials offer the potential for significant performance gains and other improvements for energetic systems used in all of the main practical energetic applications, i. e. explosives, solid rocket propellants and propellant powders. Realization of attractive, practical formulations will depend on the availability, processability, compatibility and the material property spectrum of such new materials. In this section, the known material properties of each of these new energetic materials are discussed.

1.3.1 CL-20

CL-20 or HNIW was first discovered by Arnold Nielsen [1.6, 1.7] in 1986 at the laboratory formerly known as the Naval Surface Weapons Center in China Lake, California. The molecular structure of this cage-shaped molecule is shown in Fig. 1.2.

1.3.1.1 Synthesis and Availability of CL-20

CL-20 is the highest energy single component compound known to date and, in addition, has the highest density of any known organic substance. Thiokol Corpora-

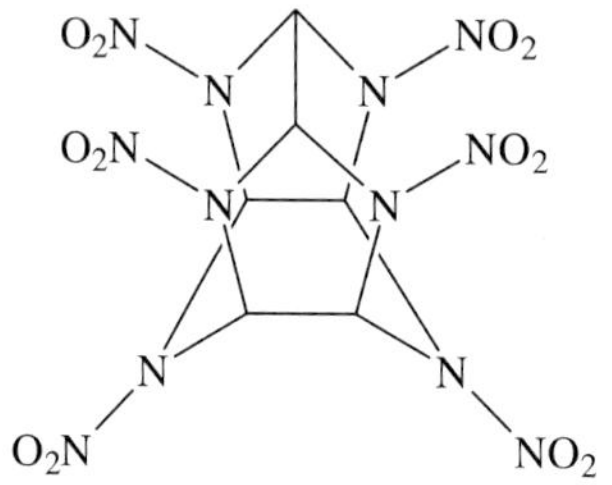

Figure 1.2. CL-20 or hexanitrohexaazaisowurtzitane (HNIW).

tion modified the original synthesis method pioneered by Nielsen [1.8–1.10] and constructed a pilot facility for small scale production of CL-20. Today, CL-20 is commercially available in kilogram quantities. It is also commercially available in various particle sizes from SNPE in France, which uses yet another modified synthesis method [1.11].

CL-20 is manufactured batchwise in lots of 50–100 kg. CL-20 has not yet been synthesized as a pure single phase, but instead exists in multiple solid phases, of which the ε-phase has the highest density. After the initial synthesis process, the raw product is recrystallized into ε-CL-20 in the presence of a suitable solvent. The resulting particle size distribution can be tailored based on the process parameters. Commercially available material is available with the following range of particle sizes:

$120 < x_{50,3} < 160$ μm;
$20 < x_{50,3} < 40$ μm;
and
$x_{50,3} < 5$ μm.

Coarse particles are produced directly via the crystallization process, whereas fine particle fractions must be obtained by grinding the coarse material.

The chemical purity [1.12] of the SNPE and Thiokol products are 98 and 96%, respectively. Impurities include residual solvent and incompletely nitrated intermediate products with benzyl, acetyl or formyl groups. It is not yet known whether and how such impurities affect the sensitivity, phase transition or other material properties of CL-20.

The quoted price of CL-20 is considerably higher than € 1000/kg. It is therefore not yet attractive for most applications, solely on account of its high cost.

1.3.1.2 Chemical and Thermal Properties of CL-20

CL-20 is a white powder. The onset of thermal decomposition occurs at ~215 °C. An example differential scanning calorimetric (DSC) curve of CL-20 is shown in Fig. 1.3 [1.13].

CL-20 is soluble in acetone, esters and ethers, but is insoluble in water and aromatic or halogenated solvents. It is compatible with most solids and binders, such as RDX, HMX, AP, PETN, nitric acid esters, isocyanates, GAP and HTPB. It is incompatible with bases, amines and alkali metal cyanides (e.g. NaCN).

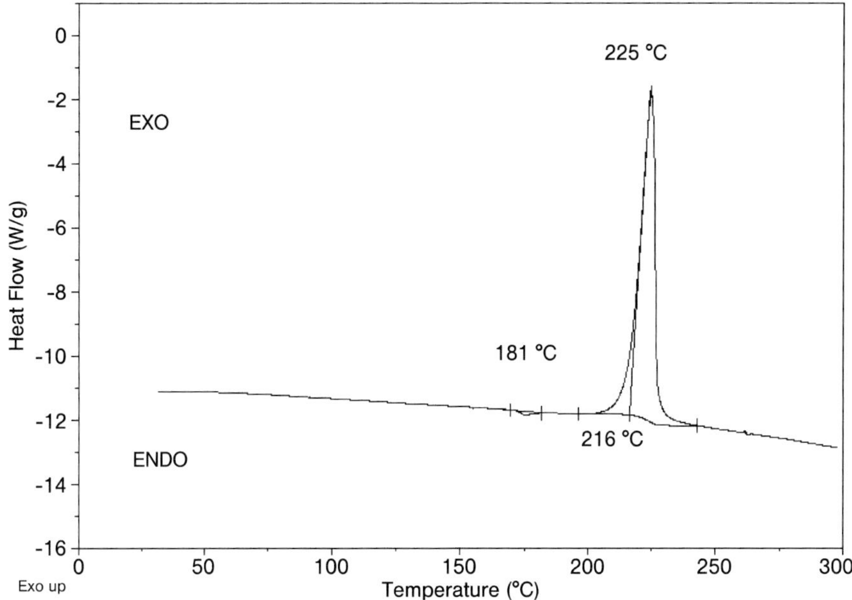

Figure 1.3. DSC curve of CL-20.

The material data sheet from SNPE lists the following properties for CL-20:

- deflagration temperature: 220–225 °C;
- decomposition temperature: 213 °C;
- maximum decomposition temperature: 249 °C;
- heat of decomposition: 2300 J/g;
- detonation velocity: 9650 m/s (experimental value);
- vacuum test, 193 h at 100 °C: 0.4 cm^3/g.

1.3.1.3 **Sensitivity and Phase Behavior of CL-20**

The impact sensitivity of CL-20 is similar to that of PETN. As a result, extensive safety precautions must be employed during processing and handling of CL-20. As a result, many international research efforts are directed at reducing the sensitivity of CL-20 through methods such as improved crystallization conditions (to produce defect-free crystals) and the development of a suitable inert coating.

Table 1.6. Properties of CL-20 phases.

Property	γ-CL-20	α-CL-20	β-CL-20	ε-CL-20	HMX
Density (g/cm^3)	1.92	1.97	1.99	2.04	1.91
Detonation velocity (m/s)	9380	9380	9380	9660	9100
Phase transition temperature (°C)	260	170	163	167	280

Data available thus far indicate that CL-20 becomes more sensitive as the particle size is reduced. Crystallized CL-20 is therefore not as sensitive as the milled material. CL-20 exists in four polymorphic structures: α, β, γ and ε (Table 1.6). The ε-phase crystallizes with a very high density and is therefore the phase of interest. All the other phases have considerably lower densities. The α-phase is a semi-hydrate and crystallizes in the presence of water. The phases can be distinguished through IR spectroscopy and X-ray diffraction.

1.3.2
Octanitrocubane

In early 2000, Mao-Xi Zhang and Philip E. Eaton of the University of Chicago, in collaboration with Richard Gilardi of the Naval Research Laboratory in Washington, DC, successfully synthesized octanitrocubane [1.14]. Octanitrocubane is a cube-shaped carbon-based molecule with nitro groups at each corner of the cube (Fig. 1.4). Its chemical formula is $C_8(NO_2)_8$. The molecule's high energetic content derives from the large number of nitro groups per carbon atom and its highly stretched ring structure. The cubic form of the molecule forces the carbon atoms to assume bonding angles of 90°.

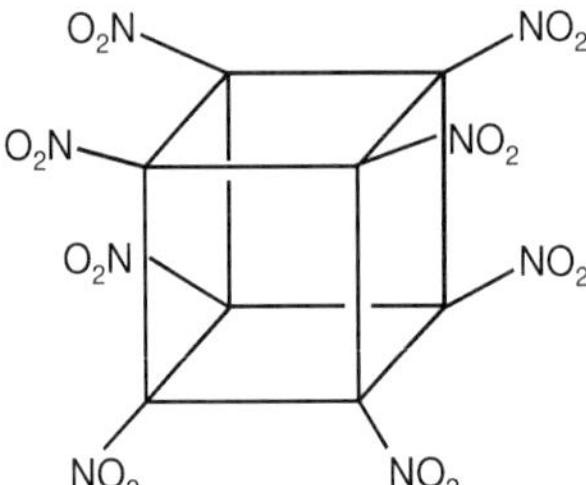

Figure 1.4. Octanitrocubane (ONC).

It could be said that octanitrocubane is a compound conceived on a drawing board. Its explosive properties were predicted years before even its precursors had ever been synthesized. Various theoretical calculations of octanitrocubane's density yielded values from 1.9 to 2.2 g/cm^3. The densities of other explosive materials are given in Table 1.1.

Compared with conventional explosive materials such as TNT, RDX and HMX, even at the lower density value of 1.9 g/cm^3, the energy content of octanitrocubane is comparable to that of HMX. If its density approaches the maximum theoretical value of 2.2 g/cm^3, it would exceed that of CL-20, which is the highest known of any organic compound.

For over 10 years, synthetic chemists have labored to add nitrate groups to carbon-based cube-shaped molecules (cubanes). A milestone in this effort was achieved in 1997 with the synthesis of tetranitrocubane (containing four nitro groups), which has explosive properties comparable to those of RDX. From a chemical development standpoint, the synthetic efforts of the research group led by Philip E. Eaton, culminating in the synthesis of octanitrocubane, represent a monumental accom-

Table 1.7. Density of nitrocubanes [1.15].

Molecule	Density (calc.) (g/cm³)	Density (g/cm³)	DSC onset tempera-ture(°C)
Cubane	1.29	1.29	
1,4-Dinitrocubane	1.66	1.66	257
1,3,5-Trinitrocubane	1.77	1.76	267
1,3,5,7-Tetranitrocubane	1.86	1.81	277
1,2,3,5,7-Pentanitrocubane	1.93	1.96	–
Octanitrocubane	2.13	1.98	–

plishment. One could follow the developments in the literature, as nearly every year an additional nitro group was incorporated into the base molecule. The recent synthesis of octanitrocubane is incredibly complex and difficult, requiring numerous intermediate steps, including the preparation of tetranitrocubane as an intermediate compound (see Table 1.7). ONC is at present still available only in minute laboratory quantities, with each milligram more valuable than gold.

Now, to answer the long awaited question: does the synthesized octanitrocubane exhibit the high density predicted theoretically, thereby fulfilling the enormous expectations of the energetic materials community? The answer is: yes and no. The measured density of octanitrocubane is 1.979 g/cm^3, which is of course high, but is near the low end of the theoretical range predicted. Hence ONC did not set a new record; instead, its density is comparable to the values exhibitied by RDX and HMX. The density of ε-CL20 remains by far the highest of the newly synthesized organics.

It is still possible of course that, like CL-20, various polymorphic forms of octanitrocubane may exist. Thus far, crystal modification of ONC has only produced an isolated hydrate compound. Future work will certainly be focused on other crystal modifications of octanitrocubane. Theoretical calculations of the density of such a still fictional compound range from 2.123 to 2.135 g/cm^3.

1.3.3 TNAZ

TNAZ (1,3,3-trinitroazetidine) [1.16], which has a performance roughly between those of RDX and HMX, was first synthesized by K. Baum and T. Archibald of Fluorochem in Asuza, California. The structural form of this compound is shown in Fig. 1.5.

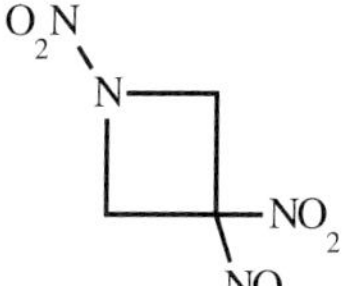

Figure 1.5. TNAZ (1,3,3-trinitroazetidine).

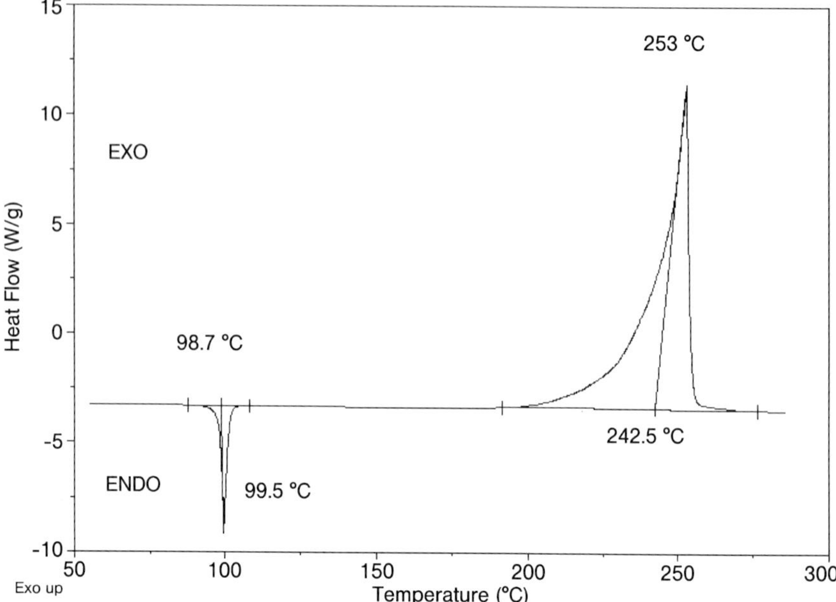

Figure 1.6. DSC diagram of TNAZ.

1.3.1.1 Chemical and Thermal Properties of TNAZ

The energy content of TNAZ is elevated owing to the tension in the four rings (the molecule has a positive heat of formation of 26.1 kJ/mol and a density of 1.84 g/cm^3). The substance exhibits outstanding thermal stability, with the onset of decomposition occurring above 240 °C (Fig. 1.6) [1.17]. A particularly attractive property of TNAZ is its low melting temperature of about 100 °C. TNAZ is thus potentially suitable for casting processes such as those in which TNT is currently used. TNAZ is compatible with aluminum, steel, brass and glass and is not hygroscopic.

A number of studies have been conducted on eutectic mixtures of TNAZ [1.18]. The results indicate that mixtures of TNAZ with RDX or HMX exhibit only a slight melting point depression of 1–5 °C. There are few data to assess the suitability of TNAZ in casting processes. Obviously, its high vapor pressure and crystallization behavior in the melt state could cause processing difficulties.

1.3.3.2 Synthesis and Availability of TNAZ

TNAZ is not commercially available. The original synthesis accomplished in 1983 had far too many steps and insufficient yield (<5 %) to make the material commercially viable. Subsequently, American researchers developed a simplified synthesis route and improved the yield to >70 %. Today, batchwise lots of ~50 kg can be manufactured. Production costs exceed € 1000/kg, primarily owing to the high cost of the catalyst required [1.19].

1.3.4 ADN

The third new component primed for application is ADN [ammonium dinitramide, $NH_4N(NO_2)_2$]. ADN is primarily of interest as an oxidizer for solid rocket propellants, particularly owing to its potential to replace ammonium perchlorate.

1.3.4.1 Synthesis and Availability of ADN

ADN was first publicly announced by Professor Tartakowsky at the ICT annual conference in 1993 [1.20]. Simultaneously, it was proposed as a new oxidizer for solid rocket propellants by Pak [1.21] in the USA.

ADN was synthesized in Moscow at the Zilinsky Institute and developed within the former Soviet Union for use in solid rocket propellants for tactical rockets. ADN was produced in ton-size quantities for this purpose. It is assumed that the manufacturing facilities for ADN were destroyed or dismantled; at any rate, it is certain that they no longer exist. The substance was unknown outside of the Soviet sphere and the Western world was neither aware of the large production facilities dedicated to ADN nor to its use in tactical rockets. Following the first publication about ADN in the West, researchers have attempted to duplicate the Russian development efforts and produce a material suitable for practical applications.

There are at least 20 different synthesis routes that successfully produce ADN, two of which have proved to be technically relevant. The first is the so-called urethane synthesis:

$$C_2H_5O_2C\text{-}NH_2 + HNO_3 \rightarrow C_2H_5O_2C\text{-}NHNO_2 + H_2O$$
$$C_2H_5O_2C\text{-}NHNO_2 + NH_3 \rightarrow C_2H_5O_2C\text{-}NNO_2NH_4$$
$$C_2H_5O_2C\text{-}NNO_2NH_4 + N_2O_5 \rightarrow C_2H_5O_2C\text{-}N(NO_2)_2 + NH_4NO_3$$
$$C_2H_5O_2C\text{-}N(NO_2)_2 + 2NH_3 \rightarrow C_2H_5O_2C\text{-}NH_2 + NH_4N(NO_2)_2$$

This synthetic method requires four synthesis steps and results in a yield of up to 60%. The processing method is not difficult and the amount of waste produced is relatively low. The urethane can be cleaned and recycled within the process. The second synthesis method uses salts of the amidosulfuric acid:

$$KO_3S\text{-}NH_2 + 2HNO_3 \rightarrow KHSO_4 + NH_4N(NO_2)_2 + H_2O$$

The nitration with amidosulfonates can be accomplished in the absence of organic solvents, but must occur in strongly acidic media, which can lead to decomposition of the dintramide. This synthesis also concludes with a purification step to reclaim ADN. This process was extensively developed by FOI and Nexplo [1.22] and today it is used for the commercial manufacture of ADN.

A number of steps are required to manufacture ADN that is useful for practical formulations. ADN crystallizes into rod or lamellar structures, which are unsuitable for further processing as energetic formulations. The raw product therefore must be

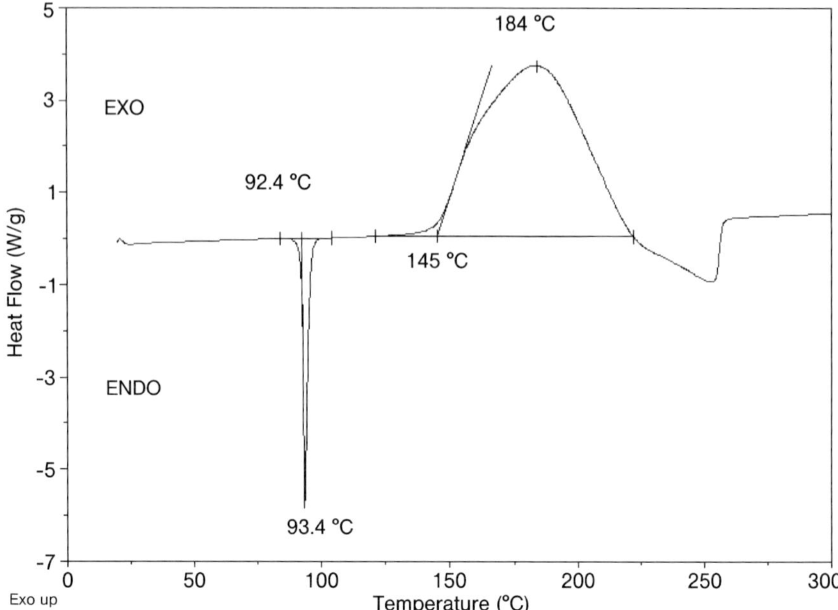

Figure 1.7. DSC diagram of ADN.

ground into spherically shaped particles of desired particle size (see Sect. 3.2.7). ADN also must be stabilized in order to attain the service life required for practical applications.

1.3.4.2 Thermal Behavior of ADN

ADN exhibits a melt temperature ranging from 91.5 to 93.5 °C (Fig. 1.7). Melt temperature values reported in the literature actually range from 83 to 95 °C. This large variation in the reported values is most likely attributable to impurities, which can have a significant influence on the ADN melt temperature even at very low concentrations. For instance, residues of water and especially of ammonium nitrate (AN) result in a clear decrease in the ADN melting temperature. Ammonium nitrate is not only a by-product of the ADN synthesis, it is also one of the primary alternative oxidizers to ADN. The AN concentration in ADN plays a key role in the long-term stability characteristics of ADN itself.

Further heating of the ADN melt (Fig. 1.7) results in exothermic decomposition. The decomposition is initiated by dissociation of ammonia and free dinitramide acid [$HN(NO_2)_2$] [1.23]. The free dinitramide acid immediately breaks down into N_2O and HNO_3. The latter condenses with NH_3 to form ammonium nitrate, which undergoes a second decomposition step, releasing N_2O and water. Other side reactions and the formation of NO, NO_2, N_2 and O_2 occur along with the reaction sequence described. ADN decomposition is an acid-catalyzed process, due to the initial step. In addition to being thermally labile, ADN is also light sensitive.

1.3.4.3 Long-term Stability of ADN

Water content

As indicated above, small concentrations of impurities have a significant effect on the long-term stability of ADN. In general, the stability of ADN increases as the amount of water impurity decreases; however, completely water-free ADN is also unstable. It has been found that ADN stability is not degraded when the water content ranges from 0.1 to 0.5%. ADN is hygroscopic and must therefore be stored and processed in a dry environment.

Ammonium nitrate

Ammonium nitrate impurities in ADN cause problems because at certain AN concentrations a lower melting eutectic compound is formed. Ammonium nitrate is a salt of a weak base and strong acid and can thus lead to acid-catalyzed decomposition of ADN. However, residual concentrations of AN below 0.5% appear to be unproblematic.

Organic impurities

Because ADN is an oxidizer, organic impurities, such as organic solvent residues, are deleterious. Such compounds apparently take part in redox reactions that accelerate ADN decomposition. In contrast, long-chain organic compounds can lower the sensitivity of ADN. Therefore, short-chain solvents should never be used to purify ADN.

Stabilization of ADN

ADN stabilizers must fulfil two basic functions: suppress the acid-catalyzed decomposition (i. e. exhibit basic properties) and remove other possible free radicals from the system (i. e. function as a radical interceptor). Substances with such properties function well as ADN stabilizers. For practical applications, the choice of stabilizer and means of incorporating it into the ADN particles must be optimized. One elegant technical solution is to incorporate the stabilizer into the raw material during the recrystallization process.

1.3.4.4 Processability of ADN

As a result of its oxidizer nature, ADN tends to oxidize organic substances. However, in general ADN does not attack C-H and C-C single bonds. Compounds with double-bonded carbons also exhibit good compatibility with ADN. However, ADN is incompatible with isocyanates, spontaneously reacting and decomposing in their presence. This means that the large number of polymers that undergo polyurethane bonding cannot be used with ADN, at least in their common form. This eliminates the possibility of using HTPB or GAP binders with isocyanate as the curing agent. Because conventional polymeric binders used in solid rocket propellants are not compatible for use with ADN, the development of future ADN-based solid propellants is severely limited by the choice of suitable binders.

Good results have been obtained, however, using PCP (polycaprolactone polymer) as a binder with ADN [1.4, 1.24, 1.25]. The formulation with this binder also exhibited the best long-term stability observed by an ADN-based material. Despite these initial encouraging results, the long-term stability and aging mechanisms of such complex formulations must still be thoroughly examined before definitive statements about the applicability of such systems can be ventured.

1.3.4.5 Safety Properties of ADN

A range of different results has been obtained for the impact and friction sensitivity of ADN (Table 1.8). These results indicate that the impact and friction sensitivity of ADN are higher than the values for neat AP. Overall ADN is somewhat more sensitive than RDX. The comparison of neat materials may be misleading. For propellant applications, the sensitivity of formulations gives more relevant data in practice. So far there are no data available with such ADN formulations.

Table 1.8. Friction and impact sensitivity results for ADN compared with AP and RDX.

Substance	Friction sensitivity (N)	Impact sensitivity(N m)
RDX	120	7.4
AP	>320	15–25
ADN	64–72	3–5

1.3.5 FOX-7 (1,1-Diamino-2,2-dinitroethylene)

This compound is a relatively new hope in the explosives world. It was introduced in the fall of 1998 by FOI collaborators in the USA [1.26, 1.27]. Three variations of the

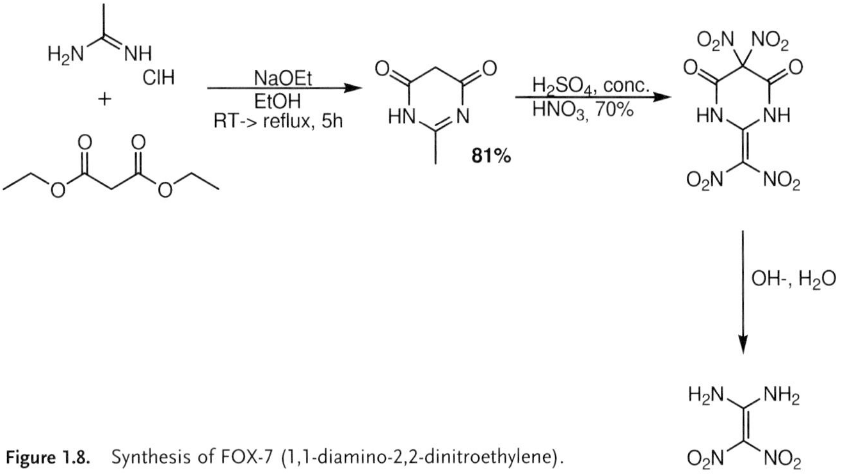

Figure 1.8. Synthesis of FOX-7 (1,1-diamino-2,2-dinitroethylene).

Table 1.9. Properties of FOX-7 (calculations with Cheetah 1.40 [1.23]) and comparison with RDX

Property	FOX-7	RDX
BAM impact sensitivity (N m)	>15	7.4
Petri friction sensitivity (N)	>200	120
Deflagration temperature (°C)	>240	230
Density (g/cm^3)	1.885	1.816
Formation energy (calculated) (kJ/mol)	−118.9	92.6
Detonation velocity (calculated) (m/s)	9040	8930
Detonation pressure (calculated) (GPa)	36.04	35.64

synthetic process have been developed [1.28–1.30], all of which involve the nitration of a heterocyclic compound followed by hydrolysis to form FOX-7. The most promising reaction path is shown in Fig. 1.8.

FOX-7 is commercially available; however, its cost of >€ 3000/kg makes it unrealistic for practical application at present.

FOX-7 has a performance in the range exhibited by RDX, but is considerably less sensitive. Little is known to date about its processability or compatibility with other materials. Such characteristics will surely be the subject of future investigations. Some properties of FOX-7 compared with those of RDX are given in Table 1.9. The very high performance of FOX-7 predicted theoretically remains to be confirmed experimentally. FOX-7 has the potential to contribute positively to the further development and improvement of insensitive munitions.

1.4 Conclusion

It is important to emphasize that the new energetic materials discussed, TNAZ, CL-20, ONC, FOX-7 and ADN, are still experimental. The material properties of these substances must be investigated further and improved before they find application in new or existing weapons systems.

Factors which must be considered when assessing whether such materials will find practical application include the following:

- availability and price;
- performance;
- sensitivity;
- processability;
- compatibility;
- chemical and thermal stability;
- temperature-dependent mechanical behavior;
- burn rate and pressure exponent.

CL-20 is currently commercially manufactured in 50–100-kg batches. As a raw product, CL-20 has a high sensitivity comparable to that of PETN, which is a significant limitation in terms of its processability. Lowering the sensitivity of the raw material by developing an improved crystallization process or coating method is a major challenge for CL-20 suppliers. When processed in a binder matrix, however, CL-20 formulations are not significantly more sensitive than RDX and HMX compositions.

CL-20 has the highest density, 2.04 g/cm^3, of any organic explosive material. It is a nitramine with good chemical and thermal stability and is compatible with most binder and plasticizer systems. CL-20 has all the characteristics required for application in all three of the energetic material sectors.

Replacing HMX with CL-20 in explosive formulations results in a 10–15% performance increase.

CL-20 combined with an energetic binder in solid rocket propellant formulations results in a reduced smoke composition that approaches the performance of conventional HTPB/AP propellants. The favorable oxygen balance of CL-20 is particularly advantageous. As with other nitramine-based energetic materials, the high pressure exponent must be reduced through the addition of burn rate modifiers.

As a result of its high energetic content and favorable oxygen balance, CL-20 is also applicable as a component of propellant powders. Performance gains of 10–15% are possible if CL-20 is used instead of traditional oxidizers.

Unlike CL-20, TNAZ is not yet commercially available. The synthesis route currently used is so expensive that it is not yet possible to examine TNAZ even in test applications. A major goal is to develop improved methods to manufacture TNAZ at a reasonable cost.

TNAZ has a performance between those of HMX and RDX. The property of TNAZ that is particularly interesting is its low melting temperature of 101 °C. TNAZ offers the possibility of functioning as a quasi-high-energy TNT. All charges that are manufactured via melt processing, such as Composition B or octol, could be formulated with TNAZ instead of TNT, resulting in compositions with considerably higher performance manufactured using the same basic technology.

ADN is commercially available. It has stability problems if the end product purity is inadequate. It crystallizes into rod-shaped structures and must therefore be ground in order to be used in practical applications. Technologies exist to grind ADN into spherical particles and stabilize it, and such methods must be incorporated into the production process. The aging behavior of ADN that has been properly stabilized is no worse than that of nitrocellulose.

Because of its strong oxidizing nature, ADN is compatabile with other systems to only a limited extent. For instance, ADN is incompatible with isocyanates, which are commonly used as curing agents in the binder matrix of energetic materials. ADN is compatabile with prepolymers, but requires a specialized curing agent. More research is required to solve the problems related to ADN compatibility.

The primary application for ADN would be for solid rocket propellants, in particular for replacing ammonium perchlorate while maintaining the high performance of AP-based systems. By using ADN in combination with an energetic binder, an impulse as high as the current space booster formulation (I_{SP} = 2600 N s/kg) is possible. Incorporating metal fuels in such a formulation, such as Al or AlH_3‘ enable even higher specific impulses to be achieved, up to I_{SP} = 2700–2800 N s/kg. ADN formulations offer the possibility of higher specific impulse propellants and corresponding longer range rocket systems.

In the areas of propellant powders and explosives, ADN is likely to only find niche or specialized applications.

FOX-7 has a performance in the range of RDX but is less sensitive. It therefore has excellent potential as a substitute in formulations that currently use RDX. As such, it could be useful in explosives, propellant powders and solid rocket propellants.

At present, however, FOX-7 is commercially available only at an exorbitant price. Reducing the cost and improving the product quality are therefore the next essential steps in order to enable FOX-7 to be tested in new formulations. Only then it will be possible to determine whether the favorable sensitivity characteristics of this ingredient will lead to IM systems with correspondingly favorable properties.

Overall it appears that new substances exist that could significantly increase the performance of energetic materials in each of the major application areas. It is therefore crucial to explore thoroughly the development potential of these substances.

1.5
Acknowledgments

The author thanks all co-workers at the Fraunhofer Institut Chemische Technologie, especially Helmut Bathelt, Indra Fuhr, Peter Gerber, Thomas Heintz, Thomas Keicher, Hartmut Kröber, Stefan Löbbecke, Klaus Menke, Heike Pontius, Dirk Röseling and Heike Schuppler, for their help and fruitful discussions.

2
Size Reduction

U. Teipel, I. Mikonsaari

2.1
Fundamentals of Size Reduction

Size reduction efficiency is largely a function of the material properties of the solid material in question. In order to split particles into ever smaller units, linear fractures that spread through the solid must develop. To create such fractures, loads must be applied to the outer surface of the particles, either from a grinding tool or from neighboring particles. These loads cause deformation that creates a stress field within the particles. Defects in the particles, such as imperfections in the particle lattice structure, promote the formation of fractures [2.1].

2.1.1
Material and Crack Behavior

The material behavior of solids can be described as linearly elastic, plastic or viscoelastic (see Chapter 12).

For a linearly elastic material, stress is proportional to the applied deformation. The proportionality factor is a characteristic material property, which is a constant value called the elastic modulus, denoted E:

$$\sigma = E \cdot \varepsilon \tag{2.1}$$

where σ is the tensile stress and ε the extensional deformation.

When a load is imposed, materials with a large elastic modulus deform very little before fracture onsets. Such materials are classified as 'brittle.' For materials with a small elastic modulus, even small stresses result in large deformations; such materials are termed 'rubber-elastic.' This behavior can be described as shown in Fig. 2.1. Here, the area under the curves is proportional to the energy input (per unit volume) required for the size reduction process. One sees that this total energy required is much less for brittle materials (a) than for elastics (b), even though the so-called fracture stress, σ_B, is considerably higher for brittle materials. Thus, brittle

Energetic Materials. Edited by Ulrich Teipel
Copyright © 2005 WILEY-VCH Verlag GmbH & Co. KGaA, Weinheim
ISBN: 3-527-30240-9

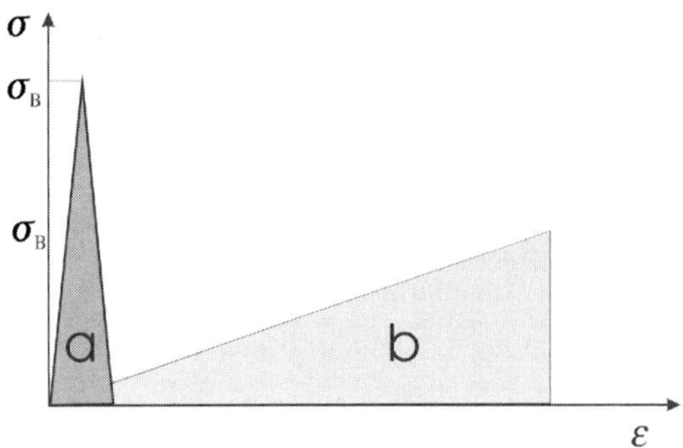

Figure 2.1. Linearly elastic material behavior.

materials can be ground using less energy than rubber-elastic materials. In practice, the material behavior is affected by the process operating parameters; for instance, size reduction of rubber-elastic materials can be effectively carried out at low temperatures.

With decreasing particle size (for quartz, starting at ~1 μm), materials show a transition from elastic to plastic behavior. At this transition, a limiting particle size for grinding is approached, because the particles' deformability increases while their probability of breaking decreases.

The way in which a solid is structured at the atomic and molecular level determines its elasticity and correspondingly its material behavior. Most energetic materials have a crystalline structure. Crystalline materials have exactly defined distances between neighboring atoms, which form a crystal lattice. This lattice is shown in Fig. 2.2 for sodium chloride, which consists of Na^+ and Cl^-. Each ion is surrounded by six oppositely charged ions.

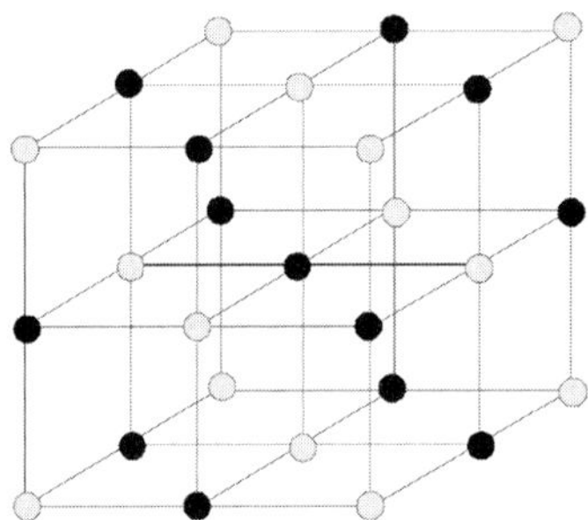

Figure 2.2. Crystal lattice of sodium chloride.

The forces involved in such a crystal lattice include attractive forces between oppositely charged ions and repulsive forces between like charged ions. Fig. 2.3 qualitatively shows the sum of these competing forces as a function of the distance between ions.

If the attractive and repulsive forces between ions are known, the strength of the crystal lattice can be calculated. The fracture of such a lattice can be described as follows (Fig. 2.4) [2.2]:

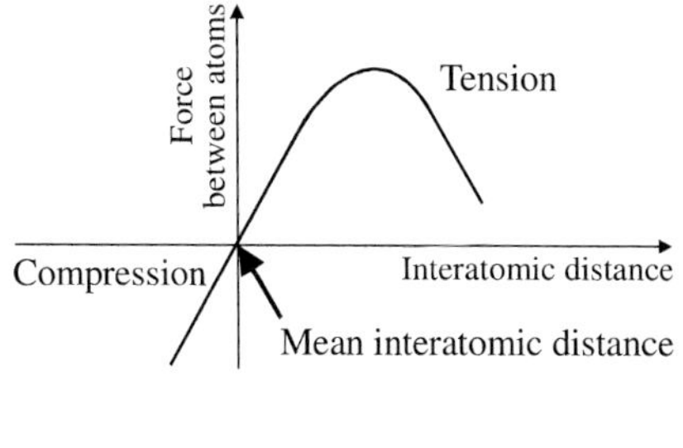

Figure 2.3. Relationships between forces in the crystal lattice.

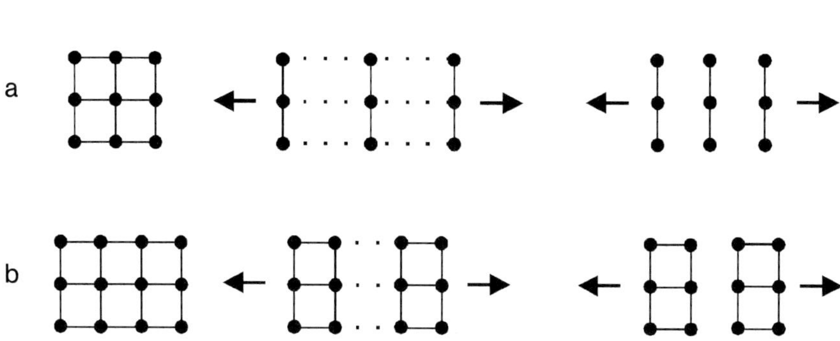

Figure 2.4. Break-up of the crystal lattice.

- All bonds in a crystal plane perpendicular to the imposed load are stretched until they all break simultaneously (case a). In this case, the theoretical energy required is much higher than the energy actually required in practice.
- If one assumes that the only bonds stretched are those in the crystal plane in which the fracture occurs (case b), then one calculates a theoretical energy for fracture much less than that actually required in practice.

Real solids have inhomogeneities such as crystal defects, cracks, voids or inclusions of foreign materials. At these sites the bonding forces are smaller and hence local stresses can arise. Such energetic effects cause cracks to develop within the solid that originate at these locations.

2.1.2 Size Reduction Energy

Several parameters are employed to characterize the size reduction process: the size reduction energy (mass and area specific), the energy utilization, the energetic efficiency and the size reduction ratio. The mass specific size reduction energy W_M is defined as

$$W_M = \frac{W}{m} \tag{2.2}$$

where W is the energy imparted by the machine and m is the mass of ground material.

The area-specific size reduction energy W_A is determined via

$$W_A = \frac{W}{\Delta A} \tag{2.3}$$

where ΔA is the area increase. The area-specific size reduction energy relates the energy expended to the size reduction obtained.

The three most important equations for describing the mass-specific size reduction energy W_M are described below. They are all based on an exponential equation of order α [2.3]:

$$\frac{1}{m} \cdot \frac{\mathrm{d}W}{\mathrm{d}x} = -\kappa \cdot d^{\alpha} \tag{2.4}$$

where x is the particle size of the ground product.

In 1867, Rittinger proposed the following equation:

$$\frac{W}{m} = \kappa \cdot \left(\frac{1}{x_0} - \frac{1}{x_W}\right) \tag{2.5}$$

where x_0 is the original particle size, x_W is the ground particle size and κ is a material constant.

The principle that the ratio of the total fracture area to the fracture energy is constant corresponds to the area-specific size reduction energy remaining constant. This holds true only for size reduction methods in which the loads are imposed on the smallest possible spaces, such as gaps. In practice, the required energy is 200–3000 times greater than this area-specific size reduction energy. Hence it is unlikely that the area-specific size reduction energy remains constant, at least for larger particles [2.1].

Kick's equation, from 1885, is written as follows:

$$\frac{W}{m} = \kappa \cdot (\ln x_W - \ln x_0) \tag{2.6}$$

Kick postulated the existence of a direct proportionality between the volume of the particles before and after size reduction [2.1]. However, his equation also appears unrealistic for a wide range of practical problems, because it predicts that the energy required to grind 10-μm particles down to 1 μm would be the same as that required to grind 100-cm particles down to 10 cm. Thus, in practice, this equation is only used for very large particle sizes.

In 1952, Bond presented the following equation:

$$\frac{W}{m} = 2 \cdot \kappa \cdot \left(\frac{1}{x_o^{0.5}} - \frac{1}{x_W^{0.5}}\right) \tag{2.7}$$

This relationship between the original and final particle sizes postulated by Bond is often used in contemporary practice. It is based on empirical data obtained in industrial processes. As a result, a large amount of empirical material data based on this formula now exists. Rittinger's equation is often used for small particles, Bond's for medium-sized particles and Kick's for larger particles. Figure 2.5 shows the

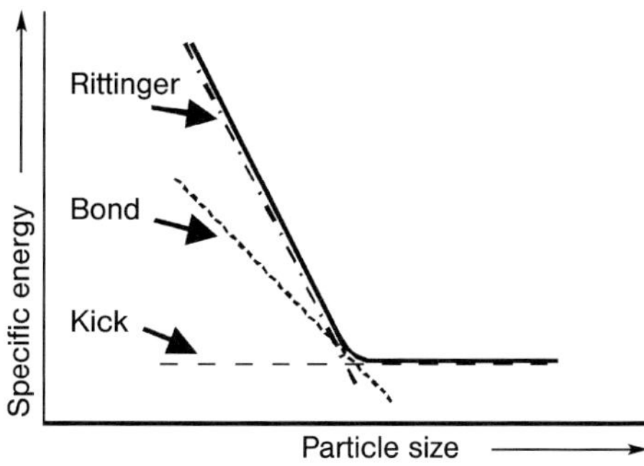

Figure 2.5. Qualitative depiction of the size reduction energy according to equations by Rittinger, Bond and Kick.

Table 2.1. Types of loads used in particle size reduction processes.

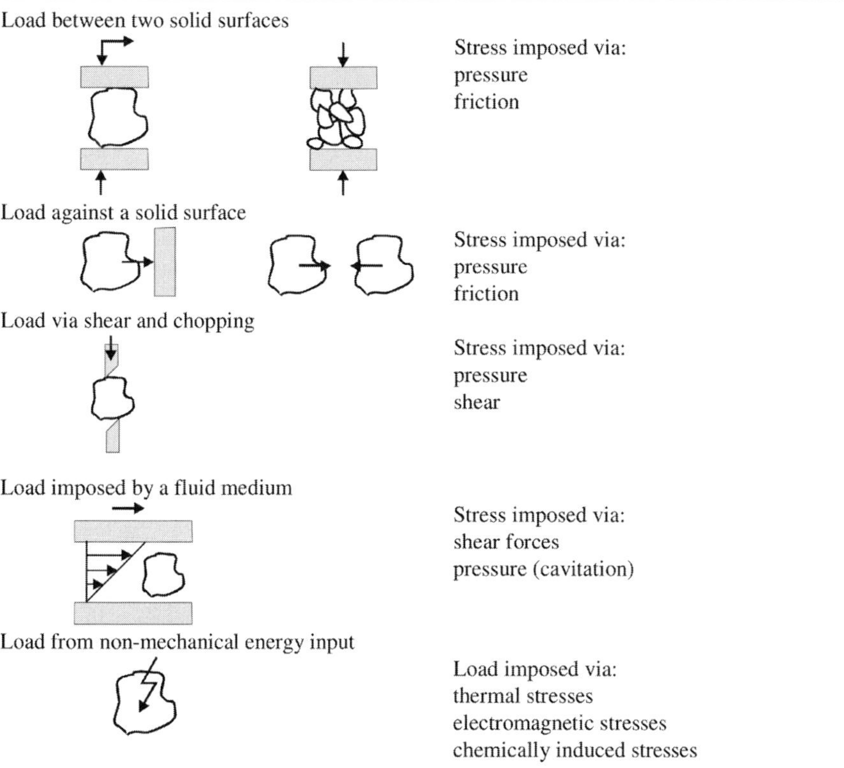

Type of load	Stress imposed
Load between two solid surfaces	Stress imposed via: pressure friction
Load against a solid surface	Stress imposed via: pressure friction
Load via shear and chopping	Stress imposed via: pressure shear
Load imposed by a fluid medium	Stress imposed via: shear forces pressure (cavitation)
Load from non-mechanical energy input	Load imposed via: thermal stresses electromagnetic stresses chemically induced stresses

specific energy as a function of the mean particle size and the size reduction energy. Table 2.1 shows different types of loads used to reduce the size of solid particles [2.4].

2.1.3 Selection Criteria for Size Reduction Processes

The most important consideration when choosing a size reduction method is the material behavior of the raw material. Table 2.2 shows the applicability of different types of loads to various material types.

Table 2.2. Material properties and type of load.

Material property	Impact	Pressure	Shear	Crushing	Chopping
Hard	++	++	–	–	–
Brittle	++	++	–	++	–
Medium hard	++	++	+	++	–
Soft	+	+	++	++	++
Elastic	–	–	–	+	++
Tough	–	–	–	++	++
Fibrous	+	–	+	+	++
Thermally sensitive	–	–	–	++	+

When selecting a size reduction process for energetic materials, special requirements must be considered. Different characteristics of the materials, such as friction and impact sensitivity, must be taken into account. Highly sensitive energetic materials such as RDX, HMX or CL 20 should be comminuted using a wet process, so that the continuous phase of the suspension can stabilize the material.

One must also take into account that dispersions of such highly sensitive materials in water are detonation sensitive at concentrations above 20 wt%. More insensitive materials, such as ammonium perchlorate (AP) or potassium perchlorate, can be ground in a dry size reduction process, e.g. with a jet mill.

Efficiency is also a criterion often used in selecting the most appropriate size reduction technique. However, size reduction efficiency remains difficult to quantify adequately [2.5]. Usually one considers the power input necessary to reduce the bulk material to the desired final particle size and compares that with the power required to grind a single particle to that size. Because the process of grinding a bulk material is so different to that of grinding a single particle, only a rough comparison can be made. Mills designed to grind coarse materials (such as a crusher) are highly effective based on this measure of merit, whereas those tailored for finer particles (such as fluid-energy and jet mills) exhibit inferior effectiveness.

2.2 Size Reduction Processes

2.2.1 Pinned Disk Mill

In a pinned disk mill, stress is applied between two concentrically oriented pins. Usually, one of the pins rotates while the other remains stationary, but in some cases both pins rotate in opposite directions. The rotor axis is horizontal so that raw material is fed into the center and discharged radially. In most designs, one cannot classify the final product by size. The final particle size is determined by the pin design, the number of rows of pins, the rotation rate and the product throughput. Rotor diameters range from 100 to 900 mm and the perpendicular velocity from 60 to 200 m/s. This type of mill finds application in the fine grinding of soft, brittle and lubricating materials in the chemical, pharmaceutical and foodstuffs industries [2.6], and also energetic materials.

A pinned disk mill built by Hosokawa Alpine, Type Kolloplex 160 Z, was used to grind various relatively insensitive energetic materials. This mill has variable rotation rates between 9000 and 14 000 min^{-1}. Table 2.3 shows typical results for comminution of ammonium perchlorate at two rotation rates (9000 and 12 000 min^{-1}); the dispersion coefficient ξ describes the particle size distribution (see Chapter 8). Other materials investigated are shown in Table 2.4.

The particle size of the AP ranged from 7 to 10 μm, which is larger than after comminution in a jet mill. The mean particle size decreased with increasing rotation rate, owing to the higher energy imparted to the material. However, an increasing dispersion coefficient ξ was observed with increasing rotation rates.

Table 2.3. Comminution of AP in a pinned disk mill.

	Rotation rate (min^{-1})	Mean particle size (μm)	ξ
AP	9000	10.3	0.72
AP	12000	7.1	0.84

Table 2.4. Various materials communited in a pinned disk mill.

Material	Mean particle size (μm)
Ammonium perchlorate	7
Ammonium dichromate	38
Ammonium nitrate	20
Nitroguanidine	4–12
Triaminoguanidine	10
Sodium nitrate	4
Potassium nitrate	6

2.2.2 Jet Mill

If a finer particle size is desired than can be obtained with a pinned disk mill, air jet mills may be used. This type of mill is used for products requiring very fine particles, such as paint pigments and graphite. Typical product throughputs range from 1 to 1000 kg/h. There are various types of jet mills, with the pancake jet mill and fluidized bed jet mill among those most commonly employed [2.7].

Size reduction of particular products in a pancake jet mill (see Fig. 2.6) is based on the principle of impact grinding. The design of the nozzle and the resulting flow patterns in the grinding chamber of the mill are especially important in achieving effective particle size reduction, because particles must be accelerated into a freely flowing stream in which particle-particle or particle-wall collisions produce the desired size reduction effect. The pancake jet mill contains a flat, cylindrical grinding chamber with nozzles situated around the periphery.

A schematic of a pancake jet mill is shown in Fig. 2.7. The bulk solid is fed via the injector (5) into a flat, cylindrical grinding chamber (7). The chamber contains a ring with a fixed number of injection nozzles, through which the grinding gas enters the chamber. A vortex flow develops that carries the bulk solid around the chamber. The particles are accelerated and impact each other at different relative velocities, resulting in size reduction.

The design of the ring of nozzles determines the velocity of the grinding gas and thus has a decisive influence on the grinding results. By using Laval nozzles, supersonic flows ($Ma > 1$) can be obtained. An advantage of Laval nozzles in jet

Figure 2.6. Pancake jet mill.

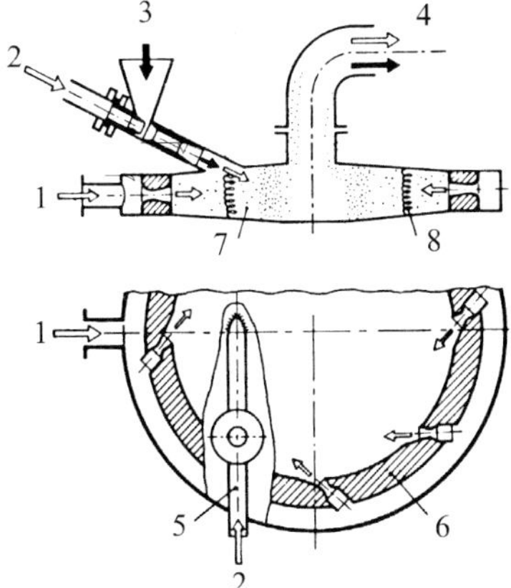

Figure 2.7. Schematic of a pancake jet mill. 1, Grinding gas; 2, injector gas; 3, bulk solid; 4, ground product; 5, injector; 6, grinding nozzle ring; 7, grinding chamber; 8, size reduction zone.

mills is the ability to achieve higher centrifugal velocities by employing the nozzles at sharper angles [2.8]. The feed material is carried centrifugally around the chamber in the gas flow and classified by size during this process. The fluid stream disturbs the rotating gas flow and vortices form at the back edges of the fluid stream. These, combined with the high relative velocity differences between particles already in the fluid stream and those just entering it, gives rise to effective particle-particle collisions, which leads to a selective size reduction of the particles via inertial forces. Coarse particles continue rotating around the grinding chamber owing to centrifugal forces, while vortex forces cause the finer particles to migrate towards the center of the chamber, where they fall into the centrifugal collector. Large particles are not carried very well in the fluid stream and as a result their collision frequency remains small. In contrast, very small particles are carried so well in the fluid stream that their probability of collision is also low, owing to their low relative velocity differences. As a result, the optimum grinding efficiency of jet mills is very dependent on the particle size [2.3].

Fluid jet mills have a number of advantages and special features that make them suitable for use with energetic materials:

- The carrier gas provides a self-cooling effect, which helps prevent temperature rise in the ground material.
- Because the particles are classified by size during the grinding operation, material build-up is effectively eliminated.
- The grinding chamber and product size can be very small and the residence time for the grinding operation is very short.
- The high size reduction energy and surface energy required to produce ultra-fine particles is applied indirectly, via the kinetic energy of the grinding gas.
- The fluid jet mill has no moving parts, such as rotating shafts or bearings.
- The equipment is subjected to very little wear and tear.

In a fluidized bed counter jet mill, the cylindrical grinding chamber is fed from below with feedstock. Two or more gas streams directed towards the center of the

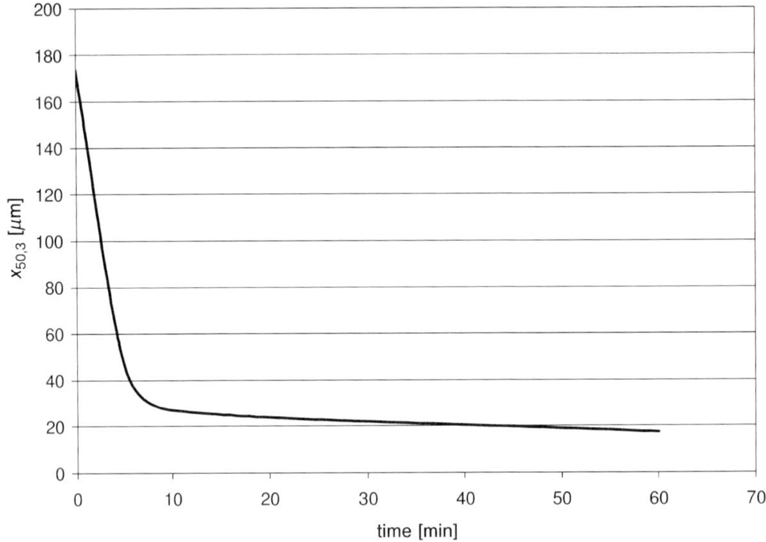

Figure 2.11. Size reduction as a function of time in a colloid mill.

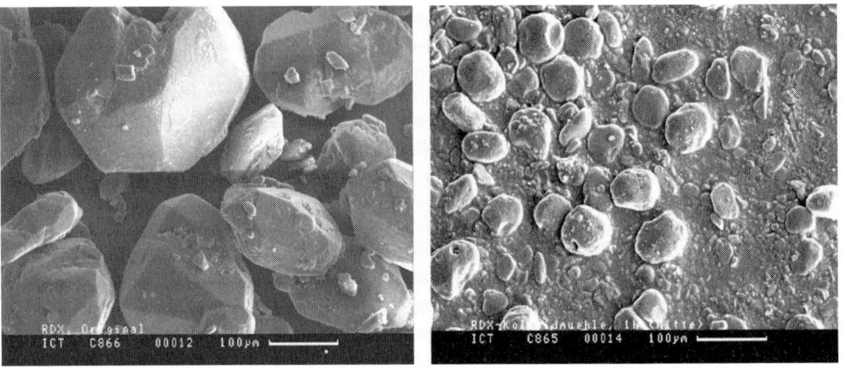

Figure 2.12. RDX ground with a colloid mill.

2.2.4 Grinding by Ultrasonic Energy

In acoustics, ultrasonic is defined as superaudible frequencies of sound, i.e. acoustic oscillations above 20 kHz [2.12]. In contrast to sounds in the audible range, where possible harm by noise has to be taken into account, high ultrasonic powers of up to several kilowatts can be generated, enabling ultrasonic energy to be used for cleaning, welding and machining of parts.

Common magnetostrictive and piezoelectric transformers are used for generating ultrasonic waves. They transform electronic oscillations into mechanical vibrations of the same frequency, which are then coupled into the medium.

There are different ways of transmitting the energy into the fluid. For example, a horn can be dipped into the fluid or energy can be transmitted by the walls of the vessel holding the fluid. In the latter case (e.g. in ultrasonic cleaning baths), ultrasonic energy is introduced into relatively large areas, resulting in a lower power intensity.

The process of size reduction of solids and emulsions by ultrasonic energy utilizes the effect of cavitation [2.13–2.16]. This effect is caused by the change in sonic pressure in the so-called suction phase of the ultrasonic oscillation. During this phase, small voids form owing to vacuum in the fluid. The suction phase is followed by a pressure phase during which the voids implode. During the short time of implosion, very high velocities occur and then immediately dissipate. The kinetic energy in a very small volume is transformed into thermal and potential energy (pressure). This leads to heating of the fluid and the propagation of a pressure shock in the form of a spherical wave. The order of magnitude of the shock wave is a multiple of the pressure caused by the sonic wave, which can be >10000 bar. In addition, local hot spots with temperatures above 5000 K can occur [2.17].

Ultrasonic grinding in an aqueous phase is ideally suited for handling energetic materials. In particular, the short stay in the comminution zone without confinement is very advantageous for these materials.

Cavitation can be characterized by the Thoma number [2.18]:

$$Th = \frac{p_i - p_V}{\frac{\rho}{2} \cdot u_i^2} \tag{2.8}$$

with fluid pressure p_i, vapor pressure of the fluid p_V, fluid density ρ and fluid velocity u_i. The necessary condition for cavitation is $Th \leq 0$.

The energy density E_m for the ultrasonic process can be determined as follows:

$$E_m = \frac{E}{m_{\text{Susp.}}} = \frac{1}{2} \cdot \omega^2 \cdot \hat{\gamma}^2 \tag{2.9}$$

where ω is the angular frequency and $\hat{\gamma}$ is the amplitude of the ultrasonic waves [2.19].

To characterize further the introduction of ultrasonic energy into the liquid phase, the intensity per effective outlet area of the ultrasonic device I is determined:

$$I = \frac{P}{A_{\text{Son}}} \tag{2.10}$$

where P is the electric power supply and A_{Son} is the effective outlet area of the ultrasonic sonotrode.

Figure 2.13 shows an experimental set-up for ultrasonic size reduction, consisting of a double-shelled vessel, a thermostat and an ultrasonic device [2.10]. The size reduction of HMX, RDX, NTO, AN and HNS by an ultrasonic device was studied at Fraunhofer ICT with the set-up shown in Fig. 2.13.

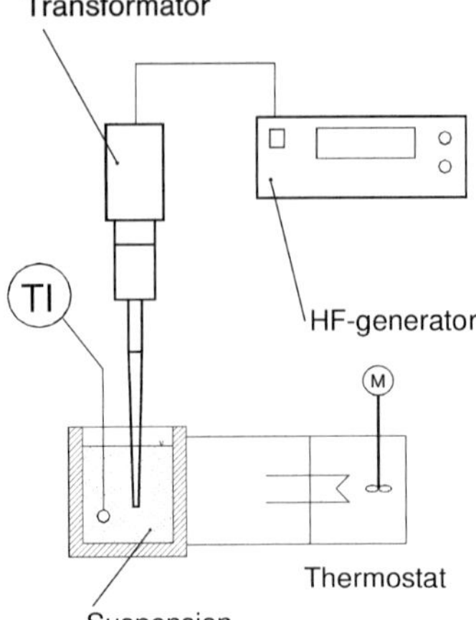

Figure 2.13. Set-up for ultrasonic size reduction.

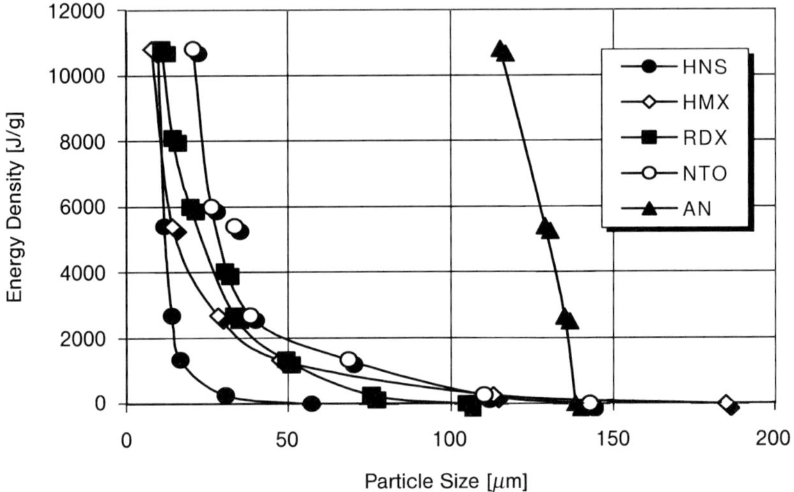

Figure 2.14. Comminution using ultrasonic energy.

Figure 2.14 shows results after grinding with an intensity of 28.2 W/mm^2. One can see that all the materials except ammonium nitrate and sodium chloride have similar comminution characteristics. These materials can be ground to a particle size of ~60 μm with a moderate amount of energy.

For the case of RDX, a curve based on the equation

$$\frac{\frac{E}{J}}{\frac{m_{Susp.}}{g}} = -\kappa \cdot \frac{1}{\alpha+1} \cdot \left[\left(\frac{x}{\mu m} \right)^{\alpha+1} - \left(\frac{x_0}{\mu m} \right)^{\alpha+1} \right] \tag{2.11}$$

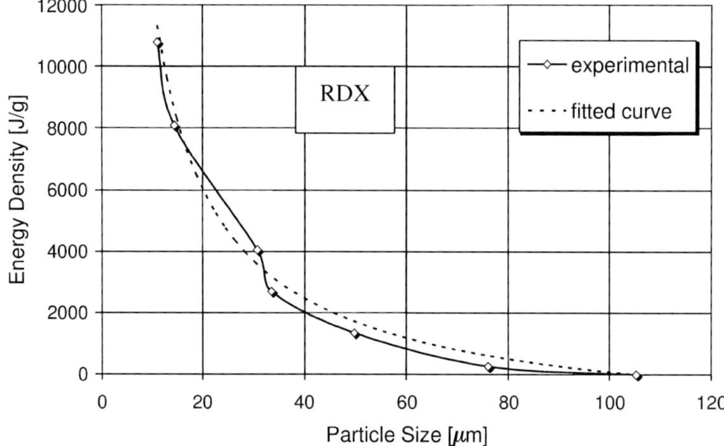

Figure 2.15. Calculated and experimentally determined energy demand for size reduction.

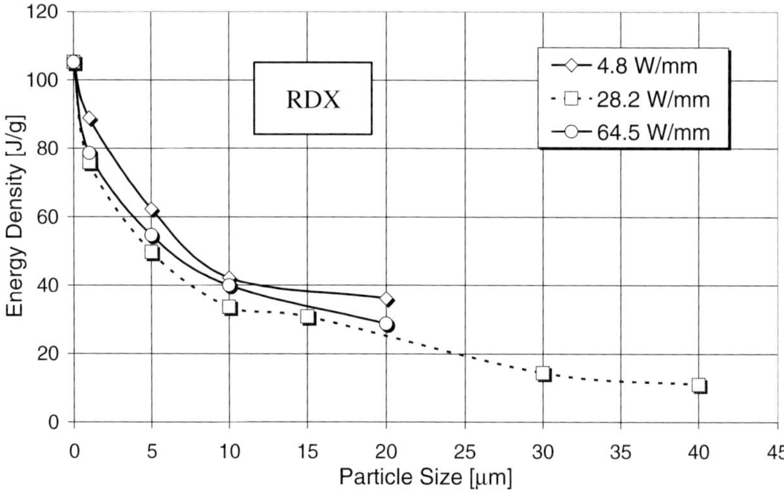

Figure 2.16. Size reduction of RDX.

was fitted to the data to determine the coefficients κ and α [cf. Eq. (2.4)]. The values obtained were $\kappa = 92\,896$ and $\alpha = -1.87$. Figure 2.15 shows the fitted curve and the measured energy demand for grinding of RDX (I = 28.2 W/mm^2).

On changing the ultrasonic intensity to 4.8 or 64.5 W/mm^2, the product particle size does not change significantly compared with use of an intensity of 28.2 W/mm^2. Figure 2.16 shows the results for size reduction of RDX in water using three different intensities. No significant change in the results was observed as a function of ultrasonic intensity.

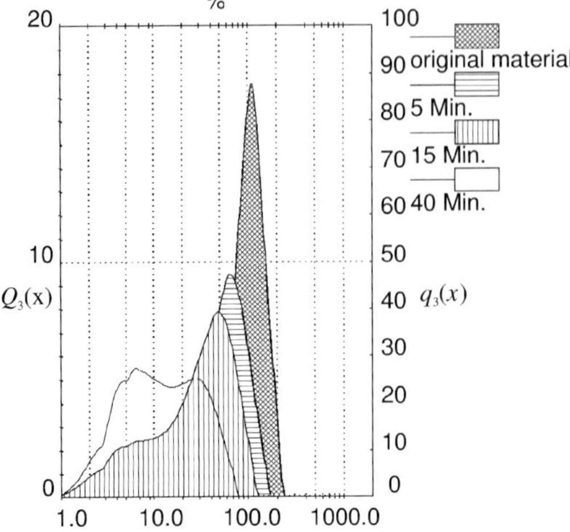

Figure 2.17. Particle size distribution of RDX during comminution.

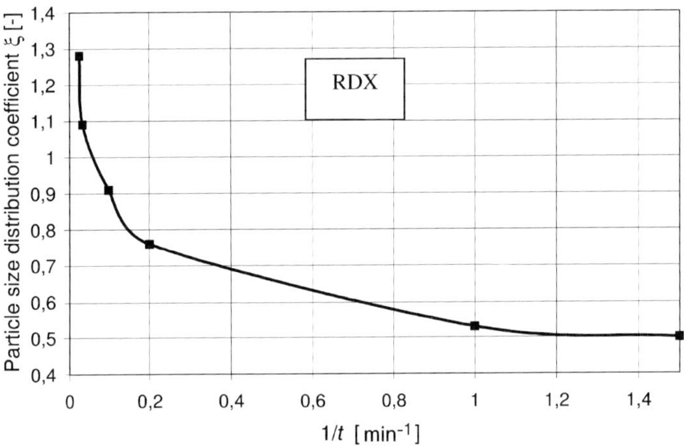

Figure 2.18. Change in the particle size distribution ξ of RDX during comminution.

Figure 2.17 shows the change in particle size distribution of RDX during size reduction for 40 min at an intensity of 28.2 W/mm^2. The width of the distribution clearly increases with time. The characteristics of the distribution are given in Fig. 2.18.

Figure 2.19 shows SEM images of RDX original material (left) and after size reduction for 40 min (right) at an intensity of 28.2 W/mm^2. A broad particle size distribution of the ground material is also evident in these images.

These results show how the product's mean particle size depends on the ultrasound intensity normalized by the effective outlet area of the sonotrode. The

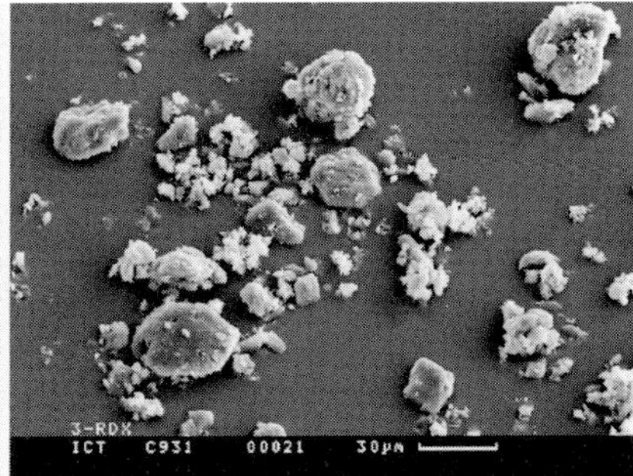

Figure 2.19. RDX original material (left) and after 40 min of size reduction using ultrasonic energy at an intensity of 28.2 W/mm^2 (right).

$I = 1$ W/mm^2 in paraffin oil

$I = 4.8$ W/mm^2 in paraffin oil

Figure 2.20. Ultrasonic jets of different intensities in paraffin oil.

transmission of ultrasonic energy into the liquid phase was investigated optically. Different intensities give rise to different cavitation behaviors in the liquid phase. Therefore, different size reduction characteristics occur. Figures 2.20 and 2.21 show ultrasonic jets of different intensity in water and paraffin oil, respectively.

The bulk viscosity of the suspension stayed constant during the experiments because the temperature was kept constant. However, hot spots formed in the liquid, causing variations in the local viscosity.

2.2.5 Rotor Stator Dispersing System

Toothed gear dispersers are high-speed rotor stator systems (Fig. 2.22). The coaxial oriented cylinder has radial indentations, which differ in geometry depending on

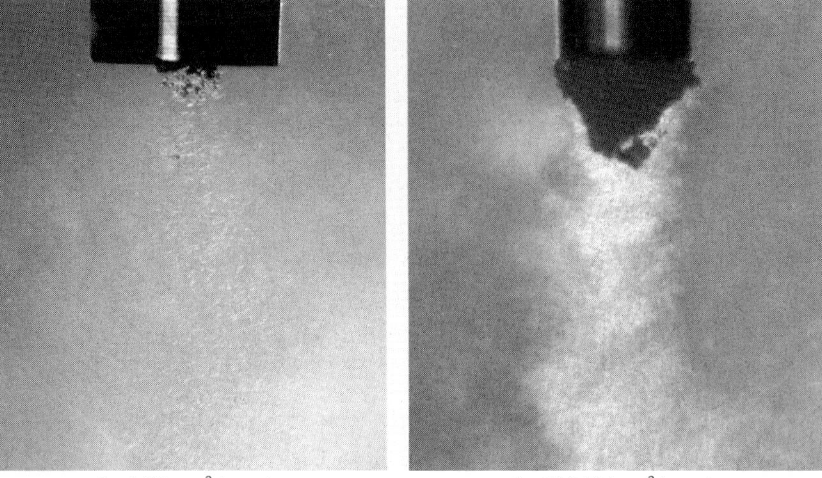

Figure 2.21. Ultrasonic jets of different intensities in water.

the industrial application. The suspension is carried into the center of the rotor stator system and exits the dispersion zone in the radial direction. As the feed suspension exits the teeth of the rotor, it is accelerated to a given radial velocity, then sheared in the gap between the rotor and stator and, upon entering the stator gap, it is decelerated to a radial velocity of zero. The size reduction in the toothed gear disperser occurs as a result of loads in the turbulent shear field [2.20–2.22]. Wiedmann [2.20, 2.21] showed that the mechanism for this size reduction is not cavitation, but instead results from the turbulent flow fields. When the process is continuous, the size reduction effect depends on the power density applied to the system and on the residence time of the material in the dispersion zone. Pedrocchi and Widmer [2.22] showed that the size reduction characteristics are also strongly influenced by the turbulent entry flow processes in the gap between the rotor and stator and the geometric features of the gap entrance.

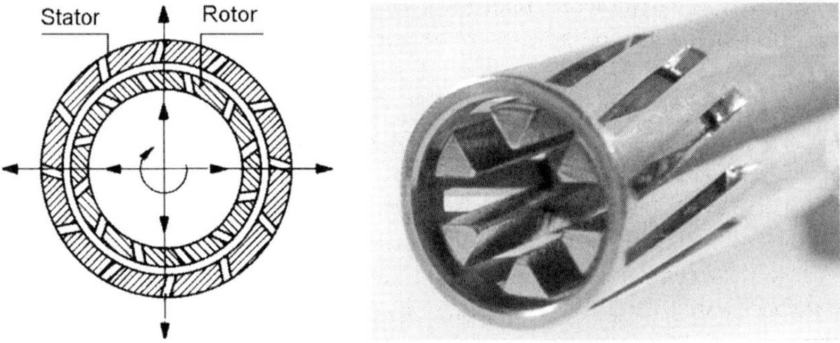

Figure 2.22. Rotor stator dispersing system.

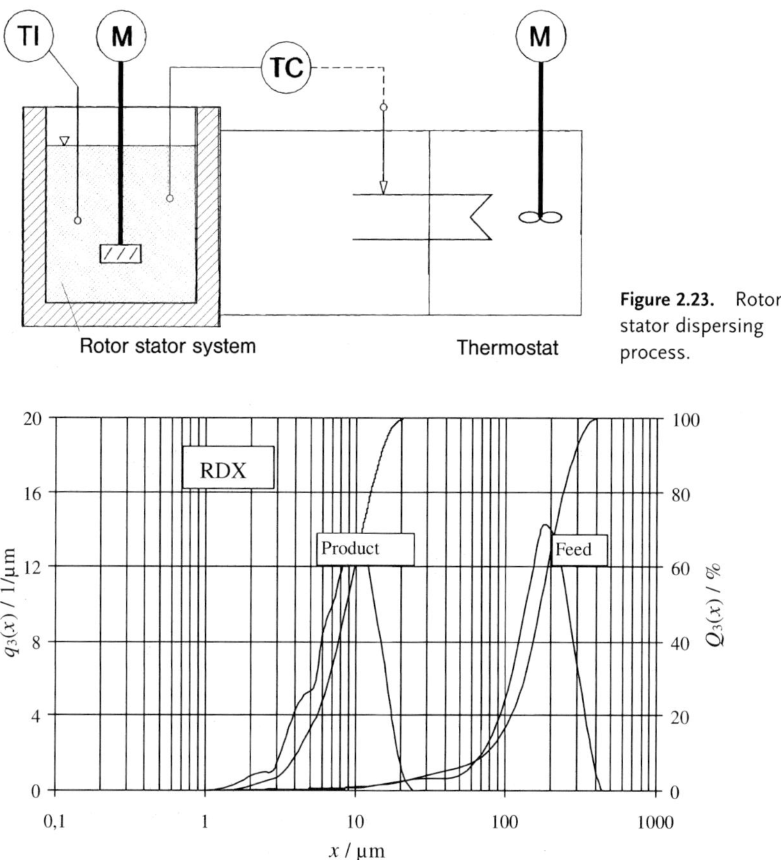

Figure 2.23. Rotor-stator dispersing process.

Figure 2.24. Size reduction of RDX using the rotor-stator system.

Depending on the size of the equipment used, rotational velocities up to 50 m/s can be obtained based on rotor revolution rates between 1000 and 20000 min^{-1}. The gap width ranges from 0.1 to 1.0 mm. A number of manufacturers offer toothed gear dispersers with multiple rotor-stator pairs oriented in series on a single axis, so that the suspension can be subjected to a multiple-step size reduction process. Because the fluid is accelerated centrifugally by the rotating rotor, toothed gear dispersers are self-feeding and a pump is thus only required for highly viscous materials. Toothed gear dispersers can be used in either batch or continuous operations; however, the batch method is usually employed with a rotor-stator dispersing system. Figure 2.23 shows a schematic design of such a process. The suspension is located in a double-shelled vessel, whose temperature is controlled during the process.

A toothed gear disperser was also used for the size reduction of crystalline energetic material [2.10, 2.23]. Two different scales were used, laboratory scale for preliminary investigations (1-l batch) and pilot scale (20-l batch). Figures 2.24 and

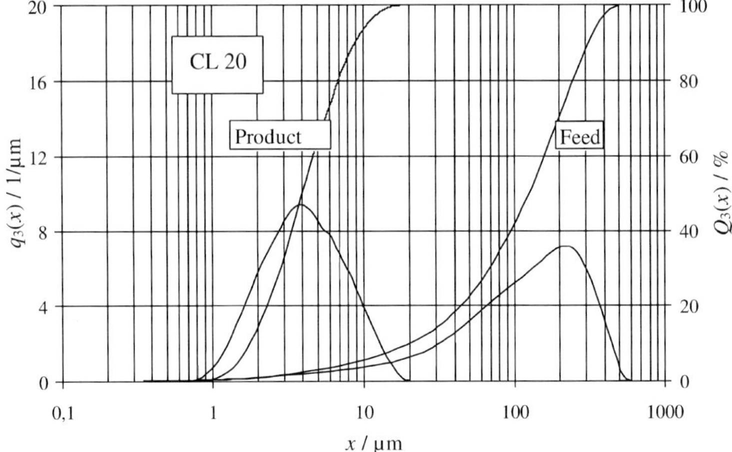

Figure 2.25. Size reduction of CL 20 using the rotor-stator system.

Table 2.5. Particle sizes and distributions after grinding.

Material	Feed $x_{50,3}$ (µm)	Product $x_{50,3}$ (µm)	Dispersivity ζ
RDX	174	8.7	0.48
CL 20	125	4.0	0.69

2.25 shows the size reduction of RDX and CL 20, respectively, after the same comminution time using the same parameters as in the laboratory-scale device. Table 2.5 shows the mean particle sizes and distribution coefficients from these experiments. The RDX product has a narrower particle size distribution but the mean particle size of CL 20 is smaller.

Figure 2.26 shows SEM images of ground RDX and CL 20 and the original materials.

Hommel [2.24] also reported on wet grinding studies of RDX and HMX using a rotor-stator disperser. He obtained mean particle sizes between 5 and 10 µm using feed materials with mean particle sizes ranging from 100 to 300 µm.

2.2.6 Agitator Ball Mill

Agitator ball mills comprise one of the largest groups of solid grinding apparatuses. Size reduction is accomplished in these mills when the bulk solid impacts freely flowing agitation balls within the grinding chamber. Agitator ball mills are used solely during wet grinding, i. e. a suspension of particles is ground by agitating balls within the grinding chamber. Such mills are used to grind colloids in the paint and coatings industry and also for pharmaceuticals and other chemicals. The feedstock size is <100 µm and the final product is <5 µm.

Figure 2.26. Size reduction using the rotor-stator system.

The grinding chamber usually consists of a cylindrical container, but in the special case of the shear gap ball mill the grinding chamber is a flat gap between two rotating bodies. The agitating medium is usually spherical. Loads are imparted on the bulk solid when they are impacted by the balls or between the balls and the chamber wall. Types of loads include impact, compression and shear. Energy to set the balls in motion can be imparted either by agitating the grinding chamber and its entire contents or by stirring the contents of the container (or in the case of the shear gap ball mill, stirring within the gap between the rotor and stator. The grinding chamber must be cooled during the operation, because the majority of the energy imparted is transformed into heat. Figure 2.27 shows a conventional agitator ball mill (a) and shear gap ball mill (b).

The agitation balls may be made from various materials; one can even use the same material as the bulk solid being ground. For relatively soft bulk solids, the balls can be made from simple, economical materials, such as glass, whereas for harder solids, substances such as ceramics are used to prevent abrasion. The size of the agitation balls is tailored to the material being ground; the finer the desired product, the smaller are the agitation balls employed. The finest size reduction obtained in such mills uses agitation balls with diameters from 1 mm to 10 μm. The grinding chamber volume of an agitator ball mill is filled between 65 and 90% with agitation

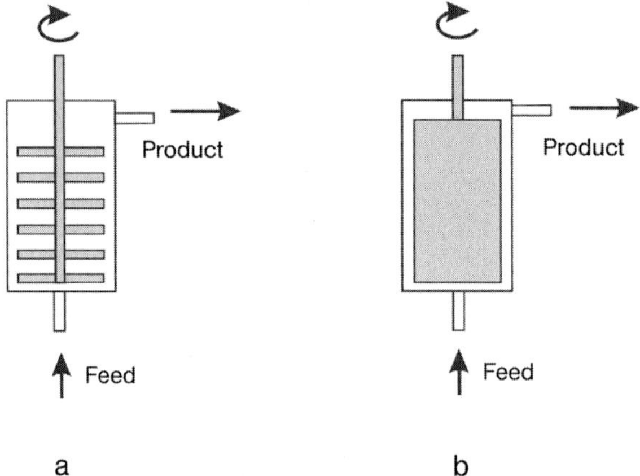

Figure 2.27. Schematic of an agitator ball mill (a) and shear gap ball mill (b).

balls. The balls are held within the chamber by a separation device, such as a sieve.

The agitators can be very different, depending on the manufacturer. They may be smooth, perforated, grooved, fluted, serrated or ribbed or they may have rotor blades. In shear gap ball mills, the rotor is simply a smooth cylinder. The agitators rotate at high speeds of 4–20 m/s, giving rise to a centrifugal acceleration of >50 g. The agitation balls are accelerated to velocities approaching the agitator velocity near the agitators and are slowed as they impact the chamber walls. In between, they should impact as many particles as possible over the largest possible surface area, in order to achieve maximum size reduction.

The power requirements of an agitator ball mill and shear gap ball mill can be quantified via the Newton number [2.25]:

$$Ne = \frac{P}{\rho_S \cdot n^3 \cdot d^5} \tag{2.12}$$

where P is the power required, ρ_S is the density of the suspension, n is the agitator rotation rate and d is the outer diameter of the agitator.

At the high Reynolds numbers and turbulent flow conditions that exist in ball mills, the Newton number is constant. It is thus a useful dimensionless number for sizing equipment; for instance, the power requirements of a laboratory mill can be used to calculate the power requirements of a production-scale machine.

Other important parameters are the hardness and particle size of the agitation balls. The product throughput can be set independently of other operating parameters. The maximum throughput limit is approached when the flow forces of the agitation balls against the suspension outlet begin to cause undesirable compression or breakup of the agitation balls themselves.

With shear gap ball mills, care must be exercised because the agitation balls are enclosed in a very narrow gap. They must therefore be no larger than one-quarter of the gap width, because otherwise wedging of the balls can occur within the gap. However, they should not be much smaller or the grinding efficiency suffers.

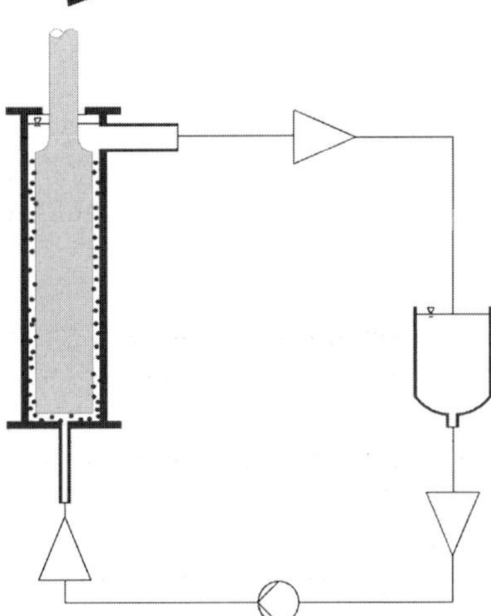

Figure 2.28. Annular gap mill.

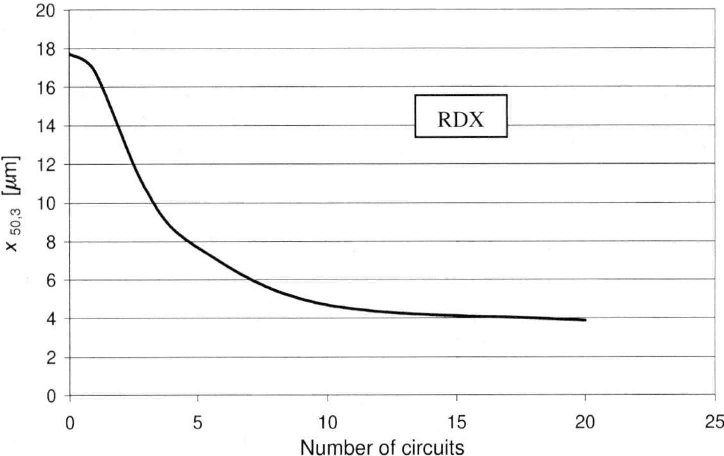

Figure 2.29. Size reduction using an annular gap mill.

The advantage of shear gap ball mills over conventional agitator ball mills is the very narrow energy spectrum required for the operation and the correspondingly small residence time distribution of the ground material. As a result, very narrow particle size distributions can be obtained. In addition, the grinding chamber volume to the radiating surface ratio is favorable, so that this method is very well suited for thermolabile substances.

Size reduction of RDX was accomplished using an annular gap mill. Figure 2.28 shows a schematic for such a process. Milling balls of zirconium oxide were used. Figure 2.29 shows the size reduction as a function of milling time. The mean

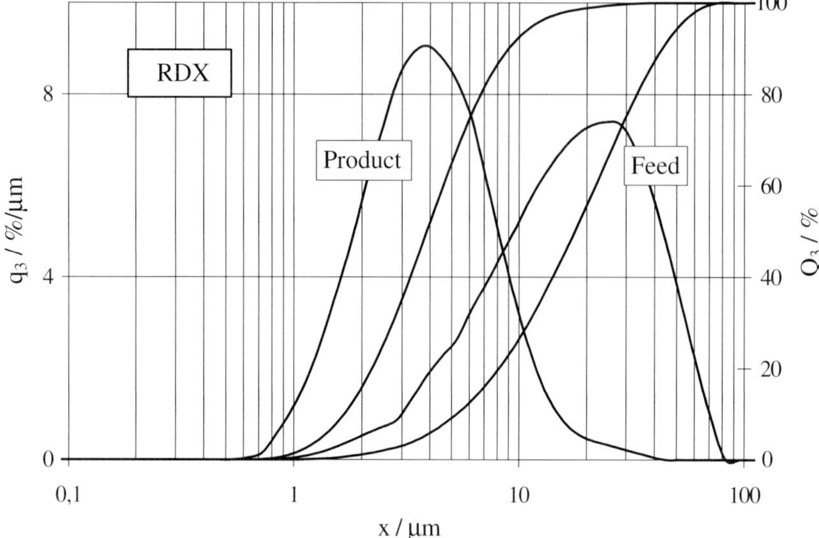

Figure 2.30. Size reduction of RDX using an annular gap mill.

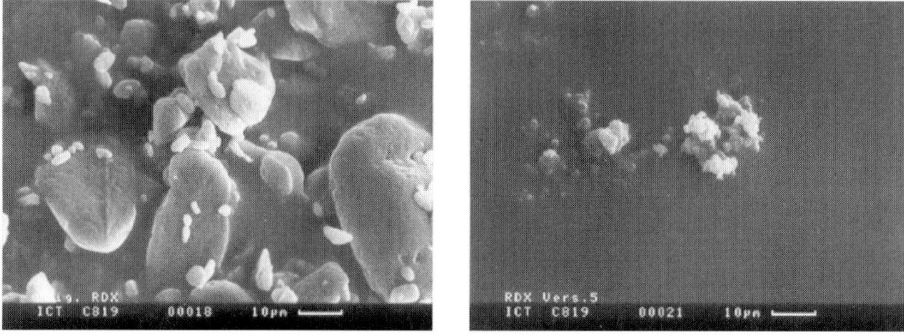

Figure 2.31. Product of an annular gap milling process.

particle size of the feed was 18 µm. The most effective size reduction occurred in the first 10 min and the product had a mean particle size of 4 µm. The particle size distribution of the product was even narrower than that of the feed material, as seen in Fig. 2.30. Figure 2.31 shows SEM images of the original and ground RDX.

In recent work [2.10], a two-step process for size reduction of energetic materials was proposed. Following pre-grinding in a colloid mill, one obtains RDX with a broad particle size distribution around a mean of 170 µm. This product has an optimum size for being ground further in a shear gap ball mill. The product from this second grinding step is a narrowly distributed RDX with a mean particle size in the sub-micron range.

Kleinschmidt [2.11] also reported on the use of an annular gap mill for size reduction of HMX, RDX and AP. The annular gap mill he used had a different

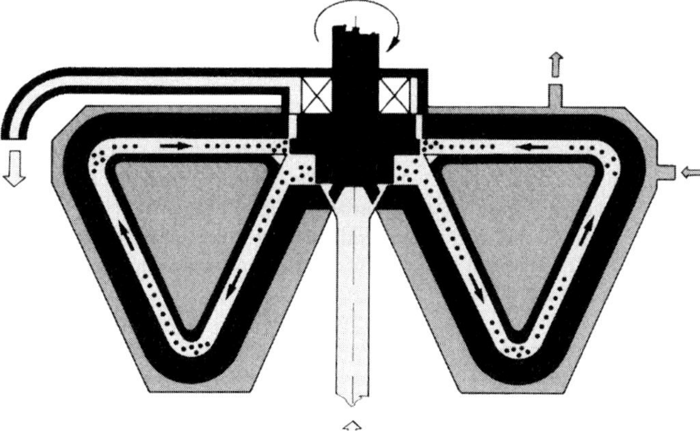

Figure 2.32. Annular gap milling process [2.11].

geometry, with a conical rotor and stator. This kind of mill is shown in Fig. 2.32. Mean particle sizes between 1 and 20 μm were obtained in Kleinschmidt's investigations.

Unfortunately, it is impossible to identify a single best size reduction process for energetic materials. The optimum process for a particular material depends on various considerations, the most important being the material properties of the feedstock and the desired product. Several publications address this topic [2.26, 2.27].

2.3 References

2.1 Prior MH (1990) Size reduction. In: Rhodes M (ed.), *Principles of Powder Technology*, Wiley, Chichester, pp. 227–297.

2.2 Rhodes M (1998) *Introduction to Particle Technology*, Wiley, Chichester.

2.3 Vauck W, Müller H (1994) *Grundoperationen chemischer Verfahrenstechnik*, Deutscher Verlag für Grundstoffindustrie, Leipzig.

2.4 Rumpf H (1965) Die Einzelkornzerkleinerung als Grundlage einer technischen Zerkleinerungswissenschaft, *Chem.-Ing.-Tech.* **37**, 3.

2.5 Stairmand CJ (1975) The energy efficiency of milling processes. In: *4. Eur. Symp. Zerkleinern, Nürnberg*.

2.6 Stiess M (1994) *Mechanische Verfahrenstechnik 2*, Springer-Verlag, Berlin.

2.7 Zogg M (1993) *Einführung in die Mechanische Verfahrenstechnik*, Teubner, Stuttgart.

2.8 Eskin D, Kalman H (2002) Engineering model of friction of gas-solid flow in a jet mill nozzle, *Chem. Eng. Technol.* **25**, 1.

2.9 Teipel U (1999) Production of particles of explosives, *propellants, explosives, Pyrotechnics* **24**, 134–139.

2.10 Mikonsaari I, Teipel U (2001) Zerkleinerung Energetischer Materialien in wäßrigen Lösungen. In: *Proc. 32nd Int. Annual Conference of ICT, Karlsruhe*.

2.11 Kleinschmidt E (1998) Mahlen von Explosivstoffen. In: *29th Int. Annual Conference of ICT, Karlsruhe*.

2.12 Bergmann L (1956) *Der Ultraschall und*

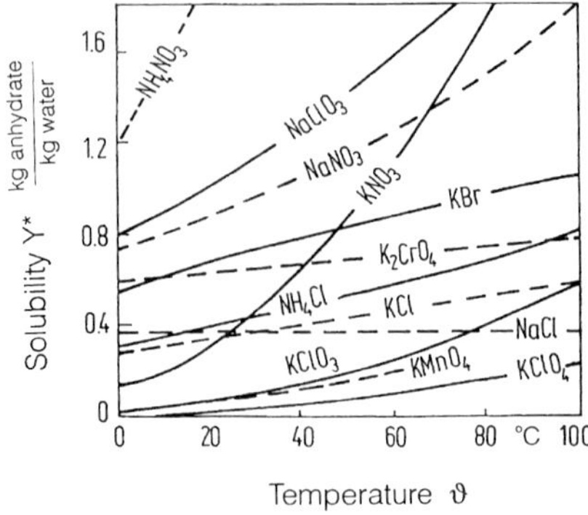

Figure 3.1. Solubility curves for different substances [3.1].

processes is achieved by the different chemical potential of the solute in the solution and solid state. The chemical potential μ_i of the solute i in the solution is

$$\mu_{i,\text{Sol}} = \mu_{0,i} + RT \ln a_i \tag{3.1}$$

For the equilibrium the chemical potential becomes

$$\mu^*_{i,\text{Sol}} = \mu_{0,i} + RT \ln a^*_i = \mu_{i\,s} \tag{3.2}$$

This gives the following relationship for the driving force:

$$\Delta\mu_i = \mu_{i,\text{Sol}} - \mu_{i,s} = RT \ln \frac{a_i}{a^*_i} = RT \ln S = RT \ln \frac{c_i}{c^*_i} = RT \ln \left(\frac{c_i - c^*_i}{c^*_i} + 1 \right) \tag{3.3}$$

Here a_i is the activity of component i, S the supersaturation and c_i the concentration of component i.

The supersaturated system then attempts to achieve thermodynamic equilibrium through nucleation and the growth of the nuclei. If a solution contains neither solid foreign particles nor crystals of its own type, nuclei can be formed only by *homogeneous nucleation.* If foreign particles (e.g. undissolved impurities) are present, nucleation is facilitated and the process is known as *heterogeneous nucleation.* Both homogeneous and heterogeneous nucleation take place in the absence of solution-own crystals and are collectively known as *primary nucleation.* This occurs when a specific supersaturation, known as *metastable supersaturation,* is obtained in the system. However, in industrial crystallizers it has been observed that nuclei occur even at very low supersaturation when solution-own crystals exist, for example, in the form of attrition fragments or added seed crystals. Such nuclei are known as *secondary nuclei.* Figure 3.2 illustrates supersaturation against solubility for several types of nucleation processes.

The growth of crystals in a supersaturated solution is a very complex process that has not been well understood up to now. The reason for this is that supersaturated solutions are composed of a variety of units and can have a certain structure. Little is known about the various species and structure. According to the model of

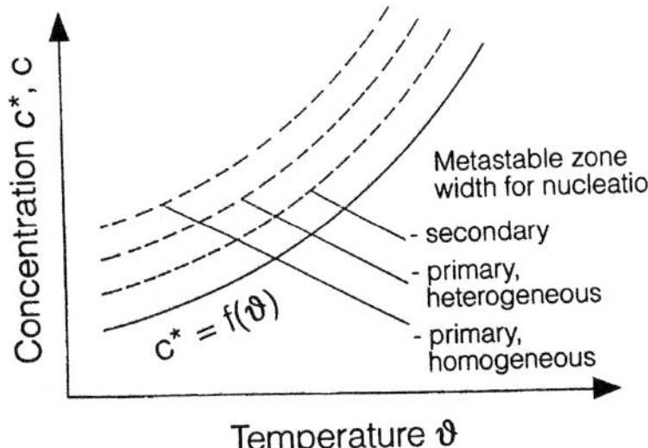

Figure 3.2. Metastable supersaturation against temperature for several types of nucleation processes [3.1].

Berthoud [3.2] and Valeton [3.3], a crystal surface grows in such a way that units in the supersaturated fluid are first transported by diffusion and convection and then integrated into the crystal structure with supersaturation being the driving force. This integration will take place preferably at the kink site positions of the crystal surface; kink sites usually provide the highest energy release since a higher number of bonds can be established between the growth unit and the crystal surface compared with an isolated position on a terrace or along a step. The integration step includes surface diffusion of a growth unit to a kink site. Depending on the system, state of flow (e.g. stirrer speed, stirrer geometry) and supersaturation the first or second step can be determined the entire process or both steps can control growth to different extents (see Fig. 3.3). Both steps can be generally described by the following kinetic statements:

$$r_D = k_D \cdot (c_\infty - c_i) = \frac{D}{\delta} \cdot (c_\infty - c_i) \tag{3.4}$$

and

$$r_r = k_r \cdot (c_i - c^*)^n \tag{3.5}$$

For $n = 1$, the overall kinetics can be written as

$$r_{tot} = k_{tot} \cdot (c_\infty - c^*) = \left(\frac{1}{k_D} + \frac{1}{k_r} \right)^{-1} \cdot (c_\infty - c^*) \tag{3.6}$$

where D is the diffusion coefficient, δ is the diffusion layer thickness, c_∞ is the concentration in the solution, c_i is the concentration at the crystal-solution interface and c^* is the equilibrium concentration (see Fig. 3.3).

At low supersaturations, the system is close to the thermodynamic equilibrium so that the crystal surface is very smooth and homogeneous. Therefore, fewer dislocations at the surface occur and the integration step is impeded; the crystal growth is limited by the integration step. At high supersaturations, the crystal surface is roughened and many dislocations and steps occur at the surface where the integration of crystal units is favored; the crystal growth is limited by the transport step.

This two-step-model is a strong simplification of the real processes that occur

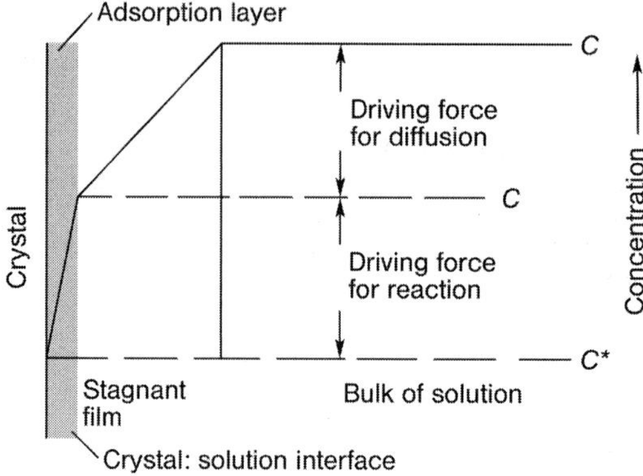

Figure 3.3. Concentration driving forces in crystallization from solutions [3.1].

during the crystal growth. Elwell and Scheel divided the growth processes of a crystal into nine individual process steps [3.4].

Several growth models based on crystal surface nucleation, followed by the spread of the monolayers, have been developed in recent years. Layer growth can be subdivided into two mechanisms: a two-dimensional and a spiral growth mechanism. The former is characterized by the fact that a two-dimensional island nucleates on the crystal surface which spreads laterally, in this way completely covering the surface before a new two-dimensional island is formed on top of the previously finished layer. A more realistic view is that islands are formed simultaneously, so also on the top of newly formed islands. These models are known as 'birth and spread' (B+S) or 'polynuclear growth' models. The growth develops from surface nucleation that can occur at the edges, corners and on the faces of a crystal. Further surface nuclei can develop on the monolayer nuclei as they spread across the crystal face. The (B+S)-model results in the following face growth velocity-supersaturation relationship:

$$v = A_1 \cdot \sigma^{5/6} \cdot \exp(A_2 / \sigma) \tag{3.7}$$

where A_1 and A_2 are system-related constants and $\sigma = (c - c^*)/c^*$.

This type of layer growth only occurs if the supersaturation is relatively high and consequently the nucleation barrier for the formation of the two-dimensional islands is relatively low. The formation of a two-dimensional nucleus is similar to that of a three-dimensional nucleus, hence also in the former case a critical nucleus – the radius of which is determined by the supersaturation – has to be formed before this nucleus is able to increase its size.

When dislocations with a screw component emerge at a crystal surface, this surface may grow layer by layer by means of a so-called spiral growth mechanism. A screw dislocation provides the constant presence of a step on the crystal surface on which this dislocation emerges. Since a step is a preferred site for growth units to integrate into the crystal structure, this step develops as a spiral on top of the crystal surface. The basic ideas for this growth mechanism were developed by Burton *et al.* and called the BCF model [3.5].

The ratio between nucleation rate and growth rate determines the size distribution of the crystals at the end of the crystallization process. A high level of supersaturation results in a high nucleation rate. Therefore, nearly all supersaturation is reduced by forming a large number of crystal nuclei and a low growth rate is achieved. Small particles are produced and the agglomeration of these fine particles is determinant for the product quality. If the nucleation rate is low, larger particles are formed and the attrition mechanism and therefore the secondary nucleation can play a decisive role in the product quality. The level of supersaturation in the system can be adjusted by the rate of cooling or evaporation. In order to guarantee a reproducible product quality in industrial crystallization, a low level of supersaturation is adjusted and the nucleation is mostly started by seeding the solution with seed crystals.

3.1.2 Crystallization Apparatus and Process

After having explained some basic principles of nucleation and crystal growth in the previous section, several crystallization techniques and equipment will now be described. Here we will focus on techniques which are most frequently applied for the crystallization of energetic materials.

Normally, continuous crystallization processes are used if there is a considerable market for the material being produced. Moreover, batch crystallization is more flexible with respect to meeting the client's wishes regarding mean size and size distribution of the particles, since variations in the process conditions leading to changes in the product characteristics can be more easily applied to batch than to continuous crystallization processes. A disadvantage is that batch crystallization is more sensitive to so-called batch-to-batch variations, which means that it is much more difficult to reproduce a batch crystallization process, even if exactly the same process conditions are applied.

The majority of the crystallzation processes involving energetic materials use the crystallization from a solvent rather from the melt of the particular material, although of course important exceptions to this rule exist (e.g. TNT, AN and ADN). The choice of the crystallization technique which is most suited for the production of a certain compound depends on, apart from other factors, the solubility of the specific substance in the solvent and the demands imposed on the final product. Another important issue is that generally energetic materials are thermally less stable, implying that high temperatures have to be avoided. The commonly used techniques will be briefly addressed in the following sections. A summary of generally applied crystallization techniques is given in Table 3.1.

3.1.2.1 Melt Crystallization

Crystallization from the melt can be operated either as a suspension growth or as a layer growth technique. In the former case, a melt is undercooled, after which nuclei appear. In the latter case, the melt solidifies on a wall, after which the material is

Table 3.1. Summary of the characteristics of a selection of generally applied crystallization techniques.

Crystallization technique	Characteristics
Cooling	High solubility of material to be crystallized, significant dependence of solubility on temperature, low supersaturation level, growth dominates nucleation, relatively large mean crystal size, moderate yield, operated both batchwise and continuously
Evaporation	Moderate to high solubility, low temperature dependence of solubility, low supersaturation level (except in boiling zone), growth dominates nucleation, relatively large mean crystal size, moderate yield, operated both batchwise and continuously
Drowning-out	Low solubility, high supersaturation level, nucleation dominates growth, high yield, generally large volumes of solvents involved, generally operated batchwise, agglomeration effects, mixing

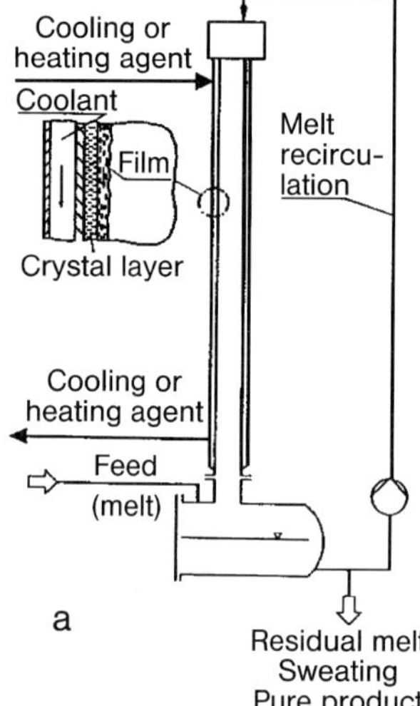

Figure 3.4. Wetted wall crystallizer (Sulzer-MWB) [3.1].

removed either by scraping it from the wall or by melting it again. This technique is mostly used for purification and no solid particles with special characteristics are formed. Figure 3.4 gives an example of an industrial layer melt crystallization apparatus.

A special melt crystallization technique is prilling: droplets of melt are sprayed in a tower after which the droplets start to fall, during which they cool and solidify. More or less spherical particles can be produced by this method. This technique will be described later in respect of the crystallization of ADN and phase-stabilized ammonium nitrate (PSAN).

3.1.2.2 Cooling Crystallization

If a substance has a high solubility in a certain solvent and if the solubility has an appreciable dependence on temperature changes, cooling crystallization is the most suitable method. As a large amount of growth units is present in the solution, sufficiently large crystals can be grown, up to mean sizes of millimeters. Since the supersaturations can be fairly low, cooling crystallization will be dominated by secondary rather than primary nucleation.

Process conditions which can be varied during cooling crystallization are the cooling rate and/or the cooling rate profile, stirrer rotation rate, stirrer type, initial and final temperature and the use of seed crystals to initiate the nucleation. In batch mode, a simple operating method consists in cooling the solution at a constant cooling rate. However, this is not optimal since at the beginning of cooling no seed surface or, after seeding, only a small seed surface is available, creating very high supersaturations followed by extensive nucleation. At the end of the cooling process, the crystal product may have a large surface, but it still grows very slowly owing to the low supersaturation. Therefore, it is beneficial to set the cooling rate so that supersaturation remains almost constant during the cooling period.

When solid material is recrystallized from a solution in a crystallizer, the suspension must be mixed and deposition must be avoided. Generally, three types of crystallizers are used in industrial applications. These three types differ with regard to the kind of circulating device which is used to avoid deposition of the crystals. The three types are (Fig. 3.5):

- fluidized-bed (FB),
- forced-circulation (FC) and
- stirred vessel (STR) crystallizers.

It must be determined whether the entire suspension including coarse particles or only that part of the flow containing crystals smaller than a special particle size (e.g. 100 μm) is to be circulated. In the former case, relatively strong attrition occurs, especially of large particles. As attrition fragments can act as effective secondary

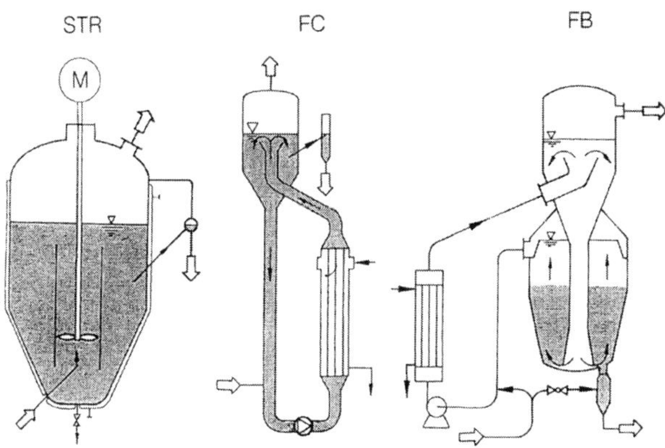

Figure 3.5. Typical industrial crystallizers [3.1].

crystal nuclei, crystal size distribution and therefore the product quality are often determined by attrition processes. The fluidized-bed crystallizer differs from the other two types of crystallizer by the fact that the suspension flow containing only small particles is conducted by the circulating pump. This is the reason why such crystallizers generally yield a coarser product than stirred vessel and forced-circulation crystallizers. FC and FB crystallizers have the advantage over stirred vessels that the ratio of the heat exchange surface to the crystallizer volume can be maintained owing to the external heat exchanger when scaling up the crystallizer.

3.1.2.3 **Evaporation Crystallization**

Materials for which the solubility in a certain solvent shows only a slight dependence on temperature cannot be crystallized very efficiently by cooling crystallization (e. g. sodium chloride in water). In this case evaporation of the solvent is much more effective. Thermally unstable substances can be crystallized reducing the pressure to evaporate the solvent. In principle, the industrial used apparatuses are similar to those used for cooling crystallization.

3.1.2.4 **Precipitation and Reaction Crystallization**

Precipitation, also called drowning-out or salting-out, uses the fact that the solubility of a substance in a certain solvent is reduced by adding a non- or anti-solvent. Upon reducing the solubility a supersaturation is created which induces the nucleation of crystals. The second solvent is selected for its good miscibility with the primary solvent. This method is usually applied to the crystallization of substances which show moderately to high solubility. Since the method can be conducted at ambient temperature, this technique is especially useful for heat-sensitive materials and very high production rates (> 90 %) can be achieved. A disadvantage is that usually large volumes of solvents are wasted, unless a recycling process is included, in order to regenerate the process streams.

In order to design and scale a precipitation process, it is necessary to know the phase behavior of the ternary system. This is more complex than the solubility curve of a binary system (Fig. 3.1). Figure 3.6 shows the triangular diagram of the system sodium sulfate, water and methanol. In the diagram the solubility line at 40 °C is included.

Because during precipitation processes two feeds are mixed, local supersaturation levels can become very high, leading to primary nucleation. In order to control the nucleation, intensive mixing of the solution is imperative. Also, careful selection of the position of the feed point with respect to, e. g., the impeller or the use of multiple feed points are methods which may be applied to improve further the controllability of the precipitation process.

In homogeneous reaction crystallization, one or more reactants react with one or more components in a liquid phase. Crystallization will take place as soon as the solubility product of the two reactants in the solution is exceeded. This crystallization technique is used for sparingly soluble substances. Hence, in order to reach

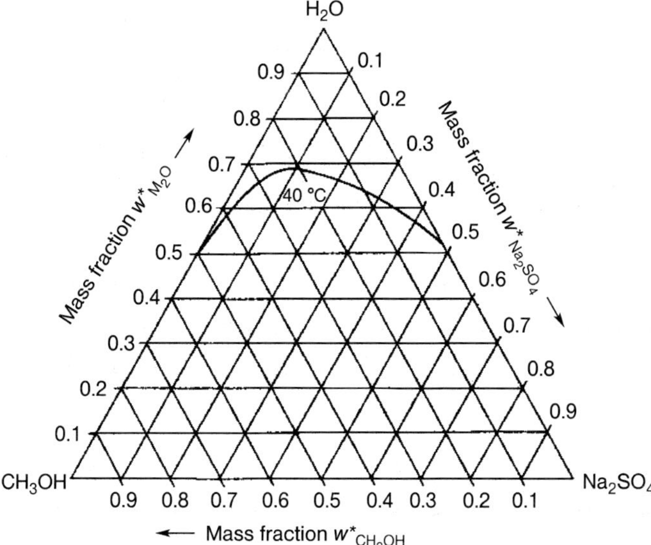

Figure 3.6. Solubility of three-component Na_2SO_4–H_2O–CH_3OH system at 40 °C [3.1].

sufficiently high yields, high supersaturation levels are required. Therefore, primary nucleation will be the dominant nucleation mechanism. As a result of the relatively high supersaturation, a large amount of nuclei will be formed which will grow to significantly smaller crystal sizes compared with evaporative or cooling crystallization. The reactants must first be macromixed and then micromixed on a molecularly disperse basis (the micromixing time is generally shorter than the macromixing time). In very fast reactions, the reaction rate and thus the product formation rate depend on the mixing time and, in very slow reactions, they depend on the reaction time.

In crystallizers for precipitation and reaction crystallization, the addition of at least one reactant or drowning-out agent is necessary. In principle, such crystallizers can be operated batchwise or continuously. In batch crystallizers, a reactant is sometimes fed into the vessel, which already contains the other reactant or the solution. It is also possible to add both reactants simultaneously. This is usually the operating mode of continuous crystallizers, with either a mixed product or classified product removal. The product characteristics depend on the mixing of the reactants. In Fig. 3.7 two different mixing devices used in industrial precipitation reactors are shown. First the reactants must be blended by macro- and micromixing. In a stirred vessel the macromixing time depends mainly on the impeller speed, whereas the process of micromixing is strongly dependent on the local specific power input. After mixing on a molecular scale the chemical reaction takes place.

High supersaturation is given in a T-mixer, assuming that the chemical reaction has finished but nucleation has not yet taken place. In stirred vessel reactors the

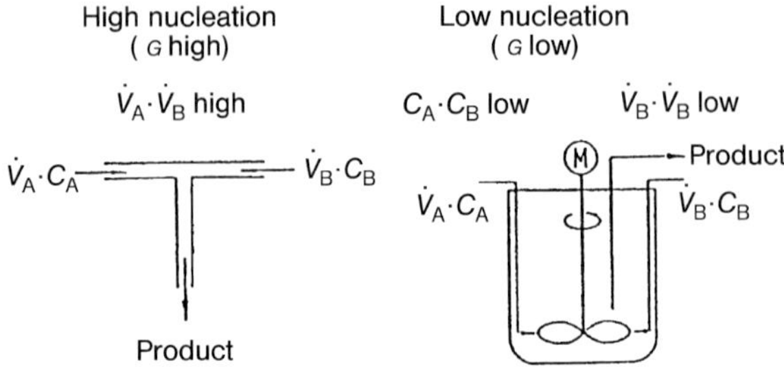

Figure 3.7. Different mixing devices: left, T-mixer; right, stirred vessel precipitation crystallizer [3.1].

supersaturation is smaller owing to a less intensive micro- and macromixing process. Studies investigating the macro- and micromixing processes with respect to precipitation can be found in the literature [3.6].

3.1.3 Crystal Defects

In each real crystal are places in which the regular lattice is disturbed. Generally such crystal defects are divided into three categories:

1. chemical impurities,
2. crystal imperfections and
3. electronic defects.

The presence of defects is indispensable in the form of edge dislocations, since thereby kinks are formed at which the crystal building components energetically favorably store themselves for the crystal growth. It should be noted that defects are characterized by an increased energy state in the lattice. In crystalline explosive components, therefore, the energy at defects, which can be supplied from the outside for detonation initiation, is lowered in relation to unimpaired crystal areas. Depending upon the type of defect, different causes are responsible for the emergence of such defects.

Electronic defects are characterized by missing or surplus electrons. These can be produced by the installation by foreign atoms, which have more or less free electrons than the regular atoms. During the production of semiconductors, this effect is used by a direct addition of foreign atoms. Chemical impurities have different effects on the crystal structure. With foreign atoms, which possess a similar chemical structure to the regular crystal building component, this leads to only a small disturbance. In other cases it can result in the formation of mixed crystals or clusters (voids), which lead to distortions and stresses in the lattice structure.

Crystal imperfections form directly during the crystal growth. They are classified in accordance with their dimensions into four classes:

- Point defects, which are called also zero-dimensional errors, are lattice dislocations. They are caused by vacancies or foreign atoms or molecules which are filled lattice sites or lattice gaps.
- Dislocations are of linear order of magnitude. One differentiates between helicoidal and edge dislocations: helicoidal dislocations are developed from lining up of edge dislocations in a 90° direction, whereas edge dislocations are formed owing to strong shear stress on the crystal. As mentioned above, the crystal growth takes place at such dislocations, since the building units which can be inserted is surrounded by up to three equal neighbors already in the 'kink' position. With an integration into an even crystal face only one neighbor would be available, which means in relation to the kink position an energetically more unfavorable status.
- One can interpret lining up of dislocations as two-dimensional, planar crystal imperfections. It also includes small-angle grain boundaries besides phase boundary surfaces, which result from a regular arrangement of edge dislocations. Boundary surfaces exist not only between different crystallites – then one speaks of grain boundaries – but also at gas or liquid inclusions.
- Such gas or liquid inclusions are three-dimensional (volume) defects, which result from irregular growth. If surface areas slow or terminate their growth, neighboring areas to them grow further, leading to the formation of inclusions, which are strengthened at dislocation lines.

For detecting such defects, different methods are available. Today, modern analytical devices offer the possibility of detecting and quantifying all types of defects. For example, high-resolution electron microscopes can even detect point defects. However, such highly sensitive methods are complex and expensive and are only rarely suitable for continuous investigations. Since these methods were developed mostly for investigations of single crystals, transfer to investigations of bulk crystals is difficult or impossible, so that in these cases a defect analysis can be carried out at the most qualitatively. Detecting chemical impurities is relatively unproblematic, since for this analytical chemistry methods can be applied. In addition, chromatographic (gas or liquid) and spectroscopic methods can be used to examine crystals for their purity. These methods are material specific and permit quantitative characterization in relation to a defined standard.

For detecting of gas or liquid inclusions, two competing methods are available. With the help of optical microscopy, inclusions can be made visible in transparent crystals. Since with this method only a limited number of crystals can be examined, a density measurement should be carried out for the determination of quantitative data. If the enclosed material differs in density from the pure crystal, larger quantities of crystals can be characterized regarding to the presence of inclusions.

Grain boundaries, and thus area defects, can be proven microscopically after complex sample preparation. Methods that use reflection or diffraction of X-rays supply information about lattice structures, from which conclusions can be drawn on the crystal modification present and possible lattice distortions.

For the formation of 'hot spots' in energetic materials, defects with a minimum size of 5 μm must be exist. Therefore, the characterization of the defects can be restricted to volume defects, e.g. inclusions and clusters. The other crystal errors specified above usually possess substantially smaller expansions and can therefore be ignored.

Inclusions are a frequent source of problems in industrial crystallization. They can cause caking of stored crystals by the seepage of liquid if the crystals become broken. Inclusions in energetic materials increase the sensitivity against shock initiation and fast heating so that it is of great interest to minimize the amount of inclusions in the crystalline material.

Inclusions can be classified as primary (formed during growth) or secondary (formed later). Primary fluid inclusions constitute samples of fluid in which the crystals grew. Secondary inclusions give evidence of later environments and are often formed as a result of crystals cracking owing to internal stresses created during growth, incorporating mother liquor by capillary attraction and resealing later. Inclusions readily form in a crystal that has been subjected to dissolution: rapid growth occurs on the partially rounded surfaces and entraps mother liquor. The rapid healing of dissolution etch pits will do the same. These regeneration inclusions, which usually lie in lines, i.e. along the former crystal edges, are characteristic of crystals grown from seeds.

Generally solvents are incorporated as inclusions or 'pockets' within the crystal structure. Agglomeration and attrition of particles due to collisions may favor the formation of inclusions. Solvent can be trapped between the spaces of individual particles in an agglomerate. Attrition of small parts of a crystal due to collisions with other crystals or (moving) hardware in the crystallization vessel may lead to damaged parts of the crystal which, owing to subsequent growth, may 'heal'. During this healing process, inclusions are usually formed [3.7]. Therefore, additional purification steps are sometimes required in order to obtain a product of sufficient purity.

Interrupted growth generally leads to inclusions. Brooks *et al.* attributed the formation of inclusions in ADP and $NaClO_4$ crystals to the introduction of a sudden upward step-change in supersaturation [3.8]. They concluded that the development of an inclusion at any point is governed by local conditions, more specifically by the concentration gradient along the height of a growth step. A critical step height was postulated beyond which a layer of solution can be trapped.

Large crystals and fast growth increase the likelihood of inclusions. Denbigh and White found that crystals of hexamine grew regularly when they were small but after they had grown larger than about 70 μm cavities began to form at the center of the faces and these were later sealed over to produce a regular pattern of inclusions in the crystals [3.9]. Cavities appeared to form only if the growth rate exceeded a certain value when the crystal had reached its critical size. Similar conclusions concerning critical size and growth rate criteria have also been recorded for ammonium perchlorate [3.10] and terephthalic acid [3.11].

3.2 Crystallization of Energetic Materials

3.2.1 Introduction

The crystallization of any material, including energetic materials, is subject to increasingly stringent demands regarding crystal size distribution, particle shape, purity, filterability, free-flowing properties and storability. For energetic materials, additional demands are placed on their hazardous properties, such as impact and friction sensitivity, thermal stability and compatibility with other chemical substances. The majority of these properties are influenced by the technique with which these materials are produced. Also, the solvents used during the production process may affect these properties. Therefore, it is of great importance to design and improve production processes and equipment in such a way that the product demands are satisfied.

Another important issue related to energetic materials is the fact that the decision as to whether or not a newly synthesized energetic material is of interest for further development is based on the thermal stability and sensitivity of the material. The assessment of these properties is often carried out with particles which are not yet optimized with regard to particle shape, mean size and purity and these properties might influence the thermal stability and sensitivity results and hence the decision as to whether the energetic material will enter a subsequent development step [3.12, 3.13]. For this reason, identification of a suitable production technique, where often a crystallization is the final process, is important not only to improve further already existing energetic materials, but also newly developed materials.

Compared with size reduction, in which the particles are subjected to high levels of mechanical stress, crystallization procedures for the preparation of particles of explosives offer substantial advantages. The crystals can be grown relatively slowly and free of stress and they have a well-defined crystalline shape and morphology.

3.2.2 Crystallization and Product Quality

This section deals with the influence of crystallization processes on the quality of the resulting product. In the following sub-sections, first a definition is given of which kind of product characteristics are comprised by 'product quality' and how these properties may affect, interfere with or even conflict with production-related processes. Different product quality aspects of energetic materials will be described, in addition to the effects of crystallization conditions and crystallization techniques on product quality. Examples and references to the literature will be given in order to illustrate the different aspects.

3.2.2.1 Definition of Product Quality

'Product quality' may be defined as the properties with which a product can be characterized with regard to its physico-chemical aspects, also taking into account the application of the product. The application of the product is explicitly included since the product quality usually determines (the limits of) the applicability of a material.

The characteristic product properties for energetic particulate materials comprised by the term 'product quality' are listed in Table 3.2. Most of these properties can be influenced or controlled by changing the process conditions during production. For instance, the particle shape can be influenced by changing the crystallization solvent and/or the growth kinetics during crystallization [3.14]. The hazardous

Table 3.2. Characteristic properties determining the overall 'product quality': for each property, several examples are given, assuming that the particulate product is processed in formulations such as PBXs or propellants; generally this involves high solid load mixtures of these particles with a prepolymer, after which curing of the prepolymer is initiated by the addition of certain chemicals.

Property	Processability	Final application	Remarks
Crystal size distribution	Rheology, hazard, processing technique	Solid load, shock sensitivity, ballistics, mechanical properties	Mean size, width of distribution, mono/polymodal
Particle shape	Breakage of particles, rheology, hazard	Solid load, shock sensitivity, ballistics, mechanical properties	Needle, rod, sphere, platelet, agglomerate
Chemical purity	Hazard, reactivity	Thermal stability, compatibility, hazard, storability (shelf-life)	Residual solvent, by-products from synthesis
Defect content	Breakage of particles, hazard	Shock sensitivity	Structural defects such as inclusions, dislocations, grain boundaries
Filterability	Solid-liquid separation after production, minimize amount of adhering solvent(s)		
Free-flowing	Dust formation (fine particles), caking, electrostatic charging, hazard		
Storability	Shelf-life, caking, hygroscopicity	Shelf-life, hygroscopicity	Caking, hygroscopicity
Hazard	Hazard during processing	Sensitivity	Impact/friction sensitivity, ESD
Thermal stability	Limits process conditions (e.g. processing or curing temperatures)	Storability (shelf-life), aging	T_m, T_{dec}, AIT
Compatibility	Chemical reactivity during processing	(Thermal) stability, storability (shelf-life), hazard	Chemical reactivity with other materials

properties, thermal stability and compatibility can be classified as more or less 'intrinsic' material properties. However, it should be noted that sometimes also aspects such as mean particle size and shape may influence, e.g., the impact and friction sensitivity [3.13, 3.15–3.17].

The criteria corresponding to these product properties are determined by the product demands, which in turn depend largely on the application of the product, not only with respect to, e.g., processability, but also with regard to the characteristics of the final product, such as its ballistic and mechanical properties or sensitivity against shock initiation. Focusing on energetic particulate materials, their usual application is in the field of plastic bonded explosives (PBX), booster charges, (gun) propellants, gas generating compositions, etc.: a polymeric binder is added to obtain a highly viscous solid-liquid mixture or a wax is added to obtain a 'granulate'. Depending on the properties of the resulting mixture, either casting, pressing or extrusion of the mixture is applied to obtain the final shape of the composition.

Generally the production of any material leads to an added value of the product because it meets certain requirements or product demands, necessary for the final application. A large part of the product quality aspects, as mentioned in Table 3.2, are determined or can be controlled by the way in which the particles are produced in the first place. Therefore, it is of the utmost importance to put sufficient effort into the selection of the most suitable production technique, the solvent(s) used and the optimization of the process conditions and the equipment. In case the required demands cannot be realized during this crystallization process, a post-treatment has to be identified with which the demands can be achieved. Apart from these aspects, there will also be an economically driven goal with regard to the cost-effectiveness and competitiveness of the production processes in relation to alternative production routes, if available.

3.2.2.2 **Process Problems and Product Quality**

As already mentioned in the previous section, the demands on the product generally result from the criteria which the final application has to meet, combined with the processing technique (casting, pressing, extrusion) in order to make the required formulation or composition. The production process of the particulate materials is usually also subject to all kind of restrictions regarding, e.g., the ranges of size which can be covered or the purity of the final product that can be achieved by the selected production technique.

The examples mentioned in this section illustrate that by selecting the proper crystallization technique and/or by optimizing the process conditions during crystallization, part of the desired product characteristics can be realized. However, sometimes the demands imposed on the final application interfere or are in conflict with the demands that would be beneficial to the crystallization process and other solutions have to be found in order to comply with the demands on product quality. For instance, the mean particle size and shape play an important role during filtration since these determine the porosity of the filtration bed; the particle properties required for good filterability may be in conflict with the demands on the

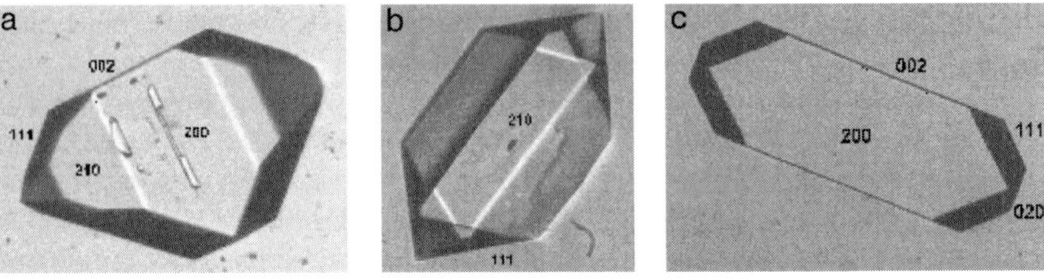

Figure 3.9. RDX grown from (a) acetone, (b) γ-butyrolactone and (c) cyclohexanone, showing several solvent-dependent growth shapes of RDX [3.27–3.29].

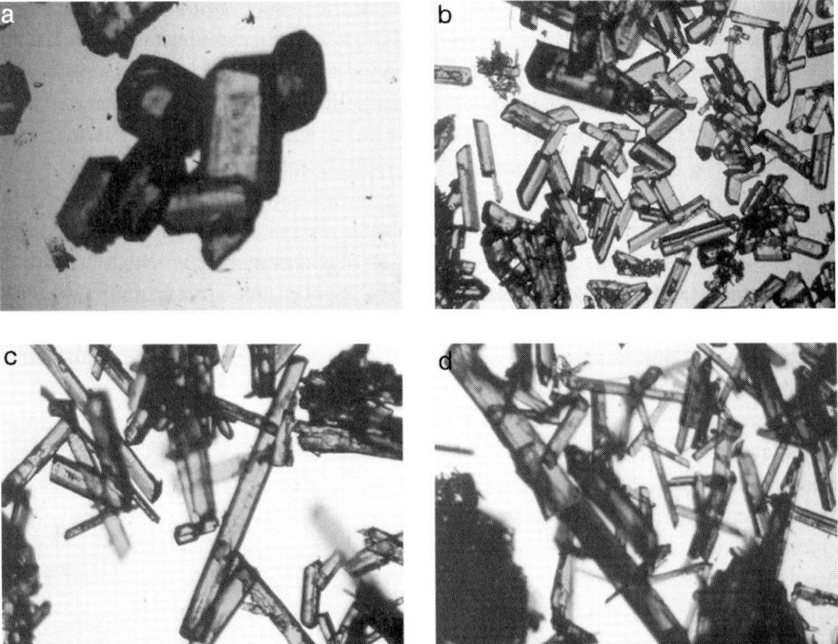

Figure 3.10. HNF crystals grown from (a) methanol, (b) ethanol, (c) 1-propanol and (d) 2-propanol. The mean aspect ratio of the crystals varies from 2–3 (methanol), 3–5 (ethanol) to 5–10 (both 1- and 2-propanol) [3.15].

can be reduced or blocked, a more isomorphic crystal will be formed. The use of habit modifiers is an interesting option, but generally these have to be developed, depending on the specific material the habit of which has to be changed.

The use of ultrasound during crystallization (sonocrystallization) is a relatively new application in the field of energetic materials [3.17, 3.31]. The use of ultrasound during crystallization may lead to different effects on the particle size distribution and other characteristics of the product. It is beyond the scope of this section to go into details of sonocrystallization, so only a few examples will be treated here. For

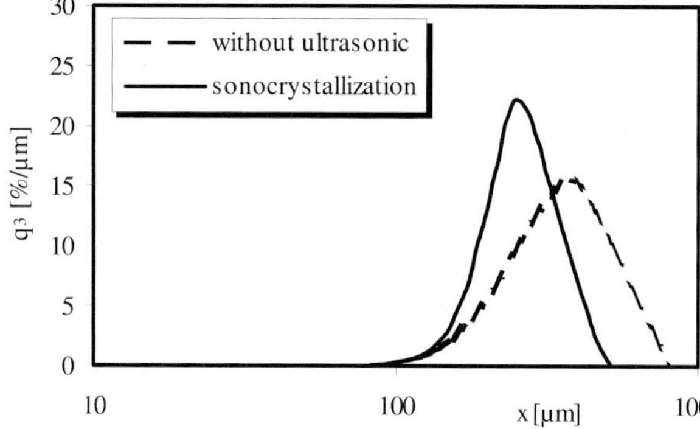

Figure 3.11. Particle size distribution of NTO particles: influence of ultrasound during nucleation.

instance, ultrasound might be used to induce (primary) nucleation to occur at lower supersaturation levels in this way using it as a kind of 'seeding' technique or it may be applied, depending on the intensity of the ultrasonic field, to break-up particles in this way reducing the mean size of the product.

Sonocrystallization has been applied to HNF [3.17]. An interesting finding was the improvement of the thermal stability of the sonocrystallized material compared with blank experiments (no ultrasound). More detailed information can be found elsewhere [3.15–3.17]. Over the past decade, new and optimized techniques have been developed to reduce the aspect ratio of the HNF crystals. This resulted in an increase in the solid load of HNF in a polymeric binder. In HTPB, solid loads of HNF up to ~80 wt% have been realized, using a monomodal HNF grade [3.32]. It is generally known that the use of well-balanced bi- or polymodal size distributions of the solid ingredients in a propellant may lead to significant increases in solid load. Therefore, it is expected that further improvements in the particle shape of HNF, the use of bimodal size distributions and/or the addition of aluminum may well lead to solid loads of 85–88 wt%, which are – according to theoretical calculations – required to obtain maximum performance for an HNF/Al/HTPB-based formulation [3.32, 3.33]. Kröber and Teipel investigated sonocrystallization of NTO (3-nitro-1,2,4-triazol-5-one) from aqueous solutions [3.31]. They compared the sonocrystallized material with material crystallized without ultrasound. It was seen that the nucleation starts earlier (ΔT = 6.5 K) when ultrasonic initiation was applied compared with the experiments without ultrasound (ΔT = 12 K). In Fig. 3.11 two crystal size distributions are presented. The continuous line represents the sonocrystallized particles and the dashed line the particles without ultrasound.

The sonocrystallized material is distinguished by a narrow crystal size distribution and by a smaller mean particle size ($x_{50,3}$ = 254 μm) in comparison with the normal crystallized material ($x_{50,3}$ = 345 μm). Microscopic pictures show that the use of ultrasound in crystallization processes can result in much regular crystals (Fig. 3.12).

For applications of energetic particles in a PBX, the desired shape of, e.g., RDX and HMX crystals is a practically isomorphic or rounded shape. In this respect, the example shown in Fig. 3.13 is undesirable. This picture shows an HMX twin crystal,

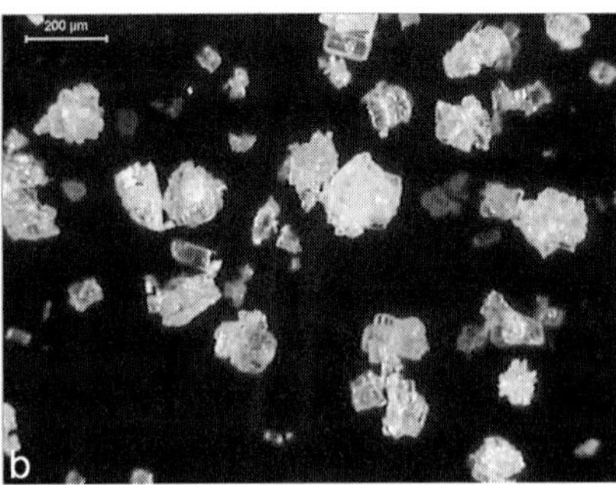

Figure 3.12. Microscopic pictures of NTO particles from sonocrystallization (a) and cooling crystallization (b).

grown from cyclohexanone [3.34]. Twins may develop spontaneously if on a certain crystallographic plane two sites are present with practically the same gain in energy for a growth unit when being built into the crystal lattice. One of these sites continues normally, whereas the other position develops in a new crystallographic direction, in this way forming a twin.

Another type of twin is a so-called transformation twin which is formed during a phase transition. One symmetry element (e.g. a mirror plane or two- or more-fold axis) of the crystallographic structure is lost during the phase transition, which is compensated by the formation of a twin. The twin plane and/or axis coincide with the lost symmetry element. Although HMX has four crystallographic structures (α, β, γ and δ), of which the α- and β-phases have been reported to be stable (at ambient temperature), only the β-phase is formed under ambient conditions [3.35]. Therefore, it is concluded that the formation of the twins occurs according to the first mechanism. An increase in the supersaturation during crystallization, which implies an increase in the crystal growth rate, enhances the formation of the former type of twins (spontaneous twin formation). Possibly impurities and the choice of

Figure 3.13. Example of an HMX twinned crystal grown from cyclohexanone as observed by means of optical microscopy [3.34]. In the center of the crystal structural defects such as cracks and inclusions can be observed.

solvent may influence this process by means of this mechanism of enhanced growth. With this mechanism of enhanced growth, less time is available to correct the energetically less favorable twin position. It was found, however, that from a solution mixture of acetone and γ-butyrolactone, the formation of twins is suppressed compared with cyclohexanone and cyclohexanone-γ-butyrolactone mixtures [3.36].

Generally, crystallization can be regarded as a purification method since the crystallization process itself is very selective with respect to the incorporation of impurities. Repeated crystallization and dissolution in a fresh solvent lead to further purification of the product. However, the uptake of impurities can never be prevented completely so a small amount of contaminants will always be present in the final product.

The purity of a crystalline material is generally expressed as the weight percentage of material which has the desired chemical composition, all other species present being referred to as impurities or contaminants. During the synthesis of the raw product or during its (re)crystallization, impurities such as by-products of the synthesis process or solvents may be incorporated into the crystal lattice. Depending on the size and the valence (charge) of the impurity, the incorporation can either be interstitial (in the available space between the atoms, ions or molecules), as a substituent replacing an atom, ion or molecule or as an inclusion. Synthesis by-products, usually differing to only a small extent compared with the desired molecule, may be incorporated as a substituent. The incorporation of such by-products may lead to a modification of the habit of the crystal [3.29]. The presence of impurities during the growth of a crystal growing by means of a layer-by-layer mechanism may lead to piling up of several layers (macrosteps), which may, in the end, lead to the formation of inclusions.

A peculiar effect was found when growing RDX from cyclohexanone with and without water present in the solution. An example of the latter case is shown in Fig. 3.14, whereas Fig. 3.9(c) shows an RDX crystal grown from cyclohexanone with water. In this case the presence of water eliminates or suppresses the formation of macrosteps, inclusions [Fig. 3.14(a)] and new crystals appearing at the sides of the original crystal [Fig. 3.14(b)]. Very likely this effect is related to the presence of a small quantity of a by-product of the RDX synthesis. This by-product undergoes a condensation reaction with cyclohexanone, resulting in impurities. These impurities might be responsible for the effects observed during crystallization of RDX, such as those shown in Fig. 3.14. This condensation reaction is an equilibrium reaction and water shifts this equilibrium to the side of the unreacted by-product and cyclohexanone. Hence the addition of water prevents or suppresses the formation of these impurities, resulting in a better defined shape of the crystals [see Fig. 3.9(c)]. More details can be found elsewhere [3.29].

Structural defects are a common feature in crystalline materials. Defects, such as inclusions (which also influence the product purity), dislocations or grain boundaries, may be introduced into the material during growth of the crystals. Especially conditions resulting in a relatively high supersaturation – involving a high growth rate – and 'severe' hydrodynamics will favor the formation of structural defects,

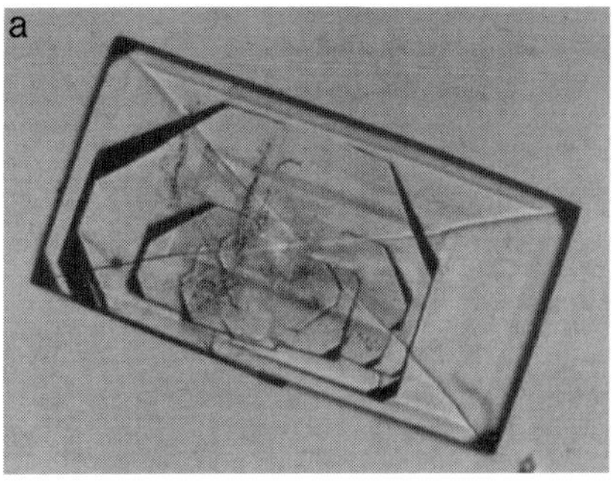

Figure 3.14. RDX grown from cyclohexanone without water (optical microscopy). (a) The formation of macrosteps and inclusions can be observed. (b) At higher supersaturations secondary growth effects can be clearly observed where several new crystals appear which start to grow from the original crystal at different crystallographic orientations [3.29].

either growth-induced or induced by mechanical damage to the particles due to collisions between the particles and the hardware of the crystallizer (stirrer blades, vessel walls) and subsequent 'healing' by growth of the damaged parts of the crystal [3.7]. The presence of these defects may influence the shock sensitivity of PBXs in which these particles are embedded [3.37, 3.38]. The defect content may be reduced when growing, e.g., under mild conditions with regard to supersaturation level and hydrodynamic conditions.

The size of (solvent) inclusions may range from macroscopic inclusions of several tens or hundreds of microns in diameter to tiny, micron-sized inclusions. The former type of inclusions can be detected fairly easily by means of optical microscopic techniques [3.36, 3.39–3.41], whereas the latter type of very small inclusions require more sophisticated optical (e.g. CSLM) or SEM techniques [3.42]. An example of CSLM applied to HMX crystals is shown in Fig. 3.15.

The size of the area is $139.6 \times 93.0\ \mu m^2$ with a thickness of 15 μm. The black spots point at local areas of refractive index differing from the surrounding area. With SEM these spots could be attributed to micro-inclusions with a mean size of $\sim 2\ \mu m$.

Figure 3.15. Slice with a thickness of 15 μm of an HMX crystal observed with CSLM revealing tiny spots pointing at locations in the crystal with a different refractive index. SEM pictures have shown that these tiny spots correspond to micron-sized voids in the crystal [3.42].

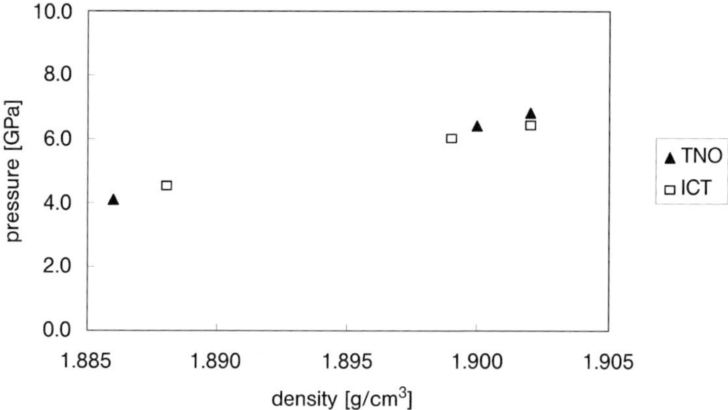

Figure 3.16. Shock initiation pressure vs crystal density of HMX embedded in a 70 wt% solid load HMX/HTPB based PBX [3.37, 3.43]. Data refer to the water gap test as carried out independently by TNO Prins Maurits Laboratory (TNO-PML, The Netherlands) and the Institut für Chemische Technologie (ICT, Germany). The theoretical crystal density of HMX is 1.903 g/cm^3.

If inclusions are present in a material, this will lead to a small but generally measurable lowering of the density, since the voids are either filled with solution or gases having a lower density than the solid material. The density of the crystals can be regarded as a measure for the internal perfection of the crystals, i.e. the closer this value is to the theoretical maximum density, the more perfect is the material. For HMX, a clear relationship between mean crystal density and shock initiation pressure was found [3.37, 3.38, 3.43]. An example is shown in Fig. 3.16 [3.37, 3.43]. The shock initiation pressure is seen to increase from ~4 GPa (reference batch) to ~6 GPa (recrystallized HMX) with increasing density of the HMX crystals. This is a significant improvement of the shock sensitivity. Shock initiation experiments using samples consisting of a liquid mixture in which the HMX particles are suspended, rather than embedding them in a polymer matrix, yielded similar results [3.38].

The liquid mixture was prepared such that its density matched that of the HMX crystals to be suspended. In this way, a possible influence of density differences when using a standard PBX on the shock initiation characteristics could be avoided. The experiments were carried out with a flyer impact set-up. Shock initiation pressures of 11 and 12 GPa were found for sieve fractions of recrystallized HMX, compared with 7 GPa for the reference material [3.38]. Again, there is a considerable difference in shock initiation pressure, which could be related to differences in internal quality of the crystals used.

In Fig. 3.13, showing a twinned HMX crystal, cracks can be observed in the center of the crystal. Such defects can also be observed in Fig. 3.17, also showing HMX crystals from a batch crystallized on a bench scale. The twinned crystals generally show one or several internal cracks or other imperfections, oriented along a diagonal crystallographic plane. Possibly the crystallographic plane where the two orientations of the twin meet lead to internal mechanical stresses at or near this

Figure 3.17. Optical microscopic picture of HMX crystals immersed in a liquid of practically the same (mean) refractive index as HMX. Defects can be clearly observed in the crystals, appearing as black spots [3.36].

boundary. The formation of cracks might then be explained as a way to release the stresses built up near to this boundary plane [3.36].

The presence of inclusions and cracks generally also leads to mechanically weaker crystals, which may break more easily under mechanical loads. This may lead to (undesired) changes in, e.g., particle shape and crystal size distribution during handling and processing of such particles. Furthermore, as illustrated above, this may lead to worse shock sensitivity aspects.

3.2.3
Crystallization of HMX and RDX

The important energetic materials cyclotrimethylenetrinitramine (RDX, hexogen) and cyclotetramethylenetetramine (HMX, octogen) belong to the family of nitramine explosives, since they contain C-N-NO_2 groups. One of the advantages of HMX over RDX is its higher density, resulting in a slightly higher detonation velocity and pressure and hence a higher performance. In contrast to RDX, HMX crystallizes in four different polymorphic forms. Apart from the four polymorphs which have been experimentally identified for HMX, also a large amount of crystal structures have been found consisting of a complex of HMX molecules with solvent molecules. Examples for such complex structures are HMX-NMP (*N*-methylpyrrolidone) and HMX-DMF (dimethylformamide), for which crystallographic data are available [3.44, 3.45].

The influence of the crystallization solvent on the crystal morphology has been studied for both HMX and RDX by several groups. Kinetic data on the crystallization of RDX or HMX in different solvents are available from only a few studies.

Xijun *et al.* presented kinetic data determined by microcalorimetry [3.46]. They concluded that the measurement of the heat produced and the rate of heat production during the crystallization process is possible in a microcalorimeter and from that data the growth kinetics can be determined. The crystallization kinetics of RDX and HMX from dimethyl sulfoxide (DMSO) and cyclohexanone can be expressed by the BCF dislocation theory model. Moreover, it was shown that the addition of seed crystals of RDX results in an increase of nuclei in the seeded solution compared with the unseeded solution.

Duverneuil *et al.* investigated the crystallization of HMX from DMSO and cyclohexanone [3.47]. They concluded that the crystal growth is limited by the diffusion step when cyclohexanone is used as solvent and the integration of the growth unit into the crystal structure is the limited step when DMSO is used. In this study, the influence of the solvent on the crystal morphology was also investigated. It was shown that the crystals of HMX from cyclohexanone are more or less spherical and twinned. Moreover, in certain crystals the presence of cyclohexanone in inclusions was also observed. In contrast, the HMX crystals were sufficiently regular and symmetric and did not contain any inclusions when the crystallization was carried out from DMSO or from a DMSO-acetone mixture. The authors explained the differences by solvent-solute interactions, which was confirmed by the calculation of the entropy factor of the surface, defined as:

$$\alpha_\mathrm{s} = \frac{4 \cdot \sigma_\mathrm{s} \cdot h_\mathrm{c} \cdot l_\mathrm{c}}{k_\mathrm{b} \cdot T} \tag{3.8}$$

where h_c and l_c are the height and width of a monomolecular growth layer, respectively, σ_s is the surface energy, k_b is the Boltzmann constant and T is the absolute temperature. The calculation of this surface factor of entropy for the investigated systems showed differences between cyclohexanone and DMSO. For cyclohexanone this factor is 0.60 and for DMSO it is 0.26, which shows that there is an absence of ideality which is more pronounced in the solution of DMSO.

Svensson *et al.* investigated the crystallization of HMX from γ-butyrolactone by cooling and precipitation [3.48]. They found that HMX can be crystallizaed from γ-butyrolactone in the desired β-modification without any detectable amount of α-HMX. The crystals are well shaped and have a relatively low tendency to form crossed crystals or other irregularities. For the crystallization they used different cooling programs. With natural cooling, that means with a high cooling rate at the beginning, the resulting mean particle size was found to be consistently around 150 μm. With a more efficient cooling system it is possible to obtain mean particle sizes of 60–70 μm.

When long crystallization times are used, coarser particles can be produced. Using a cooling period of 6 h, a mean particle size of around 500 μm was found, whereas a 3-h cooling period results in a product with a mean particle size of around 320 μm. Seeding the solution changes the particle size to a coarser product (900 μm).

Very fine particles can be obtained by precipitation of HMX from γ-butyrolactone by adding water to this solution. The mixing can be done either in a static mixer or directly in the crystallization vessel. The crystals obtained usually have mean particle sizes between 5 and 35 μm, depending on the exact conditions used. Similar results can be obtained by the crstallization of RDX from γ-butyrolactone.

The production of superfine RDX particles was described by Ruijun *et al.* [3.49].

They used solvent-non-solvent techniques and the mean particle size was around 0.2 μm. The RDX was dissolved in dimethylformamide and water was used as non-solvent. Two process variations were studied, resulting in different mean particle sizes. Both variations are distinct in mixing patterns. In the first process, the solution is fed into the non-solvent, which results in a very rapid dispersion and a high nucleation rate owing to the high supersaturation. The particles are very fine, the particle size distribution is narrow and the BET surface area is up to 6 m^2/g. The second way in which the non-solvent is fed into the solution is distinguished by larger particles with a BET surface area of about 4 m^2/g due to the lower supersaturation and therefore a lower nucleation rate. The agitation rate had a tremendous influence on the particle size. On increasing the rate from 300 to 1300 rpm, the mean particle size decreased from 20 to 1 μm and the particles became better distributed and agglomeration was prevented. The best results were obtained if the temperature of the water was as low as possible because the growth kinetics rose with the temperature increase. The temperature of the solution influences the shape of the RDX particles. At low temperatures (15 °C), needle-like particles were obtained whereas spherical particles were formed at high temperatures (60–70 °C). Between these minimum and maximum temperatures, plate-like particles were formed.

The crystallization of HMX from different solvents was studied by Kröber *et al.* [3.43]. The objective of this study was to find out whether a relationship exists between the solvent and experimental conditions and the amount of defects in the crystals (see the previous section). The experimental work was carried out in a batch crystallizer with a volume of 1 l and linear cooling rates were realized by a controlled thermostat. Five different solvents were used in this study. Before starting with the crystallization experiments, it was necessary to determine the solubility curves of these binary systems. The following equation describes the results of this investigation. The constants *a*, *b* and *c* are summarized in Table 3.3 for the different solvents.

$$\frac{X^*}{g_{\mathrm{HMX}}/g_{\mathrm{solvent}}} = a + b \cdot \frac{\vartheta}{°\mathrm{C}} + c \cdot \left(\frac{\vartheta}{°\mathrm{C}}\right)^2 \tag{3.9}$$

Owing to the very low solubility of HMX in acetone and the very high solubility of HMX in DMSO at low temperatures, batchwise cooling crystallization is not the preferred technique for these systems so no experiments were carried out.

Cyclohexanone

The crystallization of HMX from cyclohexanone was characterized by a very low yield of crystals. Theoretically, about 90% of the material should be recrystallized when the solution is cooled from 80 to 5 °C. In the experiments, only 15% was found, which can be explained by a very strong inhibition of the nucleation kinetics. Seeding of the solution was not carried out because of the disadvantages of the seed crystals with respect to the quality of the final product (maybe inclusions in the seed crystals will influence the sensitivity of the final product). Therefore, some evapor-

Table 3.3. Summary of the constants *a*, *b* and *c* in Eq. (3.9) for different solvents

Solvent	*a*	*b*	*c*
Acetone	-6.0×10^{-5}	1.0×10^{-5}	0
N-Methylpyrrolidone	5.8×10^{-3}	-9.0×10^{-4}	7.0×10^{-5}
N,N-Dimethylformamide	1.9×10^{-2}	-2.0×10^{-4}	2.0×10^{-5}
Propylene carbonate	-1.6×10^{-2}	1.8×10^{-3}	0
Dimethyl sulfoxide	0.5	3.7×10^{-3}	0
Cyclohexanone	4.0×10^{-3}	7.0×10^{-4}	0

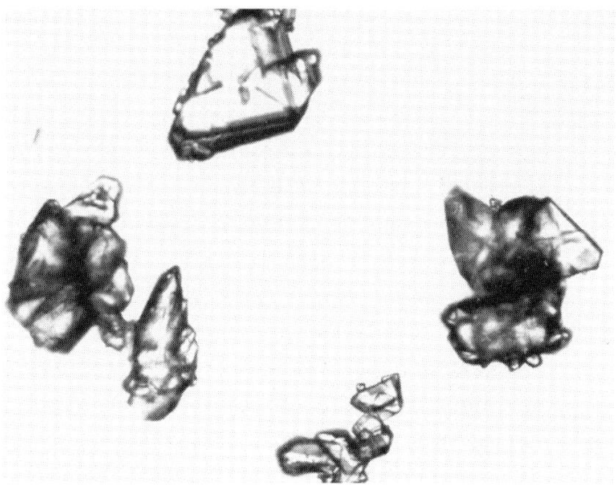

Figure 3.18. HMX crystals recrystallized from cyclohexanone.

ative crystallization experiments were carried out with cyclohexanone. The yield was slightly better than with cooling crystallization (increasing from 15 to 30%), but the crystal shape was very bad, so further experiments with cyclohexanone were not carried out. Figure 3.18 shows some crystals from these experiments, which are very irregular with no clear morphology.

N-Methylpyrrolidone (NMP)

The crystallization of HMX from NMP resulted in crystals of bad quality. The surface of the crystals was very rough and the particle shape was irregular (see Fig. 3.19). The quality of these crystals was not increased by decreasing the cooling rate, which normally results in a slower growth rate and a more regular growth.

The mean particle size increases on decreasing the cooling rate. Furthermore, some crystals were opaque which can be interpreted by solvent inclusions in the crystals. This interpretation can be supported by the measurement of the particle density. These measurements show a particle density of 1.602 g/cm^3, which is much lower than the theoretical density of HMX (1.903 g/cm^3) and of the density of the starting material (1.871 g/cm^3). A lower density can be a quantitative measure of the amount of inclusions because the density of the crystalline material is higher than

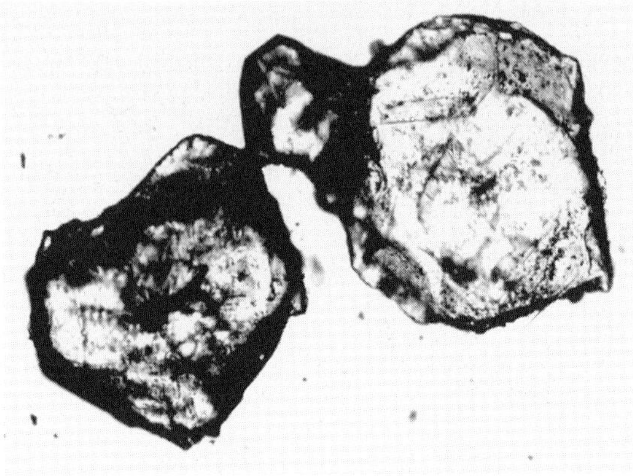

Figure 3.19. HMX crystals recrystallized from *N*-methylpyrrolidone.

the density of the solvent or a gas. However, it is possible that the low density results from the formation of an HMX-NMP complex during the crystallization.

N,*N*-Dimethylformamide (DMF)

The particles formed by cooling crystallization experiments from DMF are distinguished by a spherical shape and a smooth surface (see Fig. 3.20). The influence of the stirrer speed on the mean particle size was investigated. It was shown that an increase in the stirrer speed results in a slight increase in the mean particle size but also in a spreading of the particle size distribution. Both results can be explained: a higher stirrer speed improves the transport of growth units to the crystal surface, which results in a higher growth rate. On the other hand the crystal-stirrer collisions are intensified when the stirrer speed is increased, so that attrition is strengthened and the formation of fine particles is enhanced. The influence of the cooling rate on the particle size can be neglected.

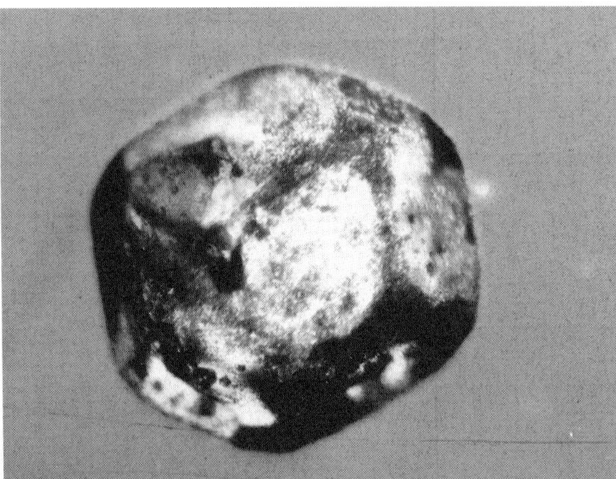

Figure 3.20. HMX crystals recrystallized from *N*,*N*-dimethylformamide.

In spite of the improvement in particle shape and surface quality, no further experiments from DMF solutions were carried out because during this crystallization HMX-DMF complexes are formed, which have a much lower density (1.612 g/cm^3) than the theoretical value. GAP-tests carried out with this recrystallized material showed a much higher shock sensitivity compared with the sensitivity of starting material.

Propylene Carbonate

The crystallization of HMX from propylene carbonate was successful with respect to an improvement of the shock sensitivity of PBX filled with this recrystallized material (see Fig. 3.16). During the crystallization, the beginning of the nucleation was very uncontrolled and high supersaturation was caused, so that a large amount of fine particles was obtained. Therefore, it was decided to start the nucleation at a certain supersaturation ($S = 1.26$) by adding a small amount of water to the supersaturated solution (drowning-out effect). Crystals were formed which were distinguished by a very compact and regular shape (Fig. 3.21(a)).

However, some twinned crystals were produced which showed one or several internal cracks or other imperfections, oriented along a diagonal crystallographic plane (Fig. 3.21(b)). Possibly the crystallographic plane where the two orientations of the twin meet led to internal mechanical stresses at or near this boundary. The crystals were mostly transparent, which indicates a low amount of internal inclusions. This assumption is supported by the very high particle density, 1.895 g/cm^3, and therefore close to the theoretical density but much higher than the density of the starting material. Again, an increase in the stirrer speed resulted in an increase in the mean particle size due to the improvement of the mass transfer from the bulk to the crystal surface. Decreasing the cooling time, which means increasing the cooling rate, resulted in a significant smaller mean particle size and the presence of

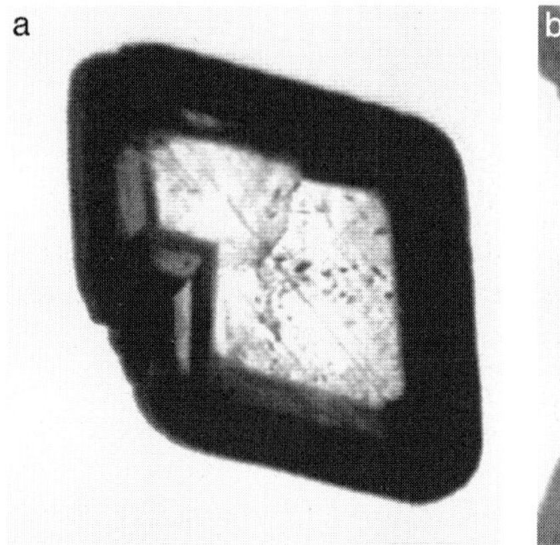

Figure 3.21. HMX crystal recrystallized from propylene carbonate.

roughness on the crystal surface, which are signs of dendritic growth caused by higher supersaturation.

3.2.4
Crystallization of CL 20

CL 20 (2,4,6,8,10,12-hexanitro-2,4,6,8,10,12-hexaazaisowurtzitane, HNIW) was first synthesized by the group of Nielsen [3.50] in 1989. Its high density, detonation velocity and pressure make it a suitable candidate for replacing HMX as a high-energy constituent in shock-insensitive explosive compositions for deformable warheads or as a constituent in propellant formulations. At least four polymorphs of this substance exist. The ε-phase is the preferred polymorph as it is morphologically stable at room temperature and has the highest density. The material purity, the nature of any impurities, crystal polymorphism, morphology and crystal quality can greatly influence the sensitivity of energetic crystals. Therefore, crystallization processes are suitable techniques to improve the quality of CL 20, especially to decrease the sensitivity so that it can be handled and processed.

The solubility of CL 20 in different solvents was determined by von Holtz *et al.* [3.51]. They showed that CL 20 is very soluble in solvents with carbonyl groups, such as ketones, esters and amides. The solubility decreases if non-carbonyl materials such as alcohols, ethers and nitroalkanes are used. CL 20 is insoluble in hydrocarbons, halogenated hydrocarbons and water. Table 3.4 summarizes the solubility data in six solvents.

The solvents investigated were characterized by a very low or very high solubility power but no solvent was found which is distinguished by a mean solubility. The solubility of ε-CL 20 in acetone and ethyl acetate is independent of temperature. Therefore, cooling crystallization cannot be applied to process CL 20 from these solvents. Owing to the high solubility, evaporation crystallization and precipitation by an antisolvent are suitable techniques to crystallize CL 20. Precipitation with an antisolvent has the advantage of crystallizing at a low temperature so that the transition from the ε-phase into the γ-phase is not favored. On the other hand this technique is characterized by a high nucleation rate, which should favor the formation of the kinetically most stable polymorph which is the β-phase.

The crystallization of CL 20 with the aim of improving the product quality was described by Wardle *et al.* [3.52]. The crude material is often polycrystalline with a

Table 3.4. Solubility data for ε-CL 20 in different solvents [3.81].

Solvent	Solubility (g/100 ml) = $f(T, °C)$	R^2
Acetone	74.8	
Ethanol	$0.778 - 0.021T + 0.0004T^2$	0.9955
Ethyl acetate	40.6	
Ethylene glycol	$1.30 - 0.027T + 0.0005T^2$	0.9873
Methylene chloride	0.043	
Water	<0.005	

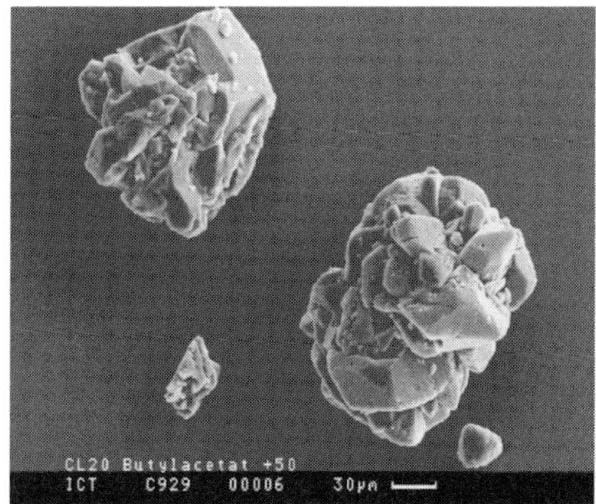

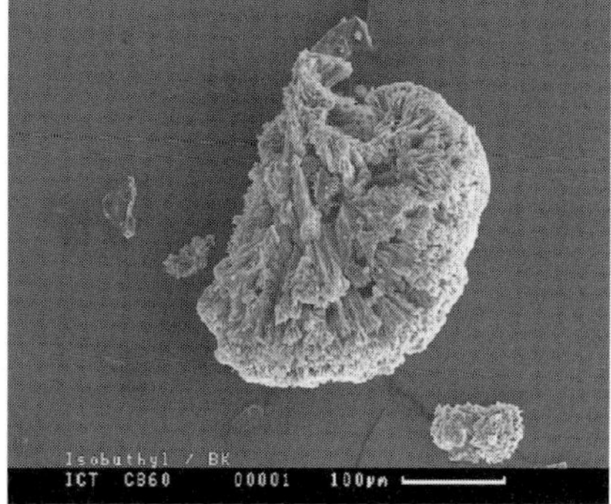

Figure 3.22. SEM pictures of CL20 particles recrystallized from acetic acid (left) and isobutyl methyl ketone (right).

broad particle size distribution and some internal defects. The particle surface is very rough and sharp corners can be seen. These imperfections and a high content of residual acid species from the synthesis result in a very high sensitivity of the raw material. The recrystallized material is not polycrystalline, has rounded edges and the ε-phase is obtained in a reproducible method. However, no details about the solvent and crystallization technique were given.

A great number of solvents have been tested at the Fraunhofer ICT [3.53]. CL 20 was dissolved in a glass vessel, tempered and the solvent was evaporated. These experiments showed that it is difficult to recrystallize pure ε-phase. Depending on the solvent used, a larger or smaller amount of other phases was formed. For example, evaporation from acetone gave a ratio between ε- and β-phases of 52/48, from acetic acid the ratio between ε- and α-phases was 87/13 and if isobutyl methyl ketone was used α-, β-, ε- and γ-phases were formed (13/20/35/32). SEM pictures show the influence of the solvent on the particle morphology (Fig. 3.22).

The differences in morphology are evident. The crystallization from acetic acid results in compact crystals whose surfaces are covered by small particles (surface nucleation), whereas the particles from isobutyl methyl ketone consist of many needles which are grown to spherical agglomerates.

In general, the α- and β-phases crystallize as plate-like crystals and the ε- and γ-phases as compact crystals with more or less sharp edges. The differences in morphology of the four phases can be explained by the different space groups in which they crystallize. The α- and β-phases crystallize in the orthorhombic space group whereas the ε- and γ-phases have a monoclinic symmetry.

3.2.5
Crystallzation of NTO

3-Nitro-1,2,4-triazol-5-one (NTO) is an attractive explosive in terms of its insensitivity and stability [3.54–3.56]. The morphology of NTO crystals coming from the

Figure 3.23. Crude NTO crystals coming from the reaction process.

reaction process has typically a jagged rod-like shape, which readily agglomerates and ultimately becomes sensitive to a sudden shock (see Fig. 3.23).

One of the ways to lower the sensitivity towards sudden shock is to control the crystal morphology to be close to spheres. NTO was typically recrystallized in solvents such as water and alcohols [3.57–3.59]. NTO recrystallized from these solvents has a needle-like or plate-like shape with agglomerates (see Fig. 3.24). The irregular crystals have a low density and eventually break easily. This causes mixtures of explosive formulations with NTO to be highly viscous and difficult to process and to pour. As a result, the amount of NTO which can be used in a processable explosive composition is limited, the performance of the explosive is reduced and the explosive eventually becomes very sensitive to unintentional shock.

On the other hand, the mother liquors contain additional components from the chemical reaction, in particular formic acid, nitric acid, sulfuric acid and 1,2,4-triazol-5-one [3.55, 3.56]. Some of these are trapped in the crystals and are present even after washing [3.55, 3.59]. A purification process is necessary to remove these

Figure 3.24. NTO crystals recrystallized in solvents: (a) water and (b) methanol.

impurities. One attempt to solve these problems is by recrystallization of the NTO in an appropriate solvent. The recrystallization offers explosive crystals with specifications of acceptable crystal size and crystal shape and narrow crystal size distribution. These can be optimized by selecting the proper operating conditions of crystallization process.

Spherical crystals of explosives can improve insensitivity towards a sudden shock, performance, processablity and packing density. The sensitivities of crystals of spherical shape were much lower than those of rod-like shape. As the bulk density of spherical crystals is much higher than that of rod-like crystals, the performance with respect to explosion per packing volume is higher.

Spherical crystals can be formed as types of agglomerates and spherulites. Agglomerated crystals are the unification of small primary particles after primary nucleation. The agglomeration technique is to transform fine crystals directly into compacted crystals during crystallization. The agglomerates have low density and low hardness compared with true crystals [3.60]. Therefore, spherical crystallization of explosives by agglomeration is not desirable. As a result, the desirable shape of explosives is a spherical spherulite. Spherulitic crystals could be obtained by cooling crystallization using a co-solvent. In cooling crystallization with a co-solvent, the cooling rate and the solvent composition were the important parameters for the control of morphology and size of the crystals [3.61–3.63]. In most cases of crystal growth, a typical primary nucleus grows into a crystallite having a discrete crystallographic orientation. Generally, this continues to develop as a single crystal until it impinges either upon external boundaries or upon other similar crystallites advancing from neighboring nuclei. However, it is a characteristic property of particular systems that primary nuclei initiate the formation of polycrystalline aggregates which are more or less radially symmetric. This type of crystal is known as spherulite. A spherulitic crystal is very different from agglomerated crystals in terms of crystallographic orientation. Keith and Padden [3.64] reported phenomenological results for spherulite crystallization. They found that high viscosity and the presence of impurities are fundamental to spherulite formation. The compact spherulites have the a similar density and hardness to true crystals. As a result, the desirable spherical shape of explosives is spherulite rather than agglomerate. Although there are many reports on the formation of spherulites of chemical compounds [3.65–3.68], spherulite of explosive compounds has never been reported. Spherulitic crystallization of NTO is presented from its mechanism to the design of a crystallizer.

3.2.5.1 Kinetics of NTO Crystallization

Solubility of NTO in Solutions

In the manufacture of explosives, crystallization from an organic solvent or mixtures of solvents is often used. The composition of the solvent determines the solubility of the substance and thus strongly influences the choice of supersatuation generation method. Furthermore, the solvent composition may have an effect on the

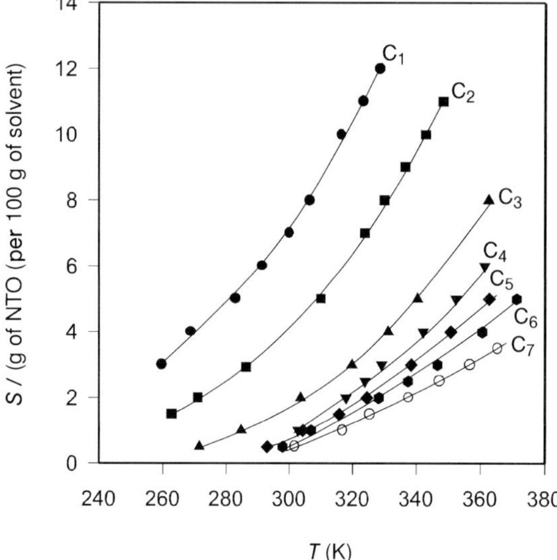

Figure 3.25. Solubilities of NTO in C_1–C_7 1-alkanols.

nucleation rate and on the crystal growth rate and thus affects the shape of the product crystals and the size distribution of the crystalline mass. The solvents for dissolving > 30% of NTO are NMP, DMSO and DMA, etc. and those for < 10% are water, alcohols, acetone, etc. [3.60].

The solubility values of NTO in C_1-C_7 1-alkanols are plotted in Fig. 3.25 [3.57]. The solubilities of NTO increase with increase in temperature and decrease with increase in the number of carbon atoms in alcohols. This means that the solubility of NTO in alcohols increases with increasing polarity of the solvent.

The solubilities of NTO in water, NMP and water-NMP at various temperatures and atmosphere pressure have been reported [3.59, 3.61] (see Fig. 3.26). The solubility of NTO in the ternary system NTO-NM-water over the temperature range 20–100 °C was determined from the phase diagram shown in Fig. 3.27. The solubility of NTO increases with decrease in temperature and the fraction of NMP. When the solubilities of NTO in NMP, water and NMP-water are compared, it can be seen that in the ternary mixture the addition of the water reduces the NTO solubility and the dependence of temperature on solubility increases with the fraction of water.

In order to calculate the enthalpy of dissolution of NTO crystals in the solvent, the interaction of a solvent and solute can be expressed as the following ideal theory [3.68]:

$$\ln x = \frac{-\Delta H_{sol}}{R}\left(\frac{1}{T} - \frac{1}{T_m}\right) \tag{3.10}$$

where x is the mole fraction of solute, ΔH_{sol} is the enthalpy of dissolution of NTO and T is the equilibrium temperature. ΔH_{sol} is equal to ΔH_{fus} for an ideal system and $\Delta H_{fus} + \Delta H_{mix}$ for a non-ideal system. The enthalpy of mixing, ΔH_{mix}, is a

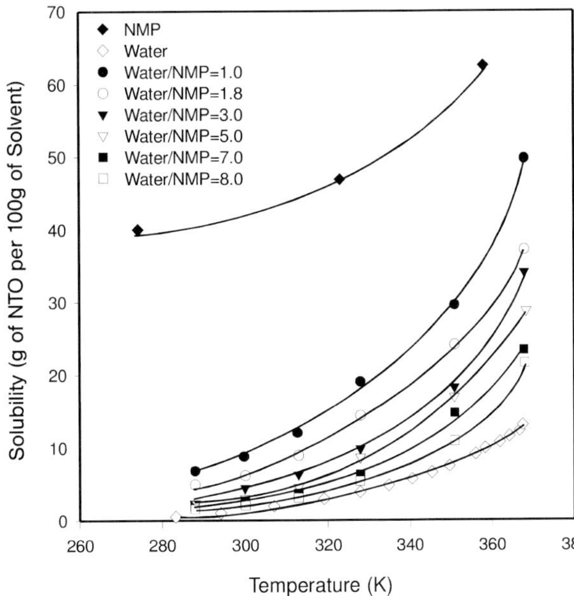

Figure 3.26. Solubilities of NTO in water-NMP mixtures.

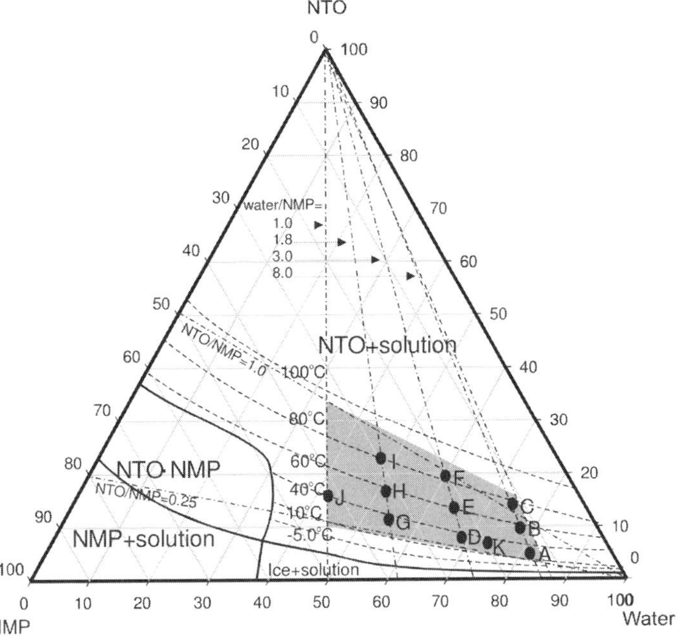

Figure 3.27. Ternary solid-liquid phase diagram for NTO-water-NMP.

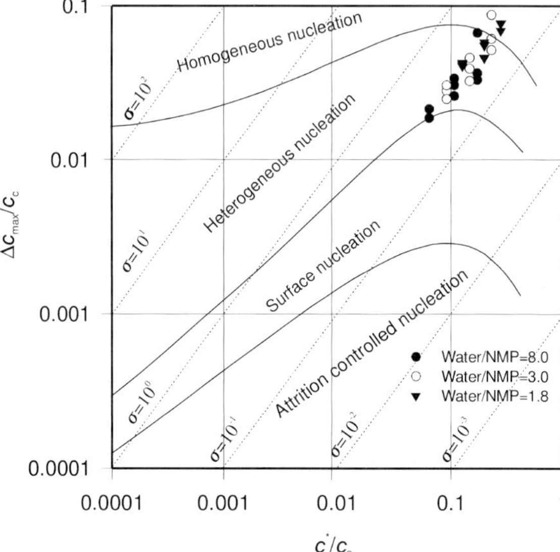

Figure 3.29. Mersmann's nucleation criterion.

Interfacial Energy

The interfacial energy can be obtained by the classical primary nucleation theory. It suggests that the classical primary nucleation rate, characterized by considering the probability of formation of a critical nucleus corresponding to the maximum free energy change, can be expressed as [3.14]

$$J = A\exp\left[\frac{-16\pi\gamma^3 v^2}{3\kappa^3 T^3 (\ln S)^2}\right] \tag{3.14}$$

A commonly used parameter to characterize the crystallization process is the induction period, t_{ind}, defined as the time that elapses between the achievement of supersaturation or subcooling and the formation of a detectable quantity of the crystals in a given system [3.74, 3.75]. The induction period can be calculated by themetastable zone width and cooling rate.

$$t_{ind} = \frac{\Delta T_{max}}{\dot{T}} \tag{3.15}$$

The induction period can be considered as inversely proportional to the rate of nucleation [3.14]:

$$\ln t_{ind} = \ln A + \frac{B}{T^3 (\ln S)^2} \tag{3.16}$$

where

$$B = \frac{16\pi\gamma^3 v^2}{3\kappa^3} \tag{3.17}$$

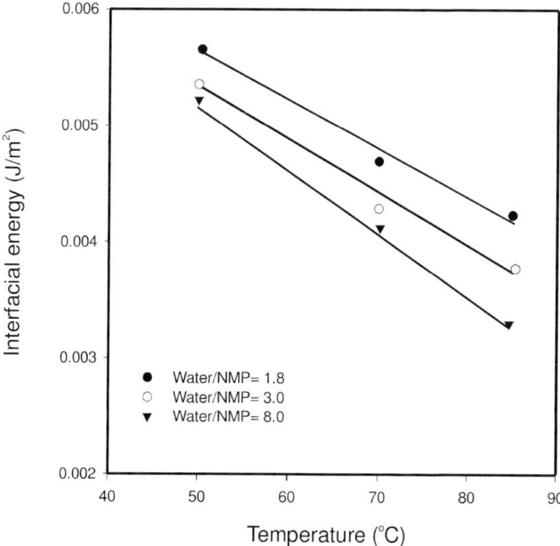

Figure 3.30. Effect of temperature and water/NMP ratio on interfacial energy.

Linear regression analysis of the experimental data in the semi-logarithmic plot of t_{ind} as a function of $T^3(\ln S)^{-2}$ gives correlations in which a straight line can be drawn for three ratios of solvents and saturation temperature [3.62]. The interfacial energy can be evaluated directly from the nucleation experiments from the thermodynamic parameter B defined in Eq. (3.17) if the experimental data can be represented by the classical nucleation theory [3.14]. In the present calculations, the molecular volume of NTO is equal to 1.12×10^{-28} m^3/molecule [3.59]. Calculated results of the interfacial energy as a function of the water/NMP ratio are shown in Fig. 3.30. A good linear dependence of the interfacial energies on the water/NMP ratio is found for all the systems. As can be seen in Fig. 3.30, the interfacial energy decreases with increasing temperature and with increasing water/NMP ratio. The interfacial energy ranges from 3.1 to 5.8 mJ/m^2 in the temperature range 50–85 °C and in the water/NMP ratio range 1.8–8.0. The interfacial energy decreases with increasing content of water in the solvent. This means that water in solvent may enhance the nucleation of spherulitic crystals for cluster aggregates, which results from the decrease in the metastable zone width. Also, the interfacial energy increases with increasing the maximum allowable undercooling and the NMP content in the solvent.

Crystal Growth Kinetics

The overall crystal growth rate including the effect of temperature can be expressed by the following expression [3.14]:

$$R_{\mathrm{G}} = k_{\mathrm{G}} \Delta c^{g} = k_{\mathrm{g}} \exp\left(-\frac{E}{RT}\right) \Delta c^{g} \tag{3.18}$$

solvent is consistent with the modifications of the physicochemical and transport properties of the system. It can be concluded that the water not only increases the diffusion in volume of the solute but also makes its integration in the crystalline structure faster. The growth of the spherulitic NTO is very likely controlled by the integration of the NTO crystals in the growth.

3.2.5.2 Control of Size and Shape by Recrystallization

Crystal Size

Crystal size depends mainly on the kinetics of nucleation and growth, which are functions of process variables such as agitation rate, feed composition and production rate. These functions are eventually related to the degree of supersaturation. The supersaturation is a main parameter for controlling crystal growth and nucleation rates.

Figure 3.33 shows the effect of solvent composition on the crystal size of spherical spherulites for NTO/NMP ratios ranging from 0.3 to 0.8. The average size of spherultic NTO increases with increasing water/NMP ratio. By addition of water to the NMP, the solubility of NTO in the ternary system can be reduced. Crystal size depends mainly on the kinetics of crystallization. It was reported that the nucleation rate decreased with increasing content of water in the solvent. It was found that growth rate increases with increasing content of water in the co-solvent. These two kinetic results suggest that the crystal size can be controlled by adjusting the water/NMP ratio. This supports the results that scale-up of a crystallizer is dependent on solvent composition and independent of production scale [3.78].

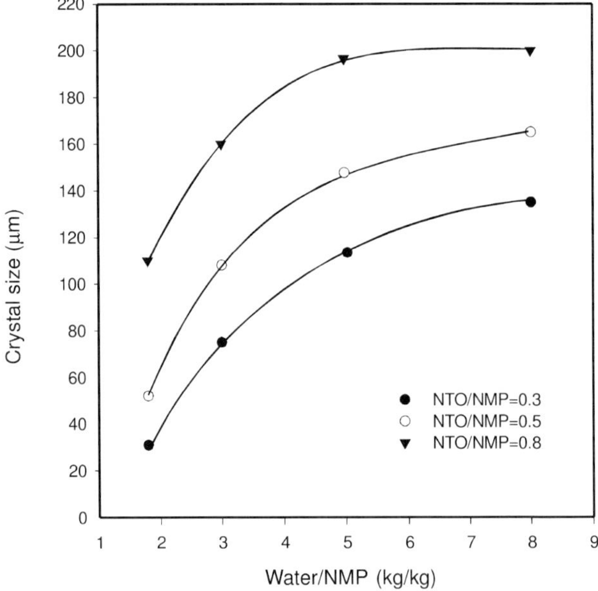

Figure 3.33. Effect of composition on the crystal size of spherical spherulites.

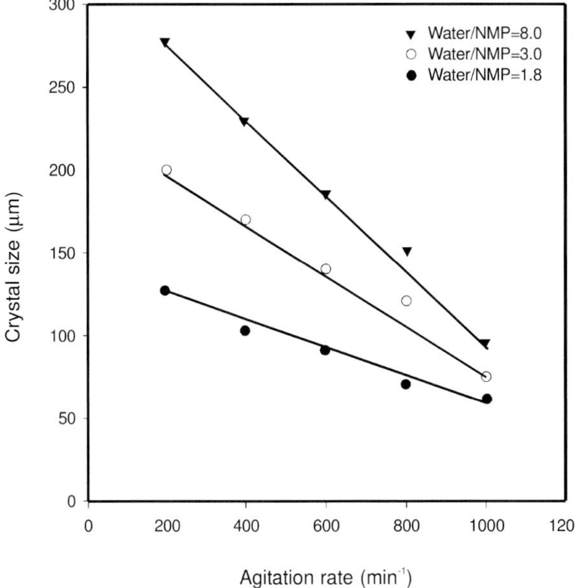

Figure 3.34. Effect of agitation rate on the crystal size at different water/NMP ratios.

Figure 3.34 shows the effect of agitation rate on the crystal size for three water/NMP ratios. The unseeded cooling crystallization was carried out at a cooling rate of 10 K/min and an NTO/NMP ratio of 0.8. Note that an increase in agitation rate resulted in a reduction in crystal size. As an illustration, the size of NTO crystals was reduced by 40% when the agitation rate was increased by a factor of 5 at the constant water/NMP ratio of 8.0. At a constant agitation rate of 500 rpm, a change from 1.8 to 8.0 in the water/NMP ratio diminished the crystal size by 50%. As the agitation rate is increased, the effect of water/NMP ratio became less important. At high agitation rates, e.g. 1000 rpm, there was little effect of water/NMP ratio and the crystal size was dictated only by the agitation conditions. Similarly, agitation had less effect on the crystal size at low water/NMP ratios.

Effect of Cooling Rate on Morphology

An almost perfectly spherical spherulite with a smooth surface formed by three-dimensional growth in all directions was crystallized by rapid cooling with water-NMP solvent. Fine rectangular tips of needles which constitute the spherulite are observed on the surface of the crystals. Figure 3.35 shows scanning electron micrographs of spherulite of NTO obtained by cooling crystallization at various cooling rates at a feed composition of 6.8 wt% NTO, 21.9 wt% NMP and 72.3 wt% water, which is indicated as point K in Fig. 3.27. Figure 3.35(a) shows NTO crystals obtained by cooling crystallization at a cooling rate of 10 K/min. The spherical spherulites of NTO were crystallized. Crystals are found to be composed of arrays of crystalline needles, arranged radially outwards from the center. Figure 3.35(b) shows NTO crystals obtained by cooling crystallization at a cooling rate of 5 K/min. The

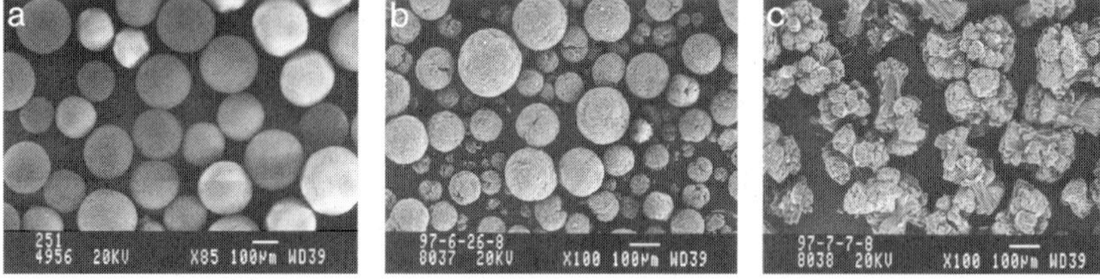

Figure 3.35. Scanning electron micrographs of NTO crystals obtained at cooling rates of (a) 10, (b) 5 and (c) 1 K/min.

crystals are a spherically spherulitic shape, but many small pores exist on its surface. This suggests that as crystal growth continued, arranging rods gradually encircled regions on either side of center, forming pores. Figure 3.35(b), an electron micrograph of the outer tips of spherulite grown at a cooling rate of 5 K/min, shows the arrangement of the needle rods. They are arranged radially outwards from the center. Figure 3.35(c) shows NTO crystals obtained by cooling crystallization at a cooling rate of 1 K/min. The spherulites are found to have an irregular shape, characterized by an elongated major axis from which needles fan out on either side, creating a sheaf-like appearance. As can be seen in Fig. 3.35(c), the crystals have a spherulitic shape, which is composed of arrays of crystalline needles, arranged radially outwards from the center. As a result, an increase in cooling rate in cooling crystallization using water-NMP solvent increases the compactness and sphericity of NTO crystals. Generally, the nucleation rate increases with increasing cooling rate [3.58, 3.59]. This suggests that nucleation of spherulite affects mainly its morphology.

Eventually, spherulitic crystals of NTO were crystallized from the ternary system with NMP-water solvent. Spherulites are composed of needle rods which radiate outwards from a common center. The rods are of approximately constant thickness and have a preferred crystal axis along the radial direction. The rod thickness increases with decreasing cooling rate and with increasing ratio of water to NMP. In addition, it was found that a spherical form of spherulitic NTO crystallized at a higher cooling rate, which resulted in a higher metastable zone width. This proves that a higher nucleation rate due to a higher metastable zone width is necessary for more compact and spherical spherulites.

Effect of Composition on Morphology

Figure 3.36 shows some SEM micrographs of the NTO crystals obtained by crystallization at a cooling rate of 10 K/min at various ternary compositions [3.61]. The compositions explored are indicated on the ternary diagram in Fig. 3.27. Figure 3.36(a), (b), (c) and (d) show the scanning electron micrographs of NTO crystals obtained for the compositions of points B, E, F and J in Fig. 3.27, respectively.

The composition of crystallization using a ternary system affects the morphology of the crystals because the interaction between solute and solvents depends on composition [3.79]. All crystals obtained were found to be spherulitic. At a constant water/NMP ratio, an increase in NTO/NMP ratio decreases the compactness of spherulitic crystals. At a constant NTO/NMP ratio, an increase in the water/NMP ratio increases the size of spherulitic crystals. Spherical spherulites of NTO were found to be crystallized at points D, E, G, H and J in the ternary phase diagram. Through further research on morphology according to composition, spherical spherulites were obtained at NTO/NMP ratios ranging from 0.2 to 0.6 and at water/NMP ratios ranging from 1.0 to 4.8 at a cooling rate of 10 K/min. At water/NMP ratios < 1.0, NTO was not crystallized, but the NTO-NMP complex was crystallized. This was also expected from the ternary phase diagram presented in Fig. 3.27. At water/NMP ratios > 4.8, with increasing water fraction of the ternary mixture the morphology of the spherulite did not change significantly, but its shape was changed to a snowball-like shape. The surface of the NTO crystals obtained at a water/NMP ratio of 3.0 (see Fig. 3.36(b)) is found to be much more compact and spherical than those obtained at a water/NMP ratio of 1.8 (see Fig. 3.36(a)).

At NTO/NMP ratios > 0.6 (see Fig. 3.36(c)), the shape of the spherulite is snowball-like with an irregular surface. On the surface of the crystals obtained at the composition of points E and H in Fig. 3.36(b), cracks were observed. This supports the view that these cracks result from the strain existing in the growth of three-dimensional spherulite. A similar result was reported for spherulite of organ-

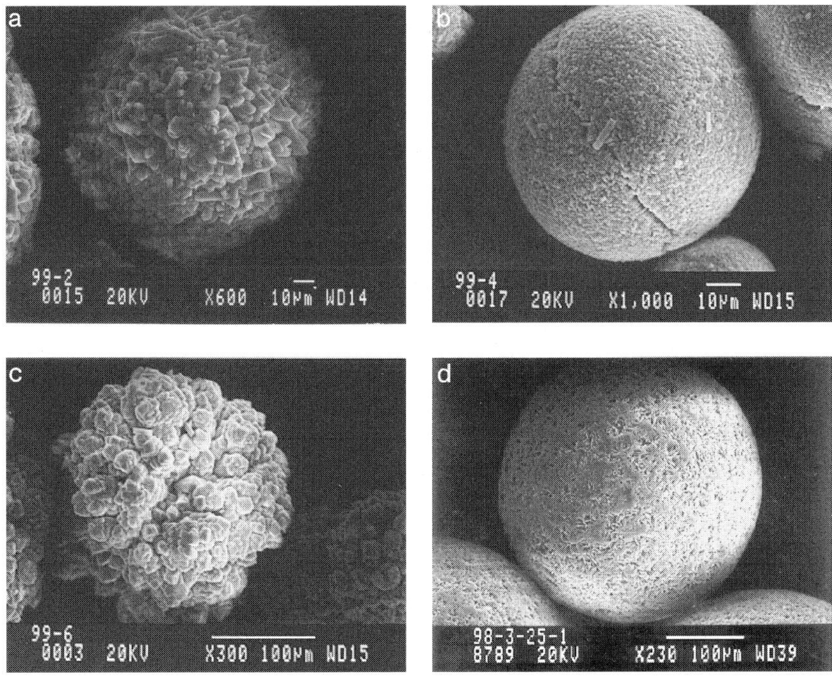

Figure 3.36. Scanning electron micrographs of crystals obtained at various compositions of water-NT-NMP: (a) point B; (b) point E; (c) point F; (d) point J in Fig. 3.27.

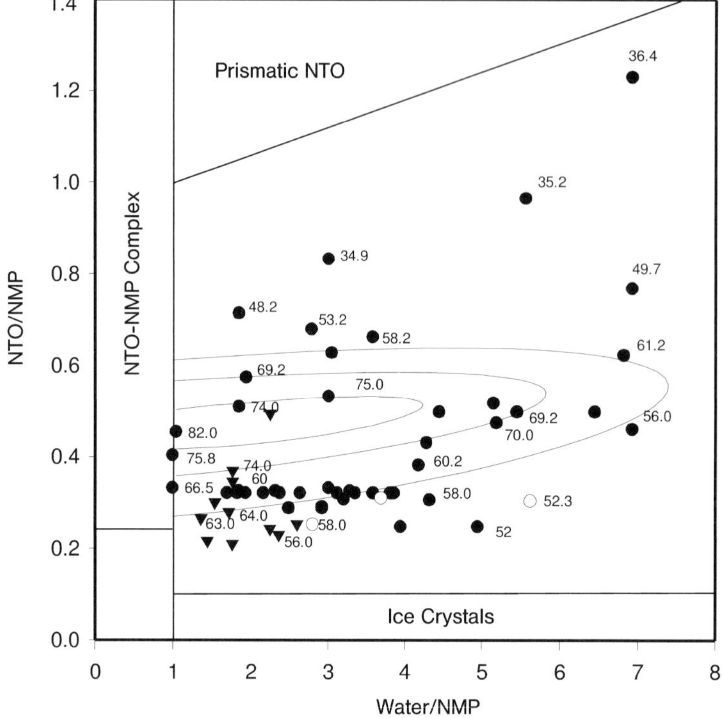

Figure 3.37. Metastable zone width at various compositions.

ics [3.66]. Cracks were formed when the spherulite could not withstand a high level of strain.

Our study revealed that the cooling rate and the composition affected mainly the morphology of spherulite. Figure 3.37 shows the status of the morphology of the NTO crystals obtained at a cooling rate of 10 K/min in the plot of NTO/NMP ratio versus water/NMP ratio. A change of morphology according to the ternary composition at the same cooling rate can result from the difference in nucleation manner, which is characterized by the metastable zone width.

3.2.5.3 Seeded Cooling Crystallization

NTO crystals obtained from aqueous solution (i. e. NTO-water) belong to the cubic system, as shown in Fig. 3.23 [3.59, 3.80]. This cubic crystal of NTO grows invariantly in the binary solution but transforms into spherulitic form during growth in the ternary system. The habit modification of NTO crystals is due to the presence of NMP in the ternary system. Figure 3.38 shows SEM photographs of NTO crystals, which were spherical seed crystals grown in water-NMP solution. Figure 3.38(a), (b) and (c) show SEM photographs of NTO crystals, which were spherical seed crystals grown in solutions with water/NMP ratios of 1.8, 3.0 and 8.0, respectively.

Figure 3.38. SEM photographs of crystals obtained in seeded crystallization at different cooling rates and water/NMP ratios.

Crystallizations were carried out at a crystallization temperature of 293 K, a cooling rate of 1 K/min and a saturation temperature of 318 K. Figure 3.38(d), (e) and (f) show SEM photographs of NTO crystals, which were spherical seed crystals grown under the same conditions except at a cooling rate of 10 K/min. With growth of the spherulitic crystals, it was found that the crystals grow radially from the center and hence have spherically spherulitic morphology.

Figure 3.39(a) and (b) show SEM photographs of NTO crystals obtained in aqueous solution at cooling rates of 1 and 10 K/min, respectively. Crystals grown from aqueous solution present faces covered with larger cubes, the edges of which show a large number of irregular needles with no regular orientation. It was found that the crystals formed in water were not spherulites but agglomerates. As the content of NMP in the solvent is increased, the number of crystallites adhering to

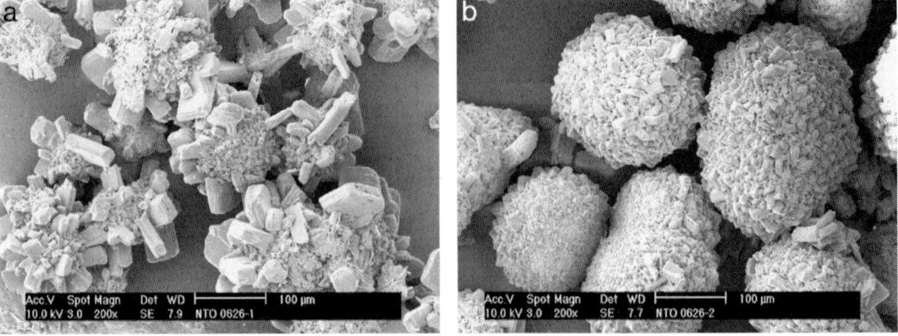

Figure 3.39. SEM photographs of crystals obtained by seeded crystallization in water at cooling rates of (a) 1 and (b) 10 K/min.

the faces and corners of the needles increases. An increased tendency for the formation of aggregates towards the radial direction parallels the content of NMP in the solvent. From the comparison between NTO morphologies obtained at two cooling rates, it is found that an increase in cooling rate in cooling crystallization increases the compactness and sphericity of NTO crystals. The thickness of needle-like crystals aggregated in the seed crystals increases with decreasing cooling rate and with increasing water content in the solvent.

In nucleation studies [3.59, 3.62], it was found that the presence of NMP increases the metastable zone width of the NTO-water systems. The slow deposition of the solute, together with a low production rate of nuclei, allow the formation of spherulitic crystals with fairly smooth surfaces. As can be seen in Fig. 3.38, the crystals grown at a lower water/NMP ratio were found to be more spherical and compact. The reasoning strongly suggests that the formation of spherulitic crystals is indeed determined by the nucleation process, while the rate of spherulite growth can be controlled by the rate of aggregation. The formation and growth of NTO spherulites in water-NMP-NTO solutions can be affected by several factors such as concentration and the type of solvent used. Solvation, selective solvation and self-association of the solute may also have an influence. From the crystal growth behavior observed in the NTO-water-NMP system, water acts as an accelerator, promoting spherulitic morphology by enhancing NTO growth.

3.2.5.4 Scale-up of Crystallizer

Theory

Mean crystal size can be expressed as a function of the growth rate G, the mass nucleation rate B_m and suspension density of crystals φ_s [3.72, 3.81]:

$$L = f\left(\frac{G\varphi_s}{B_m}\right) \tag{3.19}$$

This relation results from the fact that the mean crystal size increases with increasing growth rate and holdup of crystal and a decreasing nucleation rate.

The overall growth rate can be expressed as a power function of supersaturation [3.14]:

$$G = k_{\mathrm{G}} \Delta c^{g} \tag{3.20}$$

where G is the overall growth rate, k_{G} the overall growth constant and g the order of the process with respect to supersaturation, Δc.

A power law function of the form [3.14]

$$B_{\mathrm{m}} = k_{n} \Delta c^{n} \tag{3.21}$$

was used to fit the nucleation kinetic data.

The combination of Eqs. (3.19), (3.20) and (3.21) results in the following simple correlation for the mean crystal size:

$$L = K_{\mathrm{N}} \varphi_{\mathrm{s}} \Delta c^{g-n} \tag{3.22}$$

where

$$K_{\mathrm{N}} = \frac{k k_{\mathrm{G}}}{k_{\mathrm{n}}} \tag{3.23}$$

K_{N} is a system constant comprising the kinetic parameters of crystallization such as k_{g} and k_{n}. As the rate constants k_{n} and k_{G} in Eq. (3.23) are in a ratio and their changes with temperature or other parameters are similar, the system constant K_{N} is, relatively speaking, not too sensitive either to temperature or to intensity of mixing and to other parameters. It is therefore suitable for design calculations when changing the scale of the crystallizer.

Design of Crystallizers

The laboratory-jacketed crystallizer (300 ml, inner diameter 90 cm, height 120 cm) was made of Pyrex and was equipped with a stainless-steel marine stirrer (SUS 304) and sensors. A draft tube (diameter 60 cm, height 60 cm) and four inner baffles were included. The agitation speed was fixed at 600 min^{-1} (tip speed 1.9 m/s) to ensure a well-mixed crystal suspension.

The bench-scale-jacketed crystallizer (50 l, inner diameter 0.4 m, height 0.5 m) was made of stainless steel (SUS 304) and was equipped with a geometrically similar marine propeller stirrer (SUS 304, diameter 0.12 m, blade width 0.035 m, blade angle 25°) and temperature sensors. A draft tube (inner diameter 0.27 m, height 0.27 m), four inner baffles (length 0.23 m, width 0.12 m) and four outer baffles (length 0.34 m, width 0.037 m) were included. The agitation speed was fixed at 300 min^{-1} to ensure a well-mixed crystal suspension. To assess critically the scale-up effect of the product crystal size, the experimental conditions of the laboratory experiments were basically the same as those of the bench-scale experiments. The dimensionless group for scale-up of the crystallizer is calculated on the basis of the laboratory-scale operation: Newton number 0.5, Flow number 0.4, Reynolds number 90 000 and Froude number 0.81, at which the pumping velocity and tip speed of the agitator were 0.031 m^3/s and 1.9 m/s, respectively [3.1].

Modeling

The crystal size and crystal size distribution of energetic materials are parameters of importance in the performance of explosion in that they determine both the packing density and true density of the explosive. They are complex functions of nucleation and crystal growth rates, which are themselves functions of process variables such as agitation rate, composition of solution and production rate. The composition of the co-solvent and the temperature were also important factors in determining the spherulite size and its distribution in the crystallization of NTO in a co-solvent [3.61].

As shown in Fig. 3.33, the results of laboratory-scale tests were presented as plots of the crystal size of spherical NTO spherulites versus the water/NMP ratio for various NTO/NMP ratios. It was found that the crystal size increased with increasing water/NMP ratio and increasing NTO/NMP ratio. The crystal size increases with increasing content of water in the co-solvent. We can control the crystal size of the spherulites in the range from 20 to 210 μm by adjusting the composition of the ternary mixture at a constant cooling rate and agitation speed.

From Eq. (3.22), the mean crystal size can be estimated from the relationship between supersaturation, crystallization kinetics and suspension density. It is sensitive to the value of the difference between the kinetic exponents for growth and nucleation, $g-n$. In order to evaluate the results by application of Eq. (3.22), the data on L, φ_s and Δc for each run and the value of the exponent $g-n$ must be known. The value of φ_s, the suspension density, is calculated from the mass balance for the crystallizer. The set values of φ_s for water/NMP ratios of 1.8, 3.0 and 8.0 were 0.1, 0.048 and 0.031, respectively [3.82]. The value of the difference between orders of nucleation and crystal growth, $g-n$, can be calculated from Eq. (3.22) rewritten for the given purpose in the form

$$\frac{d\ln L}{d\ln \Delta c} = g - n \tag{3.24}$$

Supersaturation depends mainly on the cooling rate in crystallization by cooling. The previous study reported that spherical NTO crystals were obtained at cooling rates above 10 K/min and water/NMP ratios ranging from 1.0 to 8.0 and NTO/NMP ratios from 0.25 to 1 [3.61]. Supersaturation was affected by the solution composition at same cooling rate and mixing conditions.

Figure 3.40 shows a plot of supersaturation versus water/NMP ratio for various NTO/NMP ratios. Supersaturation increases linearly with increasing water/NMP ratio and decreases with increasing NTO/NMP ratio. The relationship among them can be expressed as follows:

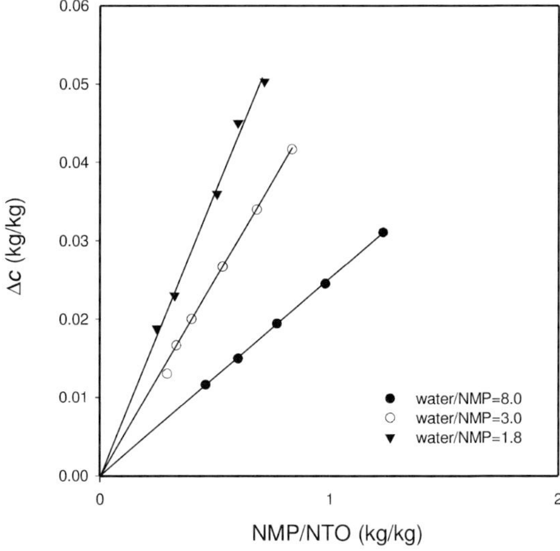

Figure 3.40. Plot of supersaturation as a function of NMP/NTO ratio for various water/NMP ratios.

$$\Delta c = 1.346X^{-0.87}Y^{1.0} \tag{3.25}$$

where X and Y are the water/NMP and NTO/water ratio, respectively. All results were correlated with Eq. (3.25), as shown in Fig. 3.41.

The relative standard deviation and the average absolute deviation between the measured supersaturation data and calculated data were 0.00008 and 0.00006 kg/kg, respectively [3.82]. Therefore, Eq. (3.25) expresses the supersaturation in which the solution composition with a constant cooling rate in such a cooling crystallizer is considered.

When the water/NMP ratio is constant, combination of Eqs. (3.22) and (3.25) gives the following relation:

$$\frac{\mathrm{dln}L}{\mathrm{dln}Y} = g - n \tag{3.26}$$

The order difference $g-n$ can be expected from Eqn. (3.26), in which a log-log plot of crystal size and NTO/NMP ratio at a constant water/NMP ratio is considered.

Figure 3.42 shows plots of L versus Y. The order difference $g-n$ can be obtained from the slope of these plots. The values of the order difference were found to be 0.828, 0.581, 0.438 for water/NMP ratios of 1.8, 3.0 and 8.0, respectively. These values are lower than the calculated value from separate measurements of nucleation and growth kinetics presented in [3.82].

The values of the system constant K_N calculated from Eq. (3.22) for each run were calculated from the slope of the curve of L versus $\varphi_s \Delta c^{g-n}$. From the NTO crystallization in water-NMP mixtures by cooling, it was found that values of K_N for water/NMP ratios of 1.8, 3.0 and 8.0 were 4889, 1539 and 324, respectively.

Despite a slight scatter of data, it can be seen that all plots st the three water/NMP

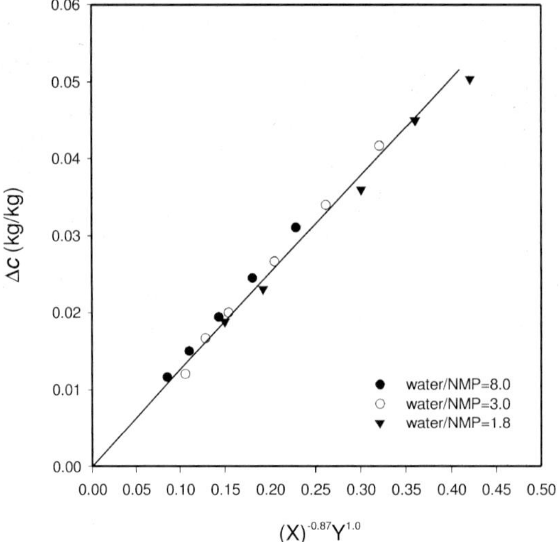

Figure 3.41. Correlation of supersaturation with NTO/NMP ratio (X) and water/NMP ratio (Y).

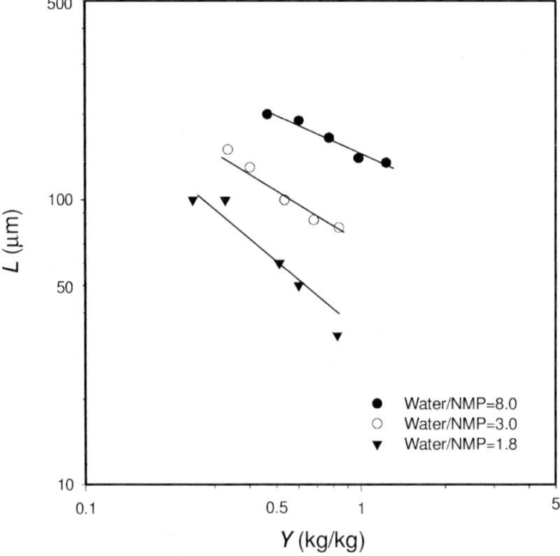

Figure 3.42. Plot of L versus Y.

ratios can be adequately fitted with a different slope. The equations found may be summarized as follows:
water/NMP = 1.8:

$$L = 324\varphi_s \Delta c^{-0.828} \tag{3.27}$$

water/NMP = 3.0:

$$L = 1539\varphi_s \Delta c^{-0.581} \tag{3.28}$$

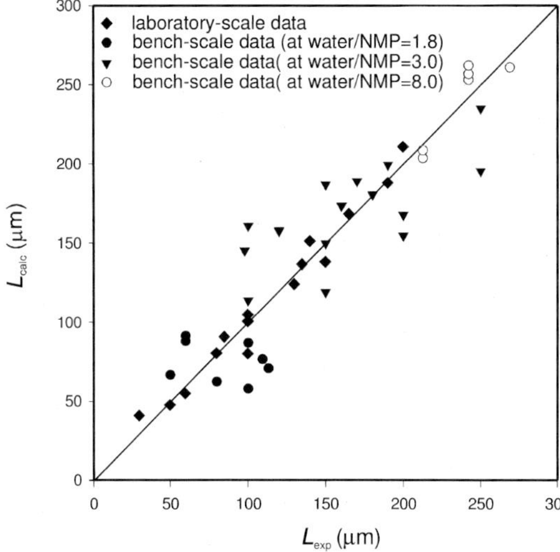

Figure 3.43. Comparison between the calculated and experimental mean crystal size for laboratory- and bench-scale crystallizers.

water/NMP = 8.0:

$$L = 4889\varphi_s \Delta c^{-0.438} \tag{3.29}$$

The mean crystal size was calculated from Eqs. (3.27) and (3.28) for the conditions of individual experiments and they were compared with experimental values. Figure 3.43 shows a comparison between the calculated and experimental mean crystal sizes obtained in laboratory- and bench-scale tests. It is obvious that the results of experiments are reasonably well approximated by Eq. (3.22). The results show that Eq. (3.22) with the parameters K_N and $g-n$ is satisfactory for representing the data from both series of measurements, which differ in the scale of the equipment used by more than two orders of magnitude. It might be concluded that further scale-up based on the above equations would be sufficiently reliable and constitute an adequate starting point for the crystallizer design.

3.2.6 Phase Stabilized Ammonium Nitrate (PSAN)

3.2.6.1 Introduction

Ammonium nitrate (AN) is used as an oxidizer in solid propellants, explosives and gas generator systems. Drawbacks are low performance, low burning rates and phase transitions that influence material properties. These drawbacks and the development of ammonium perchlorate (AP) filled composites decreased the interest in ammonium nitrate for application in solid propellants in the 1960s and 1970s. New interest was aroused when smokeless, low-sensitivity and ecologically harmless propellants were required, which could not be realized with the hydrogen chloride-emitting ammonium perchlorate composites.

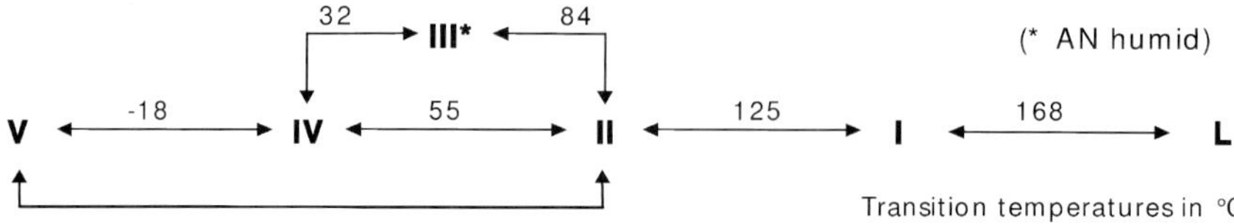

Figure 3.44. Transition paths of ammonium nitrate.

3.2.6.2 **Understanding and Measuring of the Phase Transitions**

Ammonium nitrate crystallizes in five phases. The structurally related phases I, II, IV and V appear on cooling dry ammonium nitrate from the melt after passing through order-disorder transitions. The structurally not related phase III appears only in the presence of water, which helps to nucleate the phase [3.83, 3.84]. The transition paths are represented in Fig. 3.44.

The existence of several phases and transitions in a relatively narrow temperature range together with complications caused by the presence of water make it difficult to obtain unambiguous results with classical thermal analysis methods such as DSC or TMA, as these methods measure thermal effects without identifying the phases that occur. The phases may be identified on the basis of their characteristic crystal structures if X-ray diffraction is applied (see Chapter 9).

Confronted with these problems, the first temperature-resolved X-ray diffraction experiments on ammonium nitrate were carried out by Engel and Charbit in 1978, revealing the transition paths of ammonium nitrate on heating and cooling [3.85]. Today, with modern measuring and evaluation techniques, the method has become a powerful tool in the field of thermal analysis [3.86] and it is applied extensively for the investigation of the phase behavior of energetic materials.

3.2.6.3 **Improving the Phase Behavior**

The interest in ammonium nitrate stimulated research to improve its phase transition behavior. In 1899, Müller reported the influence of potassium nitrate on the phase transitions of ammonium nitrate [3.87] and Campbell and Campbell proposed in 1946 a solid solution of potassium nitrate and ammonium nitrate to suppress undesirable transitions [3.88]. The system was investigated by Coates and Woodard using X-ray diffraction [3.89] and the crystal structures of its phases were published by Holden and Dickinson [3.90]. The proposed methods stabilize phase III; however, they fail if phase stability below −30 °C is required.

The phase transition is also influenced if cesium is incorporated into the ammonium nitrate lattice instead of potassium, which extends the stability ranges of the related phases II and V [3.91, 3.92]. The material, however, has no practical importance because of the low availability of cesium.

Another promising concept for stabilizing ammonium nitrate was developed in ICT by incorporating diammine complexes of nickel, copper or zinc into the ammonium nitrate lattice [3.93]. The materials were synthesized within a solid-state reaction by melting mixtures of metal oxide and ammonium nitrate. Samples were

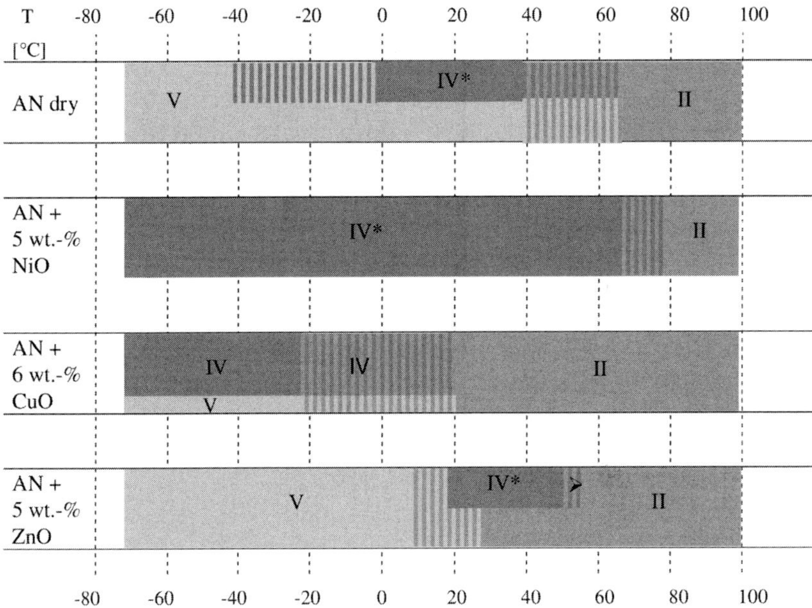

Figure 3.45. Stability ranges of ammonium nitrate phases depending on additive and temperature [3.97]. *Starting with 100% phase IV; ➢ not reversible (all measurements were started at room temperature on heating).

investigated with temperature-resolved X-ray diffraction [3.94] and applied for developing gas generators [3.95] and solid propellants [3.96]. Figure 3.45 shows the transition paths of dry ammonium nitrate and PSAN synthesized by doping with 5–6 wt% of the metal oxides. The figure was drawn using data reported by Engel *et al.* [3.97].

The investigations showed that incorporating diammine nickel(II) into the ammonium nitrate lattice stabilizes the room temperature phase IV. Compared with pure ammonium nitrate, the material changes at higher temperatures into phase II. Phase IV is still stable at –70 °C and even at lower temperatures. Propellants produced with this material show good properties during temperature cycling. However, using the material it must be taken into account that nickel and its compounds may create health problems.

Incorporating diammine copper(II) into the ammonium nitrate lattice stabilizes phase II. Above a CuO concentration of 8 wt% phase IV is suppressed so that AN is stabilized from –70 to 150 °C. As these phases are very similar, the II-V transition occurs without major volume changes. Further, the material showed excellent properties during storage over 10 years. No agglomeration was observed.

Ammonium nitrate with diammine zinc nitrate shows similarities with the material doped with diammine copper(II). The stability of phase II is extended. On cooling, however, phase V appears instead of phase IV. At room temperature phase

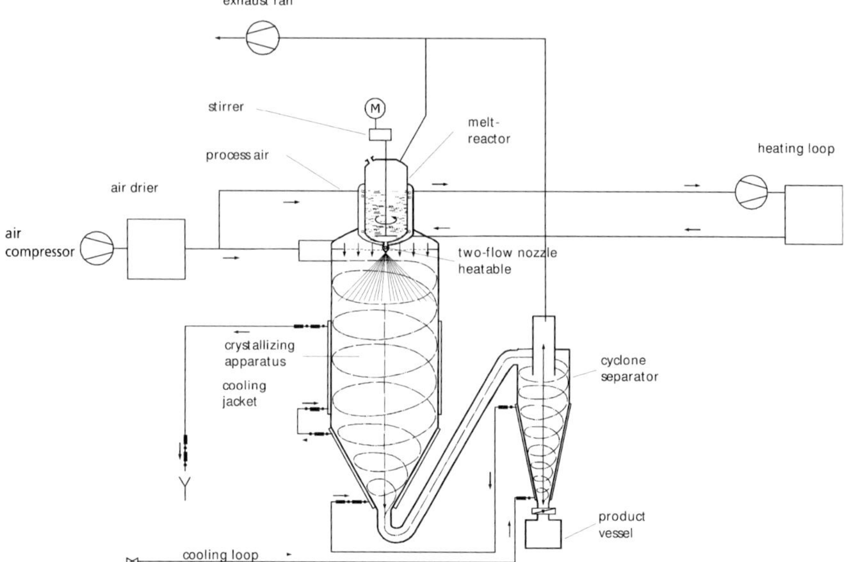

Figure 3.46. Spray-crystallization process for ammonium nitrate.

V was observed after the preparation and phase IV if the samples were stored at room temperature for more than 1–2 weeks. During the production and processing of the material water must be totally removed and excluded. The material is preferably used if catalytic activity as in the case of the copper-doped materials is not wanted.

3.2.6.4 Production Process

At the Fraunhofer ICT ammonium nitrate is produced by atomization of a melt, which leads to spherical materials especially suited for processing in propellant mixtures. Figure 3.46 shows the spray-crystallization process schematically.

Up to 40 kg of raw ammonium nitrate can be melted in the heatable reactor vessel. If necessary, phase-stabilizing additives (metal oxides) can be added to the raw material and stirred with the ammonium nitrate melt before the atomization process is started. The atomization occurs with a heatable two-flow nozzle and dried compressed air as a process gas. The recrystallization of the atomized melting droplets takes place on their spiral-shaped route through the crystallization apparatus. Subsequently the spherical product is separated in a cyclone.

The available particle sizes range from 20 to 400 μm. Standard qualities are produced with mean particle sizes of 300, 160, 50 and 20 μm (see Fig. 3.47), which allows the selection of materials suited for a variety of applications.

As phase-stabilizing additives, copper, nickel or zinc oxide at levels from 1 to 6 wt% are used. To avoid agglomeration, ~1 wt% silicon dioxide (SiO_2) or tricalcium phosphate (TCP) can be used as an anticaking agent.

Figure 3.47. Spherical ammonium nitrate produced by atomization.

3.2.7 Crystallization of ADN

Ammonium dinitramide (ADN) is an effective, non-chloride-containing solid oxidizer, which offers the potential for significant increases in performance in both propellants and high explosives [3.98, 3.99].

Raw ADN usually consists of irregularly shaped particles and some agglomerates, which complicate the processing (see Fig. 3.48). Recrystallization is required after synthesis to improve the physical properties of ADN (such as its suitability for use in the production of filled polymers), its burning characteristics and its chemical purity.

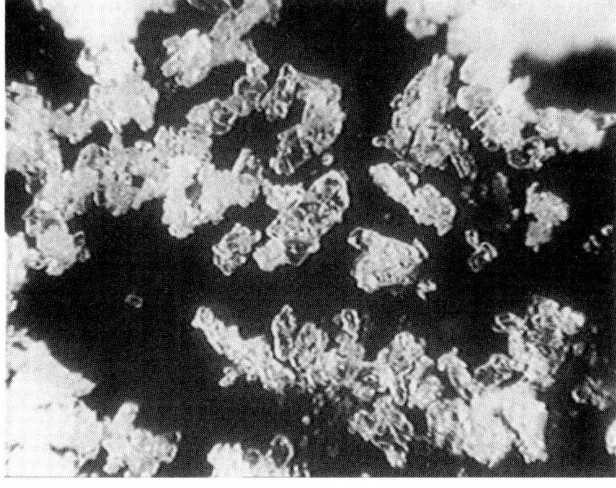

Figure 3.48. ADN as it results from synthesis.

In general, emulsion crystallization is applied to purify a two-phase system from low-melting components. The crystallization is carried out by cooling the emulsion. Impurities, in the form of eutectic mixtures, remain in the emulsion, from which they may be recovered by further cooling [3.14]. Espitalier *et al.* [3.100] described an emulsion crystallization process in which the dispersed phase is the solid product dissolved in an organic solvent. The continuous phase must not be miscible with this solution. Heat and mass transfer phenomena lead the droplets to a limit supersaturation state, allowing crystallization of the product. In contrast to this, the crystallization of ADN is carried out from molten ADN which is dispersed in a non-polar continuous phase. The crystallization behavior of dispersed melts was investigated by McClements *et al.* [3.101] in hydrocarbon-in-water emulsions. Their measurements showed that crystallization is induced in liquid droplets if solid particles of the same material are present. The solid particles in a supercooled emulsion initiate the crystallization when they collide with liquid droplets. This time-dependent crystallization mechanism can also be observed in the emulsion crystallization process of ADN.

For the formation of spherical ADN particles, two different technologies have been developed. Highsmith *et al.* [3.102] describe a prilling technique, which allows the molten droplets of ADN to fall by gravity into a tower of flowing inert cold gas. The other ADN formation process is an emulsion crystallization technique [3.103, 3.104]. In this process, production of spherical ADN particles is carried out in two stages:

1. Preparation of a W/O emulsion with liquid ADN as the disperse phase and paraffin oil as the continuous phase. The emulsion is prepared with a stirrer in a batch process. The stirrer geometry and speed must be set as appropriate to the system.
2. Crystallization of ADN droplets to spherical, solid particles. Liquid ADN exhibits a strong tendency toward supercooling, so in addition to cooling below the melting-point, crystallization must be initiated by mechanical energy input or particle-particle interaction. To avoid droplet deformation, the energy input must be kept below a certain threshold.

After successful crystallization and solid-liquid separation, the particles are washed and dried. Figure 3.49 shows the process flow chart for batch production of ADN particles by emulsion crystallization.

The continuous phase is first admitted into a temperature-controlled vessel. When the process temperature is reached, pusher S1 introduces raw ADN from intermediate storage. The ADN melts and is dispersed with the desired droplet size by the stirrer. A temperature gradient is then imposed on the system by switching three-way valves V1 and V2 over to the cooling loop and the crystallization process is initiated.

Figure 3.50 shows the spherical ADN particles that result after recrystallization in the emulsion crystallization process. This process makes it possible to produce particles with a median size in the range between 20 and 500 μm [3.105].

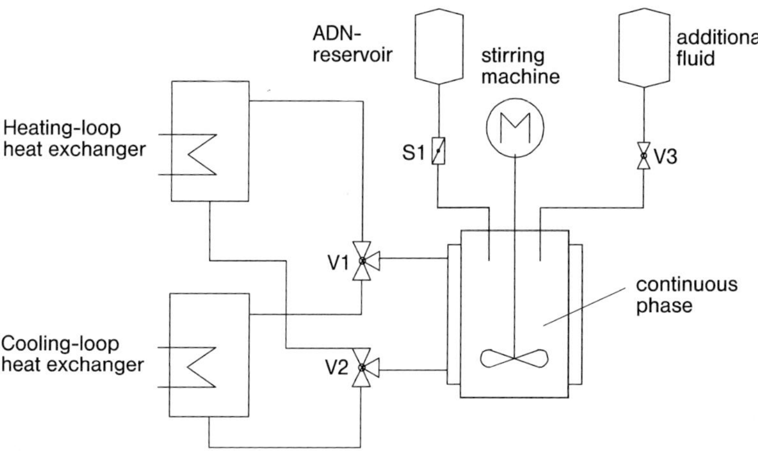

Figure 3.49. Emulsion crystallization process flow chart.

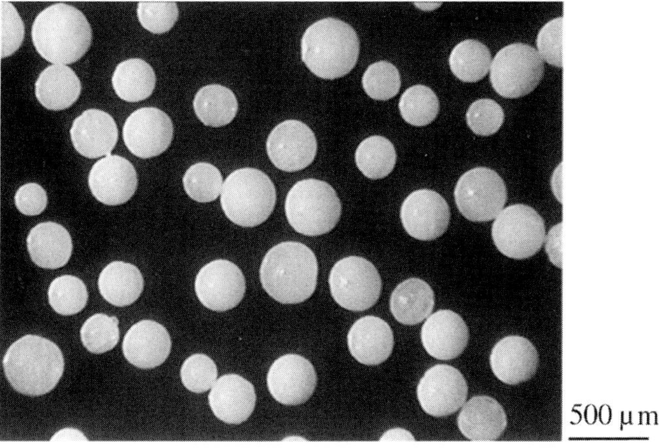

Figure 3.50. Recrystallized ADN (emulsion crystallization process).

The crystallization of dispersed ADN droplets can be initiated by stirring energy, ultrasonic energy, the rheology behavior of the continuous phase or by seed crystals. The nucleation of molten ADN in the presence of seed crystals is shown in Fig. 3.51. Within 180 s the ADN droplet is fully solidified. In corresponding experiments without seed crystals no nucleation occurred even though the temperature was decreased to 20 °C.

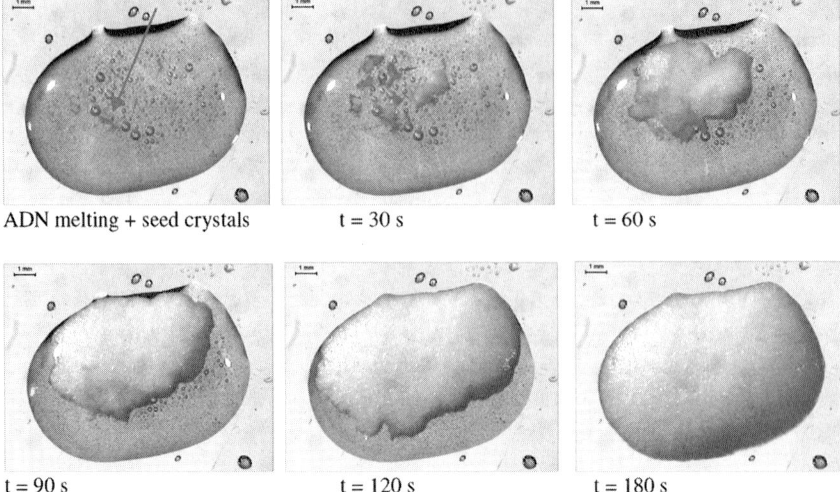

Figure 3.51. Recrystallization of molten ADN initiated by seed crystals.

3.3 Simulation

3.3.1 Introduction

The growth process of crystals in the crystallizer takes place at the interface between the crystal and the solution, the crystal surface. The growth is determined by the interactions of the molecules in that region. Hence although industrial crystallization is employed on large scales, the actual crystal growth processes in the crystallizer still take place on a molecular scale. Recently, attention has therefore become more focused on the extent to which molecular modeling can be used to explain and predict industrial crystallization behavior.

Molecular modeling is a collection of computer simulation techniques by which processes on a molecular scale can be studied. This section focuses on the application of molecular modeling to tackle crystallization problems which arise from product quality demands (crystal size distribution, crystal shape, crystal purity). A brief introduction is given to molecular modeling, after which its effectiveness for the investigation of crystallization problems is discussed.

Since crystallization is a separation and purification process, the mother liquor generally contains more compounds than the solute. These foreign compounds can be divided into solvents, additives and impurities. Foreign compounds can have a significant effect on the crystal quality (shape, inclusion content, etc.) and the process handling (filterability, scaling, caking, etc.). Molecular modeling is an effective tool for some of the effects of foreign compounds on crystallization properties.

Any attempt to understand the chemistry of energetic materials leads naturally to an interest in the structure and relative energies of the molecules involved. Energetic materials, by their very nature, possess unusual characteristics. They often contain chemical groups such as nitro or nitramine, which are not commonly found in more traditional subjects, such as biochemistry, pharmaceuticals or materials science, where computational chemistry is applied. Another way in which energetic materials store energy is to use molecules with ring structures, where internal strain adds to the energy of the material. Computational chemistry provides a set of tools for exploring the molecular structure of such molecules, exploring their conformational stability and also the way in which the molecules can interact with each other to form crystals.

One of the product quality demands for an energetic crystalline material might be the demand for a specific crystal shape. In the case of RDX, a demand exists for an isometric (rounded) crystal shape in order to obtain large crystal packing densities in the product.

3.3.2 Molecular Modeling of Energetic Materials

This section outlines some of the computational methods that are available for calculating the properties of energetic molecules and their crystals. Examples are given of the calculation of the crystal structures of RDX and HNIW (CL 20), the calculation of surface energies and approaches to the understanding of how their surfaces interact with binders and processing aids. The effect of the solvent on the RDX crystal shape is discussed and two methods are given to predict this effect.

3.3.2.1 Molecular Structure of Energetic Materials

The starting point for any calculations on the electronic structure of a molecule is the solution of Schrödinger's equation using a quantum chemistry package such as MOPAC [3.106–3.108], GAMESS–UK [3.109] or Gaussian [3.110]. The input to such packages is usually the starting point for a geometry optimization of the molecule and the output is an energy and an associated geometry, together with the electronic wavefunction, from which all manner of molecular properties can be calculated. The background to the various methods is outlined in several books [3.111, 3.112] on the subject and is not discussed in any detail here.

A short description of the various algorithms used to solve the quantum mechanical molecular electronic structure problem is given below. Quantum mechanics has limited application to large problems because the computational effort required to solve the equations expands increasingly rapidly with its size. Molecular mechanics and dynamics are the methods of choice for such systems and a brief outline of these methods is also given.

Semi-empirical Methods

Semi-empirical methods, such as those implemented in the MOPAC program, simplify the equations considerably by neglecting many terms, but then compensating for this by parameterizing some of the terms so that the calculations agree with any experimental information on, for instance, the heat of formation. Once the various approximations have been made, the molecular properties to which the parameters are fitted and the molecules used in the fitting define a model Hamiltonian, of which the most commonly used are the AM1 and the PM3 Hamiltonians found in MOPAC. A major advantage of semi-empirical methods is the speed of computation. Molecules of several hundred atoms can be treated using a semi-empirical method. However, the accuracy of a semi-empirical calculation depends on the groups that have been included in the fitting procedure.

Ab Initio Methods

Ab initio methods usually begin with a solution of the Hartree-Fock equations, which assumes that the electronic wavefunction can be written as a single determinant of molecular orbitals. The orbitals are described in terms of a basis set of atomic functions and the reliability of the calculation usually depends on the quality of the basis set being used. Basis sets have been developed over the years to produce reliable results with a minimum of computational cost. For example, double zeta basis sets such as 3–21G [3.113], 4–31G [3.114] and 6–31G [3.115] describe each atom in the molecule with a single core 1s function and two functions (hence the term double zeta) for the valence s and p functions. Such basis sets are commonly used, as there appears to be a cancellation of errors, which fortuitously allows them to predict fairly accurate results.

The Hartree-Fock method neglects the instantaneous correlation between electrons that may be incorporated by a variety of algorithms, including configuration interaction (CI), multi-configuration self-consistent field (MCSF) and Møller-Plesset second-order perturbation theory (MP2). Accurate calculations using such methods require basis sets that include polarization functions. Polarization functions, such as *d*-functions for the first-row atoms, are available for standard basis sets and are often referred to as, for example, 6–31G** [3.116]. The two stars indicate *p*-functions have been included on hydrogen as well as *d*-functions on the other elements. These correlated methods produce very reliable results for energies, geometries and vibrational frequencies. It is common to see much larger basis sets being used and in this work a triple zeta valence plus polarization basis set (TZVP) will be used [3.117], which allows for even greater flexibility in the core and the valence region of the atom.

In principle, *ab initio* methods can calculate an energy or geometry to any required level of accuracy. In practice, the more accurate methods require larger basis sets and this rapidly increases the computational cost of the calculation to a point where they are prohibitively expensive.

Density Functional Methods

Density functional theory (DFT) is a recent newcomer to the field of computational chemistry. Rather than calculating the wavefunction, they calculate the electron density directly. Although much of the computational machinery is similar, these techniques should not be regarded as related to Hartree-Fock or even post-Hartree-Fock methods. Kohn and Sham [3.118] provided the theoretical foundations for the approach. The method assumes a functional form for the contribution of the electron density to the electron correlation and to the electron exchange energies. The functional may be local (Local Density Approximation functional), in which case it depends only on the value of the electron density in space. More recent and more accurate functionals are non-local and they depend on both the values of the electron density and its first derivative (gradient-corrected functionals). In the work described here, we will be using the B3LYP gradient-corrected density functional. This uses the Becke non-local gradient correction, exact exchange using the Becke three-parameter exchange functional [3.119–3.122] and the non-local correlation functional of Lee, Yang and Parr [3.123]. DFT has proved to be as computationally efficient as Hartree-Fock methods whilst providing results with a reliability as good as some of the post-Hartree-Fock methods such as MP2. Together with the chosen basis sets, this should produce results of chemical accuracy [3.124].

Quasi-harmonic Approximation

Where possible, it is important to verify the type of turning point that the optimization process has found. Very often the minimization has been performed with an implicit symmetry constraint and the result may not be a true minimum. Diagonalization of the mass-weighted second derivatives of the energy with respect to atomic coordinates provides information on the nature of the minimum. A true minimum will have all positive eigenvalues, whilst a transition state may have one or more negative eigenvalues.

In addition to providing information on the nature of the turning point, the second-derivative matrix can be used within the quasi-harmonic approximation to estimate the vibrational contributions to the total energy, providing a so-called thermal correction.

Molecular Mechanics

The molecular mechanics method represents the total energy of a system of molecules with a set of simple analytical functions representing the interactions between bonded and non-bonded atoms. The bonded atom interactions are usually split into bond stretching, bond bending and torsion twisting terms. The non-bonded interactions are represented by three main contributions. There is a short-range repulsion term, which prevents atoms from overlapping. This is often represented by an r^{-12} term as in the Lennard-Jones potential, but a term with an exponential dependence on r is also often used. The two other non-bonded terms are longer range in nature. The first is the van der Waals interaction, which has an r^{-6} dependence and is usually included with the repulsion term described above, to give a single function. The last term is the electrostatic interaction, which usually has an

r^{-1} dependence (occasionally a distance-dependent dielectric constant is used to decrease the range of interaction to r^{-2}). The parameterization of these functions (the force field) represents the chemistry of the species involved. Various force fields have been derived, but in this work we use the force fields provided within the Cerius^2 package [3.125]. In particular, we make use of the Dreiding force field [3.126], which is capable of modification. The electrostatic interactions are represented by the interaction between charges centered on the atoms. A feature of the Cerius^2 package is its treatment of electrostatic interactions in periodic systems. Cerius^2 calculates the electrostatic interactions accurately using the Ewald method [3.127], which ensures convergence of the infinite sums over the infinite lattice. The derivation of the values of atomic centered charges is a very important requirement for the reliable prediction of crystal packing. In this work, we use the potential-derived charge (PDC) method [3.128] to assign suitable charges. In the PDC method an *ab initio* wavefunction is analyzed to give the electrostatic potential around the van der Waals surface of the molecule. The method optimizes the charges at atomic centers so as to fit this electrostatic potential, whilst constraining the charge to reproduce the total dipole moment of the molecule. This procedure ensures that the long range (dipole-dipole) interactions are reasonable, whilst ensuring that the local interactions between charge groups near the van der Waals surface of the molecules are well represented. It is interesting that there appears to be a useful cancellation of errors when using the Hartree-Fock method to calculate the charges. It is well known that larger basis sets tend systematically to predict too large a dipole moment when compared with the experimental gas-phase results. However, it is believed that this is more appropriate for molecules in the solid state, where additional charge separation is likely to lead to increased stabilization.

The reliability of molecular mechanics calculations hinges entirely on the validity and range of applicability of the force field. The force field itself can be validated against experimental and *ab initio* results and some examples of this will be shown in the present work. Because of the relative speed of molecular mechanics calculations, it is possible to consider calculations of crystals and their surface, something that is not routinely possible with *ab initio* methods. There has been some recent work [3.129–3.131] on RDX and related energetic molecules using solid-state Hartree-Fock methods, where the electronic structure problem is solved using 3D periodic boundary conditions.

Molecular Dynamics

The methods discussed above are normally restricted to geometry optimization. The introduction of molecular dynamics allows the computational chemist to include the effects of temperature and vibration in the simulation. The method essentially integrates Newton's laws of motion for each of the atoms in the system, moving along a potential energy surface defined by the molecular mechanics force field. Calculations can be done in a variety of ensembles including NVT (constant number, volume and temperature) and NPT (constant number, pressure and temperature). Calculations reported here are carried out in the NPT ensemble using the Cerius^2 molecular dynamics package. Typically, the calculation is started by assigning velocities to each of the atoms according to the temperature and a Maxwell distribution. The system is allowed to equilibrate for a period of time, usually around 20 ps, before information is collected for averaging.

Calculation of Crystal Habit

Several computational methods are available for calculating the expected shape of a growing crystal. The simplest of these is the method of Bravais-Friedel and Donnay-Harker (BFDH) [3.132], which predicts the morphology based solely on the crystal symmetry. This approach assumes that the binding energy between crystal planes is inversely proportional to the inter-planar spacing. Since the low-index faces have the largest inter-planar spacing, this method predicts that they should be most important in the crystal morphology. Thus it is possible to predict the expected crystal morphology purely on the grounds of the crystal space group, the dimensions of the unit cell and the positions of the atoms. In practice, this method often provides a useful method to screen potentially important surfaces in a growing crystal.

The attachment energy method [3.133] predicts the relative growth rate of each face by calculating the energy released when attaching a new slice to a growing surface. This calculation involves calculating both the energy of the slice of new crystal surface, E_{slice}, and the lattice energy, E_{latt}. The attachment energy is then $E_{att} = E_{latt} - E_{slice}$. Crystal faces with low attachment energies are the slowest growing surfaces and are likely to have higher surface areas. The attachment energy method is used to predict crystal habit in those cases where kinetic control of the growing surfaces dominates.

The surface energy method [3.134] determines the morphology that minimizes the total surface area of the crystal. The energy of a surface should be calculated using the Ewald method to ensure that the long-range Coulomb potential terms converge. For a surface, this can be achieved either by modifying the original 3D method to account for the loss of periodicity in one dimension or a super cell approach may be used where the 3D periodicity is maintained, but the cell contains a slab of the crystal. Provided that the surfaces of the slab are far enough apart and the slab is thick enough, this method should give well-converged energies. The surface energy method is most reliable in predicting the final shape of the crystal, when the conditions are such that the crystal is growing slowly, close to equilibrium conditions.

One other method of significance is the Hartman-Perdok method [3.135], which computes the morphology by examining networks of strong interactions between molecules. Additionally, once the morphology has been predicted and the important growth faces identified, these surfaces can be further analyzed in the presence of additives or solvents.

3.3.2.2 Molecular Modeling of Dimethylnitramine

A useful starting point for any series of calculations is a set of benchmark calculations that can be used to estimate the errors and establish confidence in an approach. Crystal structures of dimethylnitramine (DMN) are available [3.136–

3.138] and may be used to validate the methods and basis sets being used in further calculations on nitramine systems. All experimental crystal structures reported belong to the P21/m space group, which is monoclinic. Because of the relatively small size of the molecule it has been possible to perform quite accurate *ab initio* calculations of the isolated molecule in the gas phase. These were performed to understand better the nature of the nitramine group and to transfer this learning to RDX and HNIW molecules, which we will discuss later.

Ab Initio Calculations

Several conformers of DMN, shown in Fig. 3.52, were investigated. The conformers DMN1 to DMN4 keep the atoms of the nitramine unit in a plane. Conformers DMN1 to DMN3 differ in the orientation of the methyl groups. An additional conformer, DMN5, was investigated, based on DMN3 but with the planarity constraint of the nitramine group removed.

Ab initio Hartree-Fock, MP2 and DFT calculations using the B3LYP functional were performed using double zeta 6–31G, 6–31G** and triple zeta valence plus polarization (TZVP) basis sets. The results of the absolute energies and the relative energies of the molecules, after geometry optimization, are shown in Tables 3.5 and 3.6, respectively.

The Hartree-Fock calculations predict that the conformation with the N-N-O_2 atoms all lying in the same plane is the most stable. Rotation of the $-NO_2$ group, so that the oxygens are lying in a plane perpendicular to the plane of the rest of the molecule, is energetically expensive (22.7 kcal/mol) and indicates that any rotation about the N-N bond will be very slow. The CH_3 groups, on the other hand, appear to be fairly free to rotate. No calculations on barriers to rotation have been performed, but calculations on the three different conformers of DMN, brought about by

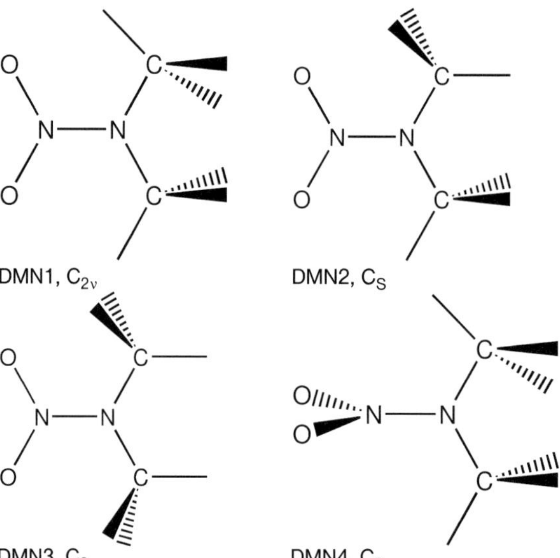

Figure 3.52. Conformers of dimethylnitramine.

Table 3.5. Absolute energies of dimethylnitramine.

Conformation	Total energy (hartree[a])			
	SCF/6–31G**	MP2/6–31G**	MP2/TZVP	DFT/TZVP
DMN1	−337.518739	−338.731985	−338.976906	−339.779191
DMN2	−337.519765	−338.732717	−338.977873	−339.779584
DMN3	−337.521364	−338.734766	−338.979572	−339.780610
DMN4	−337.486253	−338.697742	−338.947606	−339.745663
DMN5	−337.521364	−338.736557	−338.982694	−339.781328

a 1 hartree = 2625 kJ/mol.

Table 3.6. Relative energies of dimethylnitramine.

Conformation	Relative energy (kcal/mol[a])				Thermal correction
	SCF/6–31G**	MP2/6–31G**	MP2/TZVP	DFT/TZVP	(kcal/mol)
DMN1	1.6	1.7	1.7	0.9	64.0
DMN2	1.0	1.3	1.1	0.6	64.1
DMN3	0.0	0.0	0.0	0.0	64.4
DMN4	22.0	23.2	20.0	21.9	63.1
DMN5	0.00	−1.12	−2.00	−0.45	64.6

a 1 kcal/mol = 4.187 kJ/mol.

rotating these groups, indicates that they lie within 2 kcal/mol of each other. The central nitrogen connected to carbon and nitrogen could feasibly be pyramidal and the nature of the calculations so far would not show this, as they maintain the high initial symmetry provided by the starting geometry of the molecule. Consequently, calculations were performed on conformer DMN5 with lower (C_s) symmetry. Hartree-Fock calculations using the 6–31G basis converged to the more symmetrical structure DMN3 with a planar nitramine group. Although behaving poorly in this respect, the geometry predictions of this small basis set, shown in Table 3.7 show remarkable agreement with experiment.

The 6–31G basis set is known to favor planar nitrogen conformers over pyramidal conformers, so it was decided to repeat the calculations using the 6–31G** basis set and to include correlation effects through Møller-Plesset second-order perturbation theory. This routine calculation is much more expensive than standard Hartree-Fock methods, but can be expected to give more accurate bond lengths and angles, given a large enough basis set. Interestingly, this method predicts that the pyramidal nitrogen structure is now the most stable by about 1.1 kcal/mol. It is to be expected, therefore, that DMN is a fairly flexible structure whose average geometry is planar, but whose minimum equilibrium structure shows a pyramidal central nitrogen.

The MP2 calculations were repeated using the larger TZVP basis set. The major change observed was an increase in the depth of the non-planar structure DMN5 to 2.0 kcal/mol and a slight decrease in the barrier to rotation of the N-NO_2 group. DFT calculations, in this the largest basis set, showed no quantitative change in the

Table 3.7. Comparison of calculated and experimental geometries of dimethylnitramine bond lengths in ångstroms and bond angles in degrees.

Molecular geometry	Experiment	SCF/6–31G	MP2/6–31G**	MP2/TZVP	DFT/TZVP
N–O	1.23	1.24	1.24	1.24	1.23
N–N	1.33	1.32	1.36	1.36	1.36
N–C	1.44	1.46	1.45	1.45	1.46
C–C	2.56	2.59	2.58	2.58	2.59
O–N–O	124.2	124.5	126.5	126.2	125.9
O–N–N	117.9	118.0	116.7	116.9	117.0
N–N–C	117.7	117.7	116.8	117.1	117.4
C–N–C	124.2	124.6	126.4	125.8	125.2

results, although the depth of the non-planar well was reduced to only 0.45 kcal/mol. Verification of the nature of the minimum energy conformations was made by calculating the force constant matrix for each structure at the Hartree-Fock 6–31G** minimum energy. As expected, DMN5 was the only calculation to show a true minimum energy structure with no negative eigenvalues. Calculation of thermal corrections to the energies showed that at 25 °C, thermal corrections would not alter the ordering of the various minima.

Table 3.7 shows the predicted and experimental structural parameters. The table compares the structural parameters for the MP2/TZVP DMN3 structure, rather than the minimum energy MP2 structure. It can be seen that there is generally good agreement between the calculated and experimental structures at all levels of theory. The major discrepancies occur between the N-N bond length, which the *ab initio* methods are predicting to be too long by about 0.03 Å. The origin of this is not clear, as the calculations show that deviations in non-planarity tend to increase this bond length rather than decrease it. Packing interactions between neighboring molecules may be able to account for the change in the bond length, but the size of the change is unusual.

3.3.2.3 **Molecular Modeling of RDX**

RDX is one of the simplest molecules possessing the nitramine group in a ring structure and it was for this reason, along with the large amount of experimental information available about the molecule, that modeling work on this system was undertaken.

As can be seen in Fig. 3.53, RDX is a six-membered ring, with three nitramine groups positioned at alternating positions around the ring. The ring itself can exist in a chair or a boat conformation and in addition the nitro groups of the nitramine units can be orientated either axially or equatorially with respect to the ring. This latter feature implies that the central nitrogen is pyramidal in nature, something which is not absolutely clear from the experimental evidence on DMN, but which is supported by the most accurate calculations. The choice of axial or equatorial

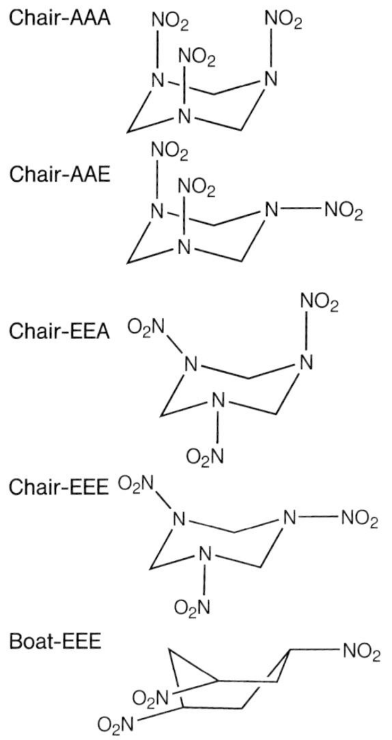

Chair-AAA

Chair-AAE

Chair-EEA

Chair-EEE

Boat-EEE

Figure 3.53. Conformers of RDX.

conformation of the ring nitrogen produces the following possible conformations for the ring: AAA, AEE, EAA and EEE, where A and E refer to axial and equatorial, respectively.

In the gas phase, electron diffraction experiments [3.139] show that RDX has a chair AAA, C_{3v} structure. In the solid state, RDX crystallizes in two polymorphs. α-RDX [3.140] is stable at room temperature and crystallizes in space group *Pbca*, an orthorhombic crystal system. The six-membered ring of RDX adopts a chair EAA structure. One of the nitramine groups lies in an equatorial position to the ring with an N-N-C-N torsion angle of 145.6°. The other two nitramine groups occupy axial positions on the ring with torsion angles close to 90°. β-RDX is extremely unstable at room temperatures, but infrared studies [3.141] indicate that it has a higher molecular symmetry than the α-phase.

Crystal packing effects obviously influence the conformation adopted by the molecule. Thus co-crystallization with tetramethylene sulfone [3.142] causes RDX to adopt an AEE configuration. Additional substitution on the ring also causes a conformation change. 2,4,6-Trimethyl-1,3,5-trinitrohexahydro-1,3,5-triazine [3.143], where one of the hydrogens of the CH_2 groups in the ring have been replaced by methyl groups, also adopts a configuration with one axial and two equatorial nitramine groups. Carbonylation of the ring results in a similar effect, with 1,3,5-trinitro-2-oxo-1,3,5-triazacylcohexane [3.143] also having one axial and two equatorial nitramine groups.

Previous calculations of the conformational energies of RDX by Rice and Chabalowski [3.144] using MP2 and DFT/B3LYP on the AAA, EEE and AAE chair conformers found the AAE conformer to be the most stable, with the EEE conformer lying highest in energy. Work by Harris and Lammertsma [3.145] using the 6–31G* and 6–311G* basis sets found five conformational minima (boat, chair-AAA, chair-AAE and chair-AEE), which are closely spaced in energy, the lowest in energy being the chair-AAE structure, in agreement with Rice and Chabalowski [3.144]. They were unable to locate the EEE chair conformer as a stable minimum. They also pointed out that the short N-N distance observed in the crystal may be due to crystal packing effects. Other theoretical work on RDX [3.146] examined the various mechanisms for degradation of RDX. A new pathway for degradation involving HONO elimination was proposed.

Ab Initio and Semi-empirical Calculations

Semi-empirical and *ab initio* Hartree-Fock calculations using both the 6–31G and 6–31G** basis sets were performed on an isolated RDX molecule. The semi-empirical calculations are very fast and success at modeling RDX strongly suggests that they would be suitable for larger nitramine-containing systems. The results are shown, together with the *ab initio* results, in Tables 3.8 and 3.9.

The semi-empirical calculations were unable to distinguish between the axial and equatorial nature of the nitramine groups and this was put down to its inability to describe the pyramidal nature of the central nitrogen. Only a single chair and a single boat conformer was found by the semi-empirical method. The nitramine

Table 3.8. Absolute energies for *ab initio* and semi-empirical calculations for RDX[a].

Conformer	MOPAC	Hartree-Fock 6–31G	Hartree-Fock 6–31G**	DFT 6–31G**
Boat	110.4	−892.004317	−892.516079	−897.4158556
Chair-AAA	104.9	−892.002137	−892.514955	−897.4165954
Chair-AAE	–	–	−892.516608	−897.4175467
Chair-EEA	110.5	–	−892.516425	−897.4153407
Chair-EEE	–	–	−892.510037	−897.408835

a MOPAC results are heats of formation in kcal/mol; Hartree-Fock results are in hartree.

Table 3.9. Relative energies (kcal/mol) of *ab initio* and semi-empirical calculations for RDX.

Conformer	MOPAC	Hartree-Fock 6–31G	Hartree-Fock 6–31G**	DFT 6–31G**	Thermal correction
Boat	0.1	0.0	0.3	1.1	107.1
Chair-AAA	0.0	1.4	1.3	0.6	106.3
Chair-AAE	–	–	0.0	0.0	106.4
Chair-EEA	5.6	–	0.1	1.4	106.5
Chair-EEE	–	–	4.1	5.5	106.3

groups were always best described by an axial conformation; no equatorial conformations were found. The boat conformation was found to be about 6 kcal/mol higher in energy than the chair.

The *ab initio* Hartree-Fock calculations, using the 6–31G basis set, were in many ways similar to the semi-empirical results. They were also unable to distinguish between the axial and equatorial conformers. The boat optimizations all converged to the same structure, which can probably be best described by EEA, whilst the chair conformers also all optimized to the same structure, namely AAA. These calculations predict the boat conformer to be slightly more stable than the chair conformer by about 1.4 kcal/mol.

The experience with DMN proved useful at this point since it is known that polarization functions are needed to represent the pyramidal nature of the nitrogen. Calculations on RDX using the 6–31G** basis set do indeed discriminate between equatorial and axial nitrogens in the chair conformation. Calculations at this level still predict that there is only one boat conformer. The overall lowest energy structure found is that observed experimentally in the crystal structure, namely the chair EEA conformer, but this is only 0.1 kcal/mol lower in energy than the next most stable structure, the EEA chair conformation. The agreement with the experimental crystal structure is probably fortuitous as there are other energetically low-lying conformers and no account has been taken of packing energies.

The DFT calculations with the 6–31G** basis set do not change the picture substantially, although the ordering of higher energy conformers is altered and the energy gaps between the conformers is increased slightly. Table 3.9 also gives the quasi-harmonic estimates of the corrections due to thermal vibrations. This indicates that the energy of the boat conformer is further destabilized, relative to the lowest energy conformer, by about 0.7 kcal/mol.

Molecular Mechanics Prediction of Crystal Packing

Taking the AAE structure calculated using the 6–31G** basis set, the electrostatic potential was used to derive PDC charges which were used in the molecular mechanics minimization of the full three-dimensionally periodic crystal. Figure 3.54 shows the electrostatic potential from the *ab initio* calculation on a contour of the electron density around the molecule. The color convention is that red represents negative electrostatic potential, green is zero and blue is positive electrostatic potential. The picture shows clearly the importance of the electrostatic potential around oxygens for controlling the energy of packing in a crystal.

The minimization of the lattice energy did not maintain the space group symmetry of the system but allowed lattice parameters and the atomic positions to adjust so as to minimize the total energy of the system. Alternatively, the molecules can be held rigid and the packing can be adjusted so as to minimize the total energy, whilst maintaining the space group symmetry of the system. Such an approach optimizes the energy with respect to the lattice parameters and the intermolecular interactions. The results for both methods are shown in the Table 3.10 and compared with experimental X-ray data. The packing approach with rigid molecules predicts a density that is too low and all lattice dimensions appear to have enlarged

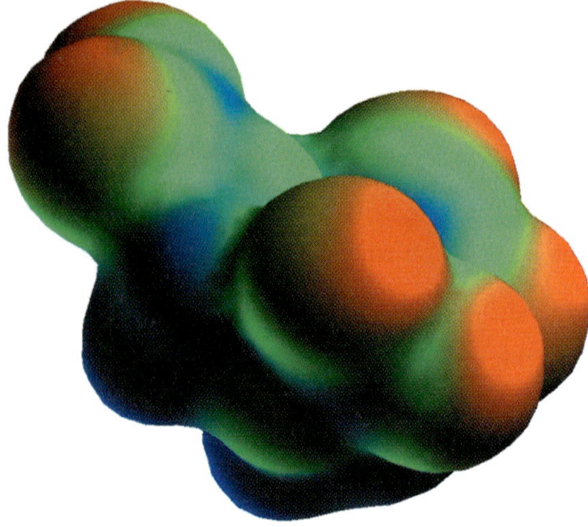

Figure 3.54. Electrostatic potential around the chair-AAE conformer of RDX.

somewhat. Full minimization leads to a lower energy and to an improved density. The agreement to within 3 % of the experimental lattice parameter is in line with the expected accuracy of these calculations. The results of this work are in broad agreement with some recent work [3.147] on extracting a rigid body, transferable potential for the nitramine group, which also includes molecular dynamics to account for the coefficient of expansion.

Various modifications to the force field were attempted in order to improve the predicted crystal parameters, whilst maintaining reasonable agreement between the *ab initio* and gas-phase molecular mechanics geometry for this conformation. This involved a minor modification of the torsional potentials associated with O-N-N-C bond term and the use of an exponential –6 non-bonding potential rather than the default Lennard Jones 12–6. The results of these modifications to the force field can be also seen in Table 3.10. The effects of the change to the non-bonded potential are dramatic. The density of the predicted crystal structure is higher than the experimental value with significant reductions in the *a* and *c* parameters of the unit cell.

Molecular dynamics calculations at constant NPT were performed on a $2 \times 2 \times 2$ super cell of the crystal. The larger simulation cell is needed for such calculations as

Table 3.10. Comparison of calculated and experimental crystal structures for RDX.

	Experiment	Calculated			
		Packing	Optimized	Optimized modified force field	MD 300 K modified force field
a (Å)	13.18	13.63	13.41	13.43	13.29
b (Å)	11.57	11.89	11.78	11.46	11.83
c (Å)	10.71	10.82	10.74	10.22	10.71
ρ (g/ml)	1.81	1.68	1.74	1.88	1.75

otherwise the small unit cell gives rise to periodic oscillations in the unit cell parameters. After equilibration for 20 ps, the average cell parameters were collected for 10 ps. The results of a simulation at 300 K show pleasing agreement with the experimental values.

3.3.2.4 **Molecular Modeling of HNIW (CL 20)**

CL 20 or HNIW is a cage molecule, which has attracted considerable interest recently because of its high energy density. The molecule appears to crystallize in many forms and little structural information is available in the open literature as to their molecular conformations.

Some examples of the possible molecular conformations of HNIW are shown in Fig. 3.55. It essentially consists of a single six-membered ring, with two nitramine groups at the 1- and 4-positions, two seven-membered rings (containing three nitramine groups) and two five-membered rings (containing two nitramine groups). The choice of an axial or equatorial conformation for the nitramine groups leads to a considerable number of feasible conformers, which we have studied using semi-

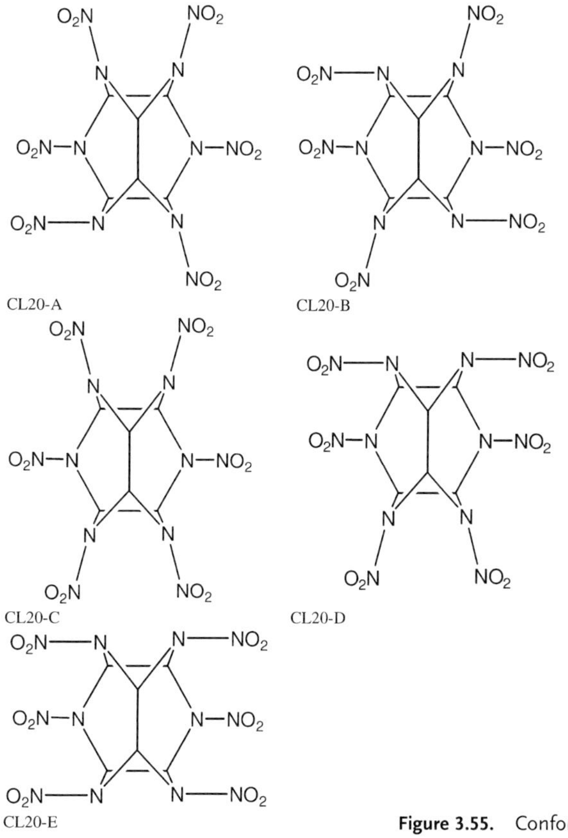

CL20-A

CL20-B

CL20-C

CL20-D

CL20-E

Figure 3.55. Conformers of HNIW.

empirical, *ab initio* quantum mechanics and molecular mechanics methods. The size of these molecules is such that it is very difficult to perform calculations of the accuracy required, given the experience described with DMN and RDX. No density functional or MP2 calculations are reported here.

Some molecular modeling work has already been published on HNIW by Pivina *et al.* using a semi-empirical approach [3.148]. In the work reported here we will use semi-empirical and *ab initio* calculations to explore the conformational flexibility of the gas-phase molecules. Other calculations include some rigid body molecular dynamics calculations of the properties of crystalline HNIW [3.149, 3.150].

Conformational Studies of HNIW

Semi-empirical calculations were performed using the MOPAC [3.106] computer program. *Ab initio* calculations were performed using the GAMESS–UK [3.109] code. Basis sets varying from STO3G, 6–31G and 6–31G** were used in the *ab initio* calculations and, where possible, symmetry was used to improve the speed of the calculations. The smallest basis sets were used to obtain suitable starting points for geometry optimizations using the larger basis and are not reported here. Tables 3.11 and 3.12 report the absolute and relative energies calculated.

Unlike the previous calculations on nitramine-containing molecules, the semi-empirical results did distinguish between the different conformers of HNIW. This can be attributed to the highly strained nature of the cage structure, which probably emphasizes the differences between axial and equatorial nitrogens. Only in the case of conformer E is the semi-empirical method unable to find a suitable minimum energy structure. The method instead changes the structure into a lower energy structure, B.

Table 3.11. Absolute semi-empirical and *ab initio* energies of molecular HNIW[a].

HNIW conformer	MOPAC	6–31G	6–31G**
A	280.4	–	–
B	278.0	–1780.490224	–1781.516381
C	283.4	–	–1781.506509
D	281.1	–1780.488718	–1781.513639
E	–	–1780.486755	–1781.502572

a MOPAC energies in kcal/mol; *ab initio* energies in hartree.

Table 3.12. Relative semi-empirical and *ab initio* energies (kcal/mol) of molecular HNIW.

CL20 conformer	MOPAC	6–31G	6–31G**
A	2.4	–	–
B	0	0	0
C	5.4	–	6.2
D	2.1	1.0	1.7
E	–	2.2	8.7

At the 6–31G Hartree-Fock level, the lowest energy structure is conformer B. Starting from conformer A, the structure optimizes to a conformer B. The next lowest energy structure is conformer D followed by E. At the 6–31G level, starting from conformer C the molecule minimizes to a structure similar to E. In agreement with earlier findings, the 6–31G basis set is not describing the pyramidal nature of the ring nitrogens well.

The 6–31G** basis set results are able to distinguish four conformers, B, C, D and E. All methods suggest that conformer B is the most stable. The next most stable appears to be conformer D. The *ab initio* calculations were unable to find a stable geometry for conformer A, as this changed to geometry B during the optimization process.

Crystalline HNIW

HNIW exists in at least five polymorphic states. Experimental X-ray structure determination of several of them has been reported in a series of papers given in [3.151–3.154]. In addition, some theoretical work has been performed on the crystal structure of HNIW using a transferable nitramine potential and rigid body molecular dynamics [3.149, 3.150]. This latter work refers to single-crystal work of Gilardi's group, the reference for which is also provided here [3.155]. Gilardi's group have also published [3.156] the synthesis of HNIW and related molecules, along with some single-crystal structure determinations. In addition, an assignment is made of some of the different molecular conformations to the different phases. The α-hydrate and ε-, β- and γ-polymorphs are stable at room temperature and pressure. The α-phase has an orthorhombic structure with *Pbca* symmetry and eight molecules of HNIW and between one and four molecules of water in the unit cell. The molecular conformation of the molecule in the α-hydrate phase is the same as that found in the γ-phase, namely conformer A. The ζ-polymorph is known to exist at high pressure, but its crystal structure is not available. The crystal structure information, including the space groups and the lattice dimensions, along with the molecular conformer associated with a particular phase, is given in Table 3.13.

HNIW Crystal Structure Predictions

Once the crystal structure is known, it provides a very useful benchmark for the force field validation. Concentrating on the ε form of HNIW, the force field for ring

Table 3.13. Crystal structure information for HNIW, from [3.157, 3.158]

Cell parameter	β	γ	ε
	$Pb2_1a$ orthorhombic	$P2_1/n$ monoclinic	$P2_1/n$ monoclinic
	$Z = 4$	$Z = 4$	$Z = 4$
A (Å)	9.6764	13.2310	8.8278
B (Å)	13.0063	8.1700	12.5166
C (Å)	11.6493	14.8760	13.3499
γ (°)	90.000	109.170	106.752
Molecular conformation	B	A	D

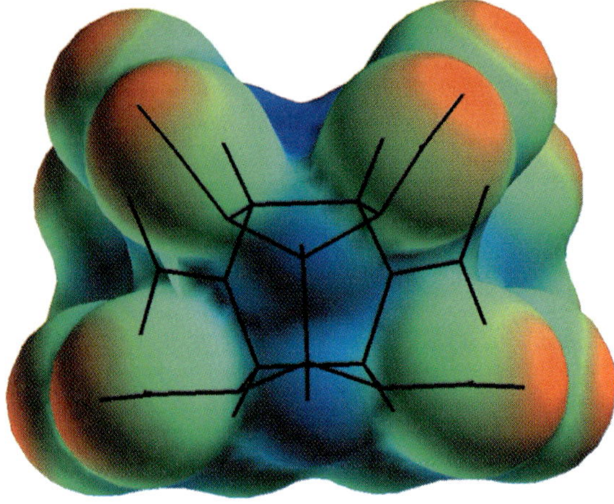

Figure 3.56. ε-HNIW molecular electrostatic potential.

molecules containing nitramine groups developed previously to describe RDX was used. The electrostatic terms in the force field were provided from atomic center potential-derived charges based on the $6-31G^{**}$ basis calculations described above, using the optimized geometry for the molecule.

Figure 3.56 shows the calculated electrostatic potential. The molecule is orientated with the five-membered rings at the top and bottom. The two nitramine groups attached to the bottom ring are equatorial and the two nitramine groups attached to the top ring are axial. The molecule has an external shell of negative electrostatic potential from the oxygens of nitrogen and internal core of positive potential.

Starting from the experimental unit cell, the crystal structure was energy minimized with respect to the cell parameters and all the atomic parameters. The minimization was carried out using the Ewald method to ensure that the long-range Coulomb sums had converged. After minimization, a molecular dynamics calculation was carried out at 300 K to establish the influence of temperature on the structure. A large unit cell consisting of $2 \times 2 \times 2$ crystallographic unit cells was constructed from the minimized structure and a constant number, temperature and pressure (NPT) calculation was carried out. The calculation involved 20 ps of equilibration followed by 10 ps of simulation during which averages of the unit cell parameters were calculated. The results of these calculations are shown in Table 3.14.

Comparing the minimized and experimental results, it can be seen that minimization produces a unit cell which is slightly too small, resulting in a too high density. In particular, the parameter *b* of the cell appears to be too small. The molecular dynamics calculations show much better agreement with experiment. There is a slight increase in the *a* and *c* lattice parameters, but *b* appears to be the most sensitive to the introduction of vibration through temperature. The final density of the unit cell is now slightly below the experimental value. The results indicate that the force field is performing very well and should be adequate for our requirements.

Table 3.14. Comparison of molecular mechanics and dynamics calculations on HNIW with experiment

Lattice parameter	Experimental	Minimization	MD 300 K
a (Å)	8.83	8.79	8.81
b (Å)	12.52	12.38	12.54
c (Å)	13.35	13.53	13.55
γ (°)	106.75	106.77	106.64
ρ (g/ml)	2.06	2.11	2.03

HNIW Morphology Predictions

The crystal morphology of ε-HNIW was predicted by the BFDH [3.132] method as described above. The method, although not predicting with absolute accuracy the shape of the crystal, does seem to be able to predict those surfaces which are likely to be important in the crystal. The predicted morphology is shown in Fig. 3.57. Several families of surfaces are predicted to be of possible importance in controlling the shape of the growing crystal. In particular, the {011}, {002}, {101} and {110} surfaces were chosen for more detailed investigation by calculating its surface energy.

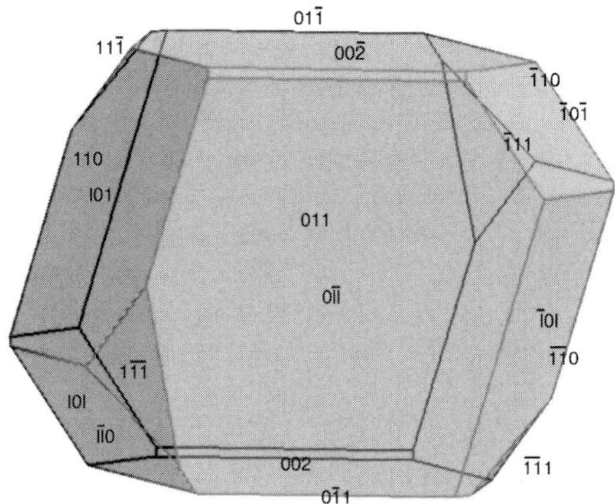

Figure 3.57. Predicted ε-HNIW BFDH morphology.

Crystal Surface Energies of HNIW

On the basis of the morphology prediction described above, there are several surfaces that may be important to the final crystal shape. An attempt wa therefore s made to calculate the surface energies of these. The surface energy is calculated by taking the bulk crystal, with known lattice energy, and cleaving it to form two crystal surfaces. This is achieved by maintaining the 3-D periodicity of the lattice but leaving a sufficiently large gap along the direction perpendicular to the surface so that the slab surfaces do not interact with each other.

The results of these surface energy calculations are shown in Table 3.15. Both

Table 3.15. Calculated surface energies of HNIW.

Surface	Area (Å^2)	Surface energy (kcal/mol/Å^2)	
		Unrelaxed	Relaxed
011	155.4	0.35	0.31
002	107.6	0.25	0.24
$10\bar{1}$	169.9	0.34	0.30
110	200.0	0.40	0.34

relaxed and unrelaxed surface energies are given. The unrelaxed energy refers to the process of cleaving and leaving the molecules at their bulk positions. The relaxed energies allow the molecules to adjust to the presence of a surface. The table shows that the {002} surface appears to be the most stable both before and after relaxation and would be expected to dominate the crystal morphology, followed by the {101} surface. The {110} surface shows the largest relaxation but is less likely to be important than the others.

3.3.2.5 Molecular Modeling of Processing Aids

Various processing aids are added to the binder and energetic filler systems to reduce the viscosity and allow the filler content to be increased. To aid the molecular understanding of how the processing aids operate, several examples have been studied using molecular modeling to understand the nature of the interaction with an RDX surface. Here we report on the bonding agent lecithin, studied using the Cerius^2 molecular modeling package. The universal force field, UFF [3.157], was used and the electrostatic potential around the bonding agent was defined using the charge equilibration method [3.158]. The potential-derived charge technique was not be applied to lecithin, as the intention is to extend this approach to other bonding agents,and the calculation of the wavefunction for extremely large molecules would be prohibitively expensive.

Lecithin is a phosphoglyceride; it has a non-polar portion, which is insoluble in water, and a polar portion, which is soluble. The structure of a representative lecithin is shown in Fig. 3.58.

The nature of the structure is such that lecithin acts as a surfactant, being able to stabilize oil/water interfaces. In order to gain a better understanding of the likely chemical interactions of this material with the surfaces of energetic materials, a simulation of lecithin on the {111} surface of RDX was performed. The surface was prepared by taking the unit cell of the RDX and slicing it along the {111} direction to create a surface cell. The slicing operation was performed in such a way that no dangling bonds were left at the surface and only whole molecules were allowed. This surface cell was then used to create a slab of the {111} surface. The slab was chosen to be wide enough in the surface plane so that lecithin would not feel the edges of the surface at a potential cut-off of 8 Å. The depth of the solid was chosen to be six molecules thick. This is not sufficiently thick to consider quantitative work but

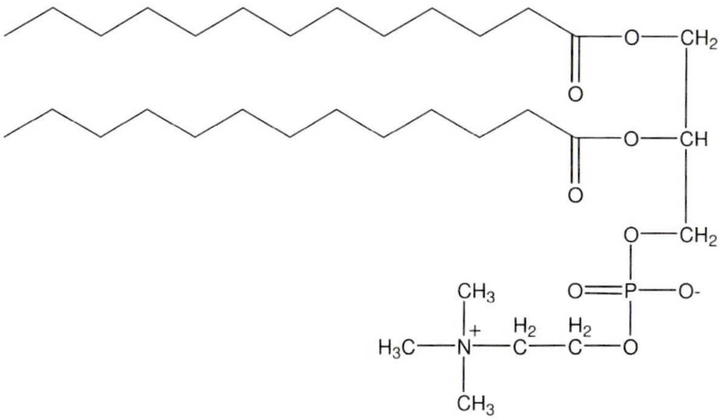

Figure 3.58. Structure of lecithin.

should be thick enough for a qualitative understanding of the bonding to the surface.

The position of the lecithin molecule on the surface was then obtained by successively running molecular dynamics and molecular mechanics calculations in order to allow the molecule to explore its own conformational space and the various sites on the surface. An example of one of the minimum energy structures of the processing aid is shown in Fig. 3.59. The figure also shows the electrostatic potential at the van der Waals surface of the molecules outlined as a collection of yellow dots. The highly polar 'core' in the center of the molecule appears to adsorb strongly on the polar surface of the RDX. The alkyl chains are not so strongly bound to the surface and would readily be displaced by any solvent that there may be in the system. The role of the alkyl chains appears to be to act as a 'lubricant' between filler particles. The 'lubrication' arises as a consequence of an entropic barrier, preventing the close approach of the particles. Any overlap of the alkyl chains between neighboring particles, which come close to each other, results in a considerable loss of conformational freedom of the chains and hence a barrier to approach.

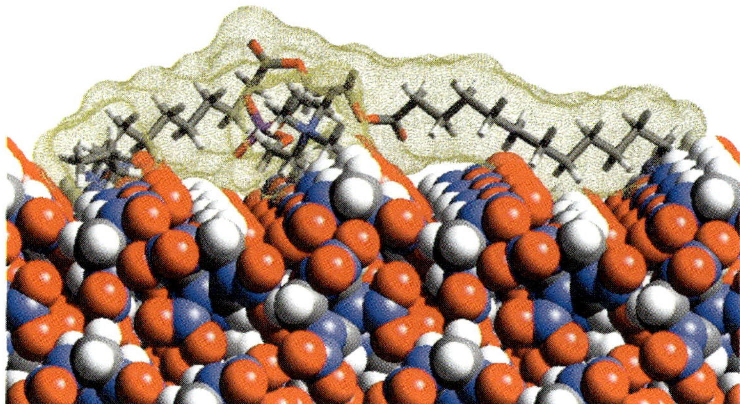

Figure 3.59. Lecithin adsorbed on the {111} surface of RDX.

The above discussion is very qualitative in nature. Such arguments can be made more substantial by consideration of different processing aids and a comparison of their strengths of adsorption on the surfaces of the energetic crystal. Such a molecular approach will not, however, lead to a complete understanding of the issues in a problem as complex as that of processing aids. This will only be achieved by a combination of molecular and mesoscale methods.

3.3.2.6 The Crystal Surface

A unit cell is the smallest repetitive unit in the crystal. It contains a number of molecules or ions and possesses a certain symmetry. In Fig. 3.60, 2 × 2 unit cells of RDX are shown. The crystal is bound by a number of different crystal faces which can be identified by Miller indices. Miller indices give the orientation of a certain crystal surface in relation to the crystal unit cell.

For one set of Miller indices such as (200), there are still several options for how the molecules are oriented in the crystal growth layer that forms the crystal surface. Each possible orientation is called a surface configuration. In the case of RDX, the crystal surface layer with the Miller indices (200) has two possible surface configurations because the crystal can be cut in the (200) orientation at two different heights in the unit cell (Fig. 3.60). The crystal surface layer has either the 200_A or the 200_B surface configuration.

The growth of a crystal surface takes place due to the addition of crystal building units at steps of crystal growth layers on the crystal surface. The thickness of a

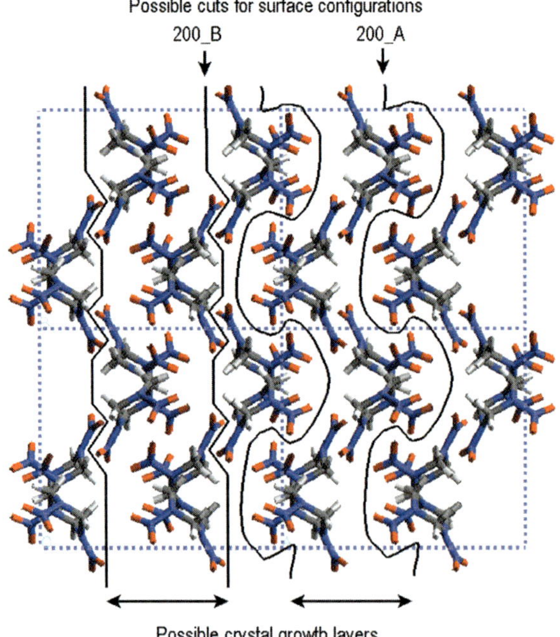

Figure 3.60. 2 × 2 unit cells of RDX, the possible cuts for the (200) surface configurations (200_A and 200_B) and the (200) crystal growth layers.

crystal growth layer is indicated in Fig. 3.60. In case of RDX the molecular orientation on both sides of a 200_A or 200_B crystal growth layer is the same due to the space group symmetry conditions in the unit cell.

The reason for emphasizing the concept of a crystal surface configuration is that the interactions of building units and foreign compounds depend on it. The orientation of the molecules in the two (200) surface configurations are different and thus a foreign compound (but also a building unit) undergoes different interactions at the two surface configurations. These interactions determine the growth rates of the crystal surface configurations, which in turn determine product quality parameters such as crystal shape and process performance parameters such as the filterability of the product. Hence a very important step in the molecular modeling procedure is the determination of the surface configurations and from thse the morphologically important crystal surfaces.

3.3.2.7 **Crystal Morphology**

The crystal morphology or crystal shape can be geometrically constructed by drawing crystal planes with an orientation $\{hkl\}$ having a distance from the origin that is proportional to the growth rate. The central volume enclosed by the set of planes is the growth form. A face with a large growth rate is positioned far from the origin and hence its surface area will be small. In contrast, a face with a small growth rate will lie close to the origin and have a large surface area (Fig. 3.61). This means that the morphology is determined by the slowest growing $\{hkl\}$ faces and that a morphology prediction can be made if all the relative growth rates of all the different $\{hkl\}$ faces are known.

Figure 3.61 also shows the effect of a growth-retarding foreign compound. Before a crystal face can grow, the adsorbed foreign compound has to be removed from the surface. This costs energy and therefore the growth rate decreases. When the interaction energies of a foreign compound are face specific, the growth retardation will vary from face to face. When the interaction of a foreign compound is very large for only one face, the relative growth rates change, as shown in Fig. 3.61 for face number 3.

In the next section a procedure to investigate the morphology and the effect of foreign compounds on this morphology is presented. From the previous sections it is apparent that such a method should include interaction energy calculations of the foreign compound with the surface in order to incorporate such a predictive behavior.

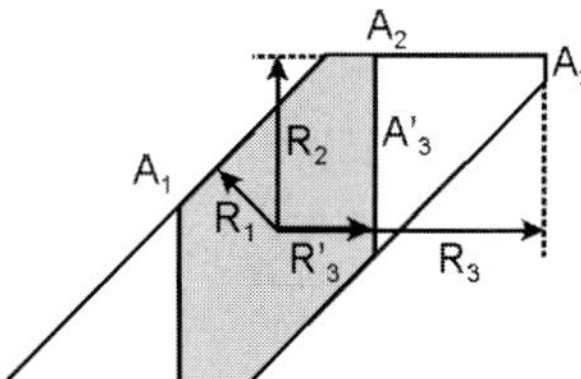

Figure 3.61. The growth rates of the different crystal faces determine the crystal morphology.

3.3.2.8 A Procedure for Molecular Modeling Simulations

A procedure to determine the effect of a foreign compound such as a solvent can be split into four parts. First comes the determination of the vacuum morphology that gives information about the potential surfaces. In the second part of the procedure it is determined which surface configurations will be used in the simulations. The third part of the procedure concerns the choice of the simulation technique that should be used to calculate the interaction of the foreign compound with the surface configurations. The fourth part uses a physical model to translate the molecular level results from the simulations to relative growth rates of the surface configurations. In the following sections each part of the procedure is discussed.

Vacuum Morphology

Nowadays a number of general prediction methods exist for determining vacuum morphologies. Vacuum here means that the influence of the solvent is not taken into account. The law of Donnay and Harker states that the importance of a crystal face decreases with the interplanar distance d_{hkl} [3.132]. A morphology can be predicted if the interplanar distance is assumed to be inversely proportional to the relative growth rate of the corresponding crystal face. A prediction method taking into account the anisotropic energies in the crystal unit cell is the attachment energy method [3.159]. The attachment energy is defined as the energy release per growth unit upon adding a surface layer with thickness d_{hkl} on to the corresponding crystal surface. The assumption that the attachment energy is linearly proportional to the relative growth rate of the corresponding crystal surface gives the morphology prediction.

Surface Configurations

Two points should be considered in order to determine which surface configurations to use in the simulations. First, a surface configuration should be able to exist. A vacuum morphology prediction discussed in the previous section is not conclusive on this point. For a more conclusive statement on whether a surface configuration will indeed develop, a periodic bond chain (PBC) analysis [3.160–3.162] can be performed.

A PBC analysis checks whether a surface layer contains a connected net. A PBC is an uninterrupted chain of strong bonds between growth units in the crystal lattice the periodicity of which is based on the unit cell parameters and symmetry and which is stoichiometric with regard to the unit cell content. Two sets of intersecting PBCs make a connected net. If a surface layer does not contain a connected net, the corresponding surface configuration cannot grow with a layer growth mechanism and becomes rough. Rough surfaces always grow fast and do not appear on the crystal shape. This means that surface configurations that correspond to surface layers consisting of unconnected chains of growth units do not have to be simulated.

The PBC analysis results in morphologically important crystal surface configurations for crystals grown from systems without boundary layer influences such as solvent interactions with the surface structures. Figure 3.62 shows the top view of

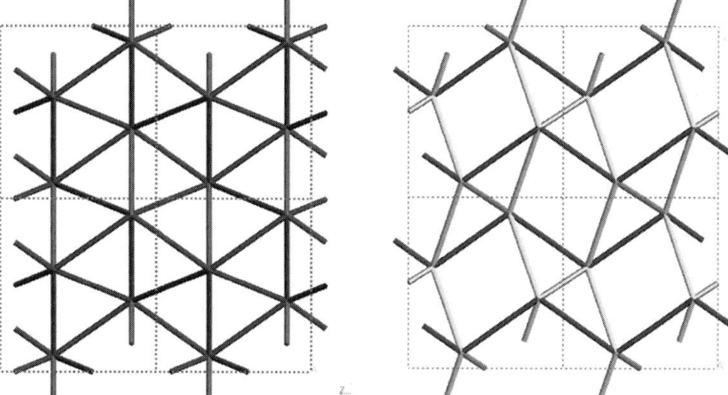

Figure 3.62. Top view of the two possible surface configurations (200_A and 200_B) of the (200) RDX crystal surface. Each intersection of lines is the center of mass of an RDX molecule in the slice.

the connected nets in the two possible RDX crystal surface configurations of the (200) surface.

The second point regarding which surface configurations to use in the simulations comes from considering all surface configurations and their energies. There is, for instance, no need to simulate the interactions with a surface configuration which will not become morphologically important even if the interaction with the foreign compound is large. Both points decrease the number of simulations and may result in a considerable time saving when a large number of molecules take part in each simulation.

Choice of the Simulation Technique

Molecular modeling [3.163] is a collection of computer simulation tools based on quantum chemical or empirical force field models that provide insight into the behavior of molecular systems. The choice as to which of these simulation techniques should be used depends on the system under investigation. Quantum mechanical simulation techniques in molecular modeling are involved with the explicit representation of the electrons. From these techniques, it is possible to derive properties that depend on the electronic distributions. They can be used to investigate chemical reactions in which bonds are broken and formed. Different from the quantum chemical methods are the empirical force field models, where electronic and nuclear motion are assumed to be separate (Born-Oppenheimer approximation). The electrons can rapidly adjust to any change in nuclear position owing to their much smaller mass compared with the atom nuclei. This means that the ground-state energy can be considered to be a function of only the nuclear coordinates and the electronic behavior can be captured in an empirical force field. Although the empirical force field models cannot provide properties that depend on electronic distributions in molecules, they can handle systems with significantly more molecules than quantum mechanical models. Some generally used empirical force field models are based upon molecular mechanics (MM), molecular dynamics (MD) or Monte Carlo (MC) simulation methods.

MM is based upon a simple model of the interactions within a system with contributions from processes such as the stretching of bonds, the opening and closing of angles, rotations about single bonds and non-bonded electrostatic and van der Waals interactions.

In MD, successive configurations of the system are generated by integrating Newton's laws of motion. At the start of each time step the forces on the atoms in the system are computed, after which the atoms are allowed to move in that time step taking into account these computed forces. The result from an MD simulation is a trajectory that specifies how the positions and velocities of the atoms in the system vary in time.

An MC simulation generates configurations of a system by making random changes to, e.g., the positions of the molecules. Each step is accepted or rejected using an algorithm based on the energy change. The configuration is accepted if there is an energy gain and there is a chance of rejection if there is an energy loss. After a number of steps the configurational energy converges and the system is assumed to be in equilibrium.

Commercial software packages nowadays have a number of tools to calculate interaction energies of foreign compound molecules on these surface structures [3.164–3.166]. Since interaction processes of a foreign compound and the crystal surface generally do not involve breaking and creating of molecular bonds, molecular modeling tools such as MM, MD and MC are suitable.

The Physical Model

From the simulation, molecular level interaction energies between surface configurations and foreign compound are obtained. These must be translated to, for instance, growth rates of the corresponding surface configurations. This can be done by assuming a physical model which interrelates the growth rate of a certain surface configuration with the calculated interaction energy between a compound and that particular surface configuration.

The models used for this relation are generally based on the assumption that the interaction between foreign compound and surface configuration reduces the growth rate. First the foreign compound has to be removed from the surface before the crystal face can grow. This costs energy and hence the growth rate decreases. An energy correction term E_s for the vacuum attachment energy E_a can be introduced which is a function of the interaction of the foreign compound and the flat crystal surface configuration [3.28]. This value for E_s is thus different for each foreign compound and each surface configuration. For a solvent E_s represents the surface configuration specific solvent effect.

$$R \propto E'_a = E_a - E_s \tag{3.30}$$

where R = growth rate of a certain crystal surface configuration; E'_a = attachment energy of a certain surface configuration corrected for the effect of a foreign compound; E_a = vacuum attachment energy of a certain surface configuration; E_s = effect of the foreign compound. Since E_a is an attachment energy, the effect of the foreign compound E_s must also be expressed as some kind of attachment.

A lot of confusion is generated owing to the definition of the sign of the energies. In this section, energies are defined as broken bond energies: the higher the interaction, the higher is the energy.

Molecular Modeling Procedure

The last few sections are used to define a molecular modeling procedure which can be used to investigate the effect of a foreign compound on the crystal morphology. The next section describes a case study of the solvent effect on the RDX crystal morphology. After a short introduction and a discussion of the experimental RDX crystal morphology from different solvents, this case study follows the guidelines of the molecular modeling procedure.

3.3.2.9 Case Study: RDX Crystal Morphology

The customer expects a certain crystal quality. In the case of RDX this is a demand for a certain crystal shape. Larger packing densities and thus higher performance in smaller volumes can be obtained if the crystal has a spherical shape rather than a needle- or plate-like shape. This means that all the crystal surfaces should have about an equal morphological importance.

The solvent used for the crystallization of an organic compound can have a large effect on the resulting morphology [3.1]. A consistent prediction method for the effect of the solvent on the crystal morphology would therefore be very helpful in order to make a solvent selection or find a suitable solvent. For this goal one should identify the determining parameters from the interface region between the crystal surface and solution in order to capture the solvent effect. These parameters can then be used in a solvent-influenced morphology prediction.

The commonly used crystal morphology prediction methods are not suitable for this purpose. These are based on crystal bulk properties such as the interplanar distance or the attachment energy. To predict the effect of a solvent on the crystal morphology, other parameters must be introduced. The actual system of a growing crystal contains three regions: crystal bulk, solution bulk and the boundary layer between crystal and solution bulk containing the interface. Since the morphology is determined by interface characteristics, information on the morphology cannot be obtained solely by calculating bulk properties. As a result, parameters from the boundary layer have to be included in a prediction method for the solvent effect on the morphology. To investigate such a parameter, a number of simulation techniques of solvent against crystal surface configurations were performed.

Experiments: the Solvent Effect

The organic compound RDX (cyclotrimethylenetrinitramine) was taken as the model compound because the solvent has a large effect on the RDX crystal morphology [3.20]. The experimental RDX crystal morphologies grown from the

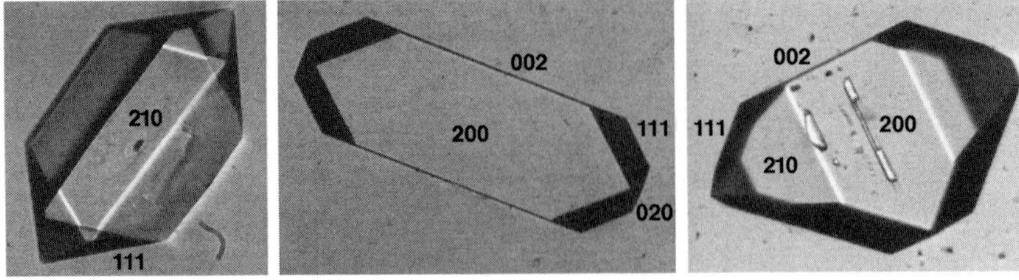

Figure 3.63. The RDX crystal morphology for RDX crystals grown from the solvents γ-butyrolactone (BL), cyclohexanone (CH) and acetone (AC).

solvents acetone, cyclohexanone and γ-butyrolactone with the aid of small-scale *in situ* crystallization experiments are shown in Fig. 3.63.

The RDX crystals grown from acetone are rather isometric. The morphology is dominated by the (200), (002), (111) and (210) faces, which are of about equal importance whereas the (020) face is smaller.

An RDX crystal grown from cyclohexanone is dominated by the large (200) crystal surface, whereas the (020), (002) and (111) faces are also morphologically important. Sometimes the (210) faces are visible.

The RDX morphology grown from the solvent γ-butyrolactone has morphologically important (210) and (111) crystal surfaces, whereas the (002) crystal surface is less important and, as in Fig. 3.63, is sometimes not visible at all. The faces (200) and (020) do not appear on crystals grown from this solvent. The (210) crystal surface is present both in the vacuum and the experimental morphology, whereas the (200) surface is absent in the experimental RDX crystal morphology grown from γ-butyrolactone. The (200) crystal surface, however, develops when RDX is crystallized from other solvents such as acetone and cyclohexanone [3.29].

Simulations

RDX crystallizes in the orthorhombic space group *Pbca* with $a = 13.182$ Å, $b = 11.574$ Å, $c = 10.709$ Å and contains eight molecules per unit cell [3.141].

Vacuum Morphology

The calculation of the attachment energies of the crystal surfaces is a relatively straightforward method to predict vacuum morphology. In Fig. 3.64, this vacuum morphology for RDX is shown. The calculated lattice energy is 25.8 kcal/mol while the RDX sublimation energy obtained from literature is 30.1 kcal/mol [3.167]. The

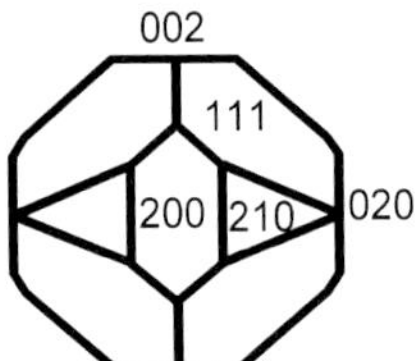

Figure 3.64. The vacuum morphology of RDX based on attachment energy calculations.

morphology shows the (111), (200), (020), (210) and (002) forms and these are therefore potential candidates for the morphologically important surfaces present on an RDX crystal in solution.

Comparison of the vacuum morphology and the experimental morphology for RDX crystallized from γ-butyrolactone indicates a strong interaction of γ-butyrolactone with the (210) crystal surface and a smaller interaction with the (200) crystal surface.

Surface Configurations

A PBC analysis for RDX results in a large number of possible surface configurations. As followed from the vacuum morphology calculations, the surface configurations of the (111), (200), (020), (210) and (002) forms must be considered to be morphologically important. These all contain two connected nets, which means that for all forms two surface configurations are possible.

To obtain a proper solvent-influenced morphology prediction for a certain solvent, the interaction of the solvent with all the surface configurations should be calculated. The more accurate simulations involve large amounts of solvent molecules, making the simulations time consuming. It was chosen to use two simulation methods. First, a simple, fast and crude method will be discussed where the interaction of one solvent molecule with all the surface configurations is calculated. Then the results from simulations of larger systems are discussed.

A Simple Tool to Calculate a Solvent Effect

The adsorption energy of a molecule on an (*hkl*) face can be calculated by a Monte Carlo simulation. The energy of the molecule in the equilibrium position can then be taken as the adsorption energy of that adsorbent. This adsorption energy is a measure of the solvent effect on that surface configuration. The higher the solvent adsorption energy, the higher is the solvent effect and the more important this surface configuration will become in that solvent.

Table 3.16 shows the adsorption energies for a solvent molecule of acetone (AC), cyclohexanone (CH) and γ-butyrolactone (BL) on all surface configurations of the (200), (020), (002), (210) and (111) faces. The morphological importance derived from the adsorption energies is also shown. The surface configuration of one (*hkl*) surface, A or B, which results in the largest adsorption energy is taken as the surface configuration which determines the growth rate and thus the morphology of RDX grown from that solvent. For instance, the adsorption energy on the surface configuration 111_A is the highest for the (111) surface and will thus determine the RDX growth and morphology from the solvent BL.

When the simulation results are compared with the experimental morphologies, the crude method does not give the exact answers, as could be expected. The morphologically important crystal faces seem to be correctly predicted for the solvent CH. The values for AC and BL give a less correct prediction of the morphological importance. However, this method is suited for a prediction in case of very strong interactions of a certain surface configuration and a given solvent. If very large differences occur between the adsorption energies of a certain solvent for

Table 3.16. Calculated adsorption energies of a solvent molecule on an RDX crystal surface configuration: the surface 020–B indicates the second surface configuration of the (020) face; the morphological importance (MI) as deduced from the adsorption energies is also given.

Surface (*hkl*)	AC (kcal/mol)	MI	CH (kcal/mol)	MI	BL (kcal/mol)]	MI
200_A	11.92	5	15.23	2	16.58	2
200_B	11.00		13.15		17.22	
020_A	12.98	1	14.76	4	17.22	3
020_B	10.28		12.99		14.92	
002_A	11.05	3	13.15	1	13.10	5
002_B	12.38		15.67		15.55	
210_A	10.73	4	12.84	5	16.71	4
210_B	12.33		14.70		15.74	
111_A	12.70	2	15.09	3	17.76	1
111_B	9.57		13.51		13.93	

the various different surface configurations, this indicates that during crystallization unwanted crystal shapes such as needles or plates will be obtained.

The Surface-induced Potential Energy Change

In this study, MD simulations of systems containing a large number of molecules were also performed. MD was chosen since rather large modeling systems are needed. These simulations provide information about the position and energy of the solvent molecules in the interfacial layer.

Calculating the solvent effect for all the possible crystal surfaces was not feasible because of the long simulation time involved. It was chosen to calculate the solvent effect of γ-butyrolactone on the (200) and (210) surface configurations.

For the (200) and (210) surface configurations [3.30], the attachment energies of the different surface configurations are given in Table 3.17. The (200)_A, (200)_B and (210)_A surface configurations have similar attachment energies (E_a) while the attachment energy of the (210)_B surface configuration is higher. According to Eq. (3.30), the solvent effect for the 210_B surface configuration should be unrealis-

Table 3.17. The vacuum attachment energy E_a, the surface slice area A_{hkl}, the number of RDX molecules in a slice N, the calculated surface induced potential energy change $E_I{}^s$, the attachment energy of the solvent layer E_s and the attachment energy E_a' corrected for this solvent attachment energy and the relative growth rate of the faces

{*hkl*}	E_a (kcal/mol)	A_{hkl} (Å^2)	N	$E_I{}^s$ (mcal/m^2)	E_s (kcal/mol)	E_a' (kcal/mol)	R_r (%)
200_A	10.76	123.9	4	9.30	1.73	9.03	125
200_B	11.04			12.75	2.38	8.67	120
210_A	11.12	285.3	8	18.22	3.91	7.20	100
210_B	13.32						

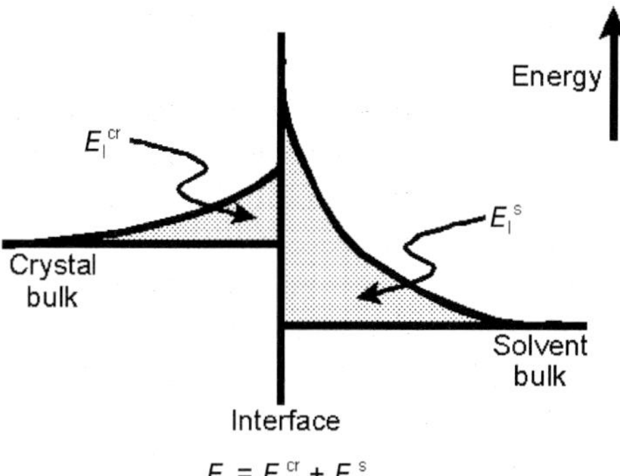

Figure 3.65. The energy profiles at the crystal interface. On the left relaxation of the crystal surface due to the solvent layer causes a potential energy change in the crystal surface E_I^{cr}. On the right a potential energy change in the solvent is induced by the crystal surface E_I^{s}.

tically high in order to transform this surface configuration into the dominant one. It was decided to simulate the solvent interaction with the (200)_A, (200)_B and the (210)_A surface configurations. Since 210_B cannot be the dominant surface configuration, it was not used in the simulations.

Consider an infinite solvent bulk and an infinite crystal bulk. As a first approximation, a bulk solvent rather than a bulk solution is considered here. When the solvent and the crystal bulk are cut in half and the resulting surfaces are stacked together, a relaxation in both the crystal surface and the solvent interface region occurs owing to their interaction. This relaxation is a surface-induced potential energy change E_I. Owing to the presence of the interface, the total potential energy is increased by an amount E_I. This energy change E_I can be divided into a solvent region part E_I^{s} and a crystal surface part E_I^{cr}. In Fig. 3.65 on the left, relaxation of the crystal surface due to the solvent layer causes a potential energy change in the crystal surface E_I^{cr}. On the right, a potential energy change in the solvent is induced by the crystal surface E_I^{s}.

In this study, it is assumed that the surface-induced potential energy change in the crystal surface E_I^{cr} is much smaller than that in the solvent interface region E_I^{s}. A first justification for this is that without the effect of a solvent, the relaxation of crystal surfaces of rigid and small organic molecules such as RDX against vacuum will be small. A second is that the interaction of the solvent will, to a considerable extent, replace the interaction of the crystal surface counterpart. Thus the relaxation of the crystal surface due to solvent interactions is expected to be even smaller than the (already minor) relaxation of a crystal surface in vacuum.

In contrast, the solvent molecules in the interface region are subjected to a considerable degree of interaction with the crystal surface. The crystal surface induces a potential energy excess in the solvent interface layer compared with the potential energy in the solvent bulk. This surface-induced potential energy change E_I^{s} is related to the interaction energy of solvent and surface: a high interaction of solvent and surface results in a large positive value of E_I^{s}.

$$E_a' = E_a - E_s = E_a - E_I^{s} A_{hkl} \frac{N_{Av}}{N} \tag{3.31}$$

where E_a' = attachment energy corrected for the solvent effect; E_S = solvent effect; E_I^S = potential energy change per unit area in the solvent layer; A_{hkl} = surface area of an *hkl* slice cut from the unit cell; N_{Av} = Avogadro's number; N = number of solvent molecules in the 3D simulation box.

From the simulations, the surface-induced potential energy change per surface area E_I^S in the solvent layer can be determined. A summation of the potential energy change over all the solvent molecules N_S in the simulation box divided by the total crystal surface area A in the simulation box results in E_I^S.

In order to derive the correction value E_S for the vacuum attachment energy E_a, the surface-induced potential energy change E_I^S has to be converted to the proper dimensions with the aid of the surface area of an (*hkl*) slice A_{hkl}, the number of RDX molecules in an (*hkl*) slice N and Avogadro's number N_{Av}. The solvent effect E_S is defined as the energy released (per mole of RDX) due to the relaxation in the solvent layer upon attaching the solvent bulk to a crystal surface. The units of $(E_I^S)_a$ are kcal per mole RDX.

The potential energy change E_I^S is 9.30 and 12.75 mcal/m^2 for (200)_A and (200)_B, respectively, and 18.22 mcal/m^2 for (210). This indicates a higher interaction of the solvent γ-butyrolactone with the (210) crystal surface and thus a morphologically more important (210) crystal surface. With aid of Eq. (3.31), the energy E_I^S can be used to correct the attachment energy values for this solvent interaction. The corrected attachment energies E_a^S are shown in Table 3.17.

Since the corrected attachment energy E_a' is smaller for (210) than for both (200) faces, its growth rate becomes smaller and thus its morphological importance is larger. A lower growth rate of the (210) surface increases the surface area of (210) but also decreases the surface area of the (200) surface. It can be calculated with aid of the unit cell parameters that the (200) surface disappears from the morphology when its growth rate exceeds 115% of the (210) growth rate. Both calculated relative growth rates of (200) are larger than this value of 115% and so the (200) faces are predicted to be absent from the crystal morphology. This agrees well with the experimental RDX crystal morphology, where (200) is not present and (210) has a large crystal surface area.

In order to compare the effect of different solvents on the morphology, the potential energy change per unit area E_I^S seems to be an adequate yardstick.

Case Study Conclusions

A procedure is proposed which is the basis for the molecular modeling simulations. First sufficient knowledge should be obtained about the morphology and the morphologically important surface configurations of the crystal compound. Then a suitable simulation technique should be chosen, after which the results from the simulations should be translated from a molecular level to the morphology level with aid of a physical model.

For the case study, two different simulation techniques were used: a crude and quick one and a more accurate and time-consuming one. The choice of which one to use obviously depends on the amount of time and money available to solve the problem. From the case study it is clear that molecular modeling can be used to determine the solvent effect and thus that crystallization phenomena can be investigated with aid of molecular modeling.

3.3.2.10 Simulation of Other Phenomena

Apart from molecular modeling as a tool to predict solvent-influenced morphologies, other applications can also be identified. The effectiveness of a *bounding agent* on a coated explosive compound can, for instance, be calculated. The function of the bounding agent is to connect the coating with the crystal. This means that the interaction of the bounding agent molecule with both the coating and all the crystal surfaces must be large. With the aid of molecular modeling the interaction energy of bounding agent and crystal surfaces can be calculated.

The effect of an *impurity* or *additive* present in small quantities (ppm level) cannot be explained by the effect of the compound on solution bulk parameters such as solubility. The impurity or additive molecules have to act on the growth process of the crystal surfaces. The adsorbed compound can impede the growth rate of the crystal surface.

If an impurity or additive and the crystal surfaces have a large interaction the impurity or additive can block the growth of these surfaces. One of the effects might be that the outgrowth of nuclei takes longer because of the growth retarding effect of the impurity or additive. This means that the induction time, i.e. the time that elapses until the outgrown nuclei are detectable, will increase. Molecular modeling calculations of the interaction between additive and crystal surface can be used to design additives which have strong effects on the induction time.

The interaction of the adsorbed additive or impurity and the crystal surface can be face specific. Only very specific faces undergo a large growth reduction and since the growth rates of the crystal surfaces determine the morphology this morphology changes. The retarded faces become morphologically more important. With the aid of molecular modeling, these retarding effects can be calculated. Also, tailor-made additives can be designed in order to impose the preferred morphology.

A substance that is capable of crystallizing into structurally different, but chemically identical crystalline forms exhibits *polymorphism*. A very common compound, which can form polymorphs, is water. Under atmospheric pressure and just below 0 °C ice is formed with a density lower than that of water, so the ice floats on the water. Under higher pressure the ice can crystallize in polymorphs with higher densities, which sink to the bottom. Common energetic compounds which possess the ability to form polymorphs are HMX (cyclotetramethylenetetramine) and CL 20 (hexanitrohexaazaisowurtzitane).

In the pharmaceutical industry, there is a strong desire to control the crystallization of polymorphs because the polymorphic form of a pharmaceutical influences its effect and its lethal dosage. The crystallization process of a certain polymorph is therefore protected by patents. Different polymorphs crystallize under different conditions. However, these conditions may be almost equal for complex organic compounds such as pharmaceuticals. It is possible, for instance, that a certain solvent influences the development of a particular polymorph.

V (1993) Ammonium nitrate: a less polluting oxidizer. In: *Proc. 24th Int. Annual Conference of ICT, Karlsruhe*, p. 3.

3.98 Chan ML, Turner A, Merwin L, Ostrom G, Mead C, Wood S (1996) ADN propellant technology. In: Kuo KK (ed.), *Challenges in Propellants and Combustion*, Begell House, Wallingford, NY, pp. 627–635.

3.99 Ramaswamy AL (2000) Study of the thermal initiation of ammonium dinitramide (ADN) crystals and prills, *J. Energ. Mater.* **18**, 39–60.

3.100 Espitalier F, Biscans B, Authelin J-R, Laguerie C (1997) Modeling of the mechanism of formation of spherical grains obtained by the quasi-emulsion crystallization process, *Trans IChemE* **75**, Part A, 257–267.

3.101 McClements DJ, Dickinson E, Povey MJW (1990) Crystallization in hydrocarbon-in-water emulsions containing a mixture of solid and liquid droplets, *Chem. Phys. Lett.* **172**, 449–452.

3.102 Highsmith TK, Mcleod C, Wardle RB (1998) ADN manufacturing technology. In: *Proc. 29th Int. Annual Conf of ICT, Karlsruhe*, p. 20.

3.103 Teipel U, Heintz T, Krause H (2000) Crystallization of spherical ammonium dinitramide (ADN) particles, *Propell. Explos. Pyrotech.* **25**, 1–5.

3.104 Langlet A, Wingborg N, Östmark H (1996) ADN: a new high performance oxidizer for solid propellants. In: Kuo KK (ed.), *Challenges in Propellants and Combustion*, Begell House, Wallingford, NY, pp. 616–626.

3.105 Teipel U, Heintz T, Leisinger K, Krause H (1999) Crystallization of spherical ammonium dinitramide (ADN) particles from emulsions. In: *Proc. 14th Int. Symp. Ind. Crystallization, Cambridge*, p. 167.

3.106 Stewart JJP (1990) MOPAC: a semi-empirical molecular orbital program, *Comput.-Aided Mol. Des.* **4**, 1.

3.107 Stewart JJP, *MOPAC Program, Ver. 5.0*, Quantum Chemistry Program Exchange, University of Indiana, Blomington, IN.

3.108 Stewart JJP (1990) Semi-empirical molecular orbital methods. In: Lipkowitz KB, Boyd DB (eds), *Reviews in Computational Chemistry*, VCH, New York, chapter 2, p. 45.

3.109 Dupuis M, Sprangler D, Wendoloski JJ (NRCC, USA), extended and modified by Guest MF, van Lenthe JH, Kendrick J, Schoffel K, Sherwood P, Harrison RJ, with contributions from Amos RD, Buenker RJ, Dupuis M, Handy NC, Hillier IH, Knowles PJ, Bonacic-Koutecky V, von Niessen W, Saunders VR, Stone AJ, *GAMESS–UK 6.2*, distributed by Computing for Science (CFS), Daresbury Laboratory, under licence.

3.110 Frisch MJ, Trucks GW, Schlegel HB, Scuseria GE, Robb MA, Cheeseman JR, Zakrzewski VG, Montgomery JA, Stratmann RE, Burant JC, Dapprich S, Millam JM, Daniels AD, Kudin KN, Strain MC, Farkas O, Tomasi J, Barone V, Cossi M, Cammi R, Mennucci B, Pomelli C, Adamo C, Clifford S, Ochterski J, Petersson GA, Ayala PY, Cui Q, Morokuma K, Malick DK, Rabuck AD, Raghavachari K, Foresman JB, Cioslowski J, Ortiz JV, Baboul AG, Stefanov BB, Liu G, Liashenko A, Piskorz P, Komaromi I, Gomperts R, Martin RL, Fox DJ, Keith T, Al-Laham MA, Peng CY, Nanayakkara A, Gonzalez C, Challacombe M, Gill PMW, Johnson B, Chen W, Wong MW, Andres JL, Gonzalez C, Head-Gordon M, Replogle ES, Pople JA (1998) *Gaussian 98*, Gaussian, Pittsburgh.

3.111 Cook DB (1998) *Handbook of Computational Quantum Chemistry*, Oxford University Press, Oxford.

3.112 Veszpremi T, Fehrer M (1999) *Quantum Chemistry, Fundamentals to Applications*, Kluwer, Dordrecht.

3.113 Binkley JS, Pople JA, Hehre WJ (1980) Self-consistent molecular orbital methods. XXI. Small split-valence basis sets for first-row elements, *J. Am. Chem. Soc.* **102**, 939.

3.114 Ditchfield FR, Hehre WJ, Pople JA (1971) Self-consistent molecular orbital methods. IX. An extended Gaussian-type basis for molecular orbital studies of organic molecules, *J. Chem. Phys.* **54**, 724.

3.115 Hehre WJ, Ditchfield R, Pople JA (1972) Self-consistent molecular orbital methods. XII. Further extensions of

Gaussian-type basis sets for use in molccular orbital studies of organic molecules, *J. Chem. Phys.* **56**, 2257.
3.116 Hariharan PC, Pople JA (1973) The influence of polarization functions on molecular orbital hydrogenation energies, *Theor. Chim. Acta* **28**, 213.
3.117 Dunning TH (1971) Gaussian basis functions for use in molecular calculations, *J. Chem. Phys.* **55**, 716.
3.118 Kohn W, Sham LJ (1965) Self-consistent equations including exchange and correlation effects, *Phys. Rev.* **140**, A1133.
3.119 Becke AD (1993) Density-functional thermochemistry. III. The role of exact exchange, *J. Chem. Phys.* **98**, 5648.
3.120 Becke AD (1993) A new mixing of Hartree-Fock and local density-functional theories, *J. Chem. Phys.* **98**, 1372.
3.121 Becke AD (1992) Density-functional thermochemistry. I. The effect of the exchange-only gradient correction, *J. Chem. Phys.* **96**, 2155.
3.122 Becke AD (1992) Density-functional thermochemistry. II. The effect of the Perdew-Wang generalized-gradient correlation correction, *J. Chem. Phys.* **97**, 9173.
3.123 Lee C, Yang W, Parr RG (1988) Development of the Colle-Salvetti correlation-energy formula into a functional of the electron density, *Phys. Rev.* **37**, 785.
3.124 Wu CJ, Fried LE (1997) *Ab initio* study of RDX decomposition mechanisms, *Phys. Chem. A* **101**, 8675.
3.125 The results published were generated using the program Cerius2, developed by Accelrys.
3.126 Mayo SL, Olafson BD, Goddard WA III (1990) DREIDING: a generic force field for molecular simulations, *J. Phys. Chem.* **94**, 8897–8909.
3.127 Ewald PP (1921) Die berechnung optischer und elektrostatischer gitterpotentiale, *Ann. Phys.* **64**, 253.
3.128 Kendrick J, Fox M (1991) Calculation of electrostatic potentials, *J. Mol. Graph. Soc.* **9**, 182.
3.129 Kuklja MM, Kunz AB (1999) *Ab initio* simulation of defects in energetic materials. Part I. Molecular vacancy structure in RDX crystal, *J. Phys. Chem. Solids* **61**, 35–44.
3.130 Kuklja MM, Kunz AB (1999) *Ab initio* simulation of defects in energetic materials: hydrostatic compression of cyclotrimethylene trinitramine, *J. Appl. Phys.* **86**, 4428–4434.
3.131 Kunz AB (1996) *Ab initio* investigation of the structure and electronic properties of the energetic solids TATB and RDX, *Phys. Rev. B: Condens. Matter* **53**, 9733–9738.
3.132 Donnay JDH, Harker D (1937) A new law of crystal morphology extending the law of Bravais, *Am. Mineral.* **22**, 446
3.133 Hartman P, Bennema P (1980) The attachment energy as a habit controlling factor, I. Theoretical considerations, *J. Cryst. Growth* **49**, 145
3.134 Gibbs JW (1928) *Collected Works*, Longman, New York.
3.135 Bennema P (1996) On the crystallographic and statistical mechanical foundations of the forty-year old Hartman-Perdok theory, *J. Cryst. Growth* **166**, 17.
3.136 Costain W, Cox EG (1947) Structure of dimethylnitramine, *Nature* **160**, 826.
3.137 Krebs B, Mandt J, Cobbledick RE, Small RWH (1979) The structure of *N,N*-dimethylnitramine, *Acta Crystallogr., Sect. B* **25**, 402.
3.138 Filhol A, Bravic G, Rey-Lafon M, Thomas M (1980) X-ray and neutron studies of a displacive phase transition in N,N-dimethylnitramine (DMN), *Acta Crystallogr., Sect. B* **26**, 575.
3.139 Shishkov IF, El'fimova TL, Vilkov LV (1992) Molecular structure of hexahydro-1,3,5-trinitrotriazine in the gas phase, *Zh. Strukt. Khim.* **33**, 41.
3.140 Choi CS, Prince E (1972) Crystal structure of cyclotrimethylenetrinitramine, *Acta Crystallogr., Sect. B* 28, 2857–2862.
3.141 Karpowicz RJ, Brill TB (1984) Comparison of the molecular structure of hexahydro-1,3,5-trinitro-*s*-triazine in the vapor, solution and solid phases, *J. Phys. Chem.* **88**, 348.
3.142 Rerat B, Berthou J, Laurent A, Rerat C (1968) Structure cristalline du complexe hexogene-sulfolane, *C. R. Acad. Sci., Ser. C* **267**, 760.
3.143 Gilardi R, Flippen-Anderson JL, George C (1990) Structures of 1,3,5-trinitro-2-oxo-1,3,5-triazacyclohexane (I) and 1,4-dinitro-2,5-dioxo-1,4-diazacyclohex-

ane (II), *Acta Crystallogr., Sect. C* **46**, 706.

3.144 Rice BM, Chabalowski CF (1997) *Ab-initio* and nonlocal density functional study of 1,3,5-trinitro-*s*-triazine (RDX) conformers, *J. Phys. Chem. A* **101**, 8720–8726.

3.145 Harris NJ, Lammertsma K (1997) *Ab initio* density functional computations of conformations and bond dissociation energies for hexahydro-1,3,5-trinitro-1,3,5-triazine, *J. Am. Chem. Soc.* **119**, 6583–6589.

3.146 Chakraborty D, Muller RP, Dasgupta S, Goddard WA III (2000) The mechanism for unimolecular decomposition of RDX (1,3,5-trinitro-1,3,5-triazine), an *ab initio* study, *J. Phys. Chem. A* **104**, 2261.

3.147 Sorescu DC, Rice BM, Thompson DL (1998) A transferable intermolecular potential for nitramine crystals, *J. Phys. Chem. A* **102**, 8386–8392.

3.148 Pivina TS, Arnautova EA (1996) Computer modeling of possible polymorphic transformations in HNIW (CL-20). In: *Proc. 27th Int. Annual Conference of ICT, Karlsruhe*, p. 39.

3.149 Sorescu DC, Rice BM, Thompson DL (1999) Theoretical studies of the hydrostatic compression of RDX, HMX, HNIW and PETN crystals, *J. Phys. Chem. B* **103**, 6783–6790.

3.150 Sorescu DC, Rice BM, Thompson DL (1998) Molecular packing and NPT-molecular dynamics investigation of the transferability of the RDX intermolecular potential to 2,4,6,8,10,12-hexanitrohexaazaisowurtzitane, *J. Phys. Chem. B* 948–952.

3.151 Ou Y-H, Jia H-P, Chen B, Xu Y, Zhang J, Liu Y (1999) Crystal structure of alpha-hexanitrohexaazaisowurtzitane, *Beijing Ligong Daxue Xuebao* **19**, 631–636.

3.152 Ou Y, Jia H-P, Chen B-R, Xu Y-J, Wang C, Pan Z-L (1999) Crystal structure of gamma-hexanitrohexaazaisowurtzitane, *Huaxue Xuebao* **57**, 431–436.

3.153 Ou Y-H, Jia H-P, Chen B, Xu Y, Pan Z, Chen J, Zheng F (1998) Structure of four polymorphs of hexanitrohexaazaisowurtzitane explosives, *Huozhayao Xuebao* **21**, 41–43.

3.154 Zhao X, Shi N (1996) Crystal and molecular structures of.epsilon-HNIW, *Chin. Sci. Bull.* **41**, 574–576.

3.155 Chan ML, Carpenter P, Hollins R, Nadler M, Nielsen AT, Nissan R, Vanderah DJ, Yee R, Gilardi RD (1995) *CPIA-PUB-625. 17P.* CPIA Abstract No. X95–07119, AD D606 761. Availability: distribution authorized to the Department of Defense and DoD contractors only: Critical Technology: March 17, 1995. Other requests should be referred to the Naval Air Warfare Center Weapons Division (4740001), China Lake. CA 93555–6001. This document contains export-controlled technical data.

3.156 Nielsen AT, Chafin AP, Christian SL, Moore DW, Nadler MP, Nissan RA, Vanderah DJ, Gilardi RD, George CF, Flippen-Anderson JL (1998) Synthesis of polyazapolycyclic caged polynitramines, *Tetrahedron* **54**, 11793–11812.

3.157 Rappé AK, Casewit CJ, Colwell KS, Goddard WA, Skiff WM (1992) UFF, a full periodic table force field for molecular mechanics and molecular dynamics simulations, *J. Am. Chem. Soc.* 114, 10024.

3.158 Rappé AK, Goddard WA (1991) Charge equilibration for molecular dynamics simulations, *J. Phys. Chem.* **95**, 3358.

3.159 Hartman P, Bennema P (1980) The attachment energy as a habit controlling factor, *J. Cryst. Growth* **49**, 145–156.

3.160 Hartman P, Perdok WG (1955) On the relations between structure and morphology of crystals. I, *Acta Crystallogr.* 8, 49–52.

3.161 Hartman P, Perdok WG (1955) On the relations between structure and morphology of crystals. II, *Acta Crystallogr.* **8**, 521–524.

3.162 Hartman P, Perdok WG (1955) On the relations between structure and morphology of crystals. III, *Acta Crystallogr.* 8, 525–529

3.163 Leach AR (1996) *Molecular Modeling: Principles and Applications*, Addison Wesley Longman, Harlow.

3.164 Myerson AS (1999) *Molecular Modeling Applications in Crystallization*, Cambridge University Press, Cambridge.

3.165 Allen MD, Tildesley DJ (1987) *Computer Simulation of Liquids*, Clarendon Press, Oxford.

3.166 Frenkel D, Smit B (1996) *Understanding Molecular Simulation*, Academic Press, New York.

3.167 Hannun, JAE (1986) Hazards of chemical Rockets and Propellents Vol 2: Solid propellents and ingredients Report chem. Popul. Inf Agency John Hopkins Univ. Laurel, MD, USA.

3.168 Aspen Technology (1985) *ASPEN PLUS Solids Manual*, Aspen Technology, Cambridge, MA.

3.169 Simulation Sciences (1991) *PRO/II Process Simulator Technical Information*, Simulation Sciences, Fullerton, CA.

3.170 Westerberg AW, Hutchison HP, Motard RL, Winter P (1979) *Process Flowsheeting*, Cambridge University Press, Cambridge.

3.171 Timar L, Simon F, Csermely Z, Siklós J, Bácskai B, Édes J (1984) Useful combination of the sequential and simultaneous modular strategy in a flowsheeting programme, *Comput. Chem. Eng.* **8**, 185–194.

3.172 Blickle T, Lakatos BG, Nirnsee BB (1993) Characterization of crystal size distribution by hyperbolic tangent distribution functions. In: *Proc. 12th Symp. on Ind. Crystallization, Warsaw.*

4
Crystallization with Compressed Gases

E. Reverchon, H. Kröber, U. Teipel

4.1
Introduction

Many product properties that are of relevance in industrial use can be adjusted by changing the particle size and particle size distribution of the powder. This statement is valid in several fields, ranging from polymers to pharmaceutical and inorganic powders.

In the case of solid explosives and propellants, small particles are as a rule required to improve the combustion process. Indeed, the attainment of the highest energy from the detonation of a solid explosive depends strongly on the particle size of the material.

Grinding and crystallization from solutions are largely used as micronization processes in industry and have been described in previous chapters. However, these processes suffer from some limitations: it is difficult to control the particle size and particle size distribution of powders, especially when very small (micron-sized) particles are required. Liquid crystallization also suffers from the problem of solvent contamination of the precipitate (crystal inclusions) and jet milling is not suitable for the treatment of shock-sensitive substances.

As an alternative to the traditional techniques, various supercritical-fluid based precipitation processes have recently been proposed. These techniques can potentially overcome all the previously described limitations of the classical micronization processes. Supercritical fluids are compressed gases that are used at temperatures and pressures higher than their critical point. At the critical point, the liquid-gas phase boundary disappears and the surface tension approaches zero. Near the critical point, even small increases in the applied pressure cause a sharp increase in the density and, correspondingly, the dissolving capacity of the supercritical fluid. Supercritical fluids exhibit properties that are a combination of liquid and gas-like characteristics (Table 4.1). The density of compressed gases is in the same range as that of organic liquids, while the viscosity is close to that of gases. Because their diffusion coefficient is larger than that of liquids, supercritical fluids have enhanced mass transfer properties, making them especially suitable for extraction operations,

Energetic Materials. Edited by Ulrich Teipel
Copyright © 2005 WILEY-VCH Verlag GmbH & Co. KGaA, Weinheim
ISBN: 3-527-30240-9

Table 4.1. Comparison of some physical data between gases, liquids and compressed gases.

	ϱ (kg/m^3)	η (10^3 Pa s)	D (m^2/s)
Gas at 0.1 MPa:			
$\vartheta = 25$ °C	0.6–2.0	0.01–0.03	$(1–4) \times 10^{-5}$
Supercritical fluids:			
T_c, p_c	200–500	0.01–0.03	7×10^{-8}
T_c, $4 \times p_c$	400–900	0.03–0.06	2×10^{-8}
Liquids:			
at $\vartheta = 25$ °C	600–1600	0.2–3.0	$(0.2–2) \times 10^{-9}$

particularly in the foodstuffs industry. As a result of these special physical properties, solid particles can be processed by supercritical fluid technologies. Supercritical fluids, in principle, do not give problems of solvent contamination because when decompression occurs, they are completely released from the solute.

The supercritical fluid of election is CO_2 since it has relatively low critical parameters ($\vartheta_c = 31.1$ °C and $p_c = 7.38$ MPa), it is not toxic, non flammable and cheap. The operating conditions for supercritical fluids processing are generally mild, thus giving no problem to process 'sensitive' materials. Also, the control of the particle size and particle size distribution of the micronized material promises to be relatively simple to obtain by continuously modulating the process conditions.

Various supercritical fluid-based precipitation processes have been proposed. Jung and Perrut summarized the experimental and theoretical work on these techniques and gave an overview of substances which were processed by these processes [4.1]. In the following part of this chapter we will discuss their major characteristics.

4.2 Rapid Expansion of Supercritical Solutions

The first micronization process based on supercritical fluids that has been proposed is the **R**apid **E**xpansion of **S**upercritical **S**olutions (RESS) [4.2–4.4]. In this process, the solvent power of the supercritical fluid is used to dissolve the compound to be micronized by fluxing the supercritical fluid in a fixed bed formed by the particles of the starting material. Then, the solution formed is depressurized down to atmospheric pressure in an atomization nozzle. The fast expansion of the supercritical solution reduces the solvent power of the fluid to nearly zero and the solute precipitates in the expansion chamber. A schematic flow sheet of a typical experimental setup is shown in Fig. 4.1. A more detailed description of the experimental setups used can be found in the literature [4.5–4.7].

Using this technique, it is possible to obtain very high supersaturation ratios that can result in the production of very small particles. Some studies on the RESS process were devoted to the identification of the process parameters that control the

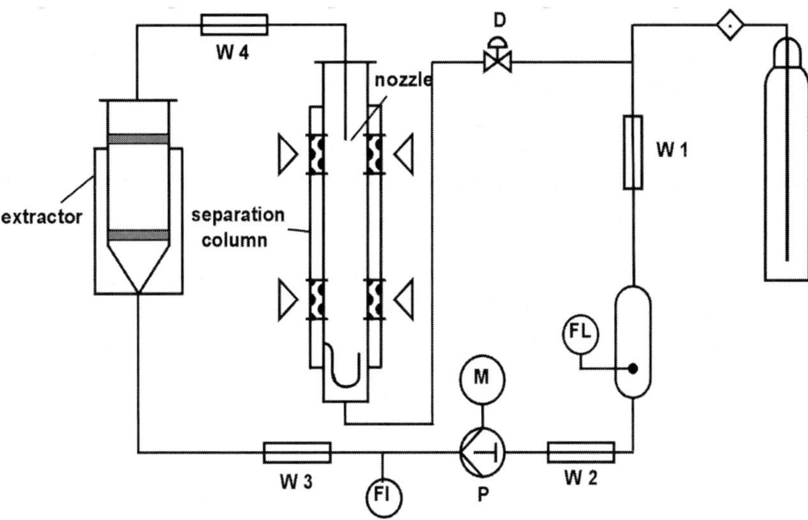

Figure 4.1. Flow sheet of the experimental setup: W, heat exchanger; P, membrane pump; FI, mass flow meter; FL, liquid CO_2 reservoir; D, pressure control.

precipitation of particles [4.4]. The main parameters that control the RESS process are pre-expansion temperature and pressure and expansion chamber temperature and pressure.

It is difficult to propose a systematic description of the investigated substances and morphologies obtained by RESS since a large variety of particle shapes and sizes have been observed and very different experimental arrangements have been used. Moreover, many different substances were investigated (organic and inorganic materials, polymers and biopolymers and biodegradable materials) which differed in their physical and chemical properties so that a comparison is difficult. Kröber *et al.* micronized benzoic acid and cholesterol as model substances for pharmaceuticals under different process conditions [4.8]. In contrast to the experimental work of other groups, their investigations were carried out with nozzles up to three-times larger than normally used so that an industrial application was simulated more realistically. The first theoretical descriptions demonstrate the complexity of the process, but only provide a rough, qualitative interpretation of experimental results. Türk showed the calculated temperature, pressure and supersaturation profiles as a function of capillary length for various experimental conditions [4.6].

4.2.1
Effect of Pre-expansion Pressure, Temperature and Concentration on RESS

The pre-expansion pressure and temperature determine the expansion path in the nozzle. This means that the achieved supersaturation and therefore nucleation rate can be influenced by varying the pre-expansion conditions. Instead, contradictory results have been reported by different authors about the influence of the pre-expansion conditions on the particle size and morphology [4.3, 4.9–4.13]. For the system carbon dioxide-naphthalene, Mohamed *et al.* found that the mean particle size increases when the pre-expansion temperature is increased whereas the pressure does not have a significant influence [4.3, 4.9]. On the other hand, RESS experiments carried out with different waxes showed a decrease in particle size when the pre-expansion temperature is increased [4.14].

Alessi *et al.* investigated the influence of the solute concentration in the supercritical solution on the product quality [4.15]. The system carbon dioxide-progesterone was investigated and an increase in solute concentration resulted in a slight increase of the mean particle size. Reverchon *et al.* showed contradictory results for carbon dioxide-salicylic acid. In this case a more concentrated solution results in smaller particles [4.4, 4.16]. Kröber *et al.* [4.5] found that the particle size of cholesterol micronized by the RESS process was more or less independent of the solute concentration.

4.2.2 Effect of Post-expansion Pressure and Temperature on RESS

The conditions in the expansion chamber can influence the spray jet behind the nozzle and the coagulation and agglomeration processes in the expansion chamber. Owing to the relative importance of these processes for product quality, the post-expansion conditions and the design of the expansion chamber should have a significant influence on the particle size and morphology. Unfortunately, investigations carried out so far do not show a uniform picture of the effect of the conditions in the expansion chamber. Some authors found that an increase in the post-expansion temperature results in larger particles and a high post-expansion pressure changes the particle morphology to a dendritic structure [4.15]. Contradictory to these results, Reverchon *et al.* described the post-expansion temperature as the main parameter which influences the particle morphology [4.10]. They found that spherical particles are formed at low temperatures whereas needle-like structures are preferred at higher temperatures. Kröber *et al.* [4.8] showed that the specific surface area increases when the post-expansion pressure is increased from 2 to 5 MPa. A higher pressure in the expansion chamber gives the opportunity to recycle the carbon dioxide in an economic way so that an industrial application of the RESS process will be more realistic.

4.2.3 Effect of Nozzle Geometry and Dimensions on RESS

The influence of the nozzle has been investigated by several groups. In principle, two types of nozzles were used: capillary nozzles with different length-to-diameter ratios and Laval nozzles with different diameters. No effect of nozzle diameter was found for the system carbon dioxide-benzoic acid by Berends *et al.* [4.17]. Kröber *et al.* [4.8] compared Laval nozzles with capillary nozzles. Again, no systematic influence on the product quality was seen. Domingo *et al.* developed a frit nozzle (sintered metal plate) simulating many parallel capillaries. The use of such a nozzle causes significant smaller particles compared with particles from normal capillary nozzles (benzoic acid and salicylic acid) [4.18, 4.19].

4.2.4
RESS Modeling

Lele and Shine [4.20] described the flow-field of the pure fluid in a capillary nozzle by the Navier-Stokes equations. The expansion in the capillary is considered as a one-dimensional, axial and steady-state flow of a real gas. The processes in the free jet behind the capillary were not investigated. The pressure and temperature in the capillary were calculated numerically as a function of different pre-expansion conditions and as a function of capillary length but no coupling with nucleation and growth kinetics were carried out.

Debenedetti [4.21] coupled the calculation of the flow pattern in a convergent nozzle with particle formation dynamics. Again, no simulation of processes in the free jet behind the nozzle was implemented. Owing to the very short residence time in the nozzle, coagulation was not considered as an important particle formation process. The particle dynamics in the nozzle are described by nucleation and condensation. Using this model, it is possible to calculate the particle size of phenanthrene in carbon dioxide for different extraction and pre-expansion temperatures. An increasing pre-expansion temperature results in larger particles whereas smaller particles are obtained when the extraction temperature increases. The highest particle size which was calculated with this model was 60 nm and therefore one magnitude smaller than the particle size measured in experimental measurements (2–7 μm).

Türk *et al.* [4.22, 4.23] calculated the flow pattern in a capillary nozzle and in the free jet behind the nozzle. Additionally, the saturation and nucleation rate were simulated as a function of the capillary length for different substances. For this reason a modified Peng-Robinson equation of state was implemented in addition to different models for the particle formation processes (nucleation, condensation and coagulation). Using this complex model, it is possible to describe the particle formation until the Mach disc by coupling the flow of a real gas with the particle dynamics. Again, the calculation results in particle sizes much smaller than the sizes measured in experimental investigations.

The reason for the differences between the calculated and measured data is the processes which take place in the expansion chamber. It present these processes are not completely understood and implemented in the models.

The major drawback of the RESS is related to the limited solubility of many of the solutes of interest in supercritical solvents. However, it is possible to increase the solubility by adding small amounts of a modifier (e.g. acetone) to the supercritical fluid before the extraction process (modified RESS). Other limitations of the process come from the analysis of the experimental results: electrostatic charging of particles occurs, owing to the high velocities of microparticles inside the precipitation chamber, and particle geometries are mainly limited to needle-like particles. Moreover, the particles obtained are not as small as one could predict from the premise of the very high supersaturation ratios that can be obtained, and as a rule particles ranging from several hundred nanometers to several micrometers are produced.

In addition to RESS, some other micronization techniques have been proposed that are based on the use of supercritical fluids. For example, the Particle from Gas Saturated Solutions (PGSS) has been proposed [4.24–4.26]. This process consists of dissolving a supercritical fluid in a liquid or a liquid suspension or solution, followed by the rapid depressurization of the formed mixture through a nozzle. The depressurization causes the formation of micrometric solid particles.

Another technique that shares some characteristics with PGSS is that proposed by Sievers and co-workers [4.27, 4.28] that is applicable to aqueous solutions. In this case supercritical CO_2 is either dissolved in or mixed with the aqueous solution. The depressurization of the non-homogeneous mixture through a capillary generates very small droplets that can produce micronic particles.

4.3
Supercritical Antisolvent Precipitation

Supercritical antisolvent precipitation (SAS) is the most promising supercritical technique since it has been demonstrated that it is capable of producing controlled micronic and sub-micronic particles [4.29]. The description of this process departs from liquid crystallization with which it is homologous in supercritical form. In SAS the supercritical fluid substitutes the liquid antisolvent and causes the precipitation of solute since it forms a solution with the primary solvent. Therefore, a prerequisite to perform SAS is that the process is performed under temperature and pressure conditions at which complete miscibility exists between the primary liquid solvent and the supercritical antisolvent. A distinctive characteristic of supercritical fluids is the diffusivity that can be up to two orders of magnitude higher than those of liquids. Therefore, the diffusion of the supercritical fluid into a liquid solvent can produce rapid supersaturation of solute dissolved in the liquid and its precipitation in micronized particles.

SAS has been performed using different process arrangements and apparatus. Since the results can be heavily influenced by the process arrangement, a brief description of the two main techniques is presented. When the precipitation vessel is loaded with a given quantity of the liquid solution and then the supercritical antisolvent is added until the final pressure is reached, a batch antisolvent precipitation is performed. This mode of operation can be referred to as a *liquid batch* or GAS process (**G**as **A**nti-**S**olvent). It is also possible to charge the precipitation chamber with the antisolvent and then to perform a discontinuous injection of the liquid solution. This mode of operation can be termed *gas batch*. The difference between these two modes of operation is that in the first case the precipitation occurs in a liquid-rich phase, whereas in the second case it occurs in a supercritical fluid-rich

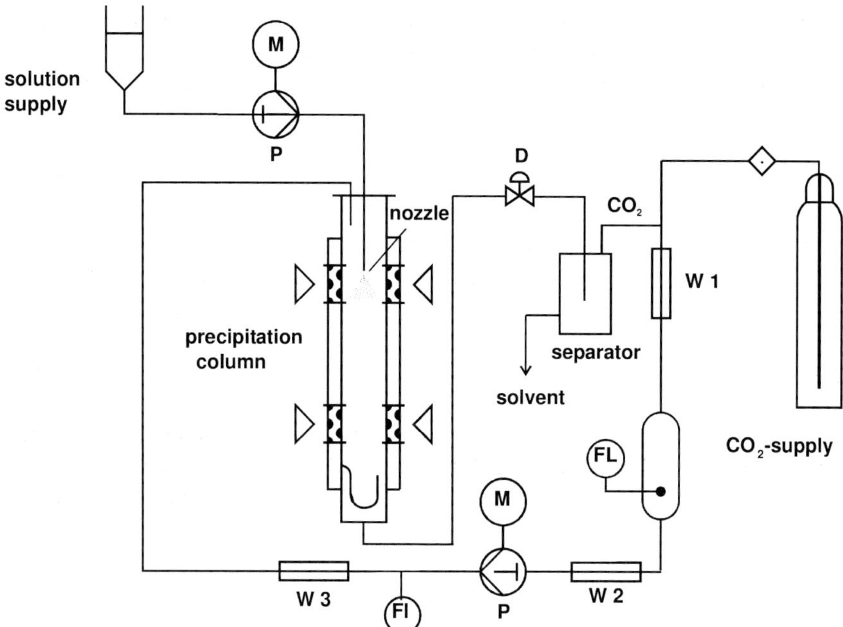

Figure 4.2. Flow sheet of the experimental setup: W, heat exchanger; P, membrane pump; FI, mass flow meter; FL, liquid CO_2 reservoir; D, pressure control.

phase. A semi-continuous SAS processing is performed when a spray of the liquid solution is continuously produced in the precipitation chamber in which supercritical CO_2 is also flowing, added from another inlet point. This mode of operation is also referred as the PCA process (**P**recipitation with a **C**ompressed Fluid **A**ntisolvent) in the literature. A typical experimental setup is shown schematically in Fig. 4.2.

Detailed descriptions of such devices have been published [4.29, 4.30]. An injector is used to produce liquid jet break-up to form small micronic droplets that expand in the precipitator. Various injection devices have been proposed. Yeo *et al.* proposed the adoption of nozzles and tested various nozzle diameters ranging from 5 to 50 μm [4.31]. Other authors used small internal diameter capillaries [4.32, 4.33] or vibrating orifices [4.32]. This last apparatus produces a spray by superimposing a high-frequency vibration on the liquid jet that exits from an orifice. Coaxial devices have also been proposed in which two capillary tubes continuously deliver the liquid solution and the supercritical antisolvent [4.34, 4.35]. The formation of small droplets in this case depends on the turbulent mixing of the two flows. Complex geometries formed by more than two capillaries and different dispositions of the liquid and supercritical fluid (inside-outside) have also been tested [4.34].

The washing step with pure supercritical antisolvent at the end of the precipitation process, either batch or semi-continuous, is fundamental to avoid the condensation of the liquid phase that otherwise rains on the precipitated powder, modifying its characteristics.

A key role in the SAS precipitation process is also played by the volumetric expansion of the liquid solvent and by its modification in the presence of different solutes for a given solvent. Expansion is due to the massive dissolution of the supercritical antisolvent in the liquid phase. Several groups have studied the volumetric expansion isotherms for various solvent-supercritical antisolvent pairs [4.36]. The percentage volumetric expansion $\Delta V\%$ can be defined as

$$\Delta V\% = \frac{V(p,T) - V_0}{V_0} \cdot 100 \tag{4.1}$$

where $V(p,T)$ is the volume of the liquid phase loaded with the antisolvent and V_0 is the volume of the pure liquid phase under atmospheric conditions. However, alternative definitions have also been suggested, based on the behavior of the molar volume of the liquid phase [4.37]. When a solution is studied, as in the case of SAS experiments, the effect of the solute at different concentrations on the expansion behavior also has to be taken into account. Indeed, the behavior of the solvent-antisolvent binary systems can be strongly modified by the addition of the solute: one or more liquid and fluid phases can be formed in these ternary systems, producing very complex vapor-liquid equilibrium (VLE) diagrams. A recent study performed on the SAS micronization of some biopolymers [4.29] confirms that to obtain micronization successfully, the process has to be performed in a region of the VLE diagram where the binary behavior of the liquid-supercritical antisolvent system has been maintained.

It is very difficult to propose a systematic description of the morphologies obtained by SAS since a large variety of particle shapes have been observed and very different experimental arrangements have been used. However, a stringent correlation between some morphologies and the liquid volume expansion has been proposed for continuous operation [4.38, 4.39], i.e. various morphologies can be explained when compared with the different solvent expansion levels at which they are generated.

In batch SAS, the rate of supercritical antisolvent addition is an important parameter in controlling the morphology and the size of the solid particles. Indeed, various authors [4.40–4.43] have tried to use different pressurization profiles. A general conclusion is that the faster is the pressurization of the precipitator, the smaller are the particles obtained.

When semi-continuous SAS is performed working at intermediate expansion levels of the liquid solution, expanded droplets (balloons) have been observed. These are dried expanded droplets and are formed by an empty shell of solute. In Fig. 4.3, the SEM image shows a landscape view of yttrium acetate balloons precipitated from DMSO. At very large expansion levels (complete miscibility), nanoparticles are produced by disintegration of the balloons. These particles can be very small (100–200 nm) and as a rule have a very narrow particle size distribution [4.38].

Some additional mechanisms have been observed that can be superimposed on that described so far of droplet expansion and explosion, which can produce more

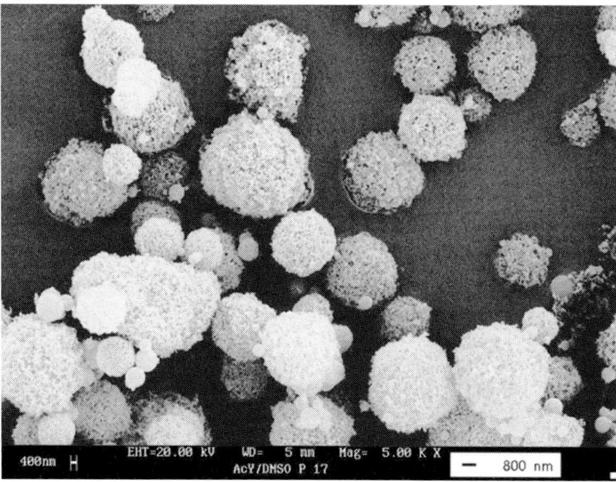

Figure 4.3. SEM image: landscape view of yttrium acetate particles produced by SAS at 12 MPa, 50 °C, 15 mg solute/ml DMSO. Very large (up to 20 μm diameter) expanded droplets (balloons) have been observed [4.38].

complex particle geometries. The first is the coalescence of nanoparticles. Two kinds of coalescence can be involved. The first can be defined as a physical coalescence and is characterized by a particle to particle interaction, for example, by impact during the precipitation process. This kind of particle can be separated by sonication or other methods. A second coalescence mechanism is chemical in nature; in this case, particles can interact with the liquid solvent and the result is the fusion of nanoparticles in groups in which the single particle no longer has a distinct identity.

The semi-continuous SAS process shares many similarities with classical spray-drying techniques. The major differences lie in spraying the solution into a compressed gas instead of into heated air as the external phase. Compressed CO_2 is miscible with the organic solvent under supercritical conditions; however, it is a non-solvent for the solute. The particle size and morphology depend on many factors, such as the jet breakup, mass transfer rates between the droplet and antisolvent phase, nucleation kinetics and particle growth rates. Especially jet breakup occurs in the compressed gas phase different from the normal breakup in atmospheric spray processes. Czerwonatis *et al.* [4.44] investigated the disintegration of liquid jets in dense gases. They used water and a vegetable oil as the liquid phase and CO_2 and N_2 as pressurized gases. With increasing jet velocity three distinct regimes of breakup can be observed: Rayleigh breakup (sinuous wave breakup) atomization (Fig. 4.4).

The boundary between the Rayleigh breakup regime and the sinuous wave regime and also the boundary between the sinuous wave breakup regime and the atomiza-

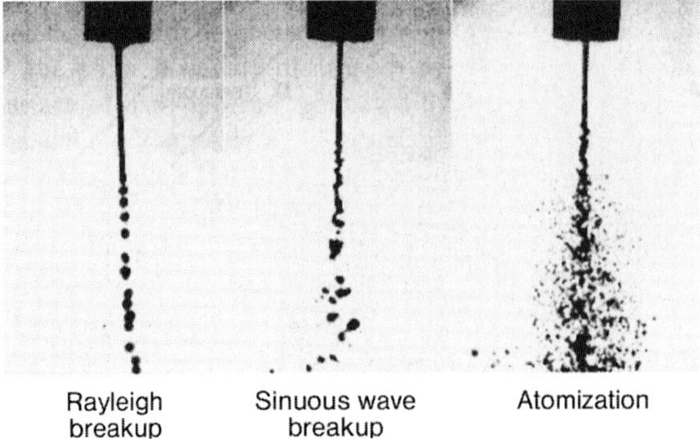

Figure 4.4. Disintegrating jet of water in pressurized N_2 at 21.6 MPa, 25 °C [4.44].

tion regime dislocate with increasing pressure, which corresponds to increasing density. With increasing gas density, the boundaries are shifted to regions of lower Reynolds numbers. This means that the sinuous wave breakup and the atomization in pressurized gases take place at lower outlet velocities of the jet compared with gases at atmospheric pressure.

Kröber and Teipel [4.45] measured the droplet size during atomization of ethanol in dense CO_2. The droplet size was relatively insensitive to the jet velocity at the nozzle outlet but increased with increasing pressure (and therefore gas density). The droplet size was measured at two different distances behind the nozzle (10 and 40 mm) which allows the determination of mass transfer properties. It was shown that the droplet size at the first measurement point was independent of the liquid that was atomized (with the exception of dimethylformamide, whose droplets were significantly smaller). At the second measurement point the droplets of toluene were distinctly smaller than the droplets of acetone and ethanol. The difference in droplet size was the lowest if dimethylformamide was atomized.

The results of both studies are relevant to the understanding of the continuous SAS process. The description of the process can be divided into two parts, both of which influence the particle size of the product: first, the formation of the droplets, and second, the precipitation of the solid particle located in the droplet.

4.3.1 Effect of Pressure and Temperature on SAS

During *liquid batch* processing, the rate of pressure increase is the most relevant parameter in controlling particle size and morphology [4.41]. Contradictory results have, instead, been obtained by different authors regarding the influence of pressure on the particle size during continuous operation [4.30–4.33, 4.46, 4.47]. For example, for a reduction in the precipitation pressure some authors found a particle size decrease [4.47], others found the process insensitive to this parameter [4.31] and others observed a particle size increase [4.30]. The same problem occurred for temperature reduction that produced a particle size decrease for some

Figure 4.5. SEM images showing the effect of concentration of the liquid solution on particle size and particle size distribution of samarium acetate nanoparticles [4.38]. (a) 5 and (b) 65 mg/ml DMSO.

authors [4.32], had no influence for others [4.31] and produced larger particles for others [4.47].

4.3.2 Effect of Concentration of the Liquid Solution on SAS

Particle size was relatively insensitive to solute concentration in the liquid for some authors [4.31, 4.32]. Large concentrations induced the formation of polymeric fibers according to other authors [4.33, 4.46].

A marked particle size increase and PSD enlargement were observed by Reverchon *et al.* in the SAS precipitation of yttrium, samarium and neodymium acetates [4.38]. This result is well illustrated in Fig. 4.5(a) and (b) that are referred to samarium acetate precipitation from DMSO at the same pressure and temperature but at 5 and 65 mg/ml concentration, respectively. SEM images give evidence of the large increase in samarium acetate particle size and particularly particle size distribution.

4.3.3 Effect of the Chemical Composition of the Liquid Solvent and of Solute on SAS

Very different behaviours can be obtained by changing the liquid solvent. For example, in the case of amoxicillin and tetracycline precipitated from DMSO and NMP, the SAS processing using the first solvent was completely unsuccessful since both antibiotics were extracted from the precipitation chamber, whereas precipitation from NMP was successful and antibiotic nanoparticles were produced [4.48]. Kröber and Teipel described the effect of different solvents during the precipitation of tartaric acid [4.49]. It was shown that the particle size decreases if acetone was used as primary solvent compared with the particles precipitated from ethanol or methanol (Fig. 4.6).

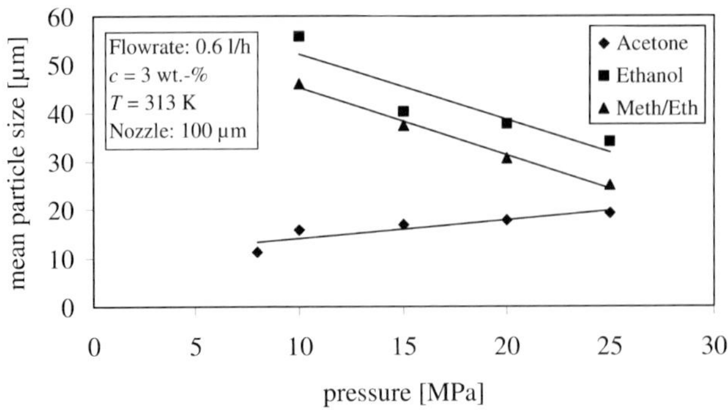

Figure 4.6. Precipitation of tartaric acid from different solvents: mean particle size affected by precipitation pressure [4.49].

4.3.4 SAS Modeling

Kikic *et al.* proposed a first attempt at modeling mass transfer mechanisms acting during the supercritical precipitation in batch and semi-continuous processing [4.50–4.52]. The initial diameter of the droplets was assumed to be equal to the nozzle diameter and hydrodynamics, mass transfer and thermodynamic considerations were embedded in the model. The composition of the two phases, the droplet diameter, the gas and liquid flow-rates and the amount of precipitated product along the axis of the vessel as a function of temperature and pressure were simulated.

Rantakyla *et al.* tried to model the supercritical antisolvent precipitation process assuming a droplet size log-normal distribution [4.53]. The particles in the droplet were described with an aerosol equation which describes the evolution of particle size distribution with time. The following assumptions (with the help of the experimental observations) were made: (a) only one agglomerated particle develops in each drop and (b) the drop size was calculated from the experimental particle size distribution. Therefore, the particle size is almost constant whereas the liquid content of particles decreases with time.

Werling and Debenedetti [4.54] concentrated their attention on the mass transfer between a single droplet of organic solvent and the supercritical antisolvent. They took into account the two-way mass transfer into the droplet and into the antisolvent, determining the radius of the droplet as a function of time. Their calculations showed that the initial interface flux is always into the droplet and causes the droplet expansion.

However, modeling of SAS is still at an early stage. The reason is that many aspects of the process are still not well understood and data required to perform the model validation are still lacking.

4.4 Precipitation of Energetic Materials by Supercritical Fluids

Although the RESS process can be used to micronize various pharmaceutical compounds [4.2, 4.9, 4.10, 4.16, 4.55], the processing of energetic materials using this technique is mentioned in only a few papers [4.56–4.58]. One reason for the lack of experimental data is the low solubility of almost all energetic substances in supercritical carbon dioxide. One exception of this behaviour is trinitrotoluene (TNT), which has a solubility of up to 2 wt% in CO_2 under moderate conditions [4.59].

The first application of the RESS process to energetic materials was proposed by Teipel *et al.* when they processed TNT [4.56]. Figure 4.7 shows two photographs of TNT particles, in both cases crystallized with supercritical carbon dioxide. The TNT particles recrystallized in a static experiment from CO_2 (Fig. 4.7, left) show the needle-like morphology which is typical of TNT. With the rapid expansion in the RESS process, the degree of development of the crystal habitus can be observed (Fig. 4.7, right). They obtained a particle diameter $x_{50,3}$ of 10 μm. Teipel *et al.* [4.57] described a slight influence of the nozzle diameter on the product quality. Results and process conditions are presented in Table 4.2.

In addition to TNT, 3-nitro-1,2,4-triazol-5-one (NTO) was micronized by the RESS process at the Fraunhofer ICT [4.57]. Owing to the very low solubility of NTO in supercritical carbon dioxide, acetone was used as modifier, which was mixed into the CO_2 just before entering the extractor. A successful test was, for example, performed under the following conditions: extraction pressure 20 MPa, extraction temperature 60 °C, CO_2 flow-rate 4 kg/h and acetone flow-rate 2.4 ml/h. Submicron particles were obtained in this case, as illustrated in Fig. 4.8, with a mean particle size of 540 nm. The agglomerate shown is very loose and consists of needle-like particles. The particles were dry and without solvent inclusions.

Hexanitrohexaazaisowurzitane (CL 20) also can be processed by the RESS process when trifluoromethane (CHF_3) is used as supercritical solvent. The solubility of CL

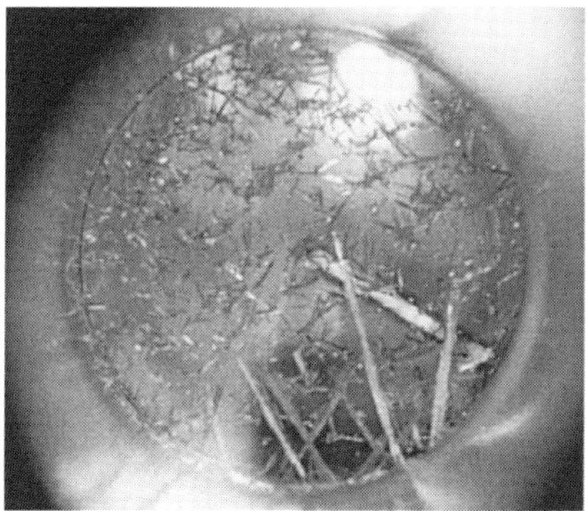

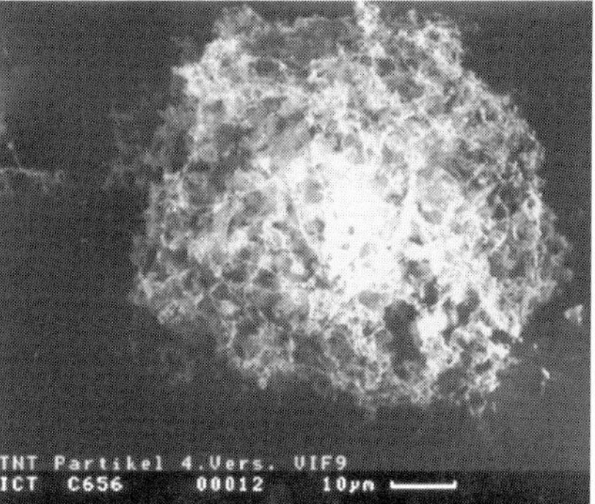

Figure 4.7. SEM pictures of TNT particles recrystallized from supercritical CO_2 (left, static experiment; right, RESS processed material) [4.56].

Table 4.2. Experimental conditions and results of the precipitation of TNT by RESS [4.57].

	Run 1	**Run 2**
Extractor:		
p (MPa)	22	22
T (K)	348	348
Nozzle:		
d (μm)	50	100
T (K)	458	458
Column:		
p (MPa)	0.1	0.1
T (K)	305	305
Mean particle size $x_{50,3}$ (μm)	14.2	10.0

20 in CHF_3 is much higher than in carbon dioxide but the critical data for both fluids are similar (CHF_3: $p_c = 4.86$ MPa; $\vartheta_c = 26.2$ °C). Figure 4.9 shows particles micronized with an extraction pressure of 15 MPa and an extraction temperature of 80 °C [4.58]. The size of these particles is between 1 and 10 μm but the larger particles seem to be agglomerates of smaller ones.

The PGSS process and related supercritical atomization techniques have been widely used to produce microparticles of polymers [4.24, 4.60], food additives [4.26] and pharmaceuticals [4.25]. No applications have been proposed so far for PGSS and CO_2 assisted atomization techniques to energetic materials, although, in principle, it is possible to use these techniques for this purpose.

SAS is the most promising supercritical-based micronization technique. It has been applied to a wide range of materials, from pharmaceuticals [4.48, 4.61–4.67] to polymers [4.684.72], colouring matters [4.47] and superconductor and catalyst precursors [4.38, 4.39, 4.73, 4.74].

Figure 4.8. SEM picture of precipitated NTO particles [4.57].

Figure 4.9. Particles of CL 20 precipitated by the RESS process [4.58].

The first application of SAS to energetic material processing was proposed by Gallagher *et al.* [4.40]. It was also the first time this kind of process was proposed. The authors used a *liquid batch* process mode to micronize nitroguanidine (NG). This explosive is not soluble in supercritical carbon dioxide or chlorodifluoromethane (CDM), but it is soluble in several liquid solvents, such as *N*-methylpyrrolidone (NMP) and *N,N*-dimethylformamide (DMF). These liquids are, in contrast, very soluble in supercritical CO_2. Therefore, if one uses a solution of NG in NMP and supercritical CO_2, the essential prerequisites for SAS precipitation are satisfied. The NG consisted of very long crystalline needles about 100 μm in length by about 5 μm in cross-section.

They found a strong influence of the rate of pressure increase on the shape and size of NG crystals. Indeed, very rapid pressure increases produced very small and regular NG microparticles of the order of a few microns in size. These particles were obtained by supercritical CO_2 injection but also by supercritical CDM injection in the liquid solution over an NG concentration range from 1 to 10% (w/w). Particles ranging from about 1 μm to hundreds of microns were obtained at very low rates of pressure increase. Large crystals with a wider particle size distribution were produced at intermediate rates. Different distributions were also obtained using a stepwise pressurization of the liquid solution. The results obtained at different rates and modes of pressure increase are summarized qualitatively in Fig. 4.10 [4.40]. Occasionally, some unusually shaped particle structures were obtained such as snowballs (agglomerates of primary particles) and starbursts (more coherent aggregates of primary particles).

Gallagher *et al.* [4.41, 4.75] also processed cyclonite (RDX) and homocyclonite (HMX). These energetic materials are produced together in an industrial process with an average content of 10 wt% of HMX in the mixture. They used acetone, γ-butyrolactone and cyclohexanone as liquid solvents and CO_2 as the supercritical antisolvent and the liquid batch mode of operation. RDX is difficult to recrystallize by conventional procedures in a form that is free of intragranular cavities. The liquid solvent inclusions in RDX crystals after evaporation generate these voids that have a

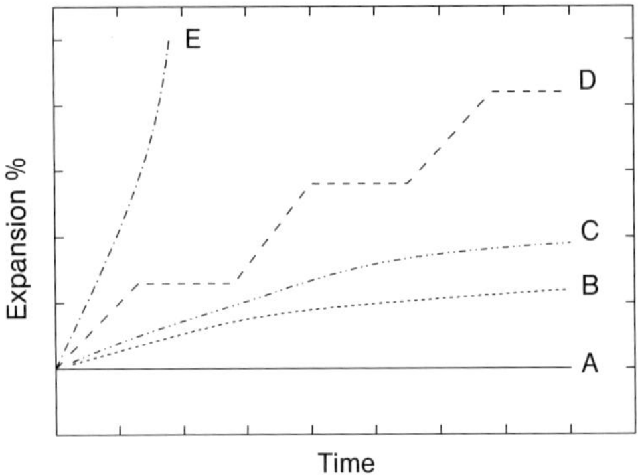

Figure 4.10. Expansion paths and their effect on particle size and morphology during batch SAS [4.40]. A, Large crystals, monodisperse; B, variable size crystals, continuous distribution; C, variable size, continuous distribution; D, variable size, discrete size distribution; E, small crystals, monodisperse.

detrimental effect on the explosive performance. The presence of HMX during the precipitation can also contribute to void formation. Therefore, in this case they had two objectives: to induce the fractional precipitation of the RDX-HMX mixture thus producing an HMX-free precipitate and to produce RDX precipitates without crystal voids. Acetone was found to be satisfactory to produce HMX-free crystals of RDX since the solubility of HMX in acetone is much lower than that of RDX: therefore, the selection of the two compounds was partly performed by the liquid solvent. Depending on various parameters (pressure, temperature, RDX concentration in the liquid solution, rate of gas introduction), a variety of shapes and sizes of crystals were obtained. Particle dimensions and regularity were also affected by the different organic solvents used. For example, the precipitation from cyclohexanone produced smaller and more regular particles than precipitation from acetone under the same operating conditions. Crystals smaller than 5 μm but also very large and void-free crystals were obtained. A smaller number of intracrystalline cavities was obtained by precipitating RDX from cyclohexanone.

HMX precipitation was particularly studied by Cai *et al.* [4.76], from acetone and γ-butyrolactone using the *liquid batch* procedure. Particles with diameters ranging from 2 to 5 μm and narrower particle size distributions were obtained when operating at 33 °C and a maximum pressure of 12 MPa. The particle size of HMX increased with increase in its concentration in the liquid solution. The recovery of HMX was > 90% in all tests performed.

Teipel *et al.* [4.77–4.79] also studied HMX precipitation, pointing out the various crystalline forms of this explosive. HMX crystallizes in four modifications: α, β, γ and δ. The most desired form is β since it has the lowest sensitivity and the highest density. This last property assures the fastest detonation rate. These authors performed liquid batch SAS precipitation from acetone and γ-butyrolactone. The experiments were performed at pressures up to 8 MPa reached in 60 s and at a temperature of 40 °C. Both the liquid solvents produced precipitation of HMX in the desired β form. The solid concentration in the solution influenced the morphology of the HMX particles since when it increases the pressure needed to obtain the onset of supersaturation decreases. Raw HMX had an $x_{50,3}$ of about 200 μm, whereas, the explosive precipitated from acetone and γ-butyrolactone had $x_{50,3}$ of 65 and 90 μm, respectively. The difference in particle sizes between the different precipitation solvents can be attributed to the different supersaturations obtained when the experiments were performed under the same operating conditions; indeed, they have different solvent capacities with respect to HMX.

Another explosive, 3-nitro-1,2,4-triazol-5-one (NTO), was processed by Lim *et al.* [4.80]. They tested in liquid batch mode its precipitation by supercritical CO_2 from DMF, dimethyl sulfoxide (DMSO) and methanol. Various morphologies were observed, but no detailed information was given by the authors.

SAS results on explosives are summarized in Table 4.3.

All SAS experiments described so far were performed using the liquid batch operation mode. Since the semi-continuous mode can produce smaller particles and continuos operation allows higher productivities, Reverchon's group decided to use this process configuration to micronize NTO. This group has previously successfully processed by continuous SAS several compounds belonging to different families: superconductor [4.38, 4.39] and catalyst precursors [4.48], pharmaceuticals [4.29] and biopolymers [4.81]. Therefore, the previously developed know-how was used as the starting point for these new experiments. Semi-continuous SAS experiments were performed using two liquid solvents, DMSO and methanol (MeOH) in which NTO is soluble. During the SAS experiments performed using DMSO (e.g., at 15 MPa, 40 °C, 20 mg/ml NTO concentration in DMSO and ratio between supercritical CO_2 and liquid solution $R = 30$), a large part of the explosive was co-extracted (co-solvent effect) and lost at the exit of the precipitator. However, the remaining part of NTO precipitated in the collection chamber in the form of networked nanoparticles that were collected at the bottom of the precipitator. The mean diameter of these particles was < 100 nm. However, we did not perform

Table 4.3. Experimental results on the supercritical antisolvent micronization of explosives and propellants. Supercritical CO_2 is the antisolvent used except where indicated otherwise.

Compound	Solvent	Process	Morphologies and particle size (μm)	Ref.
NG	DMF Cyclohexanone NMP	Liquid batch	Crystals: spheres, snowballs, starbursts 1–100	4.40
RDX HMX	Acetone γ-Butyrolactone	Liquid batch	Crystals > 200	4.41
RDX	Acetone Cyclohexanone	Liquid batch	Crystals < 5	4.75
HMX	Acetone	Liquid batch	Crystals 2–5	4.76
HMX	Acetone γ-Butyrolactone	Liquid batch	Crystals 65–90	4.77–4.79
NTO	DMF DMSO Methanol	Not available	Spheres, cubic and spherical agglomerates 0.5–20	4.80

Figure 4.11. SEM image of micronized NTO obtained by semi-continuous SAS using MeOH at 10 MPa, 40 °C, 20 mg NTO/ml MeOH and $R = 42$.

further experiments using DMSO since the yield of the micronization process was judged to be too low.

Methanol was then used as the liquid solvent. A successful test was, for example, performed under the following conditions in the precipitation chamber: pressure 10 MPa, temperature 40 °C, concentration of the liquid solution 20 and $R = 42$. Nanoparticles were obtained in this case, as illustrated in Fig. 4.11, where an SEM image of NTO nanoparticles obtained in the described experiment is shown.

The particle size distribution related to these particles is reported in Fig. 4.12; the mean diameter was 120 nm with a standard deviation of about 30 nm.

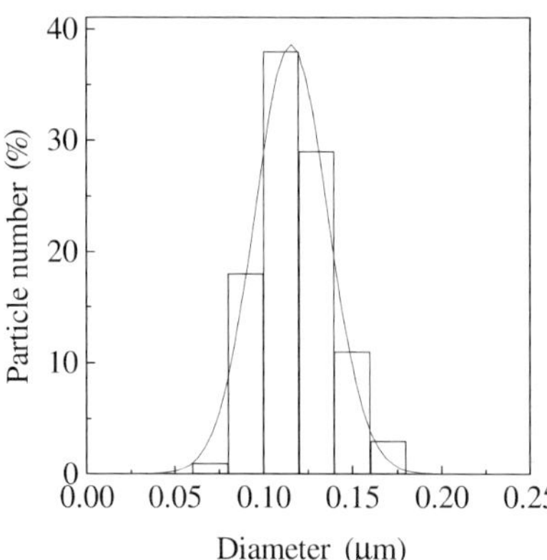

Figure 4.12. Mean particle size and particle size distribution of NTO particles produced by SAS using MeOH at 10 MPa, 40 °C, 20 mg NTO/ml MeOH, $R = 42$. Vertical bars indicate the standard deviation of the particle size distribution.

It is worth noting that during these semi-continuous SAS experiments, amorphous particles of NTO were produced, whereas batch SAS processing produces crystalline materials. This difference can be explained by taking into account that semi-continuous SAS operates in the supercritical fluid-rich phase, whereas batch SAS operates in the liquid-rich phase. Precipitation from the supercritical fluid-rich phase is expected to be faster than that from the liquid-rich phase, and therefore particles obtained using the semi-continuous operation can be amorphous.

Another explosive, hexanitrostilbene (HNS), was micronized at the Fraunhofer ICT in the semi-continuous mode using CO_2 as compressed antisolvent [4.82]. The group at the Fraunhofer ICT has previously successfully processed different materials by the semi-continuous SAS process [4.30, 4.49, 4.57, 4.83]. HNS was dissolved in a mixture of DMF and acetone (acetone acts as an entrainer to achieve full miscibility of DMF in CO_2). The solution (concentration: 1 wt%) was atomized through a two-flow nozzle with a volume flow-rate of the solution of 600 ml/h. The precipitation pressure and temperature in the autoclave were, for example, 10 MPa and 50 °C, respectively. The CO_2 flowed in co-current mode through the autoclave with a flow-rate of 8 kg/h. The HNS particles (see Fig. 4.13) were dry and had a mean particle size of 3.5 μm and a specific surface area of about 7 m^2/g (BET method). Very thin and plate-like particles were obtained.

In the same way RDX and HMX were processed by SAS. Both materials can be successfully micronized [4.57, 4.83].

The precipitation of RDX was carried out from acetone solution in which the solute concentration was 5 wt%. The precipitation pressure was 15 MPa at a precipitation temperature of 323 K. The solution was atomized through a capillary

Figure 4.13. SEM picture of micronized HNS obtained by semi-continuous SAS using DMF-acetone at 10 MPa and 50 °C [4.83].

nozzle with a diameter of 100 μm and a length of 1000 μm. The flow-rate of the solution was 480 ml/h whereas the CO_2 flow-rate was 6 kg/h. The mean particle size of the precipitated RDX was $x_{50,3} = 3.3$ μm. The influence of the nozzle type on the product quality was demonstrated by carrying out the same experiment with a two-flow nozzle. The atomization through the two-flow nozzle results in a particle size distribution which is narrower than that generated by the capillary nozzle. Additionally, the mean particle size decreases from $x_{50,3} = 3.3$ to 1.6 μm.

The precipitation of HMX was carried out from cyclohexanone solution in which the solute concentration was 4 wt%. The precipitation pressure was 20 MPa at a precipitation temperature of 313 K. The solution was sprayed through a two-flow nozzle. The flow-rate of the solution was 600 ml/h whereas the CO_2 flow-rate was 8 kg/h. The mean particle size of the precipitated HMX was $x_{50,3} = 2.3$ μm.

4.5 Conclusions and Perspectives

Crystallization with compressed gases is still in the very early stages of development. Fundamental studies are required on phase behavior and mass and heat transfer; experimental information is also required to obtain a reliable picture of the processes.

The special properties of compressed gases make them suitable for processing sensitive substances owing to the moderate operating conditions. It is possible to produce crystalline particles with a small size and narrow size distribution without spots, i.e. free from solvent inclusions. However, especially the SAS process is very promising in energetic materials processing. First results have shown the potential of this technique. Several groups were able to process nitramines such as HMX and RDX by liquid batch SAS and the semi-continuous SAS process. NTO and HNS were also micronized by antisolvent techniques and it was possible to process TNT by the RESS process. The particles generated are distinguished by particle sizes of a few microns.

Mathematical models validated on reliable sets of experimental data are also required to perform scale-up to an industrial scale.

4.6 References

4.1 Jung J, Perrut M (2001) Particle design using supercritical fluids: literature and patent survey, *J. Supercrit. Fluids* **20**, 179–219.

4.2 Matson DW, Fulton JL, Petersen RC, Smith RD (1987) Rapid expansion of supercritical fluid solutions: solute formation of powders, thin films and fibers, *Ind. Eng. Chem. Res.* **26**, 2298–2306.

4.3 Mohamed RS, Halverson DS, Debenedetti PG, Prud'homme RK (1989) Solids formation after the expansion of supercritical mixtures, *ACS Symp. Ser.* **406**, 355–378.

4.4 Reverchon E, Donsì G, Gorgoglione D (1993) Salicylic acid solubilization in supercritical CO_2 and its micronization by RESS, *J. Supercrit. Fluids* **6**, 241–248.

4.5 Kröber H, Teipel U, Krause H (2000) Manufacture of submicron particles via expansion of supercritical fluids, *Chem. Eng. Technol.* **23**, 763–765.

4.6 Türk M (1999) Formation of small organic particles by RESS: experimental and theoretical investigations, *J. Supercrit. Fluids* **15**, 79–89.

4.7 Teipel U, Förter-Barth U, Gerber P, Krause H (1997) Recrystallization of solid particles with compressed gases, *AIDIC Conference Series*, Vol. 2, ERIS C.T., Milan, pp. 231–237.

4.8 Kröber H, Teipel U, Krause H (2000) The Formation of small organic particles using supercritical fluids. In: *Proc 5th Int. Symp. on Supercritical Fluids, Atlanta, GA*, pp. 63–64.

4.9 Mohamed RS, Debenedetti PG, Prud'homme RK (1989) Effect of process conditions of crystals obtained from supercritical mixtures, *AIChE J.* **35**, 325–328.

4.10 Reverchon E, Donsi G, Gorgoglione D (1993) Salicylic acid solubilization in supercritical CO_2 and its micronization by RESS, *J. Supercrit. Fluids* **6**, 241–248.

4.11 Kim J-H, Paxton TE, Tomasko TL (1996) Microencapsulation of naproxen using rapid expansion of supercritical solutions, *Biotech. Prog.* **12**, 650–661.

4.12 Liu G-T, Nagahama K (1996) Solubility and RESS experiments of solid solution in supercritical carbon dioxide, *Ind. Eng. Chem. Res.* **35**, 4626–4634.

4.13 Liu G-T, Nagahama K (1997) Application of rapid expansion of supercritical solutions in the crystallization separation, *J. Chem. Eng. Jpn.* **30**, 293–301.

4.14 Griscik GJ, Rousseau RW, Teja AS (1995) Crystallization of n-octacosane by the rapid expansion of supercritical solutions, *J. Cryst. Growth* **155**, 112–119.

4.15 Alessi P, Cortesi A, Kikic I, Foster NR, Macnaughton SJ, Colombo I (1996) Particle production of steroid drugs using supercritical fluid processing, *Ind. Eng. Chem. Res.* **35**, 4718–4726.

4.16 Reverchon E, Della Porta G, Taddeo R, Pallado P, Stassi A (1995) Solubility and micronization of griseofulvin in supercritical CHF_3, *Ind. Eng. Chem. Res.* **34**, 4087–4091.

4.17 Berends EM, Bruinsma OSL, van Rosmalen GM (1993) Nucleation and growth of fine crystals from supercritical carbon dioxide, *J. Cryst. Growth* **128**, 50–56.

4.18 Domingo C, Berends EM, van Rosmalen GM (1996) Precipitation of ultrafine benzoic acid by expansion of a supercritical carbon dioxide solution through a porous plate nozzle, *J Cryst. Growth* **166**, 989–995.

4.19 Domingo C, Berends EM, van Rosmalen GM (1997) Precipitation of ultrafine organic crystals from the rapid expansion of supercritical solutions over a capillary and a frit nozzle, *J. Supercrit. Fluids* **10**, 39–55.

4.20 Lele AK, Shine AD (1991) Morphology of polymers precipitated from a supercritical solvent, *J. Aerosol Sci.* **22**, 555–584.

4.21 Debenedetti PG (1990) Homogeneous nucleation in supercritical fluids, *AIChE J.* **36**, 1289–1298.

4.22 Türk M, Helfgen B, Cihlar S, Schaber K (1999) Experimental and theoretical investigations of the formation of small particles from rapid expansion of supercritical solutions (RESS). In: *Proc GVC-Fachausschuß High Pressure Chem Eng, Karlsruhe*, pp. 235–238.

4.23 Helfgen B, Türk M., Schaber K (1998) Micronization by rapid expansion of supercritical solutions: theoretical and experimental investigations. In: *Proc Annual AIChE Meeting, Miami Beach, FL.*

4.24 Weidner E, Steiner R, Knez Z (1996) Powder generation from polyethyleneglycols with compressible fluids. In: von Rohr PR, Trepp C (eds), *High Pressure Chemical Engineering*, Elsevier, Amsterdam, pp. 223–228.

4.25 Weidner E, Knez Z, Novak Z (1994) PGSS (particle from gas saturated solutions) – a new process for powder generation. In: *Proc 3rd Int. Symp. on Supercritical Fluids*, pp. 229–235.

4.26 Weidner E, Petermann M, Blatter K, Simmrock HU (1999) Manufacture of powder coatings by spraying gas saturated melts. In: *Proc 6th Meeting on Supercritical Fluids*, pp. 95–100.

4.27 Sievers RE, Karst U, Schaffer JD, Stoldt CR, Watkins BA (1996) Supercritical CO_2 assisted nebulization for the production and administration of drugs, *J. Aerosol Sci.* **27**, 5497–5498.

4.28 Sievers RE, Karst U (1997) Methods for

fine particle formation, *US Patent* 5 639 441.

4.29 Reverchon E (1999) Supercritical antisolvent precipitation of micro and nano particles, *J. Supercrit. Fluids* **15**, 1–21.

4.30 Kröber H, Teipel U, Krause H (2000) Crystallization of organic substances with a compressed anti-solvent. In: *Proc 14th Int. Congress of Chemical and Process Engineering, Prague.*

4.31 Yeo S-D, Lim G-B, Debenedetti PG, Bernstein H (1993) Formation of microparticulate protein powders using a supercritical fluid antisolvent, *Biotechnol. Bioeng.* **41**, 341–345.

4.32 Randolph TW, Randolph AD, Mebes M, Yeung S (1993) Sub-micrometer-sized biodegradable particles of poly(L-lactic acid) via the gas antisolvent spray precipitation process, *Biotechnol. Prog.* **9**, 429–435.

4.33 Dixon DJ, Johnston KP, Bodmeier RA (1993) Polymeric materials formed by precipitation with a compressed fluid antisolvent *AIChE J.* **39**, 127.

4.34 York P, Hanna M (1994) Salmeterol xinafoate with controlled particle size, *World Patent* 95/01324.

4.35 Jaarmo S, Rantakyla M, Aaltonen O (1997) Particle tailoring with supercritical fluids: production of amorphous pharmaceutical particles. In: Arai K (ed.), *Proc. 4th Int. Symp. on Supercritical Fluids*, pp. 263–267.

4.36 Kordikowski A, Schenk AP, van Nielen RM, Peters CJ (1995) Volume expansions and vapor-liquid equilibria of binary mixtures of a variety of polar solvents and certain near-critical solvents, *J. Supercrit. Fluids* **8**, 205–214.

4.37 de la Fuente Badilla JC, Peters CJ, de Swaan-Arons J (2000) Volume expansion in relation to the gas-antisolvent process, *J. Supercrit. Fluids* **17**, 13–23.

4.38 Reverchon E, Della Porta G, Pace S, Di Trolio A (1998) Supercritical antisolvent precipitation of submicronic particles of superconductor precursors, *Ind. Eng. Chem. Res.* **37**, 952–957.

4.39 Reverchon E, Della Porta G, Celano C, Pace S, Di Trolio A (1998) Supercritical antisolvent precipitation: a new technique for preparing submicronic yttrium powders to improve YBCO superconductors, *J. Mater. Res.* **13**, 284.

4.40 Gallagher PM, Coffey MP, Krukonis VJ, Klasutis N (1989) Gas anti-solvent recrystallization: new process to recrystallize compounds insoluble in supercritical fluids, *ACS Symp. Ser.* **406**, 334–356.

4.41 Gallagher PM, Krukonis VJ, Botsaris GD (1991) Gas antisolvent recrystallization: application to particle design, *AIChE Symp. Ser.* **284**, 96–112.

4.42 Gallagher-Wetmore PM, Coffey MP, Krukonis VJ (1994) Application of supercritical fluids in recrystallization: nucleation and gas antisolvent techniques, *Respir. Drug Deliv.* **4**, 287.

4.43 Tai CY, Cheng C-S (1998) Supersaturation and crystal growth in gas anti-solvent crystallization, *J. Cryst. Growth* **183**, 622–628.

4.44 Czerwonatis N, Eggers R (2001) Disintegration of liquid jets and drop drag coefficients in pressurized nitrogen and carbon dioxide, *Chem. Eng. Technol.* **24**, 619–624.

4.45 Kröber H, Teipel U (2001) Experimentelle Untersuchungen von Hochdruck-Sprühverfahren. In: Eggers R, Peric M (eds), *Proc. Spray 2001, Hamburg*, P4.

4.46 Bodmeier R, Wang H, Dixon DJ Mawson S, Johnston KP (1995) Polymeric microspheres prepared by spraying into compressed carbon dioxide, *J. Pharm. Res.* **13**, 1211.

4.47 Gao Y, Mulenda TK, Shi Y-F, Yuan W-K (1998) Fine particle preparation of Red Lake C pigment by supercritical fluid, *J. Supercrit. Fluids* **13**, 369.

4.48 Reverchon E, Della Porta G (1999) Production of antibiotic micro- and nanoparticles by supercritical antisolvent precipitation, *Powder Technol.* **106**, 23–29.

4.49 Kröber H, Teipel U (2002) Materials processing with supercritical antisolvent precipitation: process parameters and morphology of tartaric acid, *J. Supercrit. Fluids* **22**, 229–235.

4.50 Kikic I, Bertucco A, Lora M (1997) Thermodynamic description of systems involved in supercritical anti-solvent processes. In: Arai K (ed.), *Proc. 4th Int. Symp. on Supercritical Fluids*, p. 39.

4.51 Kikic I, Bertucco A, Lora M (1997) Thermodynamic and Mass transfer for the simulation of recrystallization processes with a supercritical antisolvent. In: Reverchon E (ed.), *Proc 4th Italian Conference on Supercritical Fluids and Their Applications*, p. 299.

4.52 Kikic I, Lora M, Bertucco A (1997) Thermodynamic analysis of three-phase equilibria in binary and ternary systems for applications in rapid expansion of a supercritical solution (RESS), particles from gas-saturated solutions (PGSS) and supercritical antisolvent (SAS), *Ind. Eng. Chem. Res.* **36**, 5507.

4.53 Rantakyla M, Jantti M, Jaarmo S, Aaltonen O (1998) Modeling droplet-gas interaction and particle formation in gas-antisolvent system. In: Perrut M, Subra P (eds), *Proc. 5th Meeting on Supercritical Fluids*, p. 333.

4.54 Werling JO, Debenedetti PG (1999) Numerical modeling of mass transfer in the supercritical antisolvent process, *J. Supercrit. Fluids* **16**, 167–181.

4.55 Matson DW, Petersen RC, Smith RD (1987) Production of powders and films from supercritical solutions, *J. Mater. Sci.* **22**, 1919–1928.

4.56 Teipel U, Förter-Barth U, Gerber P, Krause H (1997) Formation of particles of explosives with supercritical fluids, *Propell. Explos. Pyrotech.* **22**, 165–169.

4.57 Teipel U, Kröber H, Krause H (2001) Formation of energetic materials using supercritical fluids, *Propell. Explos. Pyrotech.* **26**, 168–173.

4.58 Marioth E, Löbbecke S, Krause H (2000) Screening units for particle formation of explosives using supercritical fluids. In: *Proc 31st Int. Annual Conference of ICT, Karlsruhe*, p. 119.

4.59 Teipel U, Gerber P, Krause H (1998) Characterization of the phase equilibrium of the system trinitrotoluene/carbon dioxide, *Propell. Explos. Pyrotech.* **23**, 82–85.

4.60 Weidner E (1999) Powder generation by high pressure spray processes, *Proc. High Press. Chem. Eng.* 217–222.

4.61 Bleich J, Muller BW, Wassmus W (1993) Aerosol solvent extraction system – a new microparticle production technique, *Int. J. Pharm.* **97**, 111–116.

4.62 Bleich J, Kleinebudde P, Muller BW (1994) Influence of gas density and pressure on microparticles produced with the ASES process, *Int. J. Pharm.* **106**, 77.

4.63 Thies J, Muller BW (1996) Production of large sized microparticles with supercritical gases, *J. Pharm. Res.* **13**, 161.

4.64 Tom JW, Lim G-B, Debenedetti PG, Prud'homme RK (1993) Applications of supercritical fluids in the controlled release of drugs, *ACS Symp. Ser.* **514**, 238–252.

4.65 Winters MA, Knutson BL, Debenedetti PG, Sparks HG, Przybycien TM (1996) Precipitation of proteins in supercritical carbon dioxide, *J. Pharm. Sci.* **85**, 586.

4.66 Chou Y-H, Tomasko DL (1997) Gas crystallization of polymer-pharmaceutical composite particles. In: Arai K (ed.), *Proc. 4th Int. Symp. on Supercritical Fluids*, p. 55.

4.67 Reverchon E, Della Porta G, Pallado P (2001) Supercritical antisolvent precipitation of salbutamol microparticles, *Powder Technol.* **114**, 17–22.

4.68 Yeo S-D, Debenedetti PG, Radosz M, Schmidt H-W (1993) Supercritical antisolvent process for substituted para-linked aromatic polyamides: phase equilibrium and morphology study, *Macromolecules* **26**, 6207.

4.69 Yeo S-D, Debenedetti PG, Radosz M, Giesa R, Schmidt H-W (1995) Supercritical antisolvent process for a series of substituted para-linked aromatic polyamides, *Macromolecules* **28**, 1316.

4.70 Dixon DJ, Luna-Bercenas G, Johnston KP (1994) Microcellular microspheres and microballoons by precipitation with a vapour-liquid compressed fluid antisolvent, *Polymer* **35**, 3998.

4.71 Luna-Barcenas G, Kanakia SK, Sanchez IC, Johnston KP (1995) Semicrystalline microfibrils and hollow fibres by precipitation with a compressed fluid antisolvent, *Polymer* **36**, 3173.

4.72 Benedetti L, Bertucco A, Pallado P (1997) Production of micronic particles of biocompatible polymer using supercritical carbon dioxide, *Biotechnol. Bioeng.* **53**, 232.

4.73 Reverchon E, Della Porta G, Sannino D, Lisi L, Ciambelli P (1998) Supercritical antisolvent precipitation: a novel technique to produce catalyst precursors. preparation and characterization of samarium oxide nanoparticles. In: Delmon B (ed.). *Proc 7th Int. Symp. on Scientific Bases for the Preparation of Heterogeneous Catalysts*, Elsevier, Amsterdam, p. 349.

4.74 Reverchon E, Della Porta G, Sannino D, Ciambelli P (1999) Supercritical antisolvent precipitation of nanoparticles of a zinc oxide precursor, *Powder Technol.* **102**, 129–136.

4.75 Gallagher PM, Coffey MP, Krukonis VJ, Hillstrom WW (1992) Gas anti-solvent recrystallization of RDX: formation of ultra-fine particles of a difficult-to-comminute explosive, *J. Supercrit. Fluids* **5**, 130–139.

4.76 Cai J-G, Liao X-C, Zhou Z-Y (1997) Microparticle formation and crystallization rate of HMX using supercritical carbon dioxide antisolvent recrystallization. In: Arai K (ed.), *Proc 4th Int. Symp. Supercritical Fluids*, pp. 23–27.

4.77 Teipel U, Förter-Barth U, Kröber H, Krause H (1998) Formation of particles with compressed gases as anti-solvent. In: *Proc 3rd World Congress on Particle Technology, Brighton*, p. 189.

4.78 Förter-Barth U, Teipel U, Krause H (1999) Formation of particles by applying the gas antisolvent process. In: Poliakoff M, George MW, Howdle SM (eds), *Proc 6th Meeting on Supercritical Fluids, Nottingham*, pp. 175–180.

4.79 Teipel U, Förter-Barth U, Krause H (1999) Crystallization of HMX particles by using the gas anti-solvent process, *Propell. Explos. Pyrotech.* **24**, 195–198.

4.80 Lim G-B, Lee S-Y, Koo K-K, Park B-S, Kim H-S (1998) Gas antisolvent recrystallization of molecular explosives under subcritical to supercritical conditions. In: Perrut M, Subra P (eds), *Proc 5th Meeting on Supercritical Fluids*, pp. 271–275.

4.81 Reverchon E, De Rosa I, Della Porta G, Subra P, Letourneur D (1999) Biopolymer processing by supercritical antisolvent precipitation: the influence of some process parameters. In: Bertucco A (ed.), *Proc 5th Conference on Supercritical Fluids and Their Applications*, pp. 579–584.

4.82 Kröber H, Reinhard W, Teipel U (2001) Supercritical fluid technology: a new process on formation of energetic materials. In: *Proc 32nd Int Annual Conference of ICT, Karlsruhe*, p. 48.

4.83 Kröber H, Teipel U, Krause H (2001) Organic materials formed by precipitation with a compressed fluid antisolvent. In: Reverchon E (ed.), *Proc 6th Conf. on Supercritical Fluids and Their Applications, Maiori, Italy*, pp. 307–312.

5
Size Enlargement

E. Schmidt, R. Nastke, T. Heintz, M. Niehaus, U. Teipel

5.1
Agglomeration

5.1.1
Introduction

Size enlargement by agglomeration is a unit operation of mechanical process engineering. It is characterized by the combination of small particles brought close together to build up larger particles called agglomerates. Short-range physical forces hold the primary particles together and give the agglomerates specific product characteristics. The desired agglomerate properties together with the primary particle attributes determine the equipment and process design. Advantages of agglomerated products can be: defined size, shape, volume and mass; improved storage, handling, metering and dosing; and controlled solubility, reactivity and heat conductivity.

One of the most comprehensive textbooks on agglomeration is that by Pietsch [5.1]. The fundamentals and the great variety of techniques and applications are covered in considerable detail. More compact articles and references to related fields of mechanical process technology were given by Rumpf [5.2] and Sommer [5.3].

5.1.2
Binding Mechanisms – Interparticle Forces

The attractive forces between particles cause the size enlargement process and are essential to the strength and dispersibility of the agglomerates. Figure 5.1 gives a summary of binding mechanisms relevant to agglomeration processes.

These mechanisms were first defined and classified by Rumpf [5.2]. The presence or absence of material bridges can be considered as an overriding criterion. Material bridges can be further categorized as solid or liquid. Solid bridges include, for example, those formed by sintering, chemical reaction at contact points, hardening

Energetic Materials. Edited by Ulrich Teipel
Copyright © 2005 WILEY-VCH Verlag GmbH & Co. KGaA, Weinheim
ISBN: 3-527-30240-9

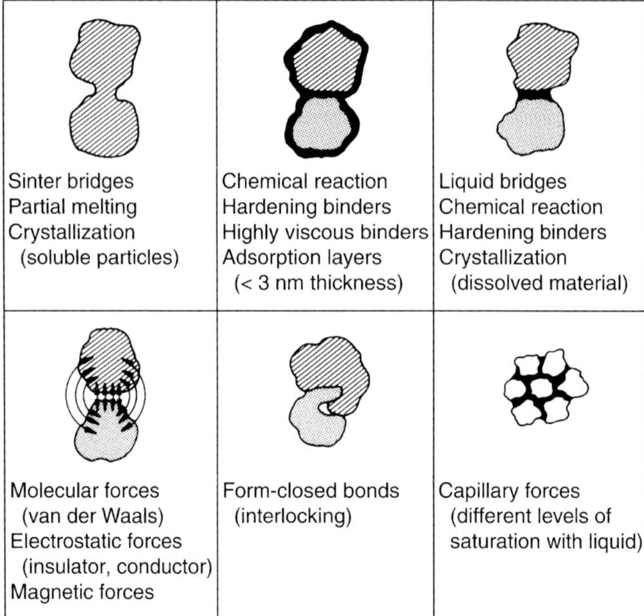

Figure 5.1. Binding mechanisms in agglomeration processes [5.1].

of binders and the crystallization of dissolved materials. Liquid bridges between the particles are formed by low-viscosity fluids owing to capillary effects or by coating with high-viscosity binders. In the case where there are no material bridges, van der Waals interaction is responsible for the principal binding force.

Attractive forces can be calculated for model systems, such as smooth, fixed, ideal spheres. Even though the values calculated for these models apply only crudely for real systems, they do indicate the effects of important parameters on the agglomeration process. To calculate the actual attractive forces between real particles is currently not possible because of their irregular shape and usually rough surfaces [5.3].

5.1.3
Growth Mechanisms and Growth Kinetics

If two particles collide and the resulting adhesive force is stronger than the effect of all other forces, they will stick together. This process can continue, causing size enlargement by agglomeration growth. However, as this operation proceeds, somewhat more complicated mechanisms evolve [5.1]. Figure 5.2 represents most of the different processes in a schematic way. It distinguishes between agglomeration and desagglomeration or size enlargement and size reduction phenomena.

Nucleation, the production of micro-agglomerates, occurs when primary particles collide and adhere. As long as more primary particles are available, they tend either

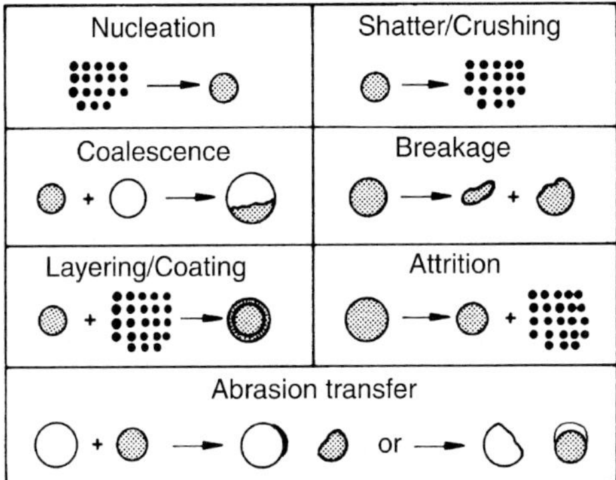

Figure 5.2. Size change mechanisms in agglomeration processes [5.1].

to adhere to each other forming more nuclei or to attach themselves to larger agglomerates. As the mass of agglomerates increases, they may break apart at structurally weaker areas, for example due to forces of impact. Abrasion will also take place, resulting in newly liberated primary particles or micro-agglomerates, which then try to attach themselves to entities offering better binding properties.

Depending on the density of the tumbling mass and the type of equipment causing agitation of the particle bed, the growth phenomena and their kinetics will differ and interfere. Nevertheless, the models given in Fig. 5.2 are the basis for population balances by which the transient agglomeration of particles, in batch and continuous processes, can be simulated.

5.1.4 Equipment and Processes

5.1.4.1 Tumble Agglomeration

Tumble agglomeration [5.1] is generally characterized by the growth of more or less spherical agglomerates after the addition of liquid binders. The wet or so-called green agglomerates are formed during suitable movement of the particulate matter containing the binder. In addition to the size of the primary particles, interfacial forces and the capillary pressure at freely movable liquid surfaces mainly cause the strength of the green agglomerates. The most commonly used apparatus for large-scale tumble agglomeration is the balling pan and the balling drum. Both employ the same size enlargement mechanisms: while the equipment rotates around its axis and a liquid binder is sprayed on to the moving bed, spherical agglomerates are built up during tumbling, rolling and sliding of the material to be agglomerated.

Figure 5.3 shows a schematic diagram of a balling pan. The movement of the material results in a distinct separation of agglomerate sizes in the bed, with the largest ones always travelling on top of the charge and near the edge of the rim. Therefore, if the process is well controlled, agglomerates of fairly uniform size can be produced. Most importantly, the recirculation of undersized material and/or the

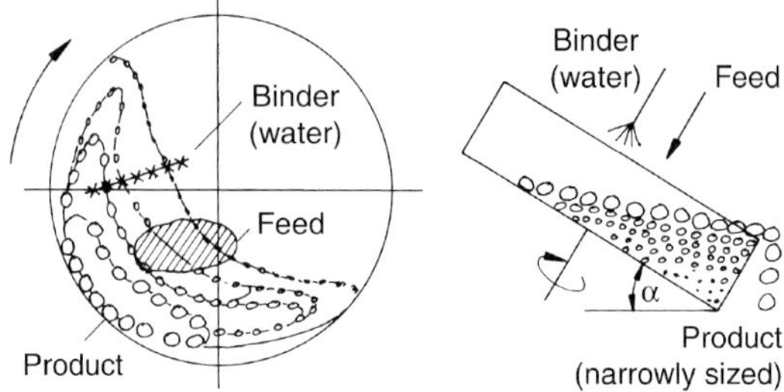

Figure 5.3. Tumble agglomeration in a balling pan (inclined rotating disc) [5.1].

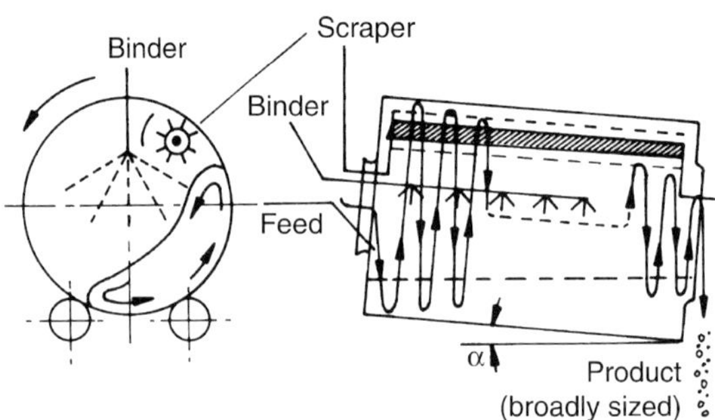

Figure 5.4. Tumble agglomeration in a balling drum [5.1].

crushing of oversized agglomerates are normally not necessary. The balling drum (see Fig. 5.4), however, always requires screening of the discharge and the recirculation of a sizeable amount to the beginning of the process.

Tumble, build-up or growth agglomeration can also be effected in a balling cone, in continuous or batch mixers with or without mixing tools, in pulsating or vibrating conveyors or troughs, in a turbulent, gas fluidized bed or in turbulently moving liquid suspensions.

5.1.4.2 **Pressure Agglomeration**

Pressure agglomeration [5.1] is sometimes characterized by very high forces acting on a mass of particulate matter within a defined volume. If fine powders of so-called plastic materials, which deform irreversible under high pressure, are pressed, no binders are required. Van der Waals forces, valence forces, partial melting and solidification or interlocking are responsible for the strength of such compacts (compare Fig. 5.1). Materials with a low melting temperature fuse at the grain boundaries to form homogeneous structures; similar results can be obtained with almost all materials during hot pressing. The prevailing high forces may activate natural components of the materials to become binders. Only a few so-called difficult materials require the addition of dry or liquid binders and/or lubricants. Pressure agglomeration is accomplished in piston, roller, isostatic and extrusion presses as well as in pelleting machines.

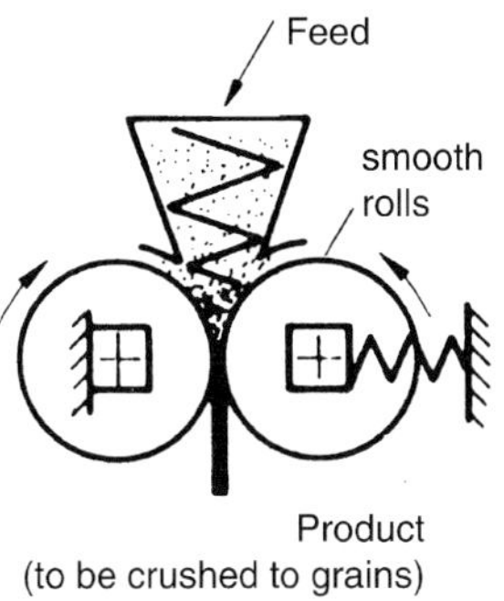

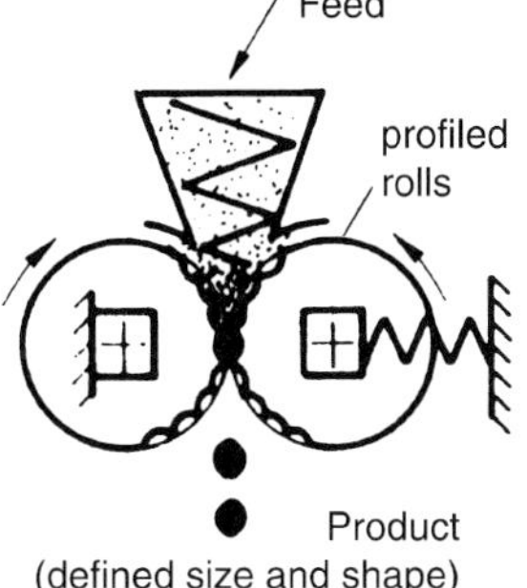

Figure 5.5. High-pressure agglomeration by a roller compacting machine (left) or a roller briquetting machine (right) [5.1].

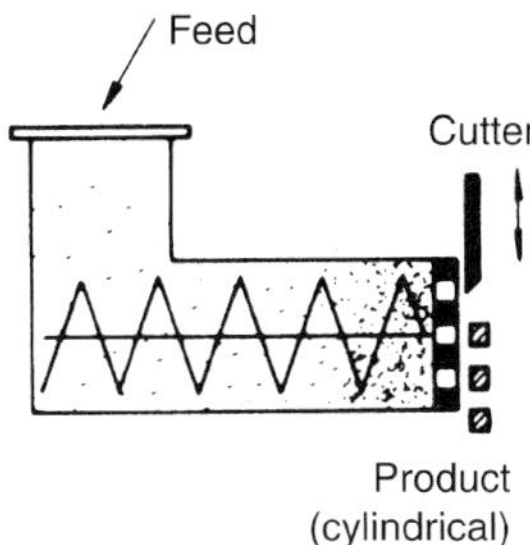

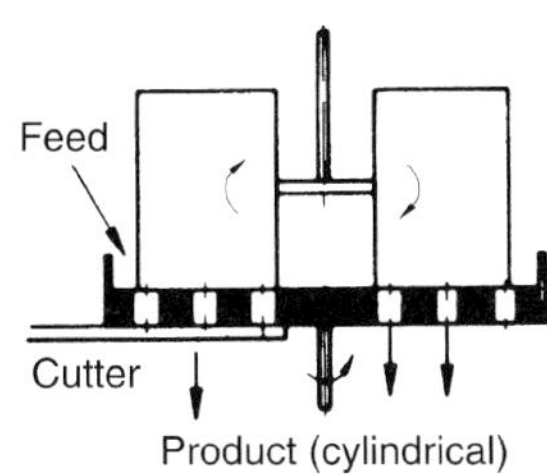

Figure 5.6. Low-pressure agglomeration by a screw extruder (left) and a machine with flat die and Muller-type press rollers (right) [5.1].

Figure 5.5 shows schematics of two different types of roller presses, which are usually employed for large capacities. The operating principle of such roller-type briquetting and compacting presses is based on two rollers of equal size with smooth or profiled surfaces, which rotate counter-currently at the same speed. The material to be compacted is normally fed from top into the nip area between the rollers and is densified and formed while it passes the nip. The rollers of briquetting machines are equipped with cups or pockets, which define the shape of the briquetted product. Compacting machines use smooth, corrugated or waffled rollers for the production of smooth or structured strips, sheets or flakes.

Extrusion presses and pelleting machines (see Fig. 5.6) make use of wall friction, causing resistance to the flow of plastic materials through bores or open-ended dies. Particularly in screw extruders extensive mixing takes place and high shear forces are applied. To obtain sufficient plasticity and product strength, binders are required in most cases. Only materials with low melting or plastification temperatures can be directly agglomerated. The agglomerates produced (which usually have to be dried) are of cylindrical shape with given diameter and variable length.

5.1.4.3 Other Methods for Agglomeration

A very old size enlargement process that is of particular importance in the iron industry is sintering. In this technology, a particle bed is heated until, by atomic and molecular diffusion, sinter bridges develop at the points of contact between the particles. After cooling, the sinter is crushed into the required particle size.

Spray drying is an agglomeration technique during which hollow spheres form in the drying tower from dissolved or suspended material. Closely related to spray drying is a process called prilling. Droplets forming from a melt at the top of the prilling tower solidify to almost spherical particles during free fall in a counter-current air stream.

5.2 Microencapsulating and Coating Processes

5.2.1 Basics of Technologies

Microcapsules consist of a core and a more or less solid shell, i.e. as a rule droplets of a liquid or particles of a solid substance will be coated with a continuous film of a polymeric material. The core may consist of substances of very different consistency, whereby when using liquid, waxed or gaseous substances single spherical patterns emerge which may form agglomerates. The microencapsulation of solid substances maintains their outer form. However, they may also exist dispersed in liquids or waxes. In such a case the liquid or waxed phase forms a matrix for the solid substance and one also obtains nearly spherical particles. For coating solid substances, in most cases the term 'coating' is used. In principle, from the point of view of application and owing to their properties, these shells have to be assigned to the microcapsules with the restriction that as a rule these shells are never closed completely. Depending on their application, microcapsules may exist as dispersed systems or free-running powders; in most cases these products are specially formulated according to their application with other auxiliary materials [5.4–5.20].

The aim of a microencapsulation of a substance may serve different purposes, including:

- controlled release behavior of a dope (sustained, controlled release)
- coating of fluid substances for the generation of free-running powders (for instance, solutions of leuco dyestuffs for copy papers)
- protection of core material against outside affects (hydrolytic decomposition, decomposition by UV radiation, ozonolysis, humidity and many others)
- reduction of solvent consumption
- reduction of volatility of easily volatilized substances (reduction of smell, release)
- reduction of the necessary quantity
- reduction of leaching in agriculture, forestry, horticulture, hygiene and in many other fields
- reduction of acute toxicity for humans and the environment, including phytotoxicity

- improvement of compatibility with other substances, (e.g. polymer compounding, fire-resistant finish, mixture of incompatible components, hydrophilization, hydrophobing and many others)
- possibilities of mixing reaction agents which later on the spot may react by pressure, heat, etc. (conversion of multi-component systems into single-component systems).

Originally, microencapsulation was developed by Green and Schleicher [5.21] of NCR Corporation in 1953 for manufacturing colorless copy paper, the so-called reaction copy paper. Since that time, the field of microencapsulation of very different substances has expanded enormously. Microcapsules are used in nearly all industrial and commercial fields.

Amongst many fields, the following can be mentioned:
- agriculture and forestry (microencapsulated pesticides, repellents, pheromones, growth regulators, attractants, depot fertilizers, metal oxides and many others
- hygiene (repellents, pesticides, odor compounds and others)
- human and veterinarian medicine (depot drugs, repellents, drugs for chemical sterilization, etc.)
- cosmetics (lipids, odor compounds, perfume oils, skin protection agents, repellents, dyestuff and pigments)
- environmental protection and environmental engineering (depot substances for nourishment of plants, repellents, oxidizing components, hollow capsules for soil processing, etc.)
- biotechnology (microencapsulated microorganisms, nutrients, regulators, biocatalysts)
- paint and lacquer industry (flakes, pigments, antifouling agents, latent hardeners)
- construction industry and building protection (pesticides used in particular for pest control in the field of timber preservation, heat accumulators, etc.)
- plastics industry (adhesives, latent hardeners, accelerators, catalysts, fire-resistant finish, compounding auxiliary materials)
- textile industry (fire-resistant finish, fixing agents, textile auxiliary material, heat and cold protection)
- printing and paper industry (inks and pigments, luminous paints, odor compounds, leuco dyestuff, agents for surface modification)
- food industry (flavors, coloring agents, yeast and other blowing agents, repellents, adhesives and many others)
- electroplating (lubricants, anti-corrosive agents, additives for galvanic baths, coloring and odor agents).

5.2.2
Introduction

Microencapsulation by using polymeric materials as the shell has been comprehensively described [5.22–5.41]. The task of microencapsulation is the inclusion of small particles of a substance (solid, fluid or gaseous) in a thin shell made by a film-forming polymeric compound. The shell isolates the particles of one or more substances from each other and from the outer environment until the contents of the capsule are released in some way. The diameter of the microcapsules may extend from fractions of micrometers up to millimeters. Usually the content of the capsules is 50–90% of the total mass, but in special cases it may be as high as 98%. These amounts are influenced by the conditions for the preparation of microcapsules. Factors that mainly exercise an influence are the preparation process, temperature, degree of dispersity, viscosity of the medium, existence of surface-active agents, protective colloids and other auxiliary agents. Using the same procedure and under the same reaction conditions, the size of the capsule within a batch, if it is not fixed by a solid substance and the content of core material of the capsule, show a Gaussian distribution.

Microcapsules with different structures are widely known. This is the reason why in addition to the usual one-shell capsules, types with a wall consisting of two or more layers can be produced and also capsules which have a second capsule inside, the two capsules containing different substances. There are also microencapsulated dispersions or emulsions or microcapsules that are in a liquid medium which are surrounded by a larger microcapsule or even dopes that are adsorbed on a matrix which then are microencapsulated as a system.

The contents of the capsule can be released by mechanical destruction of the wall (pressure, friction, increase in temperature, ultrasonics), by melting or dissolution but also by diffusion. According to the field of application, it may be necessary to bring about a release controlled by the material and the thickness of the wall or a sudden release. The wall material required is also dependent on the particular application. The shell can be either impermeable in both directions or semipermeable (from outside to inside or mostly vice versa). As a rule, the losses of capsule contents during storage are minimal. Thus it has been reported that, for instance, microencapsulated solvents (petroleum, chlorobenzene, xylene) with a capsule shell made of amino resin which had been stored for 2 years as a dispersion at room temperature had a loss of weight of core material of only 0.1–0.8%. The microcapsules as dispersion can exist in a liquid or dry in the form of a free-running powder. Also in this case the most suitable form depends on the intended application.

5.2.2.1 Preparation of the Microcapsules

Basic Material

According to the principle of the structure of microcapsules consisting of wall material and capsule contents (core), the basic material may be divided into two groups: the material for the core and the material for the wall. The choice of the former material is exclusively determined by the intended application and as far as the wall material is concerned the conditions under which the core will be released are decisive.

Without going into too much detail, in the following the most important compound classes used as core materials as taken from the specialized literature will be considered. We speak both about inorganic and about organic, monomeric and polymeric compounds of very different classes which shall be used as catalysts, stabilizers, accelerators, plastifiers, etc. Oils, liquid and solid fuels, solvents, coloring agents, pigments, insecticides and pesticides, fertilizers, drugs, flavors, odor agents, food additives and even ferments and microorganisms are microencapsulated, just to mention the most important classes. Therefore, the fields of their application also are broad.

Practically any material that has the capability to form a shell can be used for the wall. It can be synthetic or natural high molecular weight compounds, but also fusible or soluble low molecular weight substances. In this connection, organic polymers should be mentioned first (without laying claim to completeness). Of naturally occurring polymers, proteins are used (gelatin, albumin) or polysaccharides (dextran, vegetable gums, cellulose derivatives), and additionally waxes and paraffin. Amongst the synthetic polymers there is hardly a compound class that has not be used for microencapsulation. Polyvinyl compounds, polyolefins, polyacrylamides, polysiloxanes, polymaleinates, polysulfides, polycarbonates, polyurethanes, polyesters, polyamides, polyureas, epoxy, phenolic and amino resins and also many different copolymers can be mentioned here. However, even inorganic substances serve as wall material, e.g. metals, carbon, silicates, aluminates and carbides. According to the literature, in most cases, however, organic polymers were used. In medicine and cosmetics very often lipids serve as structure-forming phases for the inclusion of dopes of many different kinds.

Auxiliary Substances

In addition to wall and core materials, the materials of auxiliary substances necessary for microencapsulation also play a significant role. Such material includes solvents, precipitators, hardeners and cross-linking agents, surface-active agents, colloids and other substances that increase viscosity, and even inert packing materials and sorbents. In many cases it is essential to choose the right solvents for the encapsulation process because mostly monomers or polymers will be used as solutions. The solvent has to meet very different requirements depending on the conditions for the preparation of the capsules and their application. An important factor is the boiling temperature and in this context its volatility because residues of solvents within the wall material with insufficient volatility may influence the

The formation of fibrous materials was also observed by Bodmeier [5.51], who investigated the preparation of theophylline- and progesterone-containing poly-D,L-lactide microcapsules.

Fluidized Bed Processes

Microencapsulation may be also performed by using rotary fluidized bed processes. For this purpose, solid, finely dispersed core material is kept floating by a vertical current of air and is sprayed with the solution of the wall material or its melt. The fine droplets precipitate on to the core material, coalesce and in this way form a wall structure. This process is used advantageously for particle sizes of the core material of > 50 μm, it is very efficient and can be carried out in a single-stage manner and continuously. Some of the disadvantages include the necessity to reclaim the solvent from a huge amount of exhaust air, which is very costly (in the event of using polymer solutions as wall formers), the relatively poor quality of the capsule wall (pore formation) and the relatively large share of wall material in the total product. In the last few years, numerous different methods have been developed.

The choice of spraying methods in fluidized bed processing is based on the consideration of finished-product performance requirements and projected product volumes. Three methods of spraying are available for coating in the fluidized bed – top, bottom or tangential spraying. Tablet coating is restricted essentially to the bottom-spray method and currently small particle coating is carried out by using all types of spraying processes. In the top-spray granulator the granules are characterized by a porous wall and by an interior void space that results in increased wicking of liquid within the particles and improved disintegration or dispersibility. Another characteristic of particles produced by this technique is a bulk density that generally is lower than that produced by other techniques of coating.

The conventional top-spray method shown in Fig. 5.9 has been used for more than a decade for coating. It evolved from the fluidized bed dryers commercialized more than 30 years ago. The core material is placed in the product container (A), which typically is an unbaffled, inverted, truncated cone with a fine retention screen

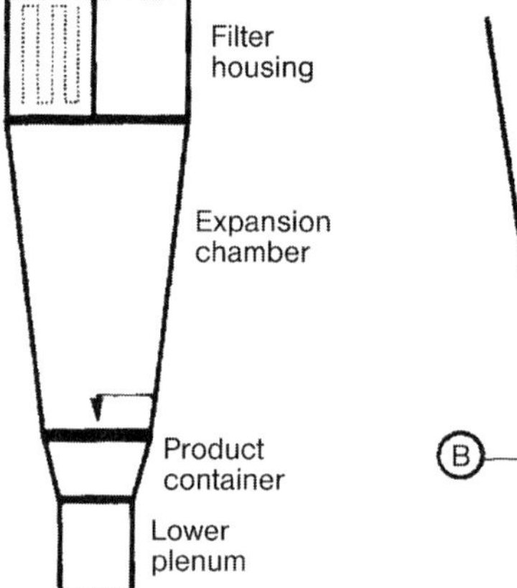

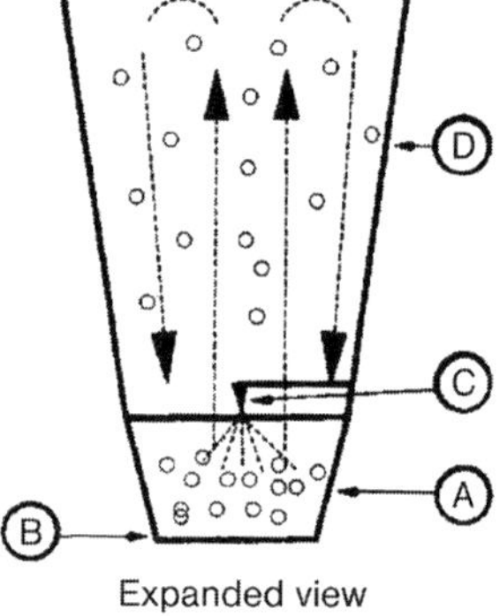

Figure 5.9. Conventional top-spray coater (Fa. Glatt, Binzen, Germany).

and an air or gas distribution plate (B) at its base. Preconditioned air is drawn through the distribution plate (B) and into the product. As the volume of air is increased the bed no longer remains static but becomes fluidized in the air stream. The point at which the bed becomes just fluidized is known as incipient fluidization. The particles are accelerated from the product container behind the nozzle (C), which sprays the coating liquid countercurrently on to randomly fluidized particles. The coated particles travel through this coating zone into the expansion chamber (D).

In 1959, Wurster [5.49] introduced an air suspension technique known as the Wurster system. Originally designed to coat tablets, nowadays the process is widely used for substrates as small as 100 μm. The components of the system are illustrated in Fig. 5.10. The coating chamber (A) is an unbaffled cylinder that contains another cylinder half its diameter known as a partition (B). At the base of the coating chamber is a fine screen and an air distribution plate (C). In the center of the plate a nozzle (D) is positioned to spray upwards. The holes in the plate in the area beneath the partition are larger in diameter than those outside. Air passes through the plate with a high volume and velocity and pneumatically transports particles vertically through the partition and the coating zone. This type of fluidization is characterized as a spouting bed.

A relatively new approach to coating is referred to as tangential spray or rotary fluidized bed coating (see Fig. 5.11), originally conceived for high-density fluidized bed granulation. This technique is used to produce high-dose pellets by placing a layer of particles on some type of seed materials. A controlled release coating can be applied subsequently.

The product container consists of an unbaffled cylindrical chamber (A) with a solid, variable-speed disc (B) at its base. The disc and the chamber are constructed in such a way that during the process a gap (C) exists at the perimeter of the disc through which preconditioned air is blown. During fluidization three forces com-

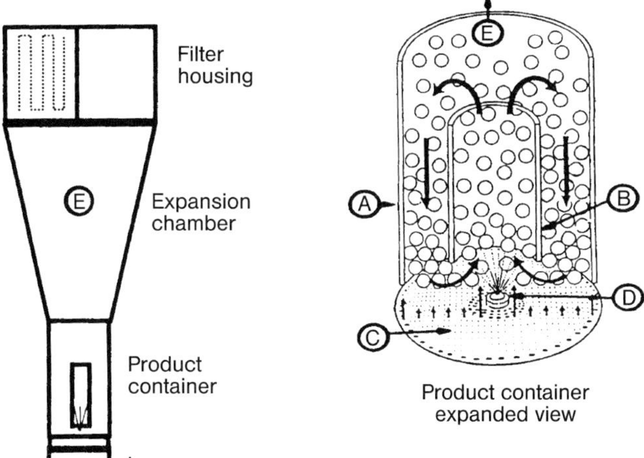

Figure 5.10. Bottom-spray coater of the Wurster technique.

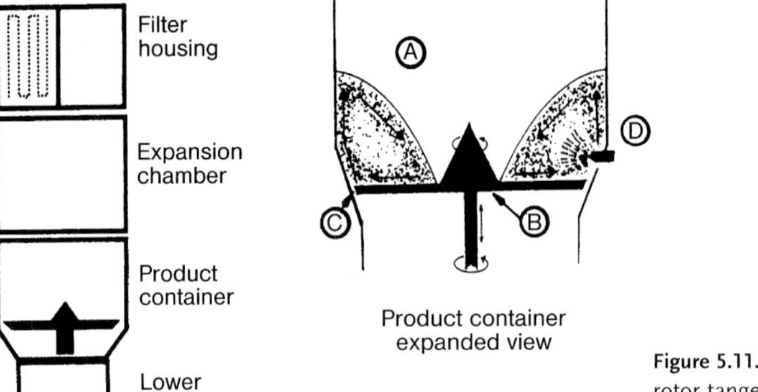

Figure 5.11. Principle of a rotor tangential spray coater (Fa. Glatt, Binzen, Germany).

bine to provide a pattern best described as a spiraling helix. Centrifugal force makes the product move towards the wall of the chamber, air velocity through the gap provides upward acceleration and gravity cascades the product inwards and towards the disc once again. The fluidization pattern may be characterized as smooth or particulate. Beneath the surface of the rapidly tumbling bed a nozzle (D) is positioned to spray the coating liquid tangentially to and concurrently with the flow of particles. The particle cycling time of this technique is very rapid, hence the films are uniform in thickness, as are those applied by using the processes which were discussed previously.

The basis of the fluidized bed process is the liquid system. In nearly all fluidized bed equipment the nozzles are binary. Liquid is supplied at low pressure through an orifice and is atomized by air. These pneumatic nozzles typically produce smaller droplets. This is an advantage when coating fine particles. As the liquid is atomized, the surface area for evaporation increases. When traveling, the droplets change rapidly in solids concentration and their viscosity increases. Some droplets may contact the surface of the substrate and fail to spread uniformly, leaving an imperfect wall. In extreme cases this produces spray drying of the coating polymer. This problem can be a severe one if an organic solvent with a low heat of vaporization is used in the spraying solution, e.g. acetone with a heat of vaporization of 0.103 kcal/ml, methylene chloride (0.118 kcal/ml) or 2-propanol (0.132 kcal/ml). The heat of vaporization of water is 0.541 kcal/ml.

A specific case of fluidized bed coating is the hot-melt fluidized bed coating process. As a rule, wax coatings are generated by spray congealing of agent/molten wax slurries. The *in vitro* release rates for drug formulations spray congealed from slurries were affected by the type of wax (especially the melting behavior), the type of nozzle and the addition of surfactants. As a rule it is applied by using a binary nozzle; liquid is supplied at low pressure and is atomized into droplets by pressurized air. A high atomization air pressure is indispensable to keep the particles small and discrete. At such high pressures the air velocity is extreme and may result in pulverization of fragile substrates. Small particle size is also achieved by the typically low viscosity of the melt and the relatively slow spray rate. The atomization air is heated to the same temperature as the spray liquid and serves to keep the molten coating at the application temperature. Major revisions of the reactor type include a lengthened expansion chamber, a conical shape of the expansion chamber and provision for alternative filter shaking without interrupting fluidization (see Figs 5.9 and 5.10).

Solvent Evaporation Technique

A typical procedure is as follows (see Fig. 5.12) [5.56]:

- manufacturing the aqueous phase by dissolving poly(vinyl alcohol) (PVA) in pure water
- dosage, e.g. of a polyester solution in methylene chloride with continuous stirring at room temperature in a vessel
- stirring at room temperature for 30 min
- stepwise increase of temperature to 303 K and solvent evaporation for 6 h
- separation of solid microparticles by sedimentation, filtration or centrifugation depending on the particle size
- washing of the isolated particles with pure water and drying under vacuum.

The loaded microcapsules were prepared by dispersing or dissolving the active agents in the organic polymer phase by means of high-speed stirrers.

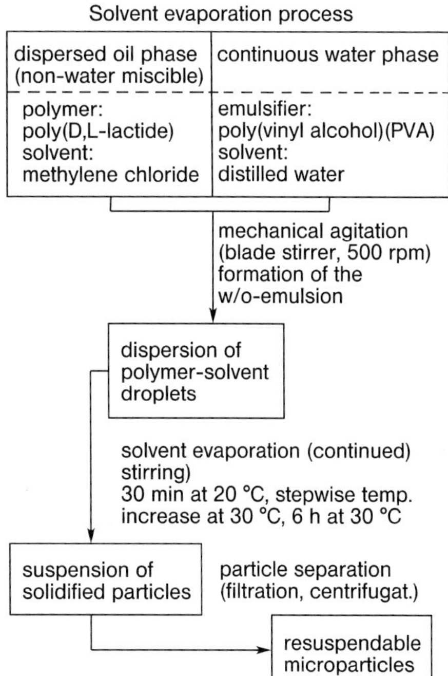

Figure 5.12. Flow scheme of the solvent evaporation technique.

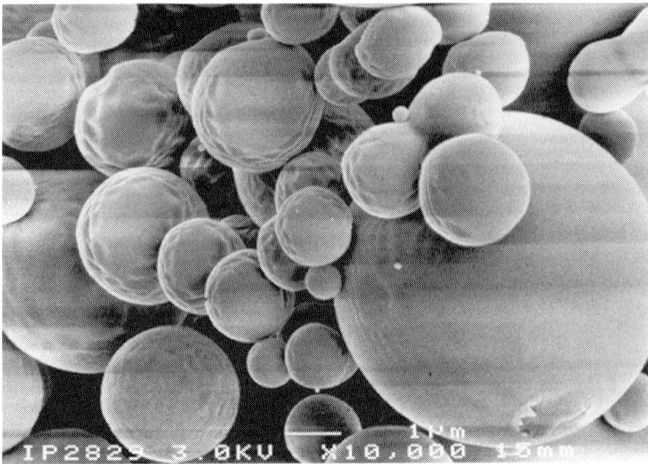

Figure 5.13. Scanning electron micrograph of microcapsules produced by solvent evaporation technique.

A scanning electron micrograph of microcapsules produced by the solvent evaporation technique is shown in Fig. 5.13.

Salting-out Processes

The salting-out method is mainly suitable for the production of nano- and microparticles. The practical use of this method requires, however, careful tailoring of the process parameters depending on the chemistry of the polymers and the active agents. Essential process parameters determining the most important particle parameters such as size, size distribution and morphology and also the release properties of active agents are the chemical structure, the concentration and the molecular weight of the protective hydrocolloid used, the hydrocolloid/polymer and salt/polymer ratios and the stirring conditions during the salting-out process. Water-insoluble or poorly soluble compounds are well encapsulated by this method. The encapsulation efficiency for water-soluble compounds is low.

The basic knowledge of this particle formation technique from preformed polymers developed on the basis of investigations on nanoparticle preparation with different polymers such as polylactides, cellulose derivatives and polyacrylates [5.57–5.59].

A typical procedure for particle preparation following a determined order of priority of the individual steps can be used as follows:

- preparation of the highly concentrated salt solution (clear and free of particles)
- preparation of the clear and highly viscous gel by slowly adding PVA to the concentrated salt solution, and, or successful particle preparation, preferably swelling the gel overnight
- progressive incorporation of the aqueous gel into the polymer solution with vigorous mechanical stirring in order to form an oil-in-water emulsion

- dilution of the resulting emulsion by adding water under conditions of continuous stirring (complete diffusion of the organic solvent in the aqueous phase)
- separation of the raw particle suspension from the homogeneous solution (organic solvent, water, PVA, salt) by centrifugation
- separation of process impurities by discharging the aqueous-organic phase and repeated resuspending of the particle sediment in pure water
- drying of the isolated particles under vacuum conditions or by freeze-drying.

The isolation of particles can be also carried out by cross-flow filtration.

The amounts and concentrations used for a typical batch for minute particle production is as follows [5.44–5.46]:

organic phase	40 ml
gel phase	50 g
polymer concentration of the organic phase	0.25 g/l
gel composition PVA-$MgCl_2 . 6H_2O$-water	10 g/90 g/60 g
water after emulsification	200 ml

Ibrahim [5.57] gave a detailed description of the influence of salt type on the phase separation of the water-acetone system. The effectiveness of acetone recovery from the aqueous solution depends on the energy of hydration of salts and it increases in the case of chlorides in the following order: $AlCl_3 < MgCl_2 < CaCl_2 < NaCl < NH_4Cl$. From the technological point of view, salts of magnesium are particularly suitable for the preparation of loaded polyester microparticles. The particle size distribution of microparticles obtained by a surfactant-free salting-out process shows a pronounced bimodal characteristic. This bimodality of particle size distribution has also frequently been found by means of other methods of microcapsule preparation, such as spray drying, solvent evaporation and coacervation [5.57–5.59]. In most cases the surface of the capsules shows microspheres (see Fig. 5.14).

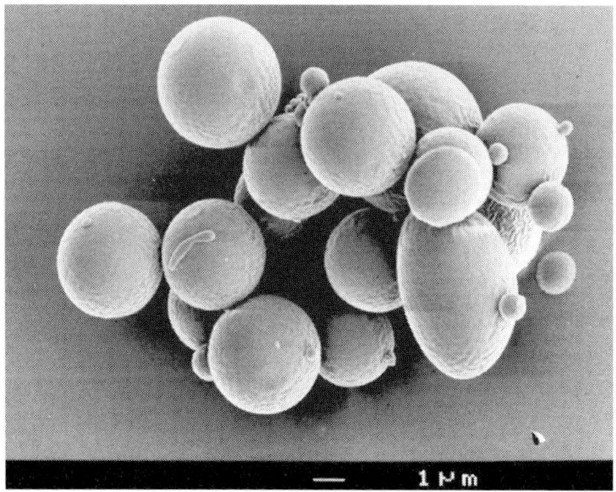

Figure 5.14. Scanning electron micrograph of phenolphthalein-loaded polylactic microparticles.

Coating by Travel Shootings

As far as this process is concerned, the solid, finely dispersed core material [here exclusively solid materials are encapsulated (or, a better expression, coated)] travel in rotating tubes, vessels, bubble caps or vibrating channels and during this process they are sprayed with melted or dissolved wall material or wetted in a suitable way, which in the case of monomeric basic material polymerizes simultaneously. In most cases of application, however, polymers or even waxed materials (e.g. paraffin waxes, higher molecular weight polyethylene glycol) are used [5.64].

5.2.3.2 Physico-chemical Procedures

Physico-chemical procedures include those which, under conditions of forming a new phase (dispersed systems in a liquid medium), are characterized by physical phase transformation, but also other procedures without a chemical transformation of the component forming the wall (e.g. coacervation, precipitation), hardening of dispersed melts and others. The technological side of all these procedures is relatively simple and hardly differs from equipment normally used in chemical technology. The physico-chemical procedures make the microencapsulation of liquid and solid substances possible, including their solutions and dispersions. They also make possible the regulation of core share, wall thickness and permeability with broad limits. As wall formers, numerous film-forming polymer materials may be used.

Most frequently the procedure of new phase formation (in an aqueous or organic medium) is used. It is characterized by high efficiency, process development with care and good reproducibility. Schematically the process may be described as follows (see also Fig. 5.15). The component to be encapsulated is added to a solution of the

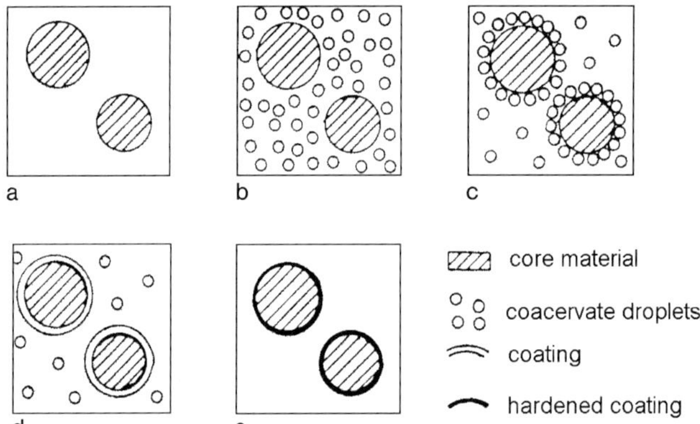

Figure 5.15. Typical steps in a coacervation method of microencapsulation [5.9]. (a) Core particles dispersed in a solution of polymer by agitation; (b) coacervation visible as droplets of colloid-rich phase; induced by one or more agents; (c) deposition of coacervate droplets on the surface of the core particles; (d) merging of coacervate droplets to form the coating; (e) shrinkage, cross-linking and drying of the coating.

wall material and dispersed up to the necessary droplet size by changing the pH or the temperature or by adding components which reduce the solubility of the wall material, which leads to the formation of a new phase enriched with the wall material. Under certain conditions the core material will be coated by this phase, which later will be hardened in one of various different ways. The resulting microcapsules are nearly or even completely spherical.

Microencapsulation by means of coacervation is also part of this technique and dates back to Green and Schleicher [5.21], who suggested encapsulating oil solutions of lyophilic dyestuffs for the manufacture of copy papers. Because the original procedure had a number of disadvantages,since then many investigations have dealt with its improvement. In addition, it was extended to solid substances as core materials.

One of the methods of the coacervation technique is complex coacervation with mutual precipitation of two solutions which are oppositely charged [5.65–5.67]. For instance, a gelatin solution is mixed with a solution of gum arabic and is diluted or the pH of the solution is slowly adjusted to ~4.5. Owing to the acidification, the gelatin, which under normal conditions is amphoteric, will be positively charged and starts to interact with the permanently negatively charged gum arabic. It is more convenient for practical work to change the pH instead of diluting the solution because one can work with smaller volumes.

5.2.3.3 Chemical Procedures

There is a distinction as to whether the user applies low molecular weightwall materials (monomers, oligomers) or already preformed polymers. In the former case the usual techniques of polymerization and polycondensation or cross-linking (use of oligomeric wall formers) may be applied. In the latter case in the limiting phase the polymer will be precipitated, forming a wall by cross-linking or other chemical reactions which result in the insolubility of the polymer in the continuing phase. During the polymerization of a dissolved monomer in one or both liquid phases, the formation of the polymeric wall material is carried out directly on the surface of the material to be encapsulated or at the interface of the phase containing the core material. The polymerization can be carried out in oil-in-water or water-in-oil systems or even in multi-phase systems. After the polymerization has been carried out and the wall structure has been formed, other layers may be built in order to change the properties of the particles. These layers can be either covalently linked to the wall material or they may also be adsorbed. For instance, another monomer can be grafted by means of initiating radiation or the polymerization is followed by a coacervation.

In the event of polycondensation, first the process of limiting phase polycondensation is applied. It is characterized by a high speed of polymer formation and the process must be carried out with care. This is the reason why a polymer with a sufficiently high molecular weight and insoluble in the acting phases is obtained (e.g. polyamide, polyester) immediately after the combination of the two phases containing the required monomers. For this reason also the range of materials which may be encapsulated is broadening. In addition, but often it is also one of the disadvantages of this procedure, one obtains by this procedure microcapsules with a small wall thickness (high dose contents) that cannot easily be obtained by means of most other procedures.

In practice when using these methods, one of the monomers will be dispersed in the continuing phase when the other components of the system are also soluble. The core material, i. e. the material to be encapsulated, must also be contained in the dispersed phase. Furthermore, in the same way as during polymerization, oil-in-water or water-in-oil emulsions may be applied. The former method offers better possibilities with respect to the regulation of the degree of dispersion of the phase to be encapsulated and also control of the polycondensation reaction itself. With this technique, one obtains copolymerizates as wall materials which are built from one hydrophilic and one hydrophobic monomer and which are insoluble in both the organic phase and the aqueous phase. For instance, ethylenediamine is soluble in water. A second solution contains terephthalic acid dichloride in toluene and also the material to be encapsulated. In practice, the toluene solution disperses under vigorous stirring in water, which preferably for the stabilization of droplets still contains a protective colloid such as PVA or polyethylene glycol, and the ethylenediamine solution is slowly added to the dispersion that contains a solution of sodium hydroxide for neutralization of the released acid. Directly with the formation of the interfaces of the two phases, which cannot be mixed with each other, an insoluble polyamide film is formed. Instead of acid chlorides isocyanates or polyisocyanates can also be used. Depending on the choice of the acid chlorides or the amines, the properties of the capsule wall can be varied. A decrease in permeability of the capsule wall is achieved, for instance, by using triamines, which cause an additional cross-linking of the polymeric material of the capsule wall.

Furthermore, there is the possibility in one of the phases of precipitating soluble polymers at the interfaces and subsequently in a chemical cross-linking reaction to convert them into their insoluble form. As example is the encapsulation of hydrophobic dyestuffs with PVA in the form of oil-in-water emulsions with subsequent cross-linking of the PVA with aldehydes. In this context, the conversion of the polymer with the aldehyde is carried out at the aqueous phase limit, and the new insoluble phase formed coats the oil droplets to be encapsulated. Higher aldehydes with high distribution coefficients between the oil and the water phase (> 1) dissolve in the non-polar phase, and their conversion with the hydrophilic polymer is the inverse, offering the possibility of encapsulation even of the polar phase. The techniques are as numerous as the compounds and the encapsulated substances used as wall materials. More detailed information can be obtained from the literature.

Another procedure is based on oligomeric or polymeric preliminary condensates which then cause further polycondensation and cross-linking of the preliminary condensate by using suitable catalysts. By means of phase splitting processes and interaction forces, concentration of the polymer at the phase limit results with coalescence of the formed particular gel particles and the final cross-linking into duromers [5.24, 5.28–5.31, 5.34].

As wall material an amino polymer is preferred, in particular based on conversion products of urea, melamine, benzoguanamine and acrylamide with aldehydes and preferably with formaldehyde, glyoxal and glutardialdehyde. However, the use of epoxy, phenolic and acrylate resins has also been described. A significant advantage of this procedure is that the wall former is contained in only one phase, preferably the aqueous phase, and in this phase there is also the catalyst. This is the reason why according to this procedure, liquid, gaseous and solid substances can be encapsulated. Another advantage is obvious because this procedure may be carried out in a batch process in a stirred reactor (one-pot process) and also by means of, for instance, cascade dispersing and a static mixer quasi-continuously [5.68].

The scientific and patent literature clearly shows that in the last few years the microencapsulation technique itself and the use of microcapsules have been developing rapidly [5.69]. As far as the encapsulation technique is concerned, new procedures and techniques have been developed and also new wall and core materials have been tested. In this connection, the concentration of research has shifted more and more from the originally controlled release applications and the synthesis of copy paper coatings to the materials themselves.

Issues such as the improvement of the compatibility of different substances are important, e.g. in polymer compounding, electronics, anti-corrosion and anti-wear agents, the development of intelligent materials, and also their utilization in the protection of plants, veterinary technology and hygiene is developing further.

5.2.4 Microencapsulation of Energetic Materials

The microencapsulation of particulate energetic materials is aimed at the improvement of the product quality regarding processing, handling and storage. Therefore, the following properties are desirable:

- insensitivity, reduction of friction and impact sensitivity
- compatibility with usual binder systems
- protection against environmental effects, e.g. humidity or radiation.

Adjusted to the properties of the core (energetic) materials, several coating materials and solvents are suitable. In this section, mostly chemical and physico-chemical microencapsulation processes (e.g. coacervation – phase separation procedures), which operate in liquid phase, are considered.

Among others the following combinations of core- and coating materials turned out to be appropriate for microencapsulation:

- ammonium dinitramide (ADN) coated with ethylcellulose
- ADN coated with cellulose acetobutyrate (CAB)
- ADN coated in two layers (first wax, second amino resin)
- hexanitrohexaazaisowurtzitane (CL20) coated with cellulose acetate phthalate
- Octogen (HMX) coated with amino resin.

In principle, all described procedures of microencapsulation may also be used for the encapsulation of explosives and fuels and of oxidizing components provided that there is compatibility of the material to be encapsulated with the procedure.

The techniques for treatment of the different components is characterized first by the properties of the components themselves. It was difficult to use explosive liquid compounds alone, such as glycerine trinitrate and ethylene glycol dinitrate, because they have high power and high sensitivity to impact and friction. In order to prevent these drawbacks, Nobel developed guhr dynamite in 1866 by absorbing glycerine trinitrate in kieselguhr, high power blasting gelatin in 1875 by absorbing glycerine trinitrate in nitrocellulose, low -freezing dynamite in 1876 by using a mixture of glycerine trinitrate and ethylene glycol dinitrate and the ammon gelatin dynamite in 1879.

Since that time, the patent literature has described procedures which are based on phlegmatisation of very dangerous liquid explosives by absortion on solid substances in fine particulate form, to procedures which protect direct microencapsulation of the material by a polymer shell according to the core-shell principle [5.70–5.78].

Microencapsulated explosive compositions can be easily obtained by microencapsulating and fixing a dangerous explosive and therefore this process is an epoch-making explosive composition. Moreover, owing to the excellent properties of this favorable form, microencapsulated explosive compositions offer new fields of utilization. As explosive liquid compounds to be used for the production of microcapsules, all liquid nitrate compounds, gelatinized products and similar materials should be mentioned. The polymer compounds to be used for the production of microcapsules in most processes include polyurethane, polyurea, polystyrene, epoxy resins and PVA. The microencapsulation procedure in these cases is as follows. An explosive liquid compound is dispersed in water together with methylcellulose and a surfactant, a monomer of a high molecular weight compound which forms the capsule wall is added to the dispersed system and a polymerization initiator or a further reactive monomer is added thereto, and the resulting mixture is subjected to a polymerization reaction. The resulting microcapsules of the explosives have the following excellent properties: high power, low sensitivity and low vapor pressure. Moreover, since the microcapsules are solids, they can be easily handled.

A special advantage of explosive formulations in the form of microcapsules is that contrary to the usual formulations, which have a content of a liquid explosive component of a maximum of 50% of their weight, microcapsule formulations may have a content of explosive of up to > 90% of their weight and simultaneously adequate safety is guaranteed. One of the disadvantages of this procedure is that products that are sensitive to hydrolysis cannot be used because of the presence of an aqueous phase. Frequently one comes across procedures that describe the simple treatment of hydrolysis-sensitive compounds with oils or coating with waxes that lead to a hydrophobic core material [5.79, 5.80].

A recent application for a patent describes a procedure for microencapsulation of particulate fuels and explosives and also of oxidizing components that are especially sensitive to humidity which involves a combination of humidity-proof impregnation with direct microencapsulation according to the core-shell principle with an amino polymer [5.81]. By dipping the particles into a melt consisting of at least one waxed material under conditions of complete wetting with exclusion of humidity in this process, these particles are coated with a thin, dense film formed by this waxed material. In order to absorb excessive waxed material, a highly active absorbent based on amino polymer hollow spheres is added and the resulting pulverized product will be microencapsulated with an amino polymer based on melamine resin. It can also be considered to be a special advantage of such a procedure that fine particulate solid substances which often are still highly porous and non-spherical resulting from their synthesis can be converted into a microencapsulated form. The typical form of the capsules of an ADN product generated by this procedure is shown in Fig. 5.16.

The experimental installation for the chemical microencapsulation processes consists of:

- reaction vessel with temperature control
- stirring apparatus with mixing propeller
- buret for exact addition of solvents, antisolvents, agents, surfactants, etc.

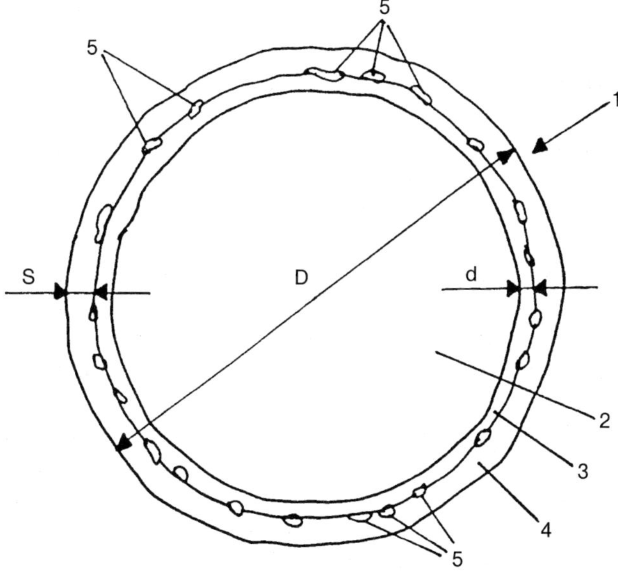

Figure 5.16. Diagram of ADN microcapsules generated by procedure of Teipel *et al.*[5.81].

The choice of materials is important for a successful microencapsulation process. The required substances are core material, coating material, encapsulation medium (carrier substance), precipitating agent and additives. Generally the carrier substance has to be a poor or non-solvent for the core material. Hence we use aqueous solutions for non-water-soluble cores such as Octogen (HMX), hexanitrohexaazaisowurtzitane (CL20) and Hexogen (RDX), whereas non-polar organic solvents are suitable for water-soluble cores such as ammonium dinitramide (ADN) and ammonium nitrate (AN).

To encapsulate spherical ADN particles produced by emulsion crystallization [5.82], the coacervation of ethylcellulose dissolved in cyclohexane can be used [5.83]. The progress of the microencapsulation process is shown in Figs 5.17–5.19.

Highly esterified cellulose acetobutyrate (CAB) has a temperature-dependent solubility in non-polar organic solvents. Therefore, a phase separation of the dissolved CAB can be achieved by cooling the solution with the suspended ADN particles. During the process the solvent-containing capsules can be deformed by the energy of the agitator as shown in Fig. 5.20. A subsequent drying process reduces the thickness of the coating.

CL20 crystals may be encapsulated in aqueous solutions of cellulose acetate phthalate (see Fig. 5.21). In this case, the coacervation is caused by variation of the pH.

Figure 5.17. Uncoated ADN particles.

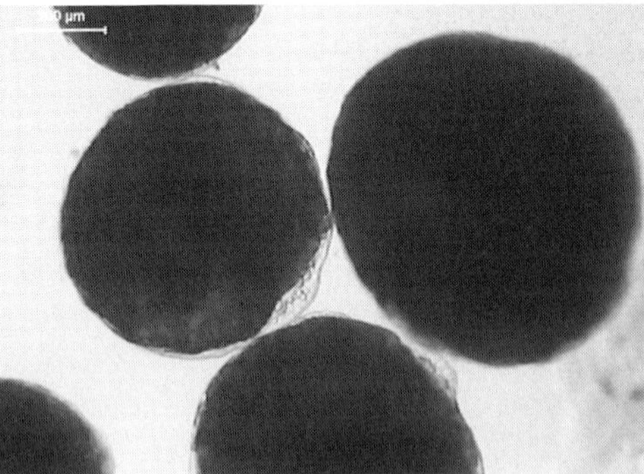

Figure 5.18. ADN particles with coacervate coating in solvent.

Figure 5.19. Encapsulated ADN particles (dried and pourable).

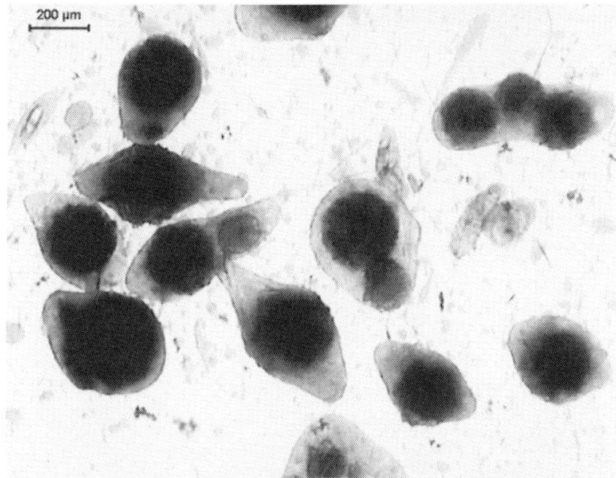

Figure 5.20. Spherical ADN particles in CAB-containing coacervate capsules.

Figure 5.21. CL20 crystals encapsulated with cellulose acetate phthalate (wet).

The coating of HMX crystals with amino resin was done by *in situ* polymerization (see Fig. 5.22). The polycondensation was started by changing the pH of the aqueous prepolymer solution.

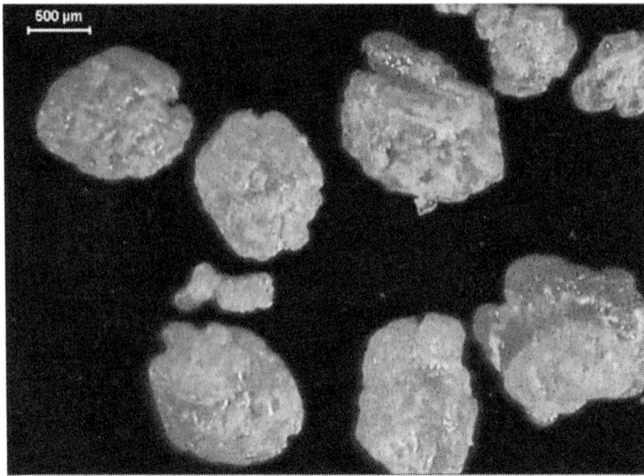

Figure 5.22. HMX crystals encapsulated with amino resin.

5.2.5 Coating with Supercritical Fluids in a Fluidized Bed

5.2.5.1 Introduction

As described before, particles can be encapsulated by spraying an organic solvent solution into a fluidized bed of particles. However, significant problems are connected with this technique. For example, organic solvents may only be applied for the coating of particles having diameters > 100 µm because otherwise strong capillary forces of the organic solvent cause defluidization of the fluidized bed [5.48, 5.49]. Also, high temperatures are necessary to perform this process, thus limiting the fluidized bed encapsulation technique to temperature-insensitive substances [5.84]. Recently, a fluidized bed encapsulation process using supercritical carbon dioxide as a solvent was developed [5.85]. The encapsulation of the thermolabile substances cyclotrimethylenetrinitramine (RDX) and pentaerythritol trinitrate (PETN) is being described. According to this study, the new process allows the encapsulation of particles having a diameter between 30 and 100 µm at ambient temperature. Another study focused on the supercritical fluid encapsulation of ADN [5.86], a substance which is also very temperature sensitive and prone to decomposition by most organic solvents. Because ADN is soluble in water, the encapsulation of this material using supercritical carbon dioxide is considered to be the most effective process.

This process exploits the capability of supercritical fluids to act as a selective solvent. Supercritical fluids are noteworthy in that their density is like that of a fluid (hence they are effective solvents) and their viscosity and diffusion coefficient are like those of gases (hence they exhibit good mass transport behavior). These properties make supercritical fluids effective as extractants. Via the RESS process (*R*apid *E*xpansion of *S*upercritical *S*olutions), particles and droplets in the submicron range can be produced by atomizing a supercritical solution through a nozzle (see Chapter 4). These fine droplets make it possible to improve the deposition on the particles, so that a thinner and more even coating layer can be achieved. In addition, because the cohesion and adhesion forces are small compared with organic solvents, the capillary forces are considerably reduced, making it possible to encapsulate particles smaller than 100 μm without forming agglomerates [5.87]. In this process, the supercritical fluid is used both as a solvent for the coating material and as a carrier fluid for the formation of the fluidized bed.

5.2.5.2 Experimental Setup

The experiments were conducted in a laboratory-scale device. A schematic view of the high-pressure fluidized coating apparatus is shown in Fig. 5.23. The apparatus consists essentially of three sections: the CO_2 supply, the extractor (in which the CO_2 becomes saturated with coating) and the high-pressure reactor (where the coating

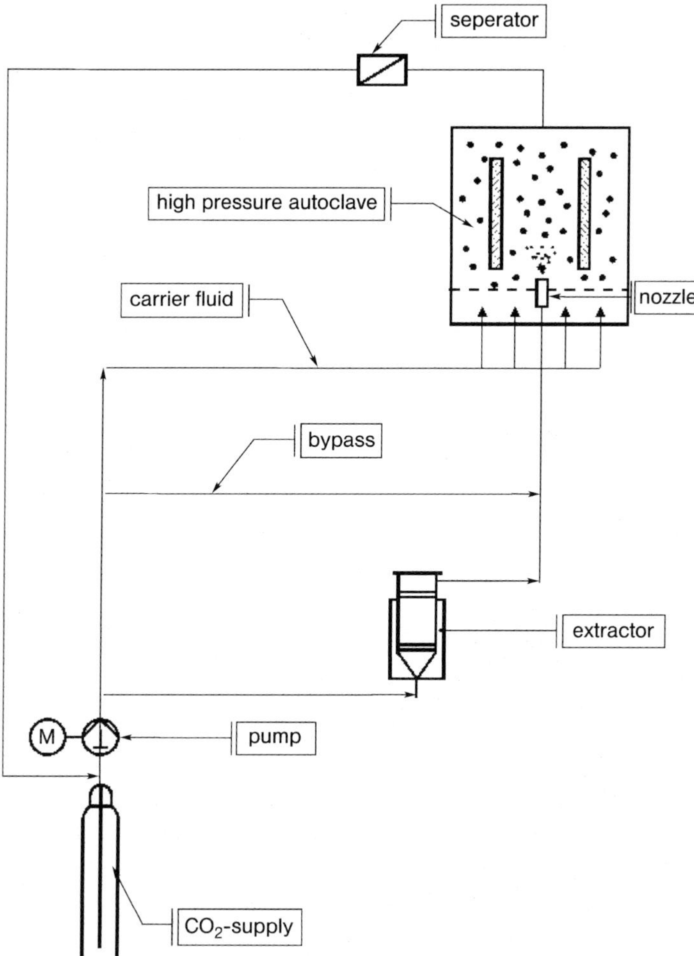

Figure 5.23. Experimental apparatus for coating particles in the presence of supercritical fluids.

takes place). The maximum allowable pressure in the device is 30 MPa at a maximum temperature of 373 K.

The first experiments were conducted using wax (obtained from Peter Greven Fett-Chemie) as the coating material and glass beads (obtained from Potter-Ballotini) as the core material. The wax was a mixture of stearyl stearate and stearyl alcohol with a molecular weight of 537 g/mol and a melting-point of 328 K. The glass beads had a high sphericity and very narrow particle size distribution, with a mean particle diameter of 50 μm.

In each experiment, the fluidized bed reactor was filled with 4–10 g of glass beads. The bed was fluidized by adding pure carbon dioxide via the line identified in Fig. 5.23 as 'carrier fluid'. The reactor floor consists of a perforated plate. When the reactor reaches thermal stability, the microencapsulation process is initiated. CO_2 flows through the extractor, saturated by the wax and then expands through a nozzle into the fluidized bed reactor. The expansion produces a fine aerosol of wax droplets that deposit on the glass particles, spread out as a film and then harden, yielding a solid layer of coating on the particles. Laval nozzles with different diameters (50, 75 and 100 μm) were used. At the end of each experiment, the connection between the extractor and nozzle was flushed by opening the bypass, at which time the product produced in the fluidized bed was subjected to post-drying.

5.2.5.3 Coating Mechanism

The mechanism of the supercritical fluid-driven encapsulation process in a fluidized bed is identical with the mechanism in conventionally operated fluidized bed coaters. Hence the encapsulation can be divided into three operational units (Fig. 5.24):

1. transport of the encapsulating substance into the fluidized bed
2. deposition of the aerosol
3. formation of the encapsulation.

Hence the encapsulation process can be described by applying models which describe phenomena occurring in the single operational units.

Transport of the Encapsulating Substance into the Fluidized Bed

The transport of the encapsulating substance can actually be divided into the extraction of the substance by the supercritical fluid (SFE: supercritical fluid extraction) and subsequent RESS into the fluidized bed.

A method for calculating the solubility of a substance comprises the calculation of the fugacity of both the supercritical phase and the phase of the solute by application of cubic equations of state (EOS). Especially the Soave-Redlich-Kwong EOS [5.88] and the Peng-Robinson EOS [5.89] have found wide use. However, this approach suffers from the need to know the values of the critical parameters p_c and T_c and the vapor pressure of the substance. Therefore, many authors [5.90–5.92] prefer to use the semiempirical model suggested by Chrastil [5.93]. This model assumes an

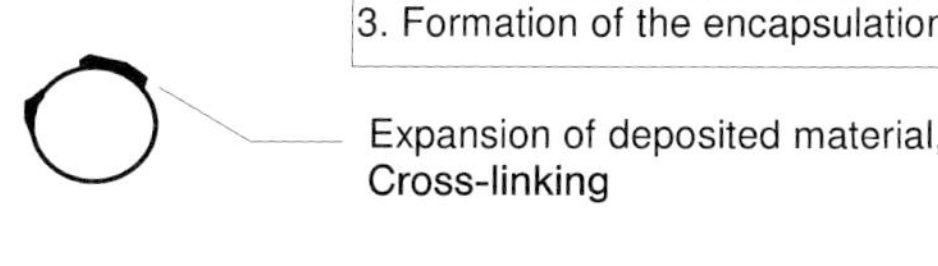

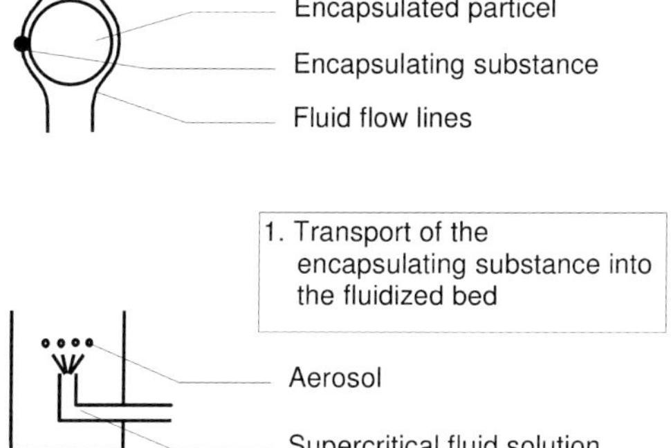

1. Transport of the encapsulating substance into the fluidized bed

Aerosol

Supercritical fluid solution

Suspending fluid

Figure 5.24. Operational units involved in the encapsulation of particles in a fluidized bed.

equilibrium between a solvato complex [AB] comprising the solute molecules A, solvent molecules B and the ambient supercritical solvent B.

Hence, with a few other assumptions, the partial density of the solute-solvent complex ρ_{AB} in the supercritical fluid may be expressed as a function of solvent density ρ_B and temperature T as follows:

$$\rho_{AB_n} = \rho_B^n \cdot \exp\left(\frac{a}{T} + b\right) \tag{5.1}$$

The parameters n, a and b have to be found empirically by regression of the experimental data. Table 5.1 lists the empirical solubility parameters of some substances for the solvent carbon dioxide.

For the wax stearyl stearate (StSt) in supercritical carbon dioxide the parameters a and b depend on the temperature of the system [5.94]. Hence the partial density of the wax in carbon dioxide within the temperature range 333–400 K and a density < 800 kg/m^3 may be calculated as follows:

$$\rho_{StSt} = \rho_{CO_2}^{8.4431} \cdot \exp(0.03489T - 65.5862) \tag{5.2}$$

Figure 5.25 shows the solubility of the coating material stearyl stearate in carbon dioxide.

It is evident that increasing the pressure at constant temperature enhances the solubility of the substance, whereas the solubility decreases on varying the temperature from 333 to 393 K at constant pressure. Most substances are extracted in a semicontinuous granular bed reactor. In this case the mass flow of the coating

Table 5.1. Solubility parameters of substances for carbon dioxide [5.93].

Substance	Formula	*n*	*a*	*b*
Stearic acid	$C_{18}H_{36}O_2$	1.821	−10664.5	22.320
Tristearin	$C_{57}H_{110}O_6$	9.750	−8771.6	−39.440
Behenyl behenate	$C_{44}H_{86}O_2$	3.250	−7328.1	0.824

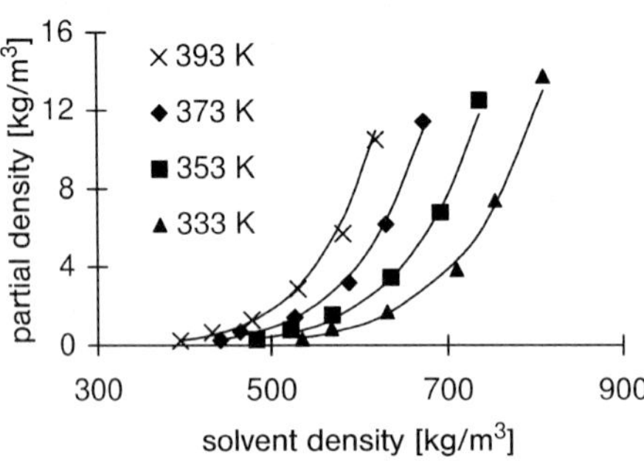

Figure 5.25. Solubility of stearyl stearate versus density of carbon dioxide.

substance may be calculated by using a one-dimensional mass balance [5.94, 5.95]. In many cases the concentration c_W of the extracted substance in the supercritical fluid is in thermodynamic equilibrium. Assuming stationary conditions, the mass balance may be simplified to

$$\dot{m}_W = \dot{m}^{out}_{CO_2} \cdot \frac{c^{Eq}_W}{\rho^R_{CO_2}} \tag{5.3}$$

where $\dot{m}_W$ and $\dot{m}^{out}_{CO_2}$ are the mass flow of the extracted substance and the mass flow of CO_2, respectively.

The mass flow of stearyl stearate as a function of the density of the carbon dioxide at 353 K is depicted in Fig. 5.26 [5.96]. It is evident that the mass flow of the wax increases with increase in solvent density. At a density < 500 kg/m³ there is no difference between the predicted and the measured mass flow of the extract. However, at a density > 500 kg/m³ a constantly increasing difference between the

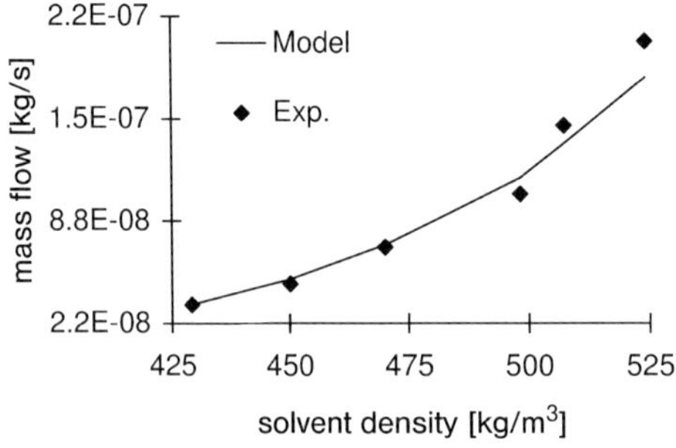

Figure 5.26. Mass flow of stearyl stearate versus density of carbon dioxide.

calculated and the measured mass flow of the extract indicates that the mass transport between the extracted substance and the solvent is the limiting step in the process.

Deposition of the Aerosol in the Fluidized Bed

Modeling the encapsulation process also has to take into account the size of the aerosol that is produced by the rapid expansion of the supercritical solution into the fluidized bed. For example, it is well known [5.97] that aerosols exceeding a certain size may cause defluidization of the fluidized bed. Also, the particle size of the encapsulating substance significantly influences the deposition kinetics and the quality of the encapsulation [5.98, 5.99].

The results in Tables 5.2 and 5.3 [5.94] indicate that a complex combination of phenomena influences the nucleation of the aerosol. Most models involving the RESS process describe the expansion of the supercritical fluid as an adiabatic process [5.100]. Hence the formation of an area where two phases, the gaseous or liquid and the solid phase, exist simultaneously depends on the temperature of the pre-expanded solvent and the pressure drop occurring at the nozzle. Empirically for the RESS of carbon dioxide the following relation between the cluster size d_N of the nucleated substance and process parameters has been found [5.101]:

$$d_N \propto p_{\mathrm{nozz.}}^{1.44} \cdot d_{nozz.}^{0.86} \cdot T_{\mathrm{nozz.}}^{-5.4} \tag{5.4}$$

According to this relation and in good agreement with experimental results [5.94, 5.102], the size of the nucleated particles grows with increasing pre-expansion pressure of the solution and with increasing diameter of the nozzle, whereas there is a reciprocal relationship between the particle size and the temperature of the solvent.

Table 5.2. Diameter of stearyl stearate particles after precipitation from supercritical carbon dioxide (T_0, 353 K; T_1, 308 K; p_0, 17 MPa; p_1, 8 MPa) [5.94].

$d_{nozz.}$ (μm)	$\dot{m}_{CO_2}$(10^{-4} kg/s)	d_P(0.16) (μm)	d_P(0.5) (μm)	d_P(0.84) (μm)
50	1.4	0.3	0.6	1.0
75	2.1	0.6	1.0	1.7
100	2.7	0.5	1.2	1.9

Table 5.3. Diameter of precipitated stearyl stearate as a function of the expansion pressure of the supercritical carbon dioxide solution ($T_{nozz.}$, 353 K; T_1, 308 K; $p_{nozz.}$, 17 MPa; p_1, 8 MPa; $d_{nozz.}$, 50 μm) [5.94].

$p_{nozz.}$ (MPa)	$\dot{m}_{CO_2}$(10^{-4} kg/s)	d_P(0.16) (μm)	d_P(0.5) (μm)	d_P(0.84) (μm)
15	1.2	0.5	0.6	0.2
17	1.4	0.4	0.6	0.3
19	1.6	3.2	4.8	0.3

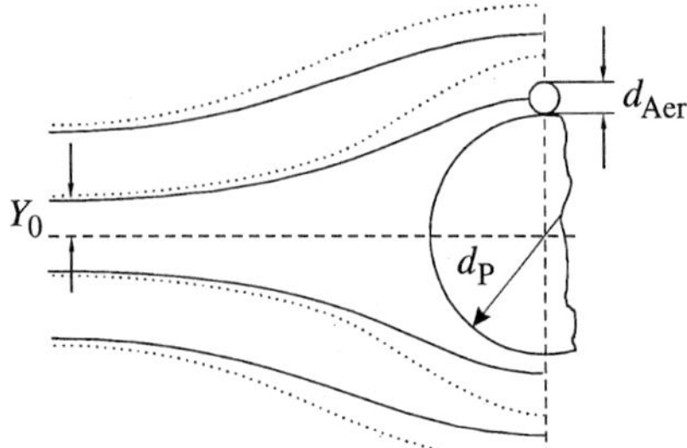

Figure 5.27. Schematic flow pattern of inertial aerosol transport to a collector particle [5.103].

As Fig. 5.27 shows schematically, aerosol particles are transported by inertia on the collector particle if the initial coordinate Y of their flow line is less than the initial coordinate Y_0 of the boundary flow line.

Hence the single particle collection efficiency η is the ratio of the projection area of the boundary flow line and the projection area of the collector particle, where r_P is the radius of the collector particle:

$$\eta = \left(\frac{Y_0}{r_P}\right)^2 \tag{5.5}$$

There have been extensive studies focusing on the calculation of the single particle collection efficiency [5.104, 5.105]. Generally, the calculation of the single particle collection efficiency requires the calculation of the aerosol stream line around the particle. Therefore, transport mechanisms such as diffusion and inertia and deposition mechanisms such as interception have to be taken into account. For example, the stochastic random motion of aerosol particles is likely to increase the probability of aerosol deposition on the collector. According to the literature [5.106], diffusion is the dominant deposition mechanism for aerosols having diameters < 0.1 µm at a flow velocity < 0.1 m/s. Studies indicate that in most cases the different deposition mechanisms are independent of each other [5.107, 5.108]. Hence the single particle collection efficiency may be calculated by adding the single particle collection efficiency maintained for the different transport and deposition mechanisms:

$$\eta = \eta_D + \eta_R + \eta_T + \eta_E \tag{5.6}$$

where the subscipts D, R, T and E indicate diffusion, inertia, interception and electrostatic forces.

Table 5.4 lists some relations for the calculation of the single particle collection efficiency. Other correlations may be found elsewhere [5.94, 5.103, 5.109].

As depicted in Fig. 5.28, the particle size of the aerosol strongly affects the deposition mechanism. At a particle size of about 1 µm the single particle collection efficiency becomes a minimum. Hence the production of aerosols having this size should be avoided because a large amount of material might be lost owing to discharge. Also, in order to achieve a smooth encapsulation, it is advantageous to

Table 5.4. Correlations for the prediction of the single particle collection efficiency.

Mechanism	Single particle collection efficiency	Validity
Diffusion [5.110]	$\eta_D = \frac{4.36}{\varepsilon} \cdot (Pe)^{-2/3}$	$165 < Sc < 600$ $Re < 55$ Fluidized bed
Interception [5.111]	$\eta_R = 1 + R - \frac{1}{1+R}$	$Re > 100$ Single cylinder
Inertia [5.112]	$\eta_T = \left(\frac{St}{St + 0.062\varepsilon}\right)^3$	$0.001 < St < 0.01$ Fluidized bed

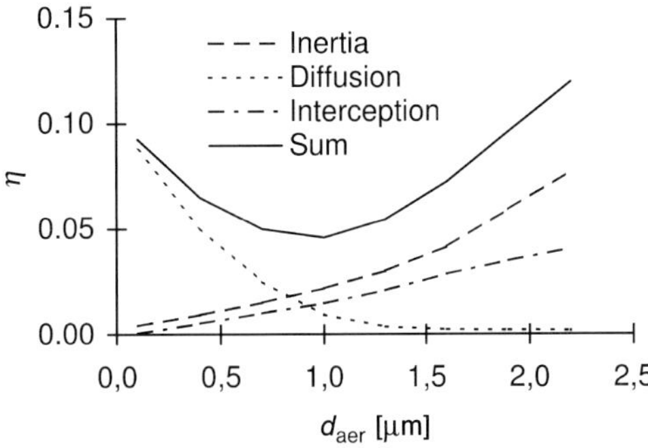

Figure 5.28. Single particle collection efficiency versus diameter of aerosol particles [5.94].

adjust the RESS process such that the particle size of the precipitated material is as small as possible.

A successful coating process requires that the encapsulation substance adheres sufficiently to the surface of the fluidized bed material. Of course, the adhesion also affects the collection efficiency. In the literature [5.94, 5.111, 5.112], the impact of adhesion on the coating process is taken into consideration by the introduction of the adhesion coefficient χ. Thus an effective single particle collection efficiency φ may be formulated:

$$\varphi = \eta \cdot \chi \tag{5.7}$$

According to literature [5.112], for liquids or very soft substances the adhesion coefficient has a value ≈ 1. Another study indicates that for solid aerosols having a diameter in the region of 0.7 μm the adhesion coefficient may vary from 0.001 to 0.01 [5.94].

Assuming homogeneous expansion of the fluidized bed, the collection efficiency E_G may be calculated as a function of process parameters, e.g. the size of the collector particle d_P or the height of the fluidized bed z and also the effective single particle collection efficiency φ and the porosity of the fluidized bed ε:

$$E_g = \frac{c_{in} - c_{out}}{c_{in}} = 1 - \exp\left(-\frac{3}{2} \cdot \frac{1-\varepsilon}{d_P} \cdot z \cdot \varphi\right) \tag{5.8}$$

According to the literature, the porosity of a homogeneously fluidized bed may be calculated by assuming that each particle is circulated by the fluid individually and that there is no collision between the particles [5.113]. There is also a correlation [5.114] which at a given porosity predicts the state of fluidization (homogeneous or inhomogeneous). The porosity ε of a homogeneously fluidized bed as a function of the velocity u of the fluid which circulates the fluidized bed material and the terminal velocity u_{term} and also the Richardson-Zaki exponent may be calculated by the following equations [5.114]:

$$\frac{u}{u_{term}} = \varepsilon^{RZ} \tag{5.9}$$

with

$$\varepsilon = 1 - \frac{V_P}{V_R} \tag{5.10}$$

$$RZ = \left(4.35 + 17.5 \cdot \frac{d_P}{d_R}\right) \cdot Re^{-0.03} \qquad 0.2 < Re < 1 \tag{5.11}$$

$$RZ = \left(4.45 + 18 \cdot \frac{d_P}{d_R}\right) \cdot Re^{-0.1} \qquad 1 < Re < 200 \tag{5.12}$$

Usually, the above equations can be applied in fair agreement with experimental data if the diameter of the reactor exceeds 15 cm. Another correlation for the expansion of a homogeneously fluidized bed has been derived for reactors having a diameter of 1.4 cm [5.94]:

$$RZ = \left(476.26 \cdot \frac{d_P}{d_R} + 1.97\right) \cdot Re^{-0.23} \tag{5.13}$$

At a given Reynolds number Re_P the porosity ε of a fluidized bed decreases with increasing particle size d_P (Figure 5.29). Figure 5.29 also shows that there is very good agreement between the values of the porosity versus Reynolds number for small particles predicted with Eqs (5.10) and (5.13). For larger particles, having a diameter of 66 and 85 μm, the agreement between predicted and measured values of the porosity becomes less accurate with increasing Reynolds number because the

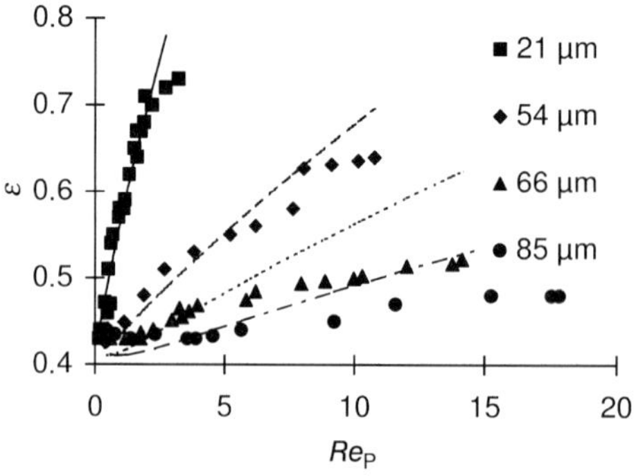

Figure 5.29. Fluidized bed porosity versus Reynolds number at different diameters of the material.

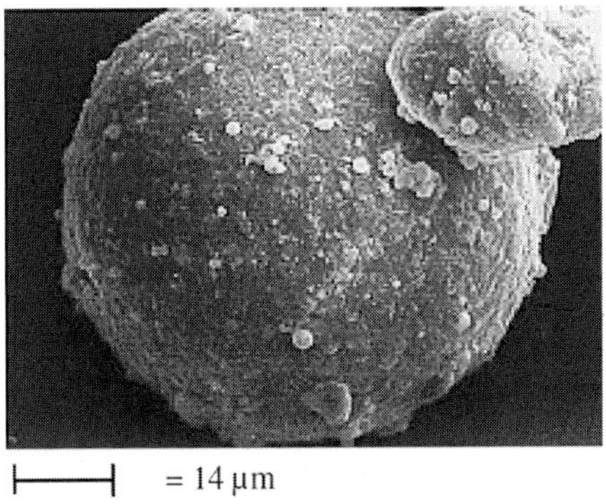

Figure 5.30. Particle encapsulated with stearyl stearate [5.94].

fluidization of the particles becomes less homogeneous. Therefore, it is essential to estimate the range of homogeneous fluidization as a function of the particle diameter and as a function of the density of the particles and the fluid if the porosity must be calculated with the Eqs (5.9)–(5.13).

Formation of the Encapsulation

Figure 5.30 presents an SEM picture of a particle encapsulated with stearyl stearate. As might be expected, the surface of the encapsulation consists of tiny dots of deposited material having an average diameter of 1.5 μm. It has been found that owing to the high kinetic energy of the aerosol generated by the RESS process, the encapsulation material deforms to small plates (see Fig. 5.31) [5.94, 5.115]. Hence a relatively smooth and tight encapsulation may be achieved.

For smooth encapsulation of monodisperse and spherical particles, the thickness th_W as a function of process parameters can be calculated according to the following equation [5.94]:

$$th_w^t = r_p^0 \cdot \left(\sqrt[3]{1 + \frac{\rho_p}{\rho_w} \cdot \frac{m_{w,g}^t}{m_{p,g}^0}} - 1 \right) \tag{5.14}$$

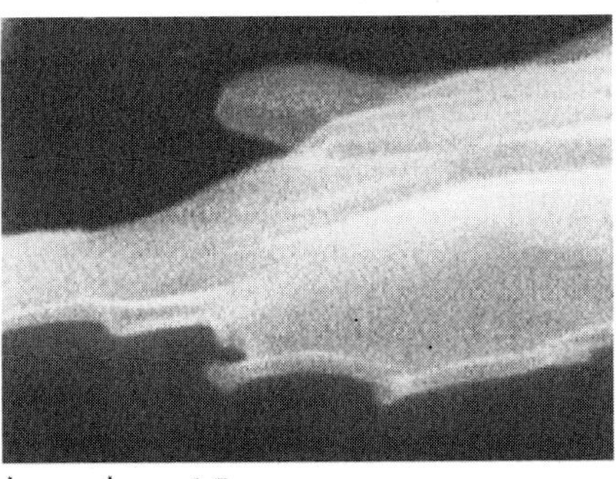

Figure 5.31. Stearyl stearate deposited on a glass plane [5.94].

where ρ_P and ρ_W are the density of the encapsulated particles and the density of the coating material, respectively, and m_P is the total mass of the fluidzed bed material. The total mass m_W of the deposited encapsulation substance can be calculated as a function of the collection efficiency E_g of the fluidized bed and as a function of time:

$$m_{w,g}^{t} = E_g \cdot \int_0^t \dot{m}_w \, dt \tag{5.15}$$

In the simplest case, stationary conditions of the mass flow of the coating substance which is dissolved in the supercritical fluid in thermodynamic equilibrium and homogeneous fluidization of the collector particles in the reactor may be assumed. Then Eqs (5.1)–(5.13) can be applied for the calculation of the deposited mass of coating material in the reactor. This makes clear that a large number of variables affect the encapsulation of particles in a fluidized bed. Because the parameters affect each other, no simple correlation between one parameter and the thickness of an encapsulation can be concluded.

For example, on the one hand the diameter of the collector particles affects the coating thickness that may be achieved with a certain amount of deposited material by the ratio of the particle surface and the particle volume. On the other hand, the size of the collector material also has an impact on the porosity of the fluidized bed. Again the porosity of the fluidized bed affects the elutriation of aerosol and hence the collection efficiency of the fluidized bed. Also, the deposition mechanism of the aerosol is strictly dependent on the geometry of the collector particles. However, in most cases a relation between the thickness of an encapsulation and the diameter of the encapsulated substance and also time may be found similar to that presented in Fig. 5.32.

Studies give evidence [5.94, 5.101, 5.115] that in a fluidized bed coating process driven by the rapid expansion of supercritical solutions, the pre-expansion pressure severely impacts the quality of the encapsulation. Not only may the smoothness of the encapsulation be adjusted by choosing appropriate expansion conditions but also the deposition kinetics are strongly influenced by the pre-expansion pressure. For example, at a given diameter of the expansion nozzle a rise in the pre-expansion pressure increases the mass flow of expanding fluid. This on the one hand also

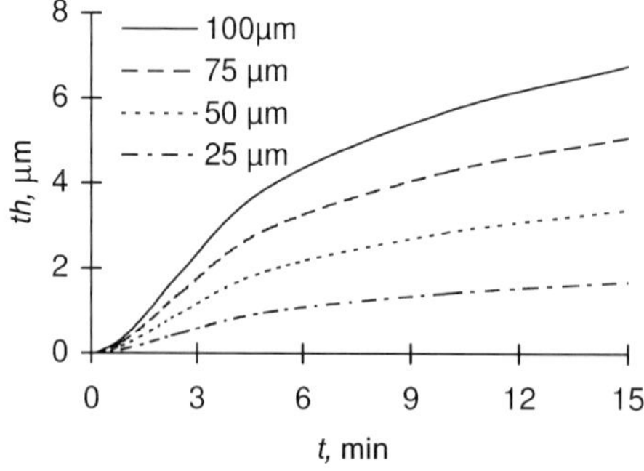

Figure 5.32. Coating thickness versus time for different diameters of the treated material.

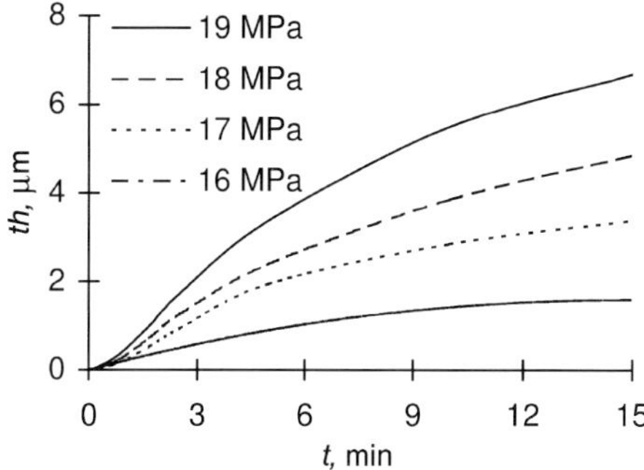

Figure 5.33. Coating thickness versus time at varying pre-expansion pressures.

increases the mass flow of the encapsulation substance. Hence the time needed to achieve a coating with a certain thickness may be reduced (Fig. 5.33).

On the other hand, results show that by increasing the pre-expansion pressure a jet of expanded fluid is formed at the nozzle which may as a consequence change the fluid dynamics from a homogeneously fluidized bed type to a jet reactor type [5.94]. In this case an axial gradient of the fluid within the reactor must be considered, thus significantly increasing the mathematical effort for predicting the collection efficiency. Additionally, the formation of a jet may contribute to enforced elutriation of the coating material, thus increasing the time needed for the achievement of an encapsulation with a definite thickness. Studies also give evidence that agglomeration may occur if the pre-expansion pressure is too large. In this case, probably the mass flow of the encapsulation substance exceeds the capacity of the reactor, thus resulting in a breakdown of the fluidized bed.

5.3 References

5.1 Pietsch W (1991) *Size Enlargement by Agglomeration*, Wiley, New York.

5.2 Rumpf H (1990) *Particle Technology*, Chapman and Hall, London.

5.3 Sommer K (1988) Size enlargement. In: *Ullmann's Encyclopedia of Industrial Chemistry*, Vol. B2, VCH, Weinheim.

5.4 Thies C (1995) *Kirk Othmer Encyclopedia of Chemical Technology*, 4th edn, Vol. 16, Wiley, New York, pp. 628–681.

5.5 Doubrow M (ed.) (1992) *Microcapsules and Nanoparticles in Medicine and Pharmacy*, CRC Press, Boca Raton, FL, pp. 73–97.

5.6 Wilkins RM (ed.) (1990) *Controlled Release of Crop Protection Agents*, Taylor and Francis, London, pp. 65–90, pp. 149–160, pp. 245–255.

5.7 Wilkins RM (ed.) (1990) *Controlled Delivery of Crop Protection Agents*, Taylor and Francis, London, pp. 279–309.

5.8 Finch CA (1990) *Ullmann's Encyclopedia of Industrial Chemistry*, Vol. A16, VCH, Weinheim, 575–588.

5.9 Deasy PB (1984) *Microencapsulation and Related Drug Processes*, Marcel Dekker, New York.

5.10 Kydonieus AF (ed.) (1980) *Controlled Release Technologies – Methods, Theories and Application*, CRC Press, Boca Raton, Fl.

5.11 Solodovnik VD (1980) *Microencapsulation*, Khimiya, Moscow.

5.12 Masters K (1979) *Spray Drying Handbook*, 3rd edn, Godwin, London.

5.13 Kondo T. (ed.) (1979) *Microencapsulation – New Techniques and Applications*, Techno, Tokyo.

5.14 Sparks R (1979) *Kirk Othmer Encyclopedia of Chemical Technology*, 3rd edn, Vol. 15, Wiley, New York, pp. 470–493.

5.15 Thies C (1979) in Mark HF, Bikolev N, Overberger CG, Menges G, Kroschwitz, JJ *Encyclopedia of Polymer Science and Engineering*, 2nd edn, Vol. 9, pp. 724–745.

5.16 Sliwka W (1978) *Ullmann's Enzyklopädie der Technischen Chemie*, Vol. 16, VCH, Weinheim, pp. 675–682.

5.17 Gutcho MH (1976) *Microcapsules and Microcapsulation Techniques*, Noyes Data Corp., Park Ridge Illinois.

5.18 Nixon JR (ed.) (1976) *Microencapsulation*, Marcel Dekker, New York.

5.19 Vandegaer JE (ed.) (1974) *Microencapsulation – Processes and Applications*, Plenum Press, New York.

5.20 Das KG (1983) *Controlled Release Technology*, Wiley-Interscience, New York.

5.21 Green BK, Schleicher LS (1957) Microcapsule system, *US Patent* 2 800 457.

5.22 Finch A (1985) Polymers for microcapsule walls, *Chem. Ind. (London)* 782–786.

5.23 Arshady R (1993) Microcapsules for foods, *J Microencapsulation* **10**, 412–436.

5.24 Dietrich K (1989) Amino resin microcapsules I, *Acta Polym.* **40**, 243–251.

5.25 Torobrin LB (1981) Centrifuge apparatus and method for producing hollow microsperes, *US Patent* 4 303 433.

5.26 Sliwka W, Mikrokapseln (1975) *Angew. Chem.* **87**, 556.

5.27 Kaye BH (1992) Microcapsulation: the creation of synthetic fine particles with specified properties, *KONA* **10**, 65–82.

5.28 Dietrich K, amino resin microcapsules II (1989) *Acta Polym.* **40**, 325–331.

5.29 Herma H (1991) Einsatz von Mikrokapseldispersionen in kosmetischen Präparaten, *Parfüm. Kosmet.* **72**, 434–438.

5.30 Nastke R (2000) Mikroverkapselung partikulärer Feststoffe. In: *Proceedings Oberhausener Umsicht-Tage.*

5.31 Nastke R, Rafler G (1999) The challenge of microencapsulation of active solid ingredients. In: *Proc. 12th Int. Symp. on Microencapsulation, London.*

5.32 Hagenbart S (1993) Microencapsulation in the food industry, *Food Product Des.* **4**, 28–38.

5.33 Hertrich B (1993) Modified hydrocolloids as covering material, *Coating*, **26**, 266–268.

5.34 Dietrich K (1990) Amino resin microcapsules IV, *Acta Polym.* **41**, 91–95.

5.35 Zessin G (1982) Zur pharmazeutischen Technologie der Mikroverkapselung, *Wiss. Z. Univ. Halle* **31**(5), 65–74.

5.36 Goto S (1984) Preparation and biopharmaceutical evaluation of microcapsules, *J Microencapsulation* **2**, 137–155.

5.37 Anon. (1988) Mikroverkapselte Textilausrüstungen, *Chemiefaser/Textil-Ind.* **90**, 446.

5.38 Aggarwal AK (1998) Microencapsulated colours, *Colourage* **8**, 10–24.

5.39 Burgess DJ (1994) Preparing animal foods, *Macromol. Complexes Chem. Biol.* 285–300.

5.40 Weiss P (2000) *CHEManager* **15**, 7.

5.41 Dietz A (2000) Mikrokapseln in der Galvanotechnik, *Galvanotechnik* **54**, 28–30.

5.42 Kondo A (1979) *Microcapsule Processing and Technology*, Marcel Dekker, New York.

5.43 Nürnberg E, Krieger M (1981) Modifizierung der physikalischen und pharmazeutischen Eigenschaften von Arzneistoffen durch Sprühtrocknung, *Chem. Ind.* **33**, 794–796.

5.44 Rafler G, Jobmann M (1994) Controlled release systems of biodegradable polymers 1, *Pharm. Ind.* **56**, 565.

5.45 Rafler G, Jobmann M (1996) Cotrolled release systems of biodegradable polymers 4, *Pharm. Ind.* **58**, 1147.

5.46 Rafler G, Jobmann M (1997) Controlled release systems of biodegradable polymers 5, *Pharm. Ind.* **59**, 620.

5.47 Jozwiakowski MJ (1990) Characterization of a hot-melt fluid bed coating process for fine granules, *Pharm. Res.* **7**, 3.

5.48 Wurster DE (1953) Method of applying coatings to edible tablets or the like, *US Patent* 2 648 609.

5.49 Wurster DE (1959) Fluid bed processes, *J. Am. Pharm. Assoc.* **48**, 451.

5.50 Christensen FN Bertelsen P (1997) Qualitative description of the Wurster-based fluid-bed coating process, Drug Dev. *Ind. Pharm.* **23**, 451–463.

5.51 Bodmeier R, Paeratokul O, Wang J (1991) Prolonged release multiple-unit dosage form. In: *Proc. Int. Symp. Control*

Rel. Bioact. Mater. **18**, Controlled Release Society, 157–159.

5.52 Thoma K, Schlütermann B (1992) Pharmazeutische Anwendungen der Mikroverkapselung, *Pharmazie* **47**, 368–372.

5.53 Mathiowitz E (1991) Spray dried polymeric microparticles, *J. Appl. Polym. Sci.* **45**, 125–132.

5.54 Schryver BB (1988) Controlled release of macromolecular polypeptides, *Eur. Patent* EP 0 251 476.

5.55 Jager KF, Bauer KH (1982) Effect of material motion on agglomeration in the rotary fluidized bed granulator, *Drugs Made in Germany*, ECV-Editio Cantor Vlg. Aulendorf (Germany) 25, 61–65.

5.56 Jobmann M, Rafler G (1998) controlled release systems of biodegradable polymers 6, *Pharm. Ind.* **60**, 979–982.

5.57 Ibrahim H (1992) Influence of salt type on the phase separation of water/acetone systems, *Int. J. Pharm.* **87**, 239.

5.58 Allemann E (1992) Nanoparticle preparation with different polymers, *Int. J. Pharm.* **87**, 247.

5.59 Allemann E (1993) Particle size distribution of microcapsule preparation process, *Pharm. Res.* **10**, 1732.

5.60 Doelker, E (1989) Process for preparing a powder of water-insoluble polymers which can be redispersed in a liquid phase and processes for preparing a dispersion, *Eur. Patent* EP 0 363 549.

5.61 Pavenetto F (1993) Evaluation of spray drying as a method of polylactide and polylactide-polyglycolide microshere preparation, *J Microencapsulation* **10**, 487.

5.62 Wang HAT (1991) J. Control. Release **17**, 23–27.

5.63 Deasy PD (1994) Microcapsulation and related drug processes, *J. Microencapsulation* **11**, 487–495.

5.64 Sliwka W (1988) Mikrokapseln. In: *Ullmanns Enzyklopädie der Technischen Chemie*, VCH, Weinheim, p. 678.

5.65 NCR Corp. (1957), Durch Druckanwendung aufbrechbare mikroskopische Kapseln, beispielsweise zur Herstellung magnetischer Aufzeichnungen, *Ger. Patent* 1 082 282.

5.66 Meierson T (1966) Treatment of capsules in liquid to inhibit clustering, *Belg. Patent* 695 911.

5.67 North B (1991) A process of the production of microcapsules, *US Patent* 5 035 844.

5.68 Nastke R, Rafler G (1999) Mit Aminoplasten mikroverkapselte partikuläre Stoffe, ein quasi-kontinuierliches Verfahren zu deren Herstellung, EP/PCT/01/11376, *Ger. Patent* 10 049 777.

5.69 Fraunhofer-Gesellschaft (1999) *Studie IAP/ICT/UMSICHT, Herstellung und Anwendung von Mikrokompositen*, Teltow, Pfinztal, Oberhausen.

5.70 Scott AC (1924) Improvements to relating to explosives, *Br. Patent* 248 089.

5.71 Whetstone (1949) Free flowing ammonium nitate and method for the production of same, *Ger. Patent* 938 842.

5.72 Shunichi TS (1995) Explosive composition and method for producing the same, *US Patent* 5 472 529.

5.73 Sparks B (1968) Method of making a hybrid liquid-solid propellant system with encapsulated oxidizing agent and metallic fuel, *US Patent* 3 395 055.

5.74 Chandler R (1974) Nitromethane explosive with a foam and microsheres of air, *US Patent* 3 794 534.

5.75 Inoue K (1976) Capsulated explosive compositions, *US Patent* 3 977 922.

5.76 Sudweeks BW (1988) Blasting agent in microcapsule form, *US Patent* 4 758 289.

5.77 Sudweeks BW (1987) Microcapsule form explosive and formation therefor, *Jpn. Patent* 1 005 990.

5.78 Sudweeks BW (1988) Blasting agent, *Eur. Patent*, EP 0 295 929.

5.79 Wahrenholz D, Kuhn H (1959) Verfahren zur Verringerung der Feuchtigkeitsempfindlichkeit von Sprengstoffen, Wasag Chemie AG, *Ger. Patent* 1 065 310.

5.80 Reichel A (1973) Verfahren zum Phlegmatisieren von körnigem hochbrisantem Sprengstoff, *Ger. Patent* 2 308 430.

5.81 Teipel U, Heintz T, Krause H, Leisinger K, Rafler G, Nastke R (1999) Verfahren zum Mikroverkapseln von Partikeln aus Treib- und Explosivstoffen und nach diesem Verfahren hergestellte Partikel, *Ger. Patent* 199 23 202 A1.

5.82 Teipel U, Heintz T, Leisinger K, Krause H (1999) Verfahren zur Herstellung von Partikeln schmelzfähiger Treib- und Explosivstoffe, *Ger. Patent* 198 16 853 A1.

5.83 Teipel U, Heintz T, Kröber H (2001) Microencapsulation of particulate materials, *Powder Handling Process.* **13**(3), 283–288.

5.84 Liu L, Litster J. (1993) Spouted bed seed coating – the effect of process variables on maximum coating rate and elutriation, *Powder Technol.* **74**, 215.

5.85 Niehaus M., U. Teipel, H. Krause (1997) Mikroverkapselung von Partikeln in einer Wirbelschicht unter Anwendung überkritischer Fluide, *Ger. Patent* 97/33019.

5.86 Niehaus M., G. Bunte, H. Krause (1996) *Überkritische Extraktion von Ammoniumdinitramid, Report Fraunhofer ICT 10.*

5.87 Sunol AK (1998) Supercritical fluid aided encapsulation of particles in a fluidized bed. In: *Proc. 5th Meeting on Supercritical Fluids, Nice*, p. 409.

5.88 Soave G (1972) Equilibrium constants from a modified Redlich-Kwong equation of state, *Chem. Eng. Sci.* **27**, 1197.

5.89 Peng DY, Robinson DB (1976) A new two-constant equation of state, *Ind. Eng. Chem. Fundam.* Vol 15 No 1.

5.90 Zehnder B, Trepp C (1993) Mass-transfer coefficients and equilibrium solubilities for fluid-supercritical-solvent systems by online near-IR spectroscopy, *J Supercrit. Fluids*, **6**, 131–137.

5.91 Del Valle JM, Aguilera JM (1988) An improved equation for predicting the solubility of vegetable oils in supercritical carbon dioxide, *Ind. Eng. Chem. Res.* **27**, 1551.

5.92 Da Ponte MN (1997) Solubilities of tocopherols in supercritical carbon dioxide correlated by the Chrastil equation. In: *Proc. 4th Meeting on Supercritical Fluids, Sendai.*

5.93 Chrastil J (1982) Solubility of solids and liquids in supercritical gases, *J. Phys. Chem.* **86**, 3016.

5.94 Niehaus M (1999) Mikroverkapselung von Partikeln in einer Wirbelschicht unter Verwendung von verdichtetem Kohlendioxid, *PhD Thesis*, University of Karlsruhe; *Wissenschaftliche Schriftenreihe des Fraunhofer ICT*, Band 23.

5.95 Brunner G (1994) *Supercritical Fluid Technology*, Deutsche Bunsen-Gesellschaft für Physikalische Chemie, Steinkopff, Darmstadt.

5.96 Niehaus M, Weisweiler W (1998) Microencapsulation of fine particles in a fluidized bed. In: *Proc 29th Annual Conf. ICT, Karlsruhe.*

5.97 Kunii D, Levenspiel O (1968) *Fluidization Engineering*, 1st edn, Wiley, New York.

5.98 Clift R, Ghadiri M, Thambimuthu KV (1981) Filtration of gases in fluidized beds, progress in filtration and separation. In: Wakeman R (ed.), 1st edn, Elsevier, Amsterdam.

5.99 Iley W (1993) Effect of particle size and porosity on film coatings, *Powder Technol.* **65**, 441.

5.100 Matson D, Petersen R, Smith R (1987) Production of powders and films by the rapid expansion of supercritical solutions, *J. Mater. Sci.* **22**, 1919.

5.101 Smith R, Wash R. (1986) Supercritical fluid molecular spray film deposition and powder formation, *US Patent* 4 582 731.

5.102 Kröber H, Teipel U, Krause H (2000) Herstellung von Partikeln im Submikronbereich durch Expansion überkritischer Fluide, *Chem. Ing. Tech.* **72**, 70–73.

5.103 Löffler F (1988) *Staubabscheiden*, 1st edn, Georg Thieme, Stuttgart.

5.104 Hähner F (1995) Trägheitsbedingte Abscheidung von Aerosolpartikeln an Einzelkugeln und Kugelverbänden, *PhD Thesis*, University of Kaiserslautern.

5.105 Ranz WE, Wong JB (1952) Impaction of dust and smoke particles on surface and body collectors, *Ind. Eng. Chem.* **44**, 6.

5.106 Lee K, Liu B (1981) Theoretical study of aerosol filtration by fibrous filters, *Aerosol Sci. Technol.* **12**, 79.

5.107 Pfeffer R, Gal E (1982) An experimental evaluation of the rotating fluidized bed filter. In: *Proc. World Filtration Congress III, Dowington, PA.*

5.108 Schmidt E (1978) Theoretische und experimentelle Untersuchungen zum Einfluß elektro-statischer Effekte auf die Nassentstaubung, *J Air Pollut. Control Assoc.* **28**, 143.

5.109 El Halwagi MM (1990) Mathematical modeling of aerosol collection in fluidized bed filters, *Aerosol Sci. Technol.* **13**, 102.

5.110 Wilson EJ, Geankoplis CJ (1966) Collection efficiency by diffusional deposition, *Ind. Eng. Chem. Fundam.* **5**, 9.

5.111 Schweers E (1993) Einfluß der Filterstruktur auf das Filtrationsverhalten von Tiefenfiltern, *PhD Thesis*, University of Karlsruhe.

5.112 Thambimuthu KV (1980) Gas filtration in fixed and fluidized beds, *PhD Thesis* University of Cambridge.

5.113 Wirth KE (1990) *Zirkulierende Wirbelschichten – Strömungsmechanische Grundlagen, Anwendung in der Feuerungstechnik*, 1st edn, Springer, Berlin.

5.114 Kunii D, Levenspiel O (1968) *Fluidization Engineering*, 1st edn, Wiley, New York.

5.115 Tsutsumi A (1995) A novel fluidized-bed coating of fine particles by rapid expansion of supercritical fluid solutions, *Powder Technol.* **85**, 275.

6
Mixing

A. C. Hordijk, A. v. d. Heijden

6.1
Introduction

In the manufacture of propellants and plastic bonded explosives, many ingredients are necessary to obtain a product which satisfies the requirements set. These requirements imply mechanical, ballistic or performance and processing characteristics. Of course, there is some interaction between these properties and in trying to improve one of them one may degrade another property. For instance, to improve the processing characteristics one may use a higher plasticizer content; however, in doing so, the mechanical strength will decrease. A number of these ingredients are listed below:

- binder
- plasticizer
- filler and filler properties (possible interaction with the binder or binder ingredients), such as shape, particle size distribution and bimodality – coarse and fine ratio
- bonding agent
- production additives (e.g. lecithin)
- curative (NCO/OH ratio)
- burning rate modifier
- stabilizers.

These ingredients have to be properly mixed, that is, the various components should be homogeneously distributed over the mix, possible agglomerates should be deleted and particle breakage prevented, so that at the end a product is obtained having the same mechanical and ballistic/explosive characteristics all over the product.

The moment when homogeneity is reached or in other words the mixing process is completed may be assessed by

- monitoring the rheological properties
- measuring the amount of energy put into the mix
- testing the mechanical and ballistic/explosive samples after curing
- characterizing samples from the cured product.

Monitoring the rheological properties and comparing these with previous measurements is one way to establish whether the batch has been prepared in a reproducible

Energetic Materials. Edited by Ulrich Teipel
Copyright © 2005 WILEY-VCH Verlag GmbH & Co. KGaA, Weinheim
ISBN: 3-527-30240-9

way. The end-of-mix and end-of-cast viscosities are well-known parameters to characterize the batch [6.1–6.3].

For a certain mix a specific amount of mixing energy has to be delivered in order to obtain a homogeneous product [6.4–6.6].

The standard procedure is to cast some test specimens together with the desired casting, in order to be able to determine some mechanical properties (e.g. tensile tests) and strand burner tests to check on the burning rate at a given pressure [6.7, 6.8].

The fourth method mentioned is perhaps the most secure one, but it also implies damaging the item. However, in case one wants to monitor the aging characteristics, this is a normal procedure [6.7, 6.9, 6.10].

6.2 Theory

Apart from the experimental approach, a more theoretical approach is also possible. A commonly used method is to use dimensionless numbers such as the Reynolds number (N_{Re}) and the Power number (N_P). Other entities such as the pumping rate (Q) or the power consumption (P) have a dimension.

$$N_P = \frac{g \cdot P}{\rho \cdot N^3 \cdot D^3} \tag{6.1}$$

$$N_{Re} = \frac{\rho \cdot N \cdot D^2}{\eta} \tag{6.2}$$

where g is the acceleration due to gravity (m s^{-2}), ρ is density (kg/m^3), N is the number of revolutions per second (s^{-1}), D is the blade diameter (m) and η is viscosity (Pa s). In *laminar flow*, the product of N_P and N_{Re} equals a constant B [6.6, 6.11]:

$$N_P \cdot N_{Re} = B \tag{6.3}$$

In laminar flow, the Reynolds number is small (< 1). For fluids with a low viscosity the relation shown in Fig. 6.1 is given in [6.12].

Dubois *et al.* [6.11] investigated Eq. (6.3) for more viscous fluids (viscosities of 15 and 175 Pa s) in a conical mixing vessel. They found that B depends on capacity, which is not surprising because of the conical vessel used and B having values of about 1900 for a 2 liter mix. Furthermore, they found that a finite element method using a chemorheological model gave results in good agreement with the experimental data.

The torque (T) needed is dependent on power and rotational speed of the mixing blade [6.13]:

$$P = T \cdot N \tag{6.4}$$

or

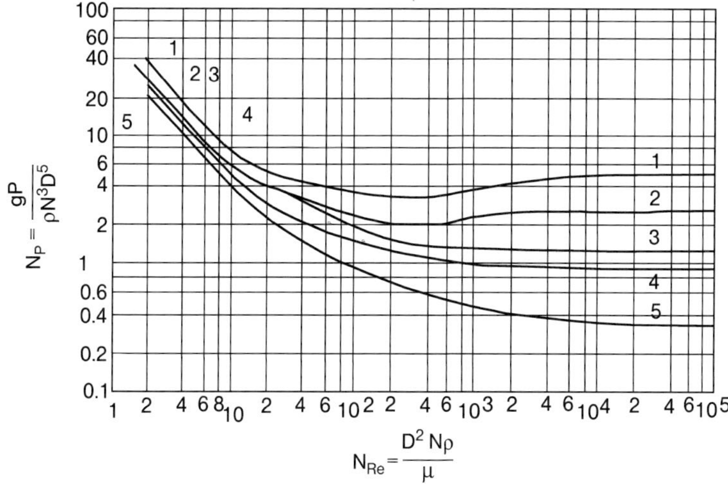

Figure 6.1. Power number as a function of the Reynolds number for low-viscosity fluids [6.12].

$$T = C \cdot N \cdot \eta \tag{6.5}$$

With constant rotational speed and measuring the torque, the total amount of work needed for homogenizing the mix is W_u:

$$W_u = 2\pi \cdot N \cdot M^{-1} \cdot \int T \cdot dt \tag{6.6}$$

to be integrated from start to end of the mixing time. M (kg) is the mass of the mix. This equation has been used in the rubber industry to study high-shear mixing. Further, the integral may be of help in the scale-up of processes and with regard to batch-to-batch uniformity.

Typical values are 80 kJ/kg for an inert propellant containing 60 vol.% of filler and 700 kJ/kg for a triple-base propellant of a much higher viscosity.

There is a relationship between the torque and the mix viscosity according to Eq. (6.5). However, the shear rate applied to the mix is proportional to N, the rotational speed and the viscosity is a function of N as well for non-Newtonian fluids. Therefore, one should keep N constant and a linear relationship between the torque measured and the viscosity of the mix will be obtained [6.6].

A considerable amount of theoretical work has been published on the description of the homogeneity of the mix. This is, however, not treated here and a summary can be found elsewhere [6.13, 6.14].

6.3 Type of Mixers

The type of mixer (the design of the mixing blades) is of considerable importance for the product quality and the required mixing time [6.15]. In Perry's Handbook [6.12], an overview is presented for paste and viscous materials. Two mixers are mentioned: batch mixers and intensive mixers. Batch mixers are:

- change-can mixers, such as vertical planetary mixers as produced by Propex;
- helical-blade mixers, as used in [6.6];
- double-arm kneading mixers, for which a number of special blades are designed, such as the Sigma blade (see Fig. 6.2) produced by Baker Perkins.

Examples of the intensive mixers are:

- the Banbury mixer, especially designed for the plastics and rubber industry with a very small clearance between wall and rotor where in effect the mixing action takes place;
- roll mills, which may provide high localized shear.

There are horizontal and vertical mixers having their own advantages and disadvantages. The larger paste mixers are usually vertical, planetary mixers.

TNO has chosen vertical mixers with planetary blades with a capacity of about 2.5 and 12.5 l of propellant mix because of their good design and mixing capability. Both mixers make use of planetary mixing blades – this is a vertical two-blade mixer, with the central blade stationary, while the outer blade rotates around the central one and simultaneously rotates around its own axis.

Smaller mixers are mostly of the horizontal type with another type of blade, so-called Z-blade or Sigma blade mixers (see Fig. 6.2 [6.12]), which may apply higher shear but are more expensive and less easy to operate than the planetary mixers during vacuum casting. The better mixing efficiency arises from the shorter time needed to obtain approximately the same rheological properties.

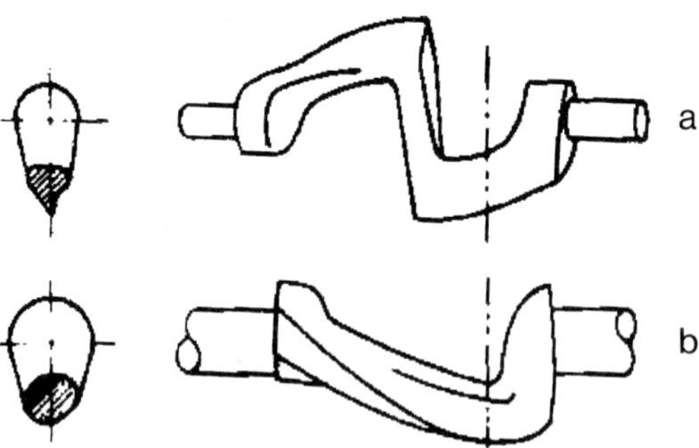

Figure 6.2. (a) Sigma and (b) dispersion blade for a double-arm kneader mixer [6.12].

In the 1980s, a study was performed on coarse AP particle breakdown by a planetary mixer. This was done by determining the effect of final mixing time on the mechanical properties and burning rate of a rocket propellant. In case particle breakdown takes place, the burning rate should increase as well as a change in mechanical properties, especially an increase in the Young's modulus [6.15]. The latter parameter is the slope of the linear part in the stress-strain relationship, representing the elastic part in which no damage is inflicted on the test specimen.

The final mixing starts after the introduction of the curing agent until the end of mixing; the final mixing time was varied between 10 and 90 min. It was found that after an initial effect of about 30 min, a stationary situation occurs with a continuous, although small, increase in both the Young's modulus and burning rate, pointing at particle breakdown.

The effect of relative rotation rates of the mixing blades has been investigated [6.1, 6.4, 6.16] in the manufacture of an 86 wt% solids composite propellant. The mixer leading (stationary) blade speed was varied from 10 to 60 rpm at five different settings. The speed of the other blade varied from 7 to 44 rpm at a fixed speed ratio. Tth temperature was set at 50 °C and the mixing cycle time was kept at 120 min.

The mixing speed appeared to have a strong influence on the yield stress and on the viscosity at unit shear rate as a function of mixing time. This effect is again due to increased grinding of the particles at higher mixing speeds. The pseudoplasticity index n (see Chapter 12) reached maximum values after 2–4 h after curative addition for all speeds. First it was concluded that a mixer speed of 25 rpm seems to be the optimum; at higher speeds too much shear could shatter the AP particles. Second, continued mixing leads to breakage of AP particles, thereby affecting the coarse to fine AP ratio and, as a consequence, also the mechanical and ballistic properties of the final composition.

Another way of mixing is to use a roller mill, as done by Sayles [6.17]. In his patent he compares the effect of mixing in a Sigma blade mixer and in a roller mill. In Table 6.1 the composition and the mixing results in terms of the mechanical and ballistic properties are given.

Table 6.1. Effect of mixing on mechanical and ballistic properties [6.17].

	Composition (%)	Sigma blade mixer	Roller mill
HTPB + AO + IPDI	9.7		
HX 752	0.3		
DOA	2.0		
AP	86.0		
Al	1.0		
Fe_2O_3	1.0		
Max. strength (MPa)		1.03	1.70
Strain at max. strength (%)		40	45
E modulus (MPa)		2.05	6.74
Burning rate at 14.2 MPa (mm/s)		4.06	7.11

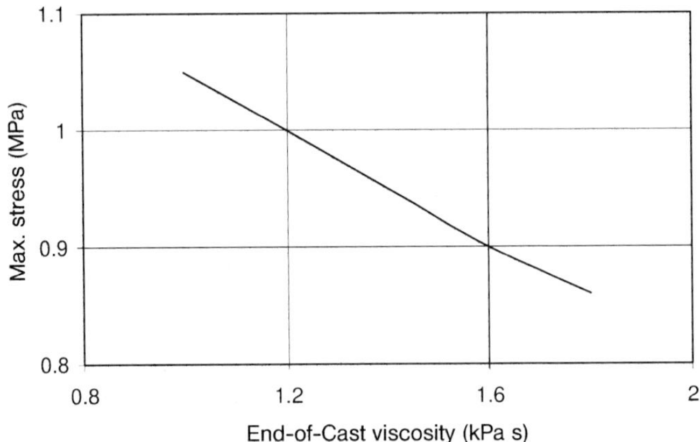

Figure 6.5. Fit between end-of-cast viscosity and maximum stress [6.2].

Muthiah *et al.* [6.16] reported on the effect of mixing speed and mixing time on the rheology of HTPB/AP composite propellants. The composition consisted of bimodal AP (68 wt%), Al (18 wt%) and HTPB/ DOA/TMP/TDI together 14 wt%.

The results of the first set of experiments have been reported in the previous section; in the second set of experiments, the mixing cycle time was varied from 60 to 240 min at a temperature of 50 °C and blade speeds of 25 and 18 rpm.

Rheological measurements were carried out 1, 2, 3, 4, 5 and 6 hours after curative addition.

The mixing speed has a strong influence on yield stress and at the viscosity at unit shear rate in time. This effect could be due to increased grinding at higher mixing speeds. The pseudoplasticity index n reached maximum values after 2–4 h after curative addition for all speeds.

It appeared that the effect of the mixing time on the viscosity at a shear rate of 1 s^{-1} can be neglected; the same holds for the yield stress. However, the pseudoplastic index n is influenced by the mixing time – a minimum mixing time of at least 120 min is necessary in order to reach a maximum value. It was concluded that mixing times longer than 180 min must be avoided because of the grinding and shattering of the coarse AP particles by blade impact. The overall conclusions were as follows:

- The rheological parameters depended only to some extent on the mixing parameters such as mixer blade speed and mixing time.
- A mixer speed of 25 rpm seemed to be the optimum – at higher speeds too much shear could shatter the AP particles.
- Continued mixing led to breakage of AP particles, thereby affecting the coarse to fine AP ratio and in the end possibly also the mechanical and ballistic properties of the propellant.

6.5 Sequence of Addition of Ingredients

The addition of some of the ingredients and their order of addition is critical; for instance, binder ingredients, which are solid and may dissolve in another binder ingredient, should give a chance of flocculation. The addition of the fines – before or after the addition of the coarse particles – determines to a (large) extent the flow behavior of the mixture and the mechanical and ballistic properties after curing.

The addition of burning rate catalysts or burning rate modifiers may also influence all of the above-mentioned properties, including the cure and cure rate.

The addition sequence is of great importance, from the viewpoints of both safety and the homogeneity of the mix. A dry mix of oxidizer and reducing agents, for instance a dry mix of AP and Al, must be prevented in view of the hazard involved (undesired ignition).

In [6.18] a solid propellant processing flow diagram is given for the composition given in Table 6.2. From this flow diagram, the following may be observed:

- First coarse AP is added followed by 5 min at atmospheric pressure and 30 min of vacuum mixing.
- The fine AP is added in two parts – the first part is mixed similarly to the mixing procedure for the coarse AP; the second part is mixed for 5 min at atmospheric pressure and 60 min under vacuum.
- The mixture is cured for 10 days at 60 °C.

The investigation concerned the effect of coarse particle break-up due to mixing, introducing more fines, which influences the burning rate and particle packing. It was concluded, however, that the break-up is very limited and that effects such as settling of the fines among the coarse in the slurry, wall effects and particle size distribution are all important and must be considered [6.18].

Marine and Ramohalli [6.8] described another propellant using aluminum. The addition sequence is somewhat changed from that given in Table 6.3.

Table 6.2. HTPB/AP composition used for mixing time experiments (plasticizer/binder ratio: 7.5% on binder).

No.	Ingredient	Content (wt%)
1	AP 200 micron	56.00
2	AP 10 micron	24.00
3	HTPB (R 45)	16.46
4	Alrosperse	0.300
5	IPDI	1.440
6	Tepanol	0.200
7	Antioxidant	0.100
8	IDP	1.500

Table 6.3. Addition sequence as used by Marine and Ramohalli [6.8].

Step	Action	Time (min)[a]	Temperature (°C)
1	Add the prepolymer, the plasticizer and burn rate catalyst	5 (atm.)	71
		10 (vac.)	
2	Add the aluminum	5 (atm.)	71
		15 (vac.)	
3	Add the AP	15 (atm.)	60
	Clean blades	30 (atm.)	
		60 (vac.)	
4	Add the curative	10 + 15 (vac.)	60

[a] atm., atmospheric pressure; vac., vacuum.

Marine and Ramohalli found that the best procedure was to store the propellant mix after step 3 for 1 day at 60 °C and then repeat 60-min vacuum mixing the next day [6.8].

The procedure for the production of a propellant, for instance the TPS (Turbo Pomp Starter) propellant used in the Ariane IV, is laid down in a SOP (Standard Operation Procedure). In such a procedure, the following items need to be treated:

- the compounds needed, the manufacturer thereof and the storage conditions;
- the number of the different production phases – production of binder, production of premix, production of propellant, casting of propellant, curing of the propellant;
- the devices needed including balances, mixer, stoves;
- safety precautions to be taken;
- detailed description of the various activities, the sequence thereof;
- detailed description of the mixing process, the time intervals between the various steps.

A first version of such a procedure is normally developed on the basis of general knowledge and experience, completed with information from the literature. The procedure is followed in small-scale mixes and adapted and improved in time until the final version is reached.

The addition sequence followed by TNO up to now for the production of propellant and PBXs is that the coarse fraction of the oxidizer/explosive is introduced first, after which it is wetted by the binder under atmospheric pressure; then the fine fraction is added in two steps, wetted and mixed under vacuum.

6.6 Scale Effects

A number of studies are dedicated to the effect of production scale on flow and final properties, which is of course of great interest to the development and scale-up of propellants and PBXs.

Marine and Ramohalli [6.8] reported an analysis on 70 batches of an AP/Al propellant based on a PBAN binder. The scaling effect, from a 4-l (1-gallon) to a 600-l mixer, was also studied. With regard to the scaling effect, the following results were obtained:

- The uncertainties within a single batch were larger in scaling up: from 2 to 13% for the various properties determined.
- The scale had a pronounced effect on the burning rate over a large range of pressures; the burning rate of the propellants produced on small scale is larger (maximum of 10%) than that of the 600-l mixes [6.2, 6.11].

6.7 Conclusions

For the production of propellants and plastic bonded explosives, planetary mixers are generally used for larger scale vacuum casting. The Sigma blade mixers show a better mixing efficiency, but are usually applied only for smaller mixers. In scaling this should be borne in mind.

Mixing times are at least 60 min after the introduction of the formulation ingredients; the torque time curve shows a minimum, in order to ensure complete homogenization of the mix. In case longer mixing times are applied, particle break-up, in particular of the coarse particles, may occur, resulting in a viscosity increase of the mix and increased power consumption and torque.

After the introduction of the curing agent, the mixing should be of at least 30-min duration in order to ensure complete homogenization.

The addition sequence is important especially for the homogeneous distribution of the bonding agent over the coarse and fine particles. Allowing the batch to be stored overnight prior to the addition of the curing agent may also lead to a more homogeneous product.

With regard to the reproducibility, it can be stated that batch-to-batch variations are within 1% and that the standard deviation of maximum stress and strain at maximum stress is about 5%.

6.8 References

6.1 Muthiah RM, Manjari R, Krishnamurthy VN (1993) Rheology of HTPB propellants: effect of mixing speed and mixing time, *Defence Sci. J.* **43**, 167–172.

6.2 Perez DL, Ramohalli NR, Rao RK, Fulian C (1991) First steps towards a scientific approach to the processing of filled polymers, *Propell. Explos. Pyrotech.* **16**, 16–20.

6.3 Hordijk AC, Bouma RHB, Schonewille E (1998) Rheological characterization of castable and extrudable energetic compositions. In: *Proc. 29th Int. Annual Conf. of ICT, Karlsruhe*, p. 21.

6.4 McKetta (ed.) Mixing and blending. In: *Unit Operations Handbook 2, Mechanical Separations and Materials Handling*, Chapter 8.

6.5 Layton RA, Murray WR (1997) The control of power for efficient batch mixing, *Propell. Explos. Pyrotech.* **22**, 269–278.

6.6 Brousseau P, Hooton I (1995) The use of a torque sensor in the processing of PBXes, report *DREV-TM-9445*.

6.7 Keizers HLJ, Hordijk AC, v. d. Vliet L (2000) Modelling of composite propellant properties. In *AIAA 2000–3323: 36th Joint Prop. Conf. and Exhibition*, AIAA/ASME/SAE/ASEE.

6.8 Marine M, Ramohalli K (1990) Processing experiments on model composite Propellants. In *AIAA 26th Joint Prop. Conf.*, AIAA 90–2313.

6.9 Keizers HLJ, Miedema JR (1996) Structural service lifetime modelling for solid propellant rocket motors. In: *AGARD Symp.*, NATO, CP-586.

6.10 Menke K, Eisele S (1997) Rocket propellants with reduced smoke and high burning rates, *Propell Explos Pyrotech.* **22**, 112–119.

6.11 Dubois C, Thibault F, Tanguy PA, Ait-kadi A (1996) Characterization of mixing processes for polymeric energetic materials. In: *Proc 27th Int. Annual Conf. of ICT, Karlsruhe*, p. 6.

6.12 Perry RH (1967) *Chemical Engineers Handbook*. McGraw-Hill.

6.13 Danckwerts PV (1953) The definition and measurement of some characteristics of mixtures, *Appl. Sci. Res.* 279–296.

6.14 Tucker CL III (1977) Principles of mixing measurement. In: Middleman S (ed.), *Fundamentals of Polymer Processing*, Chapter 12.

6.15 Ramohalli K (1984) Influence of mixing time upon burning rate and tensile modulus of AP/HTPB composite propellants. In: *Proc 15th Int. Annual Conf. of ICT, Karlsruhe.*

6.16 Muthiah RM, Krishnamurthy VN, Gupta BR (1996) Rheology of HTPB propellants: development of generalised correlation and evaluation of pot life, *Propell. Explos. Pyrotech.* **21**, 186–192.

6.17 Sayles DC (1987) A processing method for increasing propellant burning rate, *US Patent* 4 655 860.

6.18 Ramohalli K, El Helmy A (1982) Analysis of processing variables in propellant burn rate and modulus. (1) Mixing times. In: *Proc Int. Symp. Space Technol. and Sci.*, pp. 117–124.

7
Nanoparticles

A. E. Gash, R. L. Simpson, Y. Babushkin, A. I. Lyamkin, F. Tepper, Y. Biryukov, A. Vorozhtsov, V. Zarko

7.1
Nano-structured Energetic Materials Using Sol-Gel Chemistry[1]

7.1.1
Introduction

Since the invention of black powder, 1000 years ago, the technology for making solid energetic materials has remained either the physical mixing of solid oxidizers and fuels (e. g. black powder) or the incorporation of oxidizing and fuel moieties into one molecule [e. g., trinitrotoluene (TNT)]. The basic distinctions between these energetic composites and energetic materials made from monomolecular approaches are as follows. In composite systems, desired energy properties can be attained through readily varied ratios of oxidizer and fuels. A complete balance between the oxidizer and fuel may be reached to maximize energy density.

Table 7.1 is a summary of the some of the energy densities of composite and monomolecular energetic materials [7.1]. One can see that current composite energetic materials can store energy as densely as 23 kJ/cm^3. However, owing to the granular nature of composite energetic materials, reaction kinetics are typically controlled by the mass transport rates between reactants. Hence, although composites may have extreme energy densities, the release rate of that energy is below that which may be attained in a chemical kinetics-controlled process.

In monomolecular energetic materials, the rate of energy release is primarily controlled by chemical kinetics and not by mass transport. Therefore, monomolecular materials can have much greater power than composite energetic materials. A major limitation with these materials is the total energy density achievable. Currently, the highest energy density for monomolecular materials is ~ 12 kJ/cm^3,

1) This work was performed under the auspices of the US Department of Energy by University of California Lawrence Livermore National Laboratory under contract No. W-7405-Eng-48.

Energetic Materials. Edited by Ulrich Teipel
Copyright © 2005 WILEY-VCH Verlag GmbH & Co. KGaA, Weinheim
ISBN: 3-527-30240-9

Table 7.1. Energy densities for several monomolecular and composite energetic materials (adapted from [7.10]).

Energetic material	Energy density (kJ/cm^3)
ADN/Al max.	23
Compression moldable	19–22
Strategic propellants	14–16
CL-20 (neat)	12.6
Tritonal	12.1
HMX (neat)	11.1
LX-14	10
TATB (neat)	8.5
Composite C-4 as used	8
LX-17	7.7
TNT (neat)	7.6

High energy

Low energy

about half that achievable in composite systems. The reason for this is that the requirement for a chemically stable material and the available synthetic procedures limit both the oxidizer-fuel balance and the physical density of the material. Therefore, it is desirable to combine the excellent thermodynamics of composite energetic materials with the rapid kinetics of the monomolecular energetic materials. Hence the development of methods for the physical mixing of oxidizers and fuels at the nanometer scale is needed for tailorable energy and power. In addition, it is known that decreasing the scale of the grain structure can dramatically decrease shock sensitivity at low pressures and, in some cases, permits more rapid initiation at high pressures.

The field of nanoscience has grown immensely in the last 10 years [7.2–7.5]. Nanoscience can be loosely defined as the synthesis, processing, characterization and application of materials whose dimensions are in the nanometer-size range (typically defined as 1–100 nm). A primary reason for the great interest in nanoscience is the properties of materials with nanoscale dimensions, relative to properties of the bulk. It is well known that the mechanical, electronic, optical and catalytic properties of a material are significantly altered when the material is reduced to the nanometer size region. The alterations of properties occur because properties of materials have a threshold dimension size scale below which the physics of the property begin to change [7.6]. For example, the melting point of 20-Å diameter elemental gold particles is depressed nearly 800 °C relative to the bulk and nanocrystalline copper is five times harder than micrometer-sized copper [7.7]. One area of nanoscience that is currently undergoing extensive active research at many institutions is the development of nanocomposites.

A composite is a material that is a mixture of two or more phases in intimate contact with one another. Composites typically display properties that are not attainable with any one of the individual components. This is due to the considerable amount of surface contact between the constituent phases. A nanocomposite is a mixture of at least two different materials where at least one of the phases has one or more dimensions (length, width or height) in the nanosize range, defined previously. Nanocomposites very often display new, interesting and useful properties that the conventional materials lack. The special properties displayed by nanocomposites are a direct result of their nanometer-sized building blocks. The very small particles of the constituent phases have immense surface areas, which means that there is an enormous amount of surface interfacial contact between them. Hence the large amount of surface contact influences the material's properties to a great extent. In conventional composites the sizes of the constituent materials are at the micrometer size scale, with much less surface area and many fewer surface contacts and, therefore, less influence on the overall properties of the composite.

Nature has made very efficient use of a variety of nanocomposites, two of which are bone and mollusk shell. Both materials are nanocomposites of a crystalline inorganic component and an organic component that display remarkable mechanical properties [7.2]. Researchers have prepared synthetic nanocomposites for use as structural materials, coatings and catalysts and in electronics and biomedical applications. Consumers are just starting to see the effects of nanoscience in their everyday lives as commercial products in automobiles and home and personal care products utilizing nanocomposites have recently been developed. However, one field with much less research in the areas of its nanoscience and nanocomposites than the aforementioned fields is that of energetic materials.

We have been actively investigating the application of nanoscience to the field of energetic materials for several years. Our main interest in this area is the application of the sol-gel synthetic approach to the preparation of energetic materials [7.8, 7.9]. Sol-gel chemistry is an approach to creating structures at the nanometer scale and allows the formation of energetic materials in an entirely new way. The appeal of the sol-gel approach to energetic materials is that it offers the possibility of precisely controlling the composition, density, morphology and particle size of the target material at the nanometer scale. These are important variables for both safety and performance considerations. Such variables are difficult to achieve by conventional techniques. The fine control of these parameters allows the chemist the convenience of making energetic materials with tailored properties. We also believe that such control could permit the creation of entirely new energetic materials with desirable or potentially exceptional properties [7.10].

In addition to providing fine microstructural and compositional control, capable of producing materials which are both of high energy density and powerful, sol-gel methodology offers other advantages of safety and stability in energetic material processing. For example, ambient temperature gelation and low-temperature drying schemes prevent the degradation of the energetic molecules and the water-like viscosity of the sol before gelation allows easy casting to near net shapes, which is preferred over the hazardous machining alternative. Although sol-gel chemistry is a well-known means of producing nanostructured materials, until recently it had not been used to process energetic materials [7.8, 7.9].

Table 7.4. Summary of N_2 adsorption/desorption results for dry Fe_2O_3 gels made in ethanol.

Gel type	Precursor salt	BET surface area (m^2/g)	Pore volume (ml/g)	Average pore diameter (nm)
Xerogel	$Fe(NO_3)_3 \cdot 9H_2O$	300	0.22	2.6
Aerogel	$Fe(NO_3)_3 \cdot 9H_2O$	340	1.25	12
Aerogel	$FeCl_3 \cdot 6H_2O$	390	3.75	23

7.1.5.3 Surface Area, Pore Size and Pore Volume Analyses

Table 7.4 summarizes the surface areas, pore volumes and average pore sizes of some Fe_2O_3 aerogels and xerogels. In general, all of the materials listed in Table 7.3 have high surface areas and pore diameters whose dimensions are in the micro- to lower mesoporic (2–20 nm) region. Note that the xerogel solid has a comparable total surface area to the aerogel material made under identical conditions. However, the pore volume and average pore diameter of the xerogel sample are significantly smaller than that of the aerogel sample (0.22 ml/g and 2.6 nm compared with 1.25 ml/g and 12 nm, respectively). In addition, the adsorption-desorption isotherm of the xerogel solid is a Type I isotherm, indicative of a microporous solid, whereas the isotherm for the aerogel sample is a Type IV isotherm, indicative of a mesoporous solid. The differences in the results for these two samples is most likely due to the different processing conditions (evaporation and supercritical extraction, respectively) to which each was subjected. Simply put, the evaporation of the ethanol from the xerogel sample exerted substantial capillary forces on the gel's pore structure, which resulted in significant shrinkage of the pores, relative to the aerogel sample. The surface areas reported in Table 7.4 for the Fe_2O_3 gels (300–400 m^2/g) are significantly higher than those reported for other sol-gel preparations of Fe_2O_3 solids (10–80 m^2/g) [7.21, 7.22]. These results, along with those for HRTEM,

Table 7.5. Summary of syntheses of different transition and main group metal oxides from their salts by the epoxide addition method.

Precursor	Oxide	Gel?
$Fe(NO_3)_3 \cdot 9H_2O$	Fe_2O_3	Yes
$Cr(NO_3)_3 \cdot 9H_2O$	Cr_2O_3	Yes
$Al(NO_3)_3 \cdot 9H_2O$	Al_2O_3	Yes
$In(NO_3)_3 \cdot 5H_2O$	In_2O_3	Yes
$Ga(NO_3)_3 \cdot xH_2O$	Ga_2O_3	Yes
$NiCl_2 \cdot 6H_2O$	NiO	Yes
$SnCl_4 \cdot 5H_2O$	SnO_2	Yes
$HfCl_4$	HfO_2	Yes
$ZrCl_4$	ZrO_2	Yes
$NbCl_5$	Nb_2O_5	Yes
$TaCl_5$	Ta_2O_5	Yes
WCl_6	WO_3	Yes

indicate that Fe_2O_3 prepared by this method is made up of nanosized building blocks with very high surface area and significant pore volume and, depending on the process conditions, one can control the pore volume and size. This suggests that dry Fe_2O_3 gels made by this method are unique materials that should be evaluated as components of energetic materials as well as other applications utilizing Fe_2O_3.

With our demonstrated successes in the syntheses of Fe_2O_3 porous materials using the epoxide addition method, it was a logical progression to apply it to the preparation of other metal oxide aerogels. We attempted to synthesize aerogels of several different metal oxides using the epoxide addition method, the results of which are summarized in Table 7.5.

One can see that this method is applicable to the syntheses of a variety of metal oxides. Some of these same metal oxides also undergo very energetic thermite reactions according to Table 7.2. Similarly to the Fe_2O_3 materials, these metal oxides have been extensively characterized and shown to be made up of particles with nanometer-sized dimensions and very large surface areas. These findings should allow other thermitic nanocomposites to be readily prepared.

7.1.6 Iron Oxide-Aluminum Nanocomposites

The synthesis of ultrafine grain (UFG) powders of nanosized aluminum with narrow size distributions was first achieved by researchers at the Los Alamos National Laboratory (LANL) [7.23, 7.24]. The UFG aluminum powders are synthesized using dynamic gas condensation. In this technique, resistive or radio-frequency (rf) induction heating is used to heat an Al metal source to ~1300 °C. This results in the volatilization of aluminum metal atoms into a reactor chamber containing an inert gas at low pressure. In the reactor chamber aluminum atoms collide, nucleate and then grow. UFG particles of aluminum then condense in cooler parts of the reactor. Changing the reactor conditions can alter the rate of nucleation and growth and thus the subsequent size of the aluminum particles. Before recovery, oxygen is slowly introduced into the reactor chamber to passivate the UFG aluminum with a layer of Al_2O_3. This process has also been used to produce UFG powders of MoO_3, which when mixed with the UFG aluminum (~30-nm diameter particles) produces a thermitic composition that is called metastable interstitial composite (MIC) [7.25–7.27]. The MIC material reacts more than 1000 times faster than conventional powdered thermite mixtures. This observation is mainly due to the fact that the diffusion distances between reactants are much smaller in the MIC materials. The large-scale production of UFG aluminum is currently being implemented at several private and governmental sites. An HRTEM picture of nanosized aluminum made by this method is shown in Fig. 7.6. It is clear that the method produces very small particle sized aluminum.

In addition, it appears as though the particle size distribution is relatively narrow. However, the powder is not entirely made up of discrete spheres. The HRTEM image contains several examples of what looks like coalesced particles. In addition, analyses have indicated that the particles contain a passivating coating of aluminum

The products from the combustion of selected nanocomposites were analyzed using powder X-ray diffraction (PXRD) (see Chapter 9). The major products identified were metallic Fe and Al_2O_3. Some minor products consisting of iron aluminates were also observed, indicating that the reaction was either not complete or the reaction was not stoichiometric. However, the tentative results indicate that we are observing the thermite reaction in experiments such as that shown in Fig. 7.8.

The sol-gel approach also allows the relatively simple incorporation of other metal oxides into the metal oxide matrix to make a mixed metal-oxide material. Different metal oxide precursors can be easily mixed into the M^{n+} solution, before the addition of the epoxide. Dilution of the thermitic material with inert oxides such as Al_2O_3 (from dissolved $AlCl_3$ salt) or SiO_2 (from added silicon alkoxide) leads to a pyrotechnic material that is not as energetic as a pure iron(III) oxide-aluminum mixture. We have performed such syntheses. Qualitatively, the resulting pyrotechnics have noticeably slower burn rates and are less energetic. Alternatively, one could add metal oxide components that are more reactive with Al(*s*) to increase the energy released. Finally, this would also permit the addition of metal oxide constituent(s) that provide a desired spectral emission to the energetic nanocomposite. This type of synthetic control should allow the chemist to tailor the pyrotechnic's burn and spectral properties to fit a desired application. This is advantageous because the compositional changes can be made in solution and would not require the dry blending of additional inert or energetic components.

7.1.7 Gas-generating Energetic Nanocomposites

The thermitic nanocomposites described so far release their energy in the form of heat and light, as they are gasless reactions. However, a desirable quality of many energetic materials is their ability to generate gas for applications where pressure-volume work is needed. The sol-gel approach to nanostructured energetic materials described here is versatile enough to incorporate gas-generating materials into the process. There are two methods to do this. In one case a gas generator, such has an organic polymer, may be added to an otherwise thermitic sol-gel material. A second approach is to make carbon-based gels that are oxidized to gaseous species.

7.1.8 Hydrocarbon-Ammonium Perchlorate Nanocomposites

As shown previously, sol-gel chemistry can be used to prepare energetic nanocomposites where the skeleton of the material is the oxidizer and the fuel particles reside in the pores. However, we can also make energetic nanocomposites where the solid skeleton is the fuel and the oxidizer resides in the pores; subsequent reactions produce all gaseous products. Porous organic-based solids can be derived from a sol-gel procedure reported by Pekala [7.28]. These materials are synthesized by the polycondensation of resorcinol (1,3-dihydroxybenzene) with formaldehyde (RF).

Figure 7.9. Transmission electron micrograph of resorcinol-formaldehyde aerogel with ammonium perchlorate crystallized in the pores.

The result of this reaction is the formation of a porous organic-based ($-CH_2-$) solid that is made up of nanometer-sized clusters linked together in a fashion very similar to that shown in Fig. 7.2. Crystallization of an oxidizer into the nanometer-sized pores of this material would result in an energetic nanocomposite. We have prepared such materials where ammonium perchlorate (AP) was the oxidizer deposited in the pores [7.9].

Transmission electron microscopy (TEM) was performed on a nanocomposite using the procedures described above. Figure 7.9 shows a solid structure composed of interconnected clusters of nanometer-sized primary particles and crystallites, generally smaller than 20 nm. Near-edge X-ray absorption on the same material was also done. This technique creates an image by scanning a monochromatic X-ray beam from a synchrotron source across the sample and recording the near-edge X-ray absorption intensity of nitrogen. As the only source of nitrogen is from AP, the distribution of the oxidizer in the nanostructured material can be examined. It was shown that the nitrogen was uniformly distributed in the material on a scale less than 43 nm, which was the limit of resolution for the instrument.

The characterization described above showed that the RF-AP materials are nanostructured. To determine whether the material was energetic, differential-scanning calorimetry (DSC) was performed. DSC plots for an RF-AP nanocomposite and neat ammonium perchlorate are compared in Fig. 7.10 (a) and (b). The trace for the RF-AP nanocomposite showed an exotherm at about 250 °C, indicating that it is indeed energetic; pure AP has significantly less exothermicity and integrated decomposition enthalpy in the absence of the fuel gel skeleton (RF).

7.1.9 Summary

We have presented a summary of our work on the preparation and characterization of energetic nanocomposites using sol-gel methods. We have shown that sol-gel methods can be used to prepare pyrotechnic and explosive composites that are intimately mixed and whose constituents are on the nanometer size scale. Through both qualitative and quantitative characterization we have shown that the materials

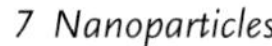
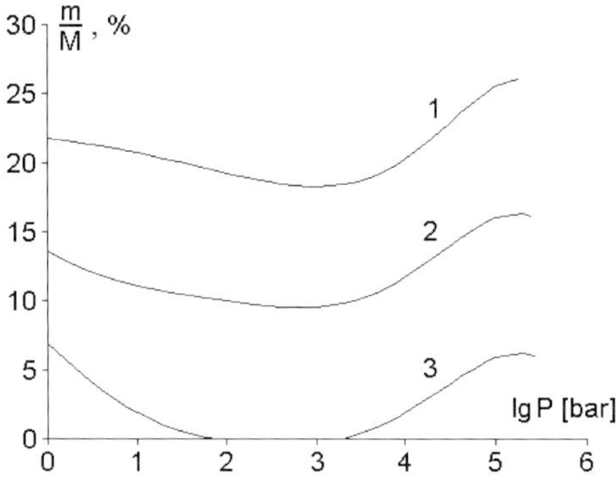

Figure 7.13. Amount of condensed carbon at the expansion isentrope of detonation products. 1, TNT; 2, TNT-RDX 50/50; 3, RDX ($\rho_0 = 1.6$ g/cm^3).

yield with TNT is justified by another reason – the incomplete transformation of amorphous carbon into diamond in the zone of chemical reaction.

There is an additional proof in favor of the proposals made relating to the character of losses. Thus, for example, the diamond yield for TNT-RDX 50/50 can constitute 10% whereas the calculations give 9.5%. The value of 10% was obtained with the charges placed in a massive ice envelope. In the case of TNT, the diamond yield in the regime of the overcompressed detonation equals 15% [7.29], whereas the calculated value is 18%.

The maximum yield of diamond for a given HE composition is reached when the parameters of preservation provide DP expansion conditions close to the isentropic regime. The diamond losses in this case will be called the basic losses. On expansion, the DP are subjected to the action of the shock wave (SW), except in the idealized case of the unlimited expansion into a vacuum. The more intense the SW, the greater is the deviation of the state of the DP from the C–J isentrope and the larger are the diamond losses. Hence there are additional losses in the SW dependent on the external conditions.

In gas dynamics calculations, the influence of the charge mass and surrounding gas pressure on the diamond yield was studied, with the composition of TNT-RDX 50/50 and gases (CO_2 and N_2) as an example. Figure 7.14 (a)–(d) represent four consecutive phases of the DP course. The vertical marks on the horizontal axis show the demarcation line between the DP and the external gas. After the HE detonation, the SW in the surrounding gas and the inner SW that spreads in the DP (Fig. 7.14 (a)) are formed. The appearance of the inner SW is attributed to the specificity of the spherical expansion. The velocity of the inner SW in relation to the substance is smaller than that of a flow and this is why the SW is carried away by the flow towards the wall of a chamber. Having been reflected from the wall, the SW comes up to the DP-gas demarcation line (Fig. 7.14 (b)) and after tangential rupture enters the DP region. The successive compression of the DP by the inner SW and by the reflected SW is close to the isentropic one, so the additional losses are minor. Thereafter the reflected SW catches up with the inner wave and a strong SW is formed (Fig. 7.14(c)). As a result, the DP is divided into a comparatively rarefied hot inner nucleus and a denser cold external region preliminarily compressed with the

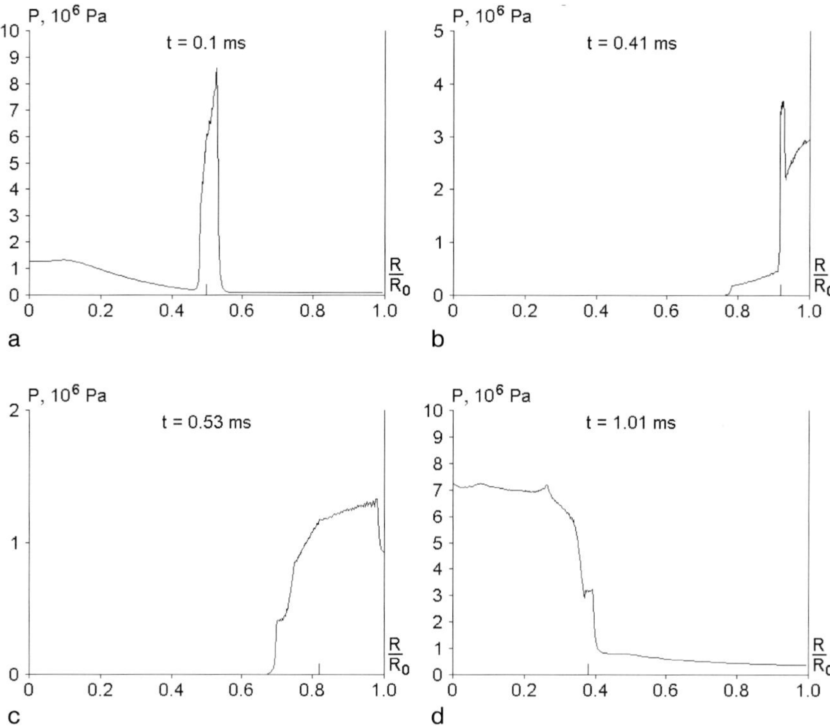

Figure 7.14. Pressure profiles in the explosion chamber (2 m^3) at times (a) 0.1, (b) 0.41, (c) 0.53 and (d) 1.01 ms. TNT-RDX 50/50 (1 kg); CO_2 1 atm.

inner SW. The calculations showed that the hot nucleus contributes the greatest additional losses of the diamond (see Figs 7.18 and 7.19). Subsequently, the wave is reflected from the symmetry center and reaches the DP-gas contact boundary (Fig. 7.14(d)). The progression of the process is hindered by the interaction of the SW with the demarcation boundary between hot and cold DP. The further circulation of the SW proceeds with a much more lower intensity and does not result in changes of the calculated values reached (the local diamond yield and peak temperatures).

Figures 7.15 and 7.16 display the corresponding dependences of the local diamond yield and peak temperature on the Lagrange charge coordinate. In the frames of the initial assumptions, the losses of diamond connected with annealing were not observed. Further, when compared, the figures show that the peak temperatures in the presence of CC do not exceed 1500 K. This suggests that there is no annealing of the diamond in reality, since the value of 1500 K is considerably lower than the reasonable evaluations of the temperature of annealing, and its losses are entirely due to the chemical reactions.

As the pressure of the surrounding gas increases or the mass of a charge decreases, the intensity of HE and the size of the hot nucleus decrease and the

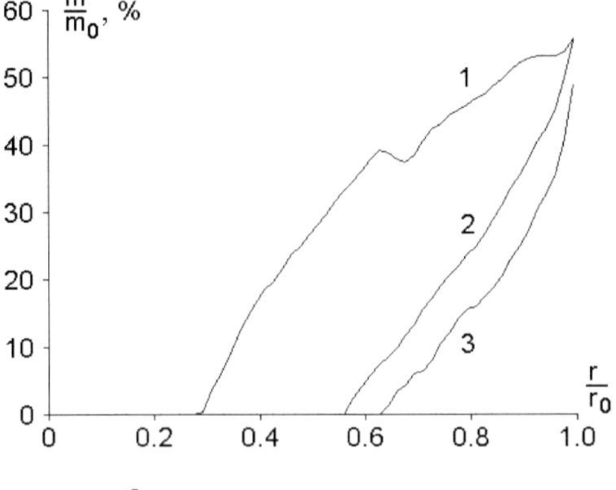

Figure 7.15. Local diamond yield (m). m_0 = Diamond yield in C-J plane; r = Lagrange coordinate; r_0 = charge radius. TNT-RDX 50/50; CO_2; volume of explosion chamber = 2 m^3. (A) Pressure of gas = 1 atm; charge mass: 1, 250; 2, 500; 3, 1000 g. (B) Charge mass = 1000 g; pressure of gas: 1, 4; 2, 2; 3, 1 atm.

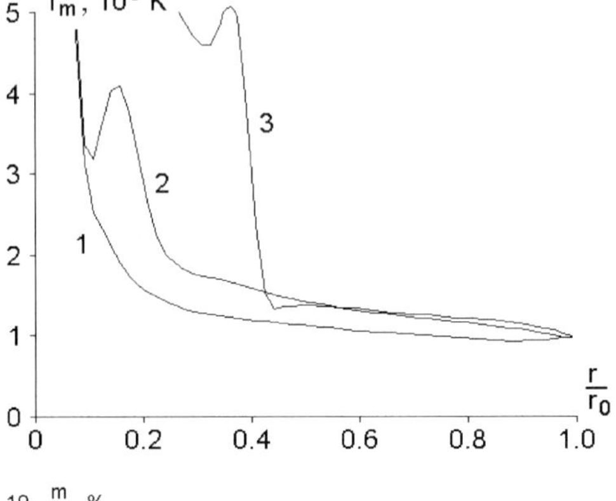

Figure 7.16. Local peak temperatures (T_m). r = Lagrange coordinate; r_0 = charge radius. TNT-RDX 50/50; CO_2; volume of explosion chamber = 2 m^3. (A) Pressure of gas = 1 atm; charge mass: 1, 250; 2, 500; 3, 1000 g. (B) Charge mass = 1000 g; pressure of gas: 1, 4; 2, 2; 3, 1 atm.

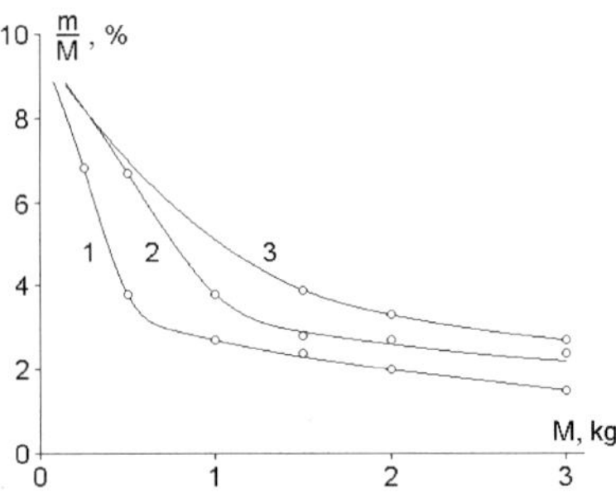

Figure 7.17. Diamond yield (m) versus mass of TNT-RDX 50/50 (M); Explosion chamber = 2 m^3. Pressure of CO_2: 1, 1; 2, 2; 3, 4 atm.

nucleus can disappear. Hence the synthesis parameters have an effect on the value of the final diamond yield. The respective dependences are shown in Fig. 7.17. We should mention that if the chamber is filled with nitrogen, the diamond yield under the same parameters (pressure and charge mass) is noticeably smaller.

Some similarity of the calculated values was established. The diamond local yield and the peak temperature, i.e. their dependence on the reduced Lagrange coordinate, are determined by the ratio between the charge mass and gas pressure (Figs 7.15 and 7.16). Hence the conclusion is that the integral diamond yield under other invariable conditions is expressed as follows:

$$m = M \cdot f(P/M) \tag{7.2}$$

where m = the diamond yield, M = the HE mass, P = the pressure of the surrounding gas and f = some function. A dependence of this kind, referring to [7.41], was observed experimentally.

The experimental diamond yield data vary in character and do not agree with the calculation. The exception is a study [7.42] of the effect of the nitrogen pressure in a chamber on the diamond yield. Figure 7.18 presents the data obtained [7.42] and the results of calculations for these conditions of synthesis. Taking the model simplification into consideration, one can regard the correspondence with experiment as satisfactory.

The results of calculations of gas dynamics compared with the experimental data suggest a weak influence of the phase of the DP and the surrounding gas (non-reacted with CC) mixing on the diamond yield. The chemical adsorption is considered to occur at the mixing phase and as a result protection of the CC, diamond included, takes place. In any case this occurs at the ordinary parameters of the synthesis. Diamond synthesis is also feasible in air, in which case the charge is thickly covered with ice or water. On expansion under these conditions, the DP is cooled strongly and consequently the diamond remains safe on interaction with air.

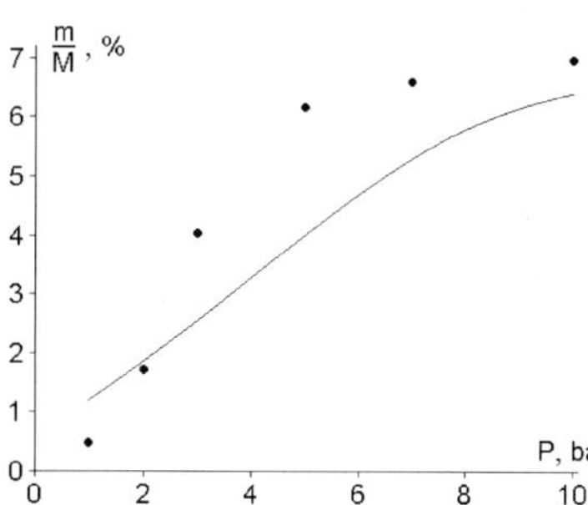

Figure 7.18. Diamond yield (m) versus pressure (P) of nitrogen in chamber. Mass of TNT-RDX 50/50 = 100 g (M); volume of explosion chamber = 0.175 m^3. Solid line, calculation; circles, experiment [7.42].

7.2.4 Properties and Application of Ultrafine Diamond Particles

The condensed product of detonation (CP) consists of graphite, diamond (about half of the CP mass) and a small amount of metal impurities from the chamber walls [7.44]. The electron diffraction pattern of diamond particles gives a circular picture. The measurements on electron microscopic photographs give size values of 3–12 nm.

It has been found that the width of the X-ray lines for the synthesized ultrafine diamond is an order of magnitude more than for that obtained in static conditions. The size of the coherent scattering area is 4 nm. The same size is specified by the specific surface area, which is 380–390 m^2/g determined from the adsorption of argon at the temperature of liquid helium.

The diamond powders produced from HE have the following characteristics: diamond content not less than 98%; graphite content not more than 0.5%; bulk density = 0.3–0.5 g/cm^3; specific surface area not less than 290 m^2/g; range of particle size = 2–12 nm; average particle size = 4 nm.

A small-sized particle consists of several thousand atoms and its properties are intermediate between those of a separate molecule or atom and of massive material. In this case unusual and at times unique mechanical, physical and chemical properties of the nanophase materials are realized.

A diamond film a few tens of atoms thick has been produced from nanophase diamond using modern processes of sol-gel technology. These films are now produced on various materials, such as metals, semiconductors, dielectrics and glass, which are of particular value.

Various electrochemical techniques of depositing wear-resistant coatings with diamond agents on general-purpose tools (taps, drills, milling cutters, dental borers, saw blades, etc.) have been mastered. The surface microhardness is thus increased (up to 20%), and also corrosion resistance and durability (by factors of 1.3–10).

Lapping pastes and suspensions based on diamond materials are used for superfinely machining (lapping, finishing, polishing) products from difficult-to-machine tool, hard, structural and special alloys and ceramics. The suspensions for the superfinish machining of monocrystal surfaces allow the production of high-quality surfaces (not more than 2 nm roughness) without any damage of the surface coating. The suspensions permit the treatment of high-strength materials (diamond, sapphire, silicon), wafers and substrates of semiconductors, optical pieces made of glass, quartz and glass ceramic, gold and silver ware, precious and semi-precious stones and optoelectronic products.

Plastic greases containing ultrafine diamond agents diminish the wear of friction components by factors of 2–4 and increase the limiting loading ability of friction units.

Anti-friction additives based on diamond-graphite mixtures for gearing and bearing greases have been developed. Such greases decrease the friction coefficient, diminish the wear of parts and increase the transmission-carrying ability. Also, additives for motor oils to perform green runs of internal combustion engines have been developed and manufactured. The additives reduce friction and wear of the cylinder-piston components by factors of 1.5–2. They permit petrol and oil consumption to be reduced by 5–7% and engine service life to be increased.

There have been developed some lubricating and cooling technological media with diamond-graphite mix additives to machine the metals. In the process of machining the metals, the lubricant-coolants on the basis increase the tools service life up to 4 times as long, whereas in drawing, grooving and hot die forging they decrease the friction coefficient as many as 1.5 times and increase quality of the work pieces.

Composite materials based on metals and ultra-fine diamond particles have been produced. In view of the small size of the diamonds, it is possible to have a composite disperse hardening known in material technology as the expense of crack mobility decrease, dislocations and other defects. The high hardness and modulus of elasticity ($E = 900$ GPa) of the diamond component should enhance this effect. The hard inclusions of the diamond phase penetrate under high pressure into near-surface coats of metal powders, which should hinder shift deformations of the material. The small-sized diamond particles fill the pores between large metal dusts.

Measurements of microhardness in sample cross-sections have shown an excess on some parts compared with tabular data for continuous materials by factors of 2–2.5. The following reasons for this effect are probable: disperse hardening, occurrence of hard compounds as a result of chemical reactions in the shock wave and hardening as a result of plastic deformation under high pressure with formation of a large number of defects. There is noted melting point elevation of the composite in comparison with the pure material.

7.2.5 Conclusion

The detonation synthesis of diamonds is a complex phenomenon, the detailed investigation of which is far from nearing completion. The data obtained allowed to conclude:

1. The formation of a diamond in the detonation wave proceeds through the intermediate amorphous phase. The synthesis region in the $p-T$ plane is characterized by the hysteresis line and the transient area within the boundaries of which the transformation is incomplete.
2. The diamond losses in post-detonation processes are accounted for by the secondary chemical reactions proceeding in the DP under expansion. Their amount is defined by the external conditions of the synthesis: the charge mass, the kind of surrounding gas and its pressure. The maximum diamond yield for a given HE is observed under the conditions providing an isoentropic character of the DP expansion.
3. Further investigations of the properties of ultrafine materials will allow the features of conventional materials to be improved and new materials with unique properties to be made.

7.3 ALEX® Nanosize Aluminum for Energetic Applications

7.3.1 Introduction

Aluminum powder is extensively used in solid rocket propellants because of the high energy produced during its combustion with oxidizers. Because aluminum has a specific gravity of 2.7 g/cm^3, it is particularly advantageous for increasing the *density* specific impulse, thereby reducing the size and therefore the weight of the rocket. In many cases there is interest in increasing the burning rate of aluminized propellant and also in increasing its combustion efficiency. The most desirable means of achieving such improvements is to use high surface area aluminum. The process of electroexplosion of metal wire (EEW) produces aluminum powder with particle sizes at least an order of magnitude or two smaller than the aluminum powder ordinarily used in propellants.

7.3.2 Description of the Process

EEW was first achieved by Narne in 1774 and later by Michael Faraday. They applied an electrical pulse to a wire, causing it to explode in air to form aerosols of metal oxide. The process is similar to the brilliant flash that occurs at the moment of burn-out of a filament of an incandescent bulb. The exploding bridge wire (EBW) of a detonator is also a similar process. A considerable amount of research was done on EBWs in the 1940s and 1950s and it was demonstrated that temperatures > 15000 K were reached at the moment of explosion.

Figure 7.19 shows an EEW machine. The process is suitable for producing nanopowders in any metal that is available as ductile wire. A reel of wire is enclosed in the reactor where argon at 2–3 atm is recirculated. The wire is fed through an electrically insulated baffle and when it contacts a strike plate the circuit is closed, causing a large pulse (10^2–10^3 J during 1 μs) to flow through the wire. An explosion occurs that forms a plasma. The plasma is contained by the very high field created

Figure 7.19. Reactor for producing nanopowders by electroexplosion of wire.

by the pulse. When the vapor pressure of the metal exceeds the force of the field, there is an interruption in current flow, causing the plasma to explode into clusters of metal that are projected at supersonic speeds through the argon. The condensing metal is carried by an argon flow (created by an internal blower) to a gravimetric separator where the now agglomerated particles are collected. Several hundred grams are produced per hour with the rate proportional to the specific gravity of the metal. Virtually all the metals produced by this process are combustible and several of them, such as aluminum, iron, titanium and zirconium, are either pyrophoric or nearly so. These powders are collected and immersed under hydrocarbon or under argon. In the case of Alex® the particles are passivated by exposure to dry air before removal from the chamber and are packaged as a dry powder. A variant of Alex®, called L-Alex®, is passivated with a hydrophobic organic coating.

7.3.3 Characteristics of the Powders

A scanning electron micrograph of Alex® is shown in Fig. 7.20. The particles are almost always spherical and fully dense and are ~ 100 nm in diameter. BET surface area measurements of Alex® range from about 10 to 20 m^2/g. Figure 7.21 shows a sector of an air-passivated Alex® particle with a coating ~ 2.5–3 nm thick. X-ray powder diffraction shows essentially metallic aluminum and the presence of aluminum oxide, nitride and an oxynitride. The amount of active aluminum is ordinarily 88–90 wt%. Compared with evaporation-deposition processes that produce relatively defect-free crystals, EEW powders have crystal defects, faults and twins. The disorder in the EEW aluminum crystal extends into the oxide outer layer. When ignited in air, Alex® appears to have two separate ignition steps. The first occurs at lower temperatures (400–500 °C) and a second where the powder burns white hot. The first appears to be a physical process – one of crystal transformation accompanied by a surface burning – and the second is chemical oxidation. Alex® and other electroexploded metals may be converted from their metastable form by electron irradiation as in a transmission electron microscope. The powders may be

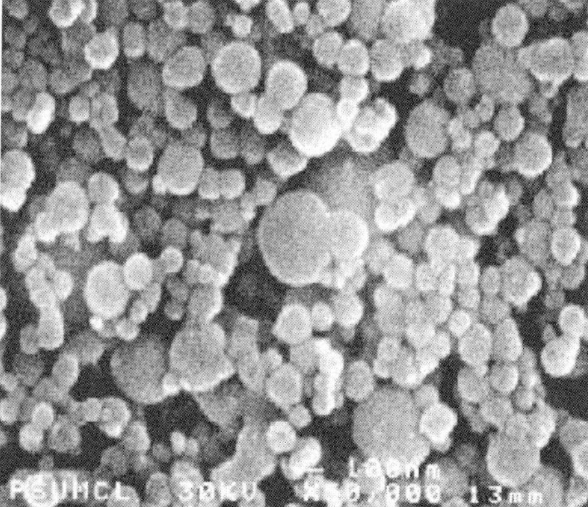

Figure 7.20. Field emission electron microscopic view of ALEX® (×50000).

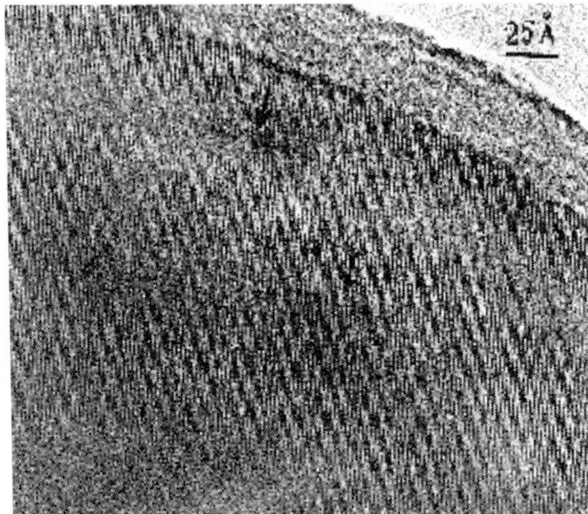

Figure 7.21. Thin alumina on the surface of spherical ALEX® particle (×400 000).

seen to twinkle under the microscope and to coalesce into larger spheres. Mench *et al.* [7.45] studied the thermal behavior (DTA) of Alex® powder and compared it with micron-sized aluminum. Their data show that Alex® powder has very sharp exotherms when heated in air, oxygen or nitrogen. These occur well below the melting point (660 °C) of aluminum, while 20-μm sized aluminum does not react with oxygen or air until about 1000 °C.

The L version of Alex® involves coating unoxidized nanosized aluminum with palmitic acid to form a chemical bond to aluminum. For particles that are ~110 nm in diameter, this organic coating constitutes ~5 wt%. It provides energy on combustion, whereas the 10–12% oxide on Alex® does not. Under accelerated aging for 40 days when both versions are exposed to conditions of low humidity (32% RH at 45 °C), neither version was affected. However, at 75% RH and 45 °C, Alex® was converted almost completely to aluminum hydroxide within 20 days, whereas L-Alex® was unaffected even after 40 days of exposure [7.46].

7.3.4
Behavior as a Solid Propellant and Hybrid Additive

Figure 7.22 shows burning rates of Alex® and ammonium perchlorate (AP) powder at 40 atm in a Crawford bomb [7.47] as a function of Al/AP ratio and compared with propellant-grade aluminum. The typical solid propellant is a castable mixture based on rubber (HTPB) binder and the non-porous nature of such mixtures leads to much slower burning rates than binderless mixtures of fuel and oxidizer. Mench *et al.* [7.48] studied the burning rates of Al-HTPB-AP blends and found that Alex® improved the burning rates by ~100% compared with micron-sized aluminum. Simonenko and Zarko [7.49] found that replacing commercial-grade aluminum with Alex® shortened the ignition delay and increased the burning rate by ~2–5 times over the pressure range 10–90 atm and with decreases in pressure exponent. They also noted more stable burning with Alex®. Several studies have indicated that Alex® when used in Alex®-HTPB-AP propellants is completely consumed at the

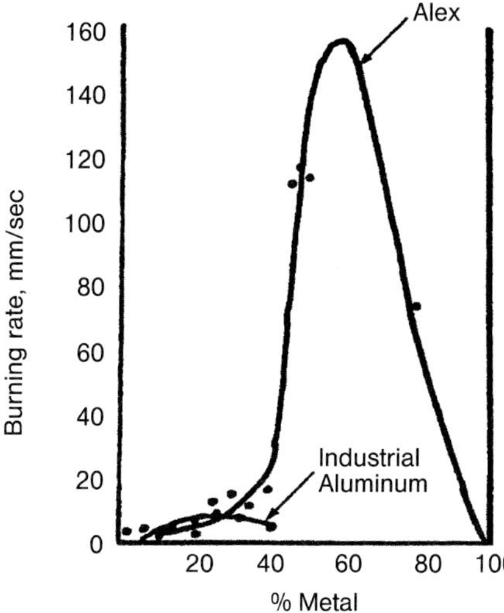

Figure 7.22. Burning rate of aluminum-ammonium perchlorate powder mixes.

solid-gas burning interface, whereas micron-sized aluminum particles continue to burn in the engine and may be ejected into the plume without being fully oxidized. More recently, Cliff [7.50] found that L-Alex® mixes more readily than Alex® into HTPB, minimizing inhomogeneity.

There have been extensive studies on modeling of aluminum combustion in solid propellant engines and the data appear to fit those models where the life of a burning aluminum particle in oxygen is inversely proportional to d^2. The molten aluminum particle reacts principally with water vapor and carbon dioxide produced by the combustion of the organic binders with AP. These models project that the life of micron-sized aluminum particles are > 1 ms, whereas a burning 100-nm particle such as Alex® is consumed in 0.1 μs, supporting experimental observations. The use of Alex® as an additive to hybrid engine fuels was studied [7.51] with additions of 4, 12 and 20% Alex® to HTPB slabs. When burned in gaseous oxygen, the 20% formulation had a 40% higher global recession rate and a 70% higher mass burning rate than pure HTPB, whereas conventional aluminum powder had a much lower regression rate. Alex® also seemed to smooth out irregularities in the surface during burning.

7.3.5
Alex® Powder as an Additive to Liquid Fuels

Ivanov and Tepper [7.47] studied the combustion characteristics of aluminized water and aluminized hydrazine as monopropellant mixtures. In both cases, hydrogen is generated with theoretical peak temperatures in excess of 3200 K with water and 3560 K in the case of the hydrazine reaction. With Alex® they found that the reactions were rapid, easy to initiate and burned uniformly to produce hot hydrogen that could further oxidize. Unfortunately, Alex® is not compatible with these fluids during long-term storage.

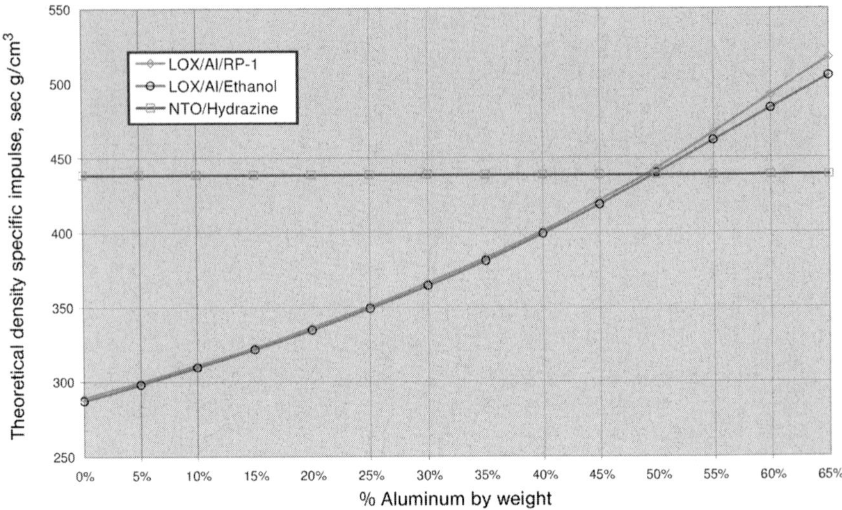

Figure 7.23. Density specific impulse (vacuum, 100:1 expansion ratio) of aluminized liquid propellants when burned with oxygen.

Figure 7.23 shows how increasing the aluminum content of kerosene or ethanol substantially improves the theoretical density specific impulse of the liquid propellant when burned with oxygen. Because of this potential, metallized gelled propellants have been studied analytically and experimentally during the last 70 years [7.52]. There have been significant efforts by NASA [7.53–7.57] to use aluminum as an additive to gelled rocket propellant (kerosene, RP-1) and also as an additive to liquid hydrogen [7.58]. However, adding micron-sized aluminum to hydrocarbon fuel is somewhat ineffective. Wong and Turns [7.59] studied the burning of aluminum slurried in jet propellant (JP-10). They found that the residence time is prolonged because of delays in burning of the aluminum, and also the coalescence of the aluminum droplets into larger ones while the hydrocarbon is burning. In 1998, NASA awarded Argonide a study contract to determine whether Alex® powder, if immersed in RP-1 kerosene, would burn efficiently.

7.3.5.1 **Formulation of Aluminized Gels**

Gels of aluminized (RP-1) kerosene were formulated [7.60] and then characterized for ignition delay as compared with gelled RP-1, neat RP-1 (no gellant) and RP-1 containing 5 μm-sized aluminum. A detailed description on the rheology of gels is given in Chapter 12. In most cases, fumed silica (Cab-O-Sil) was used as a gellant. When the Alex® content was greater than about 25 wt%, silica was unnecessary since the powder acts as a pseudo-gellant. Tween-85, a non-ionic surfactant, was added to wet and penetrate any aluminum agglomerates.

7.3.5.2 **Ignition Delay Measurements**

Figure 7.24 shows a stainless-steel combustion bomb used to measure the ignition delay of kerosene and its aluminized versions [7.60]. With this device we can inject ~0.3 cm^3 of fuel mixture into the bomb, which is pre-heated to a maximum temperature of 800 °C and with an oxidizing gas of known pressure. The injection port is water-cooled to keep the fuel at ambient temperature until injected. A water-cooled piezoelectric pressure transducer measures the pressure rise due to combustion. Time is measured from the first movement of the piston forcing the fluid into the chamber, until there is a rapid rise in pressure as measured by an internal pressure transducer. In a typical experiment, an oscilloscope simultaneously records a pressure trace together with another trace of piston movement that indicates the start of the injection. Surface thermocouples and thermocouple probes are used to determine the temperature of the bomb wall and gases inside the bomb. The ignition delay was measured at a fixed pressure of 0.8 MPa of oxygen and at 410, 460, 520 and 580 °C.

Seven different fuels were examined. The aluminized versions were compared with neat (pure) and gelled kerosene, with the latter intended to achieve equivalent viscosity and allow comparison under similar spray patterns. Figure 7.25 shows the chemical ignition delay times of neat RP-1, gelled RP-1 and aluminized gelled RP-1 in hot oxygen. The data indicate that the 25 wt% Alex gel ignites faster than gelled RP-1, but is about equivalent to that of pure kerosene. However, adding a surfactant to the 30 wt% Alex gel causes a substantial reduction in ignition delay to below that of pure RP-1 kerosene, showing that Alex® is a combustion accelerant for kerosene. The ignition delay for L-Alex® was at least as short as that of Alex®, when ignited in either air or oxygen. Gels produced from 5-μm aluminum did not show a reduced ignition delay in either air or oxygen. Small rocket engine tests of these gels were done at Pennsylvania State University [7.61], confirming the ignition delay experi-

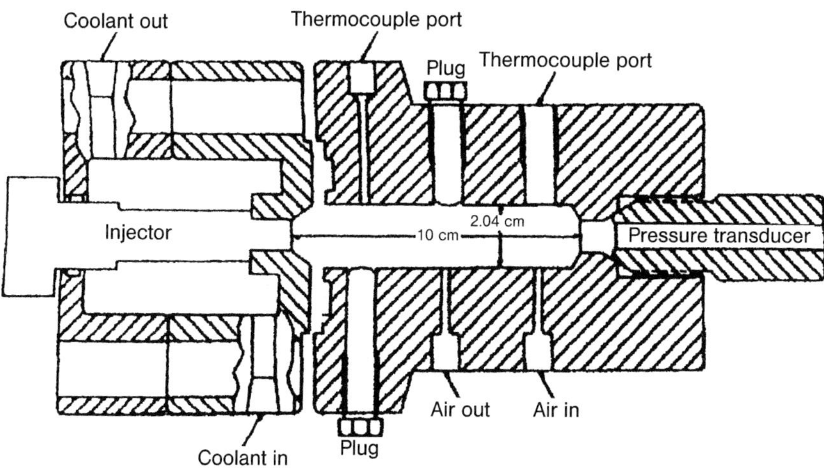

Figure 7.24. Ignition delay device.

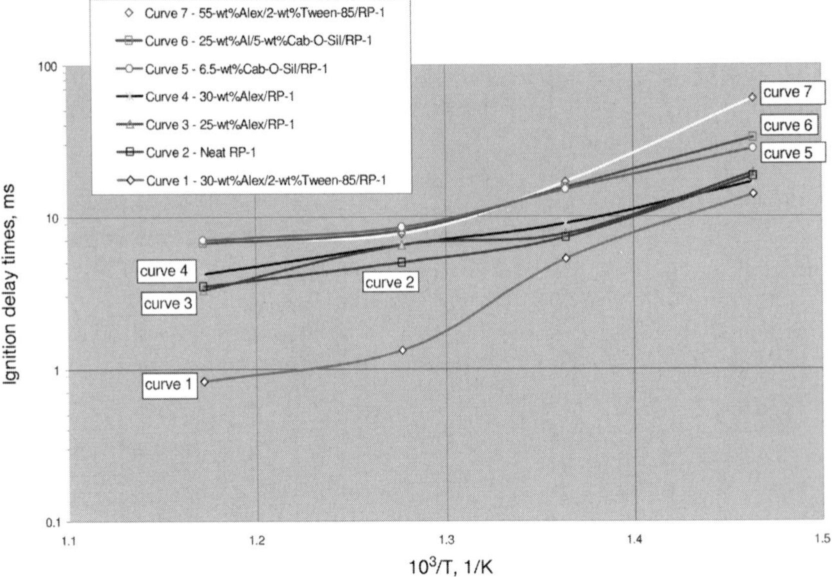

Figure 7.25. Average ignition delay in oxygen.

ments. It was found that the aluminum oxide that is produced appears to be submicron in size. We speculate that this oxide would be retained in the turbulent flow regime of the exhaust gases of the engine and would be too light to impinge through the laminar layer and deposit on the walls of the engine.

Alex® when gelled into kerosene was found to be reactive with nitrogen, suggesting its use to maintain combustion stability in oxygen-starved air that would exist in air-breathing engines such as hypersonic, combined cycle rocket and pulse detonation engines.

7.3.6 Alex® in Explosives

Aluminum is often added to organic (CHON) explosives to improve their blast force because aluminum can liberate amounts of energy several times greater than that of organic high explosives. As with solid propellants, the contribution of aluminum to explosives is limited by the low burning rates of micron-sized aluminum. Yet aluminized explosives have been under study for some years and compositions such as H-6 have been developed to produce an explosive filling based on RDX-TNT but with enhanced blast characteristics. In the early 1980s, Reshetov *et al.* [7.62] found that Alex® enhanced the detonation rate of hexogen, a high explosive. Although there was little effect on the detonation rate with low levels of additives, adding more than 50% Alex® resulted in an increase in detonation rate from about 5400 to 7000 m/s. More recently, a beneficial enhancement was demonstrated [7.63] for both velocity of detonation (VoD) and brisance for a number of TNT-based tritonal and H-6 derivatives containing Alex®. VoD enhancements of 200–300 m/s and improvements in brisance of up to 27% were observed in a number of tritonal charges when Alex® was substituted for conventional aluminum grades.

7.3.7 Alex® in Gun Propellants

The combustion behavior of Alex in gun propellants up to high pressures (280 MPa) was studied by Baschung *et al.* [7.64]. The burning rate is nearly doubled in comparison with high-caloric conventional double-base propellants. Simultaneously, the pressure exponent of Vieille's burning law decreases from > 0.8 for double-base propellant to 0.66. They proposed that Alex® would be useful as an additional accelerator and also for applications in high-pressure rocket propulsion as an igniter or as a booster.

7.3.8 Conclusions

Improvements in Alex® are currently aimed at its use in energetics, and these studies include:

1. Better characterization of the powder so as to produce meaningful specifications.
2. Improvement in the process and in particular surface modification of the particles.
3. Reducing particle size. Further benefits of reducing particle size are likely, but such benefits would not be effective without substituting a combustible organic layer for the ~3 nm thick native oxide that forms on exposure to air.
4. Alloying with other reactive metals to improve the burning rate.

7.4 Pneumatic Production Methods for Powdered Energetic Materials

7.4.1 Fundamentals and Advantages of Pneumatic Production Methods for Powdered Energetic Materials

In many industrial processes, such as pharmaceutical materials production, the precise processing of surfaces, obtaining superconductors or components of functional ceramics and the production of superfine particles of metals, specific requirements are imposed on powder technology: maintenance of size uniformity of submicron particles and their mixes, constancy of particle shape and size distribution and variation of moisture content. In view of the topicality of the problem, in

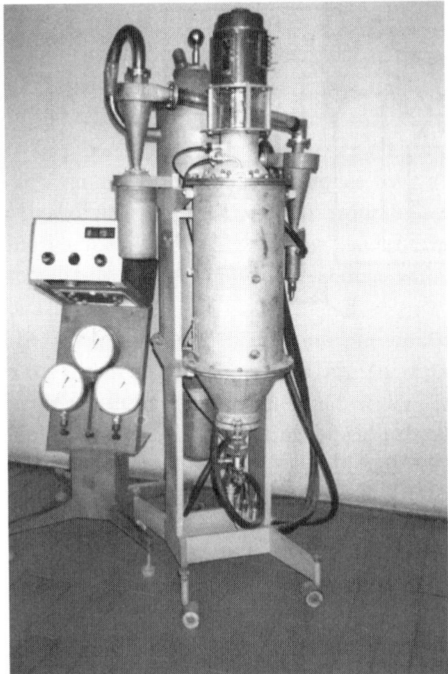

Figure 7.27. General view of the 'Combi 90' unit.

Figure 7.28. Pilot plant 'PCA-50'.

studies have shown that the process of particle self-attrition is of major importance whereas the nozzles design and installation of obstacles hardly affected the process outcome. The high purity of the end product after size reduction of hard and super-hard materials (carbides, nitrides, oxides of some metals) is also explained by the domination of particle interaction (self-attrition) instead of interaction of particles with working surfaces [7.69].

In Table 7.6, the results are presented of the analysis of impurities in beryllium oxide powder. The content of impurities was measured before and after processing. The initial powder had a specific surface area of 7200 cm^2/g and the end product 12 300 cm^2/g.

Experiments at pressures > 1 MPa (up to 10 MPa) did not give any qualitative change in characteristics or increase in process efficiency, but further investigations along this line are planned. The specific energy consumption on production of submicron powder of electric corundum, starting from powder with initial particle size $x_{50} \approx 30$ μm, with the use of complex technology, is 3–5 kWh/kg. The specific yield of the pneumatic unit is 15 kg/h and the working pressure is 0.7–0.8 MPa. Potential opportunities for improvement of the processing technology for sub-micron and ultrafine powders, e. g. for aluminum oxide and other ceramic powders, still exist. This requires a more in-depth study of gas dynamics of flow of 'gas-solid particles' in the 'Combi' unit.

The combination of Al_2O_3 pneumatic grinding with classification in multi-contour separation loops resulted in the production of abrasive powders with comparatively high-level characteristics (Table 7.7, Fig. 7.29) that has allowed an increase in the precision of manufacture of ball-bearings and the problems connected with the disposal of toxic wastes and harmful products to be overcome. The average productivity of the 'Combi' unit in obtaining such powders reaches 1–1.5 kg/h.

Table 7.6. Impurities in beryllium oxide powder on processing in the 'Combi' unit.

Element	Initial content (%)	Content in end product (%)
B	3.6×10^{-5}	9.4×10^{-5}
Si	4.7×10^{-3}	7.2×10^{-3}
Mn	3.6×10^{-4}	5.0×10^{-4}
Fe	4.3×10^{-2}	5.4×10^{-2}
Mg	5.0×10^{-3}	3.9×10^{-3}
Cr	5.4×10^{-3}	6.8×10^{-3}
Ni	1.5×10^{-2}	1.5×10^{-2}
Al	1.2×10^{-2}	1.7×10^{-2}
Cu	4.7×10^{-4}	4.3×10^{-4}
Zn	4.7×10^{-3}	4.7×10^{-3}
Ca	5.7×10^{-3}	5.7×10^{-3}
Ag	1.1×10^{-5}	1.1×10^{-5}
Li	1.1×10^{-3}	1.1×10^{-3}
Na	3.6×10^{-3}	4.3×10^{-3}

Table 7.7. Abrasive powders for the manufacture of ball-bearings

Fraction	Particle size of main fraction (μm)	Mass percentage of main fraction (%)
M 0.5B	< 0.5	65.0
M 0.7B	0.5–0.7	40.0
M 1.0B	0.7–1.0	45.0
M 3	1.0–3.0	60.0

The physical-mechanical processes realised in the 'Combi' unit considerably increase the wear resistance of powders. The viable time of their activity in technological processes exceeds by an order of magnitude the service lifetime of powders obtained by hydroclassification (sedimentation in liquid). This may be attributed to nodulizing and decreasing the total number of particle defects during treatment by underexpanded gas jets flooded in the powdered material.

Figure 7.30 gives the results of size reduction for aluminum powder with an ordinary 10–30 μm size range at a working gas pressure of 0,7–0,8 MPa and a rotation speed of the rotor of 7000 rpm. The fact of size reduction is confirmed by the presence of small-sized aluminum particles in the end product, which are

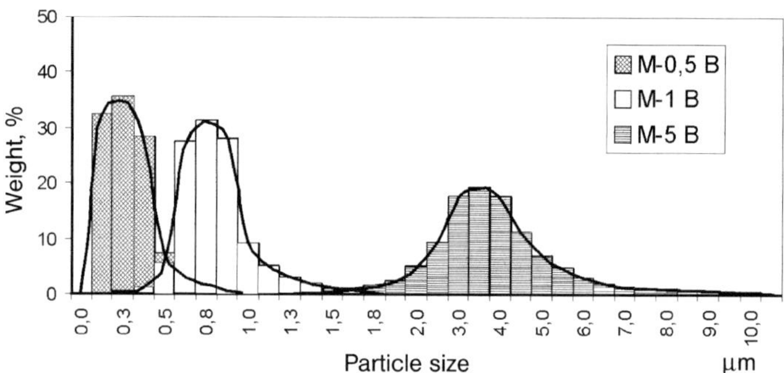

Figure 7.29. Mass size distribution of polishing powders of corundum Al_2O_3 (particle sizes determined by optical microscope).

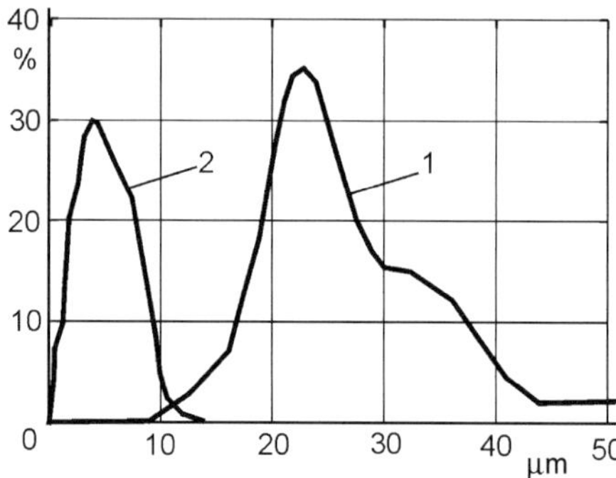

Figure 7.30. Differential function of particle size distribution for initial aluminum powder (1) and for the ground powder (2).

absent in the initial material. The computer images of fine aluminum particles, obtained with the use of an optical microscope and TV camera, are shown in Figs 7.31 and 7.32.

The use of pneumatic technology for the separation of coarse-grain inclusions in ultrafine powder of copper (Fig. 7.33), obtained by the method of electrical discharge [7.70], allowed the production of highly efficient additives to oils and lubricants.

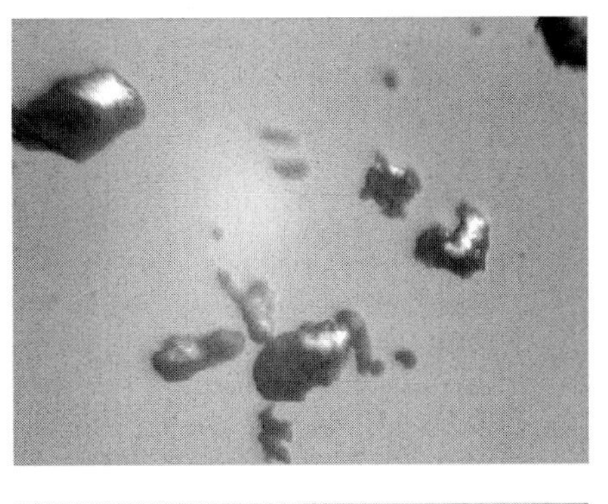

Figure 7.31. Initial aluminum powder (the maximum size of particles is 50 µm and the average size is 25 µm).

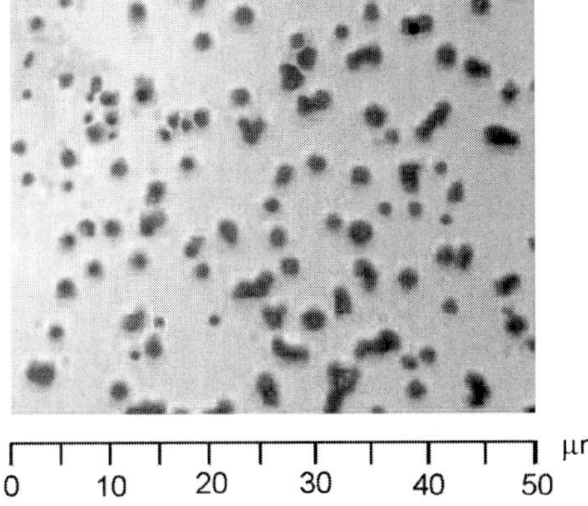

Figure 7.32. Fine fraction of end product with sizes 3 µm (average), 7 µm (maximum) and 1.5 µm (minimum).

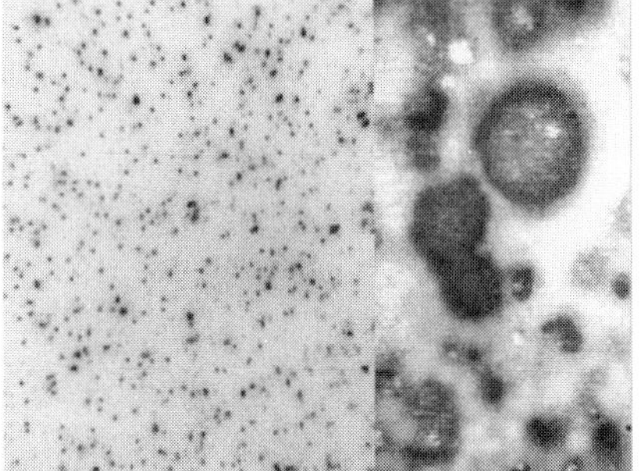

Figure 7.33. Ultrafine copper powder ($x_{97} = 0.7$ µm). On the right the initial powder is shown.

Similar powder characteristics were obtained by combining the processes of grinding and classification for a wide range of materials [organic – polyethylene, poly(vinyl chloride), nozepam, cinnarizin, sorbite, salbutamol, vitamin concentrates, vitamin meal; inorganic – aluminum, niobium, boron, tantalum, Nichrome, talc, chalk, ammonium perchlorate] [7.71].

7.4.3.2 Classification of the Particles by Sizes

Theoretical and experimental studies of air-centrifugal particle classification have allowed the formulation of some principles for providing an increase in classification efficiency and for developing on this basis a number of new processes and devices for the classification of finely dispersed materials [7.72]. The scheme of a typical design of new air-centrifugal classifier with a profiled rotating zone of separation is shown in Fig. 7.34.

The basic performance characteristics of the process of particle classification are productive capacity, boundary (limiting size) of separation and efficiency of classification. For an air-centrifugal classifier, the capacity is determined by the mass flow rate of air and by the maximum admissible (without a decrease in separation quality) concentration of particles in the carrying medium. The influence of the main operating and design parameters on the boundary size of separation can be determined from the analysis of a simplified model for single-particle motion in a swirling carrying flow [7.67].

Equation (7.3) describing this motion after appropriate transformations may be written as

$$x_b^2 = \frac{V_r \cdot R}{V_\varphi^2} \cdot \frac{18\rho_f \nu}{\rho_m} \tag{7.3}$$

where x_b is the boundary diameter of a particle, ρ_m and ρ_f are the densities of particle and carrying medium, respectively, R is the radius of the circle along which the particle rotates with a peripheral velocity V_φ, V_r is the radial velocity of carrying medium relative to the particle and ν is the kinematic viscosity of the carrying medium.

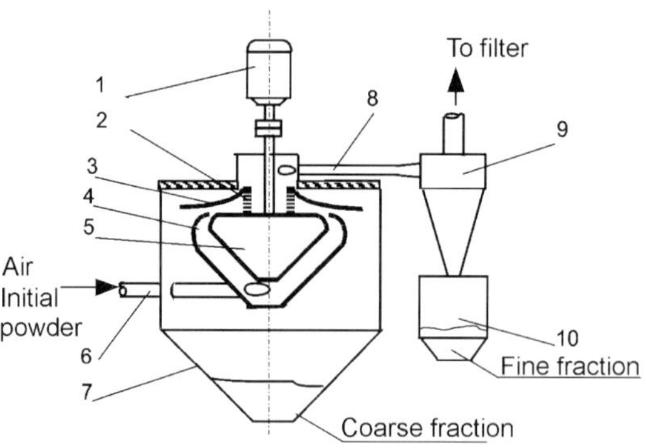

Figure 7.34. Air-centrifugal classifier. 1, Electric drive; 2, set of disks; 3, profiled disk; 4, collector funnel; 5, rotor cone; 6, inlet pipe; 7, hopper for receiving the coarse fraction; 8, outlet pipe; 9, cyclone; 10, hopper for receiving the fine fraction.

Considering that variations in gas viscosity are insignificant, the following expression holds upon classification of specific material with given density: $18\rho_f \cdot \nu / \rho_f =$ constant. Hence, to realize an equality of the centrifugal and aerodynamic forces for a particle of given (boundary) size, it is necessary that the following condition is satisfied throughout the separation volume:

$$\frac{V_r \cdot R}{V_\varphi^2} = \text{const}\,. \tag{7.4}$$

Technical implementation of the advanced method of air-centrifugal powder classification can be rather varied [7.68, 7.72], but the key point is in ensuring equal action of forces on the particle irrespective of its location in the separation zone.

The main characteristic defining the quality of the classifier operation is the curve of separation, which expresses the probability of a particle of a given size getting into small-sized (or coarse-sized) products of separation. Since the curve of separation is considered by many researchers [7.73] as the device characteristic independently of classified material properties, we made an attempt to generalize available experimental data in the form of an empirical curve of separation. Figure 7.35 presents the curve of separation constructed on accumulated data from 160 experiments, which cover the classification of various materials on air-centrifugal classifiers of several standard sizes under laboratory and industrial conditions. The data are averaged on typical diameters of particles, 75, 50 and 25% (by mass) of which enter into small-sized products of separation. The diameter of particles with a content in small-sized products of < 3% was taken as the maximum particle size of the product. The other typical sizes are normalized to this diameter:

$$\overline{X}_{0.75} = \frac{x_{0.75}}{x_{0.03}} \tag{7.5}$$

$$\overline{X}_{0.5} = \frac{x_{0.5}}{x_{0.03}} \tag{7.6}$$

$$\overline{X}_{0.25} = \frac{x_{0.25}}{x_{0.03}} \tag{7.7}$$

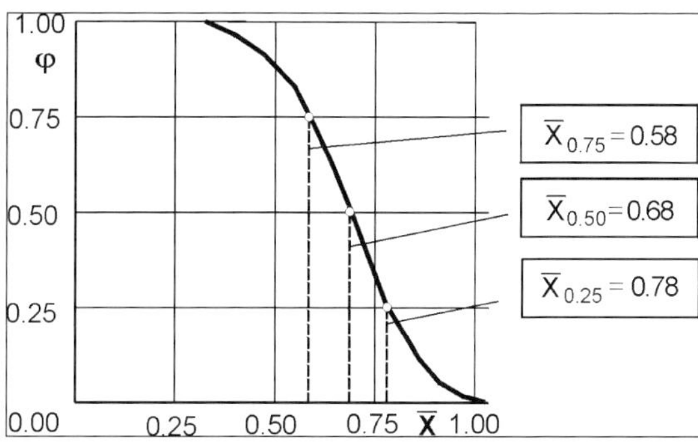

Figure 7.35. Integral separation curve for classifiers with a profiled rotating zone of separation.

Mathematical expectations of the magnitude of $\bar{X}_{0.75}$, $\bar{X}_{0.5}$ and $\bar{X}_{0.25}$ were obtained with a confidence level of 0.95. The curve drawn through these points (Fig. 7.35) may be considered as a curve of separation that characterizes the dividing ability of air-centrifugal classifiers with a profiled rotating zone of separation. It is approximated well by the following empirical equation:

$$\varphi(\bar{x}) = \exp\left(-5.18 \cdot \bar{x}^{5.31}\right) \tag{7.8}$$

The curve of separation obtained can be considered as an accessible, in principle, characteristic for a given type classifier and can be used in the design of pneumatic units and also for choosing the separation mode to produce dispersed materials with a fixed particle size distribution.

It is known that the particle size distribution of heterogeneous components of composite solid propellants (CSPs) has significant effects on both the production technology and operating parameters of solid rocket motors. The development of CSPs with high specific impulse is based on an increase in the filler volume content in the propellant mass that can be achieved by a decrease in the filler particle size. At the same time, a size reduction of ammonium perchlorate (filler) particles leads to a considerable rise in the burning rate. However, as the volume content of the filler increases and its particle size decreases, the viscosity and casting properties of the uncured propellant mass decrease. Predetermined filler particle sizes and rheology characteristics of uncured propellant mass can be attained by preparation of CSP heterogeneous components with optimal particle size distribution.

On the other hand, ensuring reproducibility of the CSP burning rate is of great importance in practice. The reproducibility depends in turn on heterogeneous component characteristics, in particular on the particle size distribution of ammonium perchlorate.

In this connection, the development of methods for the production of crystalline CSP components with a strictly standardized particle size distribution, in particular, methods of effective classification, is gaining in importance.

Processes of air-centrifugal classification of ammonium perchlorate have been investigated with the use of air-centrifugal classifiers described in this section. The goal was to choose separation modes, to determine operating parameters of the apparatus and to produce some test batches for the investigation of the kinetics of the combustion of individual AP fractions. Some results are presented in Fig. 7.36. From the plots one may see that the classification efficiency is fairly high but falls with decreasing value of the separation boundary. The reason is the high adhesive ability of finely dispersed ammonium perchlorate.

The range of separation boundaries was found to be 5–45 μm. From a source material having specific surface area $S_m = 2700\ cm^2/g$, three narrow fractions (S_m = 2400, 5500 and 8300 cm^2/g) were produced in amounts of 20 kg each. The firing tests demonstrated an increase in burning rate with decrease in the mean particle size. Air-centrifugal classifications have been carried out also for other CSP heterogeneous components (aluminum hydride, aluminum, vitan).

Thus, the high separation efficiency for particle sizes < 50 μm and the relatively

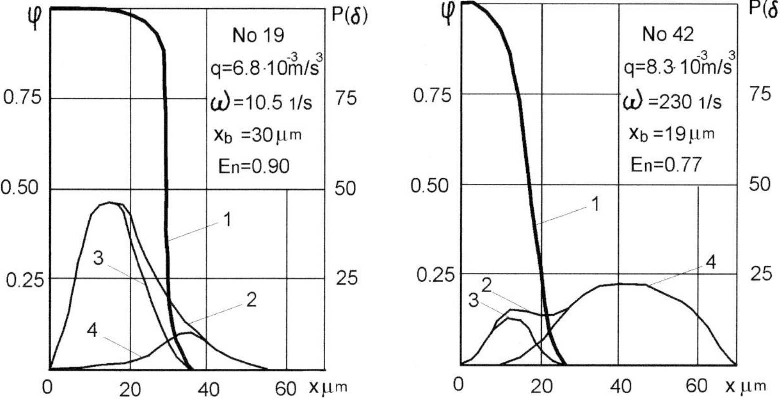

Figure 7.36. Results of air-centrifugal classifications for ammonium perchlorate powder.

high output and possibility of component classification in closed technological loops and in inert media make it possible to use the described air-centrifugal classifiers in CSP production.

7.4.3.3 Blending and Homogenization of Powders and Components of Energetic Materials

The intensity of the blend process in the 'Combi' unit is determined by the magnitude of convective transfer, namely by the function of particles residence time $f(t_{max}/t_{min})$ in the fluidized bed during a cycle. The quality of mixing on the micro level is determined by the effect of underexpanded gas jets on incoming aggregates of particles. The variation of the key component concentration (C) with the number of displacements to blend (N), needed to achieve the given heterogeneity factor (V_c = 5%), is shown in Fig. 7.37.

The results of mathematical modeling and the experimental dependence of N on t_{max}/t_{min} (ratio of maximum to minimum particle residence time) are shown in Fig. 7.38. The value of V_c is calculated by

$$V_c = \frac{100}{\bar{c}}\sqrt{\frac{1}{n-1}\sum_{i=1}^{n}(c_i - \bar{c})^2} \quad [\%] \tag{7.9}$$

Referring to Fig. 7.38, the maximum efficiency of the process of blending is reached at t_{max}/t_{min} = 2.2–2.4. In experiments, the particle velocities and trajectories were measured via use of inductive and capacitive pickups (through injection of marked particles).

On the basis of combined processes of blending and dispersion, the technology for obtaining highly homogeneous compositions with low-mass additives has been developed. An example is the production of a medicinal composition of sodium

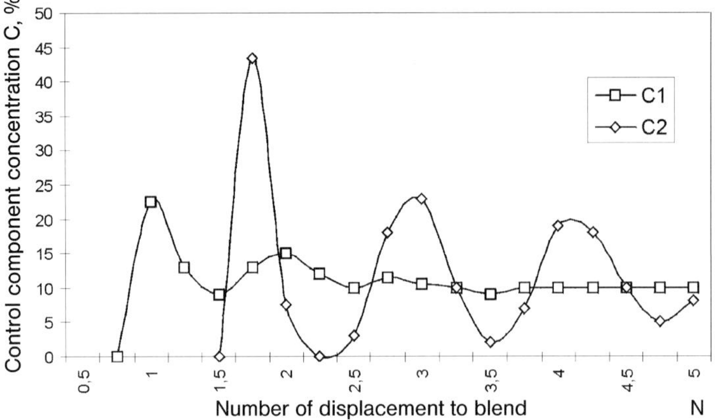

Figure 7.37. Variation of the key component concentration versus number of displacement to blend. C1, $t_{max}/t_{min} = 2.4$: C2, $t_{max}/t_{min} = 1.7$.

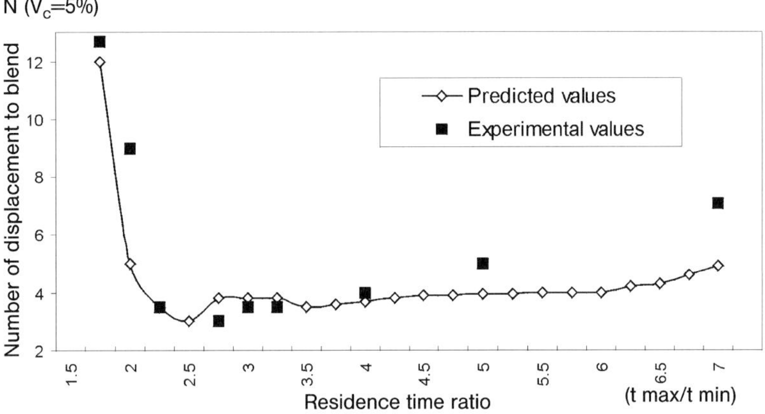

Figure 7.38. Comparison between experimental and calculated N values.

benzoate and salbutamol in the ratio 98 : 2, which contains 85 % by mass of particles with size < 5 mm and is characterized by a heterogeneity coefficient $V_c < 2.7\%$. Figure 7.39 presents the results of the tests. The work with pectin confirmed the efficiency of the proposed technology.

One of the essential technical requirements in the production of finely dispersed powder materials by pneumatic circulation equipment is to reduce the effect of adhesive and cohesive properties of loose material, which hinders the motion of the bulk bed over the internal surfaces of the equipment, causes bridging between particles and does not allow efficient control of the circulation motion. This problem arises with each specific material. The results of our studies have allowed the development of commercial plants for processing several slurry materials (moisture content up to 25 %).

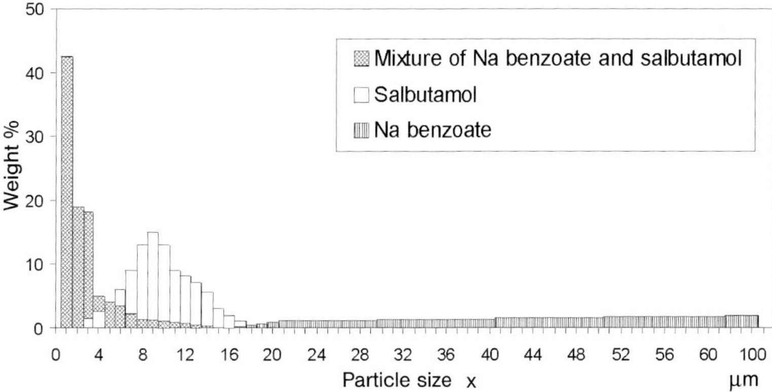

Figure 7.39. Results of combined blending and dispersion of salbutamol and sodium benzoate components. The resulting coefficient of heterogeneity is $V_c = 2.7\%$.

7.4.3.4 **Powder Drying**

Powder drying via the use of air streams is a widely used method in various technological processes [7.74]. However, extensively used fluidized-bed dryers and pipe dryers have some crucial drawbacks, which reduce their performance. For instance, it is impossible to apply fluidized-bed units in the case of high initial temperatures of a drying agent. In pipe dryers, the initial temperature can be rather high but in the drying process the moisture content in the drying agent constantly increases, which does not allow the final moisture content of a material under treatment to be reduced effectively.

A comprehensive study has been carried out of drying processes in pneumatic circulating dryers developed on the basis of the 'Combi'-type units for various materials. As an example, the results for ammoniac saltpeter drying are presented in Fig. 7.40. They show that the dryer performance (Q) is essentially improved with increase in the initial temperature of the drying agent (T_{init}). Overall, research has shown that the drying process in devices with controlled interior circulation of a dispersed material has a number of positive features:

1. The efficiency of convective heat and mass transfer is increased owing to the large difference in the velocities of gas and particles which takes place in the speed-up zone.
2. The optimum relation between wet material surface area and quantity of drying agent can be ensured by control of the component mass flow rate.
3. Interaction of the drying agent with the material in the circulation tube takes place during a rather small time span, allowing the temperature of the agent to be elevated and drying to be intensified.
4. It is possible to achieve a relatively low final moisture content of the material (down to 0.1%) owing to the continuous renewal of the drying agent.

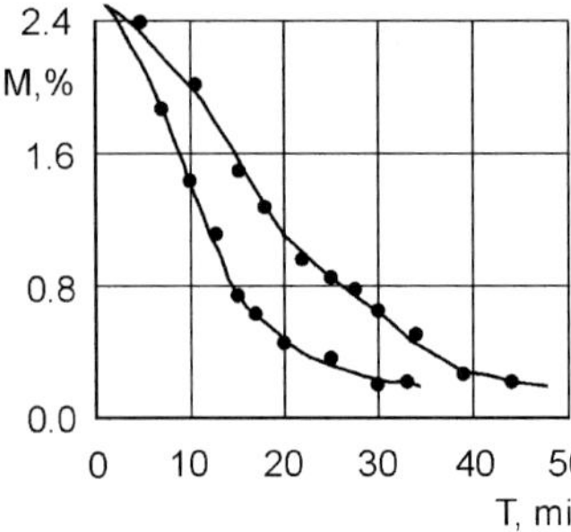

Figure 7.40. Drying of ammoniac saltpeter under different operating conditions: 1, T_{init} = 140 °C, Q = 25 kg/h; 2, T_{init} = 170 °C, Q = 56 kg/h.

5. The principle of the unit's operation allows drying to be performed in variable modes that reduce energy consumption.

'Soft' drying equipment and a process for moist and thermally unstable materials have been developed. The cooling of gas on its expansion and intensive flow around particles prevent thermal decomposition of the material to some extent. This permits the treatment of some high molecular mass compounds, organics and some medicinal materials such as polyethylene, poly(vinyl chloride), vitan, nozepam, pentoxifylline, pectin, *Acorus calamus* and *Thermopsis lanceolata*.

7.4.3.5 **Granulation**

Efforts are under way to solve the problem of granulation in a unified production line. The complexity of the problem consists in the necessity of bringing together in one working volume almost all processes of powder technology such as dispersion, classification, mixing of particles with liquids and simultaneous drying. The first encouraging results in this direction have been obtained with the 'Combi' unit.

7.4.3.6 **Pneumatic Transport of Powder Materials**

The extensive theoretical and experimental studies of hydrodynamics of two-phase flows in pipes performed in the early 1960s allowed the realization of the ideas of Shvab and the development of a new method of impulse pneumatic transport, which in many cases, and particularly for transportation on the long distances, is all that is possible [7.75].

The principle of operation of a piston transportation unit (Fig. 7.41) is the periodic delivery of transported product from the feeder into the pipeline by means of a special feeding device and compressed air (gas). Compressed air is fed additionally into the pipeline in chosen places along the transportation line by means of special pressure regulators. Therefore, the unit operation is characterized by a stable regime of movement of material portions.

The advantage of impulse pneumatic transport is determined first by the small relative velocity of the carrying medium and material, resulting in full utilization of expanded gas energy. The high initial pressure of gas compared with a 'suspended'

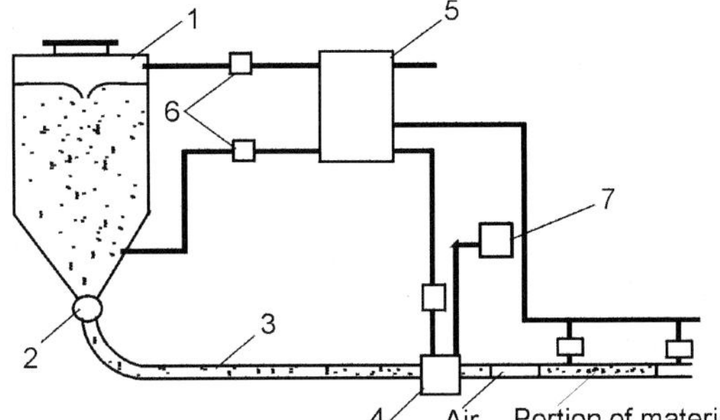

Figure 7.41. The principal of impulse powder transportation. 1, Chamber feeder; 2, stop valve; 3, transport pipeline; 4, divider; 5, receiver; 6, pressure regulator; 7, impulse device.

transportation mode allows increases in the transportation distance and mass flow concentration by 10–15-fold.

The principle of impulse transportation can be realized also under 'vacuum' conditions when the pressure in the pipeline is below atmospheric. This is of significance upon transportation of materials hazardous for the environment.

7.4.3.7 A Solution to Dust-collection Problems in Pneumatic Technology of Powders

Since the use of air as a working medium is the basis for pneumatic methods of powder technology, the problems of dust collection are of double importance. First, it is necessary to prevent the loss of processed material that is important essentially for materials with micron and submicron particle size, as they are fairly expensive. Further, it is necessary to exclude exhaust of dust into technological compartments and the environment.

Numerous experimental studies of methods of dust collection have been developed and various devices and installations have been tested. As this tooks place, the efficiency of micron-sized particle capture, the reliability of operation and also the ease of dust collection service systems (unloading, regeneration, cleaning, repair) were also estimated.

Cyclone systems with closed contours were developed for ultrafine powder materials processing. This approach allows reductions in the dust concentration in released air and the total amount of dusty air output from pneumatic units. Released air is cleaned on cloth filters in addition. For materials hazardous for human health and the environment, pneumatic units operated in a closed cycle have been designed.

7.5 References

7.1 Fried LE, Howard WM, Souers PC (1998) *Cheetah 2.0 User's Manual*, Lawrence Livermore National Laboratory, Livermore, CA.

7.2 Dagani R (1999) Putting the 'nano' into composites, *Chem. Eng. News* 77, 25–31.

7.3 Komarneni S, Parker JC, Thomas GJ

(1992) *Nanophase and Nanocomposite Materials*, Materials Research Society, Vol. 286.

7.4 Komarneni S, Parker JC, Wollenberger HJ (1996) *Nanophase and Nanocomposite Materials II*, Materials Research Society, Vol. 457.

7.5 Komarneni S, Parker JC, Hahn H (1999) *Nanophase and Nanocomposite Materials III*, Materials Research Society, Vol. 581.

7.6 Siegel RW (1999) *Report on Nanostructure Science and Technology, http://itri.loyola.edu/nano/final.*

7.7 Buffat P, Borel JP (1976) Size effect on the melting temperature of gold particles, *Phys. Rev. A* **13**, 2287–2292.

7.8 Gash AE, Simpson RL, Tillotson TM, Satcher JH Jr, Hrubesh LW (2000) Making nanostructured pyrotechnics in a beaker. In: *Proc of the 27th International Pyrotechnics Seminar, USA*, pp. 7.11–53.

7.9 Simpson RL, Tillotson TM, Hrubesh LW, Gash AE (2000) Nanostructured energetic materials derived from sol-gel chemistry. In: *Proc of the 31st Int. Annual Conference of ICT, Karlsruhe*, p. 35.

7.10 Simpson RL, Hrubesh LW, Tillotson TM (2000) Patent pending, USA.

7.11 Iler RK (1979) *The Chemistry of Silica*, Wiley, New York.

7.12 Brinker CJ, Scherer GW (1990) *Sol-Gel Science*, Academic Press, New York.

7.13 Clifford T (1999) *Fundamentals of Supercritical Fluids*, Oxford University Press, Oxford.

7.14 Ishizaki K, Komarneni S, Nanko M (1998) *Porous Materials: Process Technology and Applications*, Kluwer, Dordrecht.

7.15 Gregg SJ, Sing KSW (1982) *Adsorption, Surface Area and Porosity*, Academic Press, New York.

7.16 Goldschmidt H (1908) *Iron Age*, **82**, 232.

7.17 Fisher S, Grubelich MC (1998) Theoretical energy release of thermites, intermetallics and combustion metals. In: *Proc of the 24th International Pyrotechnic Seminar, USA*, pp. 231–286.

7.18 Wang LL, Munir ZA, Maximov YM (1993) Review: thermite reactions: their utilization in the synthesis and processing of materials, *J. Mater. Sci.* **28**, 3693.

7.19 Tillotson TM, Hrubesh LW, Simpson RL, Gash AE (2000) Patent pending, USA.

7.20 Strategic Environmental Research and Development Program Home Page (2000) *http://ww.serdp.org.*

7.21 Feng Z, Zhao J, Huggins FE, Huffman GP (1993) Agglomeration and phase transition of a nanophase iron oxide catalyst, *J. Catal.* **143**, 510.

7.22 Takahashi N, Kakuta N, Ueno A, Yamaguchi K, Fujii T, Mizushima T, Udagawa Y (1991) Characterization of iron oxide thin films prepared by the sol-gel method, *J. Mater. Sci.* **26**, 497–504.

7.23 Aumann CE, Skofronick GL, Martin JA (1995) Oxidation behavior of aluminum nanopowders, *J. Vac. Sci. Technol. B* **13**, 1178.

7.24 Taylor TN, Martin JA (1991) Reaction of vapor-deposited aluminum with copper oxides, *J. Vac. Sci. Technol. A* **9**, 1840.

7.25 Danen WC, Martin JA (1993) Energetic composites and method of providing chemical energy, *UK Patent Appl.* 2 260 317.

7.26 Danen WC, Martin JA (1993) Energetic composites, *US Patent* 5 266 132.

7.27 Dixon GP, Martin JA, Thompson D (1998) Lead-free percussion primer mixes based on metastable interstitial composite (MIC) technology, *US Patent* 5 717 159.

7.28 Pekala RW (1989) Organic aerogels from the polysondensation of resorcinol with formaldehyde, *J. Mater. Sci.* **24**, 3221.

7.29 Titov VM, Anisichkin VF, Mal'kov IYu (1989) Investigation of the ultrafine diamond synthesis process in detonation waves, *Phys. Combust. Explos.* **3**, 117–126.

7.30 Dremin AN, Pershin SV *et al.* (1989) About bend of dependence of detonation velocity versus TNT initial density, *Phys. Combust. Explos.* **5**, 141–144.

7.31 Kozirev NV, Golubeva ES (1992) Investigation of ultrafine diamond synthesis process in mixtures of TNT and RDX, HMX, PETN, *Phys. Combust. Explos.* **5**, 119–123.

7.32 Gubin SA, Odintsov VV *et al.* (1990) Influence of shape and size of graphite and diamond crystals on the phase equilibrium of carbon and detonation parameters of explosives, *Chem. Phys.* **3**, 401–417.

7.33 van Thiel M, Ree FH (1987) Properties of carbon clusters in TNT detonation products: the graphite-diamond transition, *J. Appl. Phys.* **5**, 1761–1767.

7.34 McGayer R, Ornellans D, Acst I (1981) Chemistry of detonation processes: diffusion phenomena in nonperfect explosives, In: *Detonation and Explosives*, pp. 160–169.

7.35 Anisichkin VF, Derendyaev BG, *et al.* (1990) Investigation of detonation process in condensed explosives by isotope method, *Rep. Acad. Sci. USSR* **4**, 879–881.

7.36 Kozirev NV, Brilyakov PM, *et al.* (1990) Investigation of ultrafine diamond synthesis process by labeled atoms method, *Rep. Acad. Sci. USSR* **4**, 889–891.

7.37 Kozirev NV, Sakovich GV, *et al.* (1991) Investigation of ultrafine diamond synthesis process by labeled atoms method. In: *Proc. 5th All-Union Meeting on Detonation, Krasnoyarsk*, pp. 176–179.

7.38 Pyaternev SV, Pershin SV, Dremin AN (1986) Dependence of shock induced graphite-diamond transformation pressure versus initial graphite density, hysteresis line of the transformation, *Phys. Combust. Explos.* **6**, 125–130.

7.39 Aksenenkov VV, Blank VD, *et al.* (1994) Formation of diamond monocrystal in plastically deformed graphite, *Rep. Acad. Sci. USSR* **4**, 472–476.

7.40 Bundy FP (1963) Melting of graphite at very high pressure, *J. Chem. Phys.* **3**, 618–630.

7.41 Petrov EA, Sakovich GV, Brilyakov PM (1990) Conditions of diamond conservation in process of detonation transformation, *Rep. Acad. Sci. USSR* **4**, 862–864.

7.42 Mal'kov IYu (1993) Preservation of carbon in explosion chamber, *Phys. Combust. Explos.* **5**, 93–96.

7.43 Bundy FP (1963) Direct conversion of graphite at very high pressure, *J. Chem. Phys.* **3**, 631–643.

7.44 Guschin V, Zakharov A, Lyamkin A, Staver A (1996) Carbon composition production process, *US Patent* 5 482 695.

7.45 Mench MM, Kuo KK, Yeh CL, Lu YC (1998) Comparison of thermal behavior of regular and ultra-fine aluminum powders (Alex) made from plasma explosion process, *Combust. Sci. Technol.* **135**, 269–292.

7.46 Cliff M, Tepper F, Lisetsky V (2001) Ageing characteristics of Alex® nanosize aluminum. In: *Proc. of the 37th AIAA Joint Propulsion Meeting, Salt Lake City*, p. 3287.

7.47 Ivanov GV, Tepper F (1997) Activated aluminum as a stored energy source for propellants. In: Kuo KK (ed.), *A Challenge in Propellants and Combustion*, Begell House, pp. 636–645, Stockholm.

7.48 Mench MM, Yeh CL, Kuo KK (1998) Propellant burning rate enhancement and thermal behavior of ultra-fine aluminum powders (ALEX). In: *Proc. of the 29th Int. Annual Conference of ICT, Karlsruhe*, p. 30.

7.49 Simonenko VN, Zarko VE (1999) Comparative study of the combustion behavior of composite propellants containing ultra fine aluminum. In: *Proc. of the 30th Int. Annual Conference of ICT, Karlsruhe*, p. 21.

7.50 Cliff M (2001) Personal communication.

7.51 Chiaverini MJ, Kuo KK, Peretz A, Harting GC (1997) Heat flux and internal ballistic characterization of a hybrid rocket motor analog. In: *Proc. of the 33rd AIAA Conference, Seattle*, p. 3080.

7.52 Sanger E (1933) *Raketenflugtechnik*, R. Oldenburg, Berlin, p. 5.

7.53 Rapp DC, Zurawski R L (1988) Characterization of aluminum/RP-1 gel propellant properties. In: *AIAA Propulsion Conference, Boston*, p. 2821.

7.54 Palaszewski B, Zakany JS (1996) Metallized gelled propellants: oxygen/RP-1/aluminum rocket heat transfer and combustion measurements. In: *AIAA Propulsion Conference, Lake Buena Vista*, p. 2622.

7.55 Galecki DL (1989) Ignition and combustion of metallized propellants. In: *AIAA Propulsion Conference, Monterey*, p. 2883.

7.56 Palaszewski B, Rapp D (1991) Design issues for propulsion systems using metallized propellants. In: *AIAA Conference on Advanced SEI Technologies, Cleveland*, p. 3484.

7.57 Palaszewski B, Zakany JS (1995) Metallized gelled propellants: oxygen/RP-1/aluminum rocket combustion experiments. In: *Proc. of the 31st Joint Propulsion Conference and Exhibition, San Diego*, p. 2435.

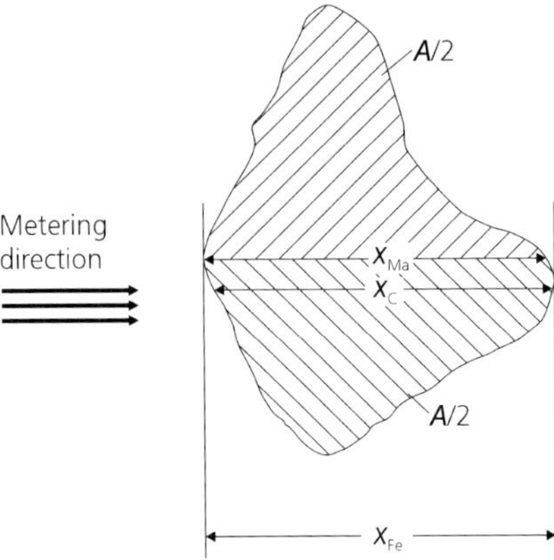

Figure 8.1. Statistical particle diameter.

which splits the projection area in half parallel to the measurement direction is denoted the Martin diameter, x_{Ma}. Another possible parameter useful for characterizing particle size is the longest chord, x_c, determined in the measurement direction [8.1, 8.2]. Yet another empirical particle characteristic is the projection area, A_m, of the particle. As a characteristic particle parameter, particle volume has the advantage that it is an unambiguous quantity, independent of the particle's spatial orientation.

Methods commonly used to determine physical size parameters of particles include sedimentation velocity in a fluid, scattering of light, attenuation of sound, diffusion coefficient, mobility in an electrostatic field, passage through a sieve opening and many others.

The concept of equivalent diameters was developed as a practical means to characterize the size and shape of non-spherical particles. An equivalent diameter is defined as the diameter of a spherically shaped particle that has the same geometric or physical properties as the irregularly shaped particle in question [8.1, 8.2].

Some commonly used equations for the equivalent diameter include the following: the diameter of a sphere of equivalent volume:

$$x_V = \sqrt[3]{\frac{6 \cdot V}{\pi}} \tag{8.1}$$

the diameter of a sphere of equivalent surface area:

$$x_S = \sqrt{\frac{S}{\pi}} \tag{8.2}$$

the diameter of a sphere of equivalent mean projection area A_m:

$$x_{Proj} = \sqrt{\frac{4 \cdot A_m}{\pi}} \tag{8.3}$$

the diameter of a sphere of equivalent sedimentation velocity (the Stokes diameter):

$$x_{ST} = \sqrt{\frac{18 \cdot \eta_L \cdot u_S}{(\rho_S - \rho_L) \cdot g}} \tag{8.4}$$

where η_L and ρ_L are the viscosity and density of the fluid, respectively, ρ_S is the particle density, g is the gravitational acceleration and u_S is the sedimentation velocity.

The equivalent diameter determined for a particle may differ significantly depending on the equivalence property upon which the measurement is based. It is therefore essential that the measurement method and/or equivalence property are documented for any type of particle size measurement.

In general, it is not sufficient to characterize non-spherical particles solely on the basis of size. The properties of such particles in many cases are highly dependent on their specific shape characteristics. With a gradually increasing deviation from spherical shape, the values of the different equivalent diameters also gradually diverge. This phenomenon provides an additional means to characterize the particle shape by defining a shape factor:

$$\Psi_{\alpha,\beta} = \frac{x_\alpha}{x_\beta} \tag{8.5}$$

where x_α and x_β are different equivalent diameters. One of the most commonly used shape factors is Wadell's sphericity [8.1]:

$$\Psi_{Wa} = \left(\frac{x_V}{x_S}\right)^2 \leq 1 \tag{8.6}$$

The Wadell sphericity is the ratio of the surface area of a sphere with a volume equivalent to that of the particle to the particle's actual surface area.

8.1.2
Particle Size Distributions

The properties of particles depend strongly on their size and shape. Characterization of a discrete dispersed particulate system or a particle collective includes classifying the system based on the concentration of particles with a given particle size x. This characterization of the particle size distribution [8.1–8.4] varies depending on the measurement method used. For instance, if particles of a given size are counted, then a number-average particle size distribution is obtained, whereas if the weight of particles of a given size is determined, then a mass-average distribution is obtained. Different methods of classification are denoted according to the index r as shown in Table 8.1.

The cumulative distribution $Q_r(x_i)$ represents the concentration of particles equal to or smaller than a given particle size x_i [8.1].

$$Q_r(x_i) = \frac{\text{Amount of particles } x \leq x_i}{\text{Total amount of particles}} \tag{8.7}$$

Values of the cumulative distribution $Q_r(x_i)$ range from 0 to 1:

Table 8.1. Designations of different particle classification methods.

Classification basis	Dimension	Index
Number	L^0	$r = 0$
Length	L^1	$r = 1$
Area	L^2	$r = 2$
Volume	L^3	$r = 3$
Mass	L^3	$r = 3$

$$Q_r(x_{\min}) = 0; \qquad Q_r(x_{\max}) = 1 \tag{8.8}$$

The density distribution $q_r(x_i)$ represents the amount of particles of a given particle size x_i, relative to the entire particle size distribution. Amounts that fall within a given particle size interval are referenced to the interval size:

$$q_r(\bar{x}_i) = \frac{\text{Amount between } x_i \text{ and } x_{i+1}}{\text{Interval size } x_{i+1} - x_i} \tag{8.9}$$

where $\bar{x}$ is the mean particle size of the interval:

$$\bar{x}_i = \frac{1}{2}(x_i + x_{i+1}) \tag{8.10}$$

If $Q_r(x)$ is a continuously differentiable function, then $q_r(x)$ can be determined via differentiation:

$$q_r(x_i) = \frac{dQ_r(x)}{dx} \tag{8.11}$$

The cumulative distribution $Q_r(x)$ and density distribution $q_r(x)$ are illustrated in Fig. 8.2.

From the measured size distribution, various parameters can be determined that characterize the particle collective, including:

- The median value $x_{50,r}$ is the particle size for which $Q_r(x_{50,r}) = 0.5$. This means that 50% of the total particle collective is smaller than this value. The median value $x_{50,r}$ is often given as the mean value of a disperse system.
- The modal value x_{mod} is the particle size at which the density distribution $q_r(x)$ exhibits a maximum.
- The mean value $\bar{x}$ of a particle size distribution of type r is determined according to the equation

$$\bar{x}_r = \int_{x_{\min}}^{x_{\max}} x \cdot q_r(x) \cdot dx \tag{8.12}$$

So, for example, the mean particle diameter for a distribution based on particle surface area (the Sauter diameter x_S) is a mean particle diameter that approximates the particle collective as having the diameter of a sphere, with the same ratio of volume to surface area as the particle collective itself:

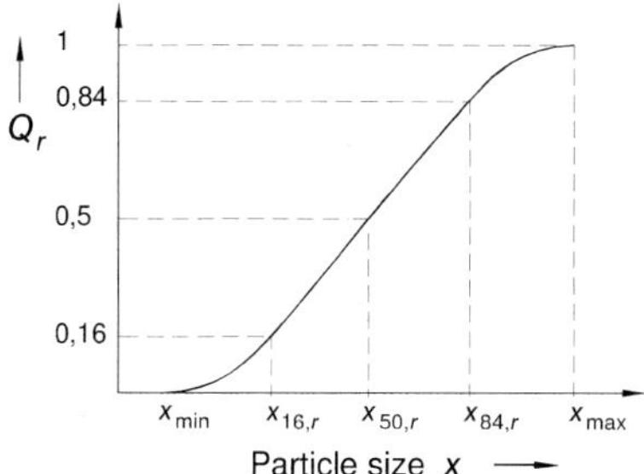

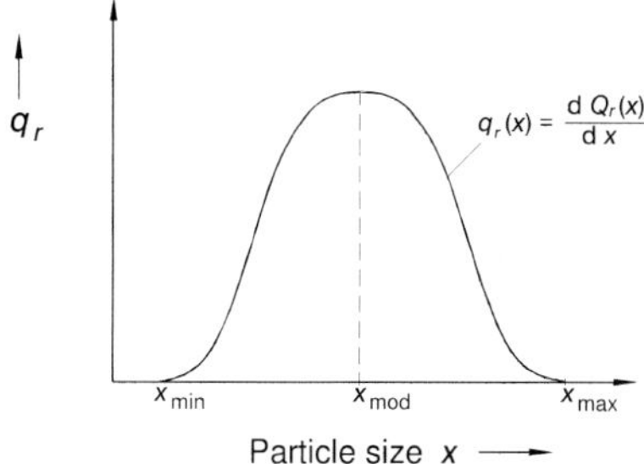

Figure 8.2. Cumulative distribution $Q_r(x)$ and density distribution $q_r(x)$.

$$x_S = \int_{x_{min}}^{x_{max}} x \cdot q_2(x) \cdot \mathrm{d}x \tag{8.13}$$

The homogeneity or dispersity of a particle collective or dispersed system is a function of the width of the particle size distribution. This distribution width can be determined simply on the basis of the maximum and minimum particle sizes or it can be calculated based on other measurable parameters of the distribution. According to specifications published by the Society of German Engineering (VDI 3491 [8.5]), the width of a size distribution can be characterized by the dispersity ξ, defined as

$$\xi = \frac{x_{84,3} - x_{16,3}}{2 \cdot x_{50,3}} \tag{8.14}$$

where

$x_{84,3}$: 84% of the particle collective is smaller than $x_{84,3}$; $Q_3\,(x_{84,3}) = 0.84$
$x_{16,3}$: 16% of the particle collective is smaller than $x_{16,3}$; $Q_3\,(x_{16,3}) = 0.16$

The dispersity of a particle collective or dispersed system can be categorized as follows [8.5]:

$\xi < 0.14$ → monodisperse;
$0.14 \le \xi \le 0.41$ → quasi monodisperse;
$\xi > 0.41$ → polydisperse.

In general, a particle collective is considered monodisperse when the diameters of the individual particles are within 10% of each other.

Different particle size distributions are only directly comparable when they have been determined based on the same equivalence property. When that is not the case, it may be possible to compare them using mathematical transformations. For instance, if the number density distribution $q_0(x)$ is determined via a counting technique, then as long as $q_0(x)$ is a continuous function, the volume density distribution $q_3(x)$ can be calculated as follows:

$$q_3(x) = \frac{x^3 \cdot q_0(x)}{\int_{x_{\min}}^{x_{\max}} x^3 \cdot q_0(x)\, dx} \tag{8.15}$$

Often, empirically determined cumulative distributions can be described via mathematical functions. Such approximation functions can be plotted on specialized coordinate systems that enables them to be portrayed linearly [8.3, 8.4, 8.6].

The power law function of Gandin and Schuhmann is the simplest approximation function. It is used to approximate the volume distribution and is defined as follows:

$$Q_3(x) = (x/x_{\max})^m \qquad \text{for } x \le x_{\max} \tag{8.16}$$

$$Q_3(x) = 1 \qquad \text{for } x > x_{\max} \tag{8.17}$$

The parameters $x_{\max}$ and m are adjusted so that the mathematical distribution function fits the empirically determined distribution $Q_3(x)$. If this function is plotted on a double logarithmic coordinate system, a linear relationship is obtained.

A logarithmic normal distribution (analagous to the well-known Gaussian normal distribution) results when the logarithm of the variable x is normally distributed. The density function of a logarithmic normal distribution $q_r(x)$ is expressed as:

$$q_r(x) = \frac{1}{\sigma_x \cdot \sqrt{2\pi}} \exp\left[-\frac{1}{2}\left(\frac{x - x_{50,r}}{\sigma_x}\right)^2\right] \tag{8.18}$$

where $x_{50,r}$ is the median value of the r distribution and σ_x is the standard deviation of the length x.

The normal distribution is one in which values of the length x are randomly distributed around a mean value $x_{50,r}$ and the individual values of the length are independent of each other.

The RRSB distribution, named after Rosin, Rammler, Sperling and Bennet, is defined as

$$Q_3(x) = 1 - \exp\left[-\left(\frac{x}{x'}\right)^n\right] \tag{8.19}$$

The fit parameters are x' and n, where x' is the particle size at which the value of the cumulative distribution function $Q_3(x')$ is 0.632.

8.1.3
Sample Collection and Sample Preparation

The first important steps when conducting analysis of a particle collective or dispersed system are collection and preparation of the samples. During these steps the product must be transferred from the process or bulk material and turned into a measurable sample. Prior to the analysis, the sample must be prepared for the measurements. The particle concentration is usually reduced; in some cases the volume concentration of particles may be diluted by up to six orders of magnitude. Additional procedures may include breaking up agglomerates, adding stabilizers and dispersing the particles in a fluid. One must take care, however, to ensure that the sample is not changed in a way that causes the measurement results to no longer be representative of the process/bulk material. During sample collection one must also ensure that the material collected is representative of the larger particle collective, i.e. product properties such as size distribution and moisture content must be the same. Randomly collecting the sample from the product stream is not sufficient to ensure a representative sample, because such a sample may have randomly different properties than that of the particle collective. Instead, every particle size analysis should use a process based on the random collection of a large number of very small samples which are combined together to create one small test sample. Good results have been obtained using a rotating sample collector, such as shown in Fig. 8.3.

Using this device, a given sample is divided into eight partial samples (for example) and a small amount of additional material is added during each revolution of the collector. This provides representatvie sample materials even if some of the material collected differs significantly from the bulk material. Mistakes made during the sample collection step may be magnified or reduced during the subsequent analysis, depending on the analysis method employed.

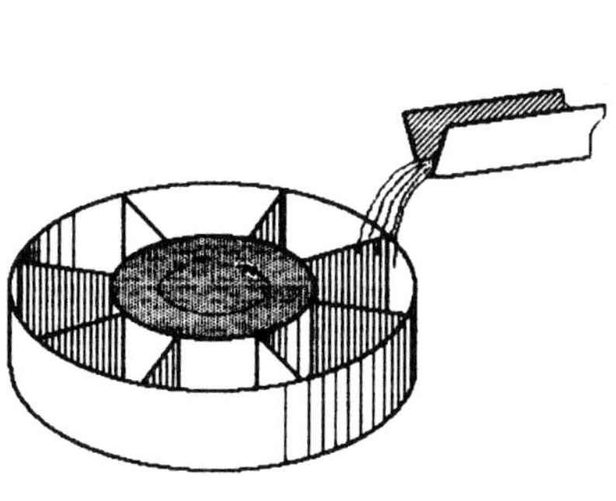

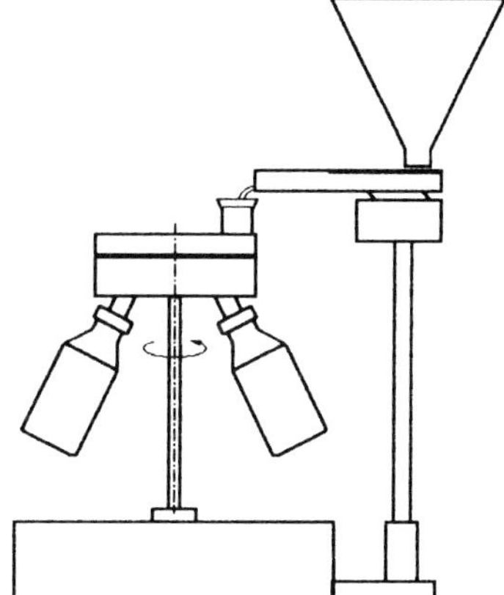

Figure 8.3. Rotating sample collector [8.2].

Sieve analysis requires multiple gram quantities of material, whereas laser light diffraction spectrometry requires only milligrams or at most a few grams.

When dispersing the particles in a gas, it is important that the particle size is not reduced as the sample is conveyed to the measurement zone. With commercially available equipment used for dry dispersion, the sample is conveyed via shear flow, either with or without subsequent impact pulverization. As the pressure applied increases (in order to increase the conveyance speed), more load is applied to the particles. The susceptibility of the particles to break apart is, of course, strongly dependent on the type of material and the particle size.

Most particle size analysis methods are disturbed by the presence of agglomerates (at least for those cases in which one does not intend to measure the size distribution of the agglomerates). Therefore, the particles need to be dispersed prior to the measurement. For effective dispersion, adhesive bridges between the particles must be broken (without destroying the particles) and the particles must be evenly distributed on to a substrate or in a fluid.

Wet preparation is advantageous in lowering the adhesion forces between particles [8.7]. The particles are suspended in a wetting but non-dissolving liquid, whereby the deagglomeration can be assisted by stirring, shaking, ultrasound or adding dispersion agents. Dispersing agents for wet preparation adsorb on the particles' surface, causing the wettability, zeta potential and van der Waals forces to be modified [8.8–8.10]. Possible dispersing agents include sodium pyrophosphate ($Na_2P_2O_7$), silicates, various surfactants, gelatine and many others.

Especially during sonification, the dispersing process should be monitored to ensure that agglomerates are completely dispersed, without destroying the primary particles. In general, one can conclude this has been successfully accomplished if the particle size distribution remains unchanged with continued dispersion.

8.1.4
Methods of Particle Size Measurement

As discussed in the previous sections, the particle size and size distribution of a particle collective are, in principle, measurable parameters. This chapter introduces various measurement methods that can be reliably employed (by trained personnel) and discusses their underlying fundamental principles. Particular special features of various methods are not addressed. In many cases the underlying physical measurement principles of the various methods are different, meaning that the particle size distribution measured is not always identical from one method to another.

Obtaining a quantitative measurement signal that can be related to the properties of the particle collective is an essential requirement in order to determine a particle size distribution using any of the particle characterization methods. Variations in the type of measurement signal detected and the means of detection also lead to differences in results produced by the various methods. Figure 8.4 provides an overview of the application range of the most common particle size analysis methods.

Particle size analysis techniques can be distinguished based on their underlying physical principle into the following broad categories: separation procedures

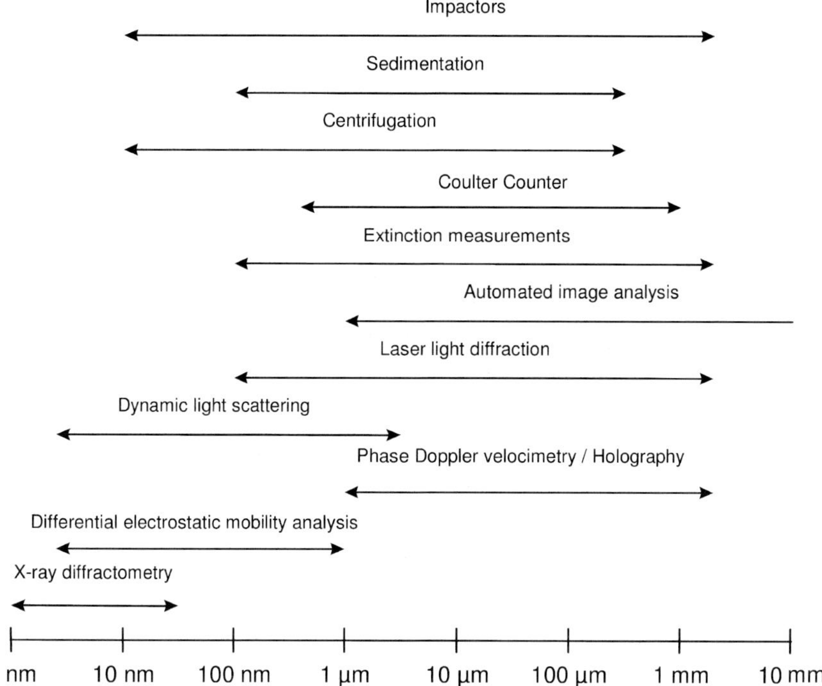

Figure 8.4. Measurement range of various particle size analysis methods.

(sieving, impactors, sedimentation), counting methods (image analysis, Coulter Counter, extinction measurement), spectrometric methods (laser light diffraction, ultrasonic, dynamic light scattering) and other methods (e.g. phase Doppler velocimetry, holography and differential electrostatic mobility analysis) [8.3, 8.4, 8.11–8.13].

8.1.4.1 Sieve Analysis

Sieving using woven wire sieves is a simple, cheap method of size analysis. Sieve analysis allows one to determine the mass fraction of particles falling within a specific size interval between the mesh size of two sieves. A particle size range from approximately 5 μm to 125 mm is accessible via sieve analysis [8.2–8.4, 8.14]. The range for dry sieving is from approximately 45 μm to 125 mm, air jet sieving is useful for ranges from approximately 10 to 500 μm and wet sieving, which is used primarily for cohesive particles and other hard to handle powders, can be used for particles ranging from 5 to 50 μm. A sieve analysis is conducted by stacking a series of sieves vertically in order of mesh size, with the larger sizes at the top. A given amount of sample is spread on the top sieve and, after a predetermined sieve time, the masses of material collected on each sieve are determined.

Sieve analysis is a particle size chacterization technique that can be used to determine the cumulative mass distribution curve. If m_i is the mass of material collected on each sieve (Fig. 8.5), then

$$\Delta Q_i = \frac{m_i}{\sum_{i=1}^{n} m_i} \text{ and } Q_i = \sum_{i=1}^{n} \Delta Q_i \tag{8.20}$$

To convey solid particles through the sieve apertures requires that the sieve moves relative to the particles. This relative motion can be effected either by moving solely the sieve or via a system in which the particles are transported in a fluid or air stream into a rotating sieve. Movement of the sieve insures that the feedstock is uniformly distributed on the mesh so that every particle has the opportunity to pass through the apertures and, most importantly, so that blocked apertures are continuously opened. For sieves with a diameter of 200 mm, 100–200 g of material are sufficient to conduct a sieve analysis. A typical sieve time is of the order of 20 min. Using too much material causes the required sieve time to be unnecessarily long and makes separation into the various size fractions more difficult. Using too little material reduces the measurement accuracy. Finally, the accuracy and comparability of sieve measurements are also affected by the type of sieves used (see Fig. 8.6).

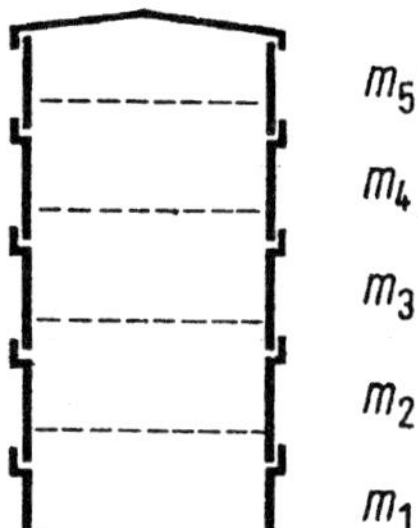

Figure 8.5. Schematic of a sieve apparatus [8.4].

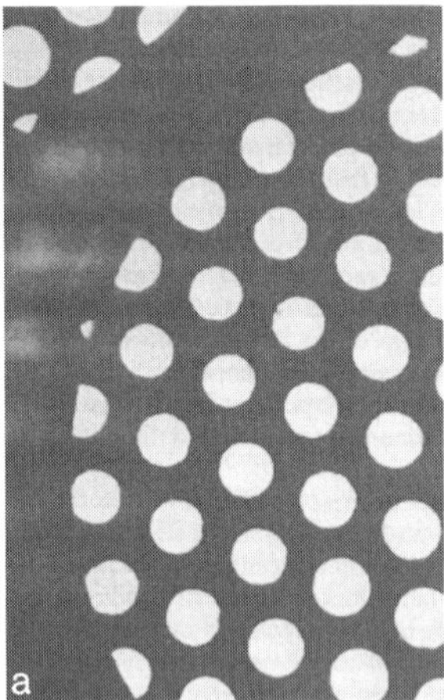

Figure 8.6. Sieve plates: (a) perforated plate and (b) woven wire.

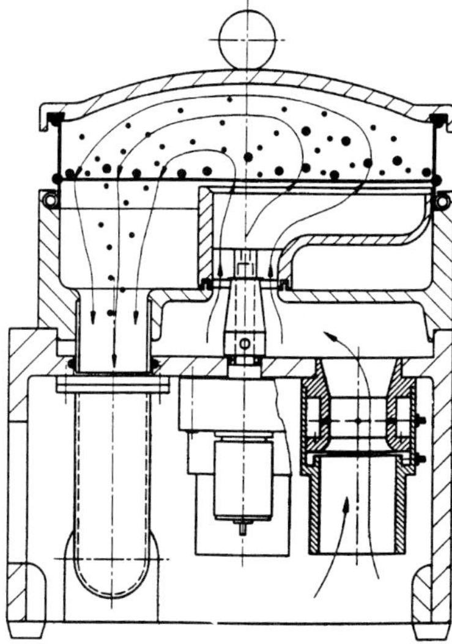

Figure 8.7. Air jet sieve [8.15].

Air jet sieving is suitable for analyzing lightly agglomerated particles and for characterizing particles with sizes up to approximately 10 μm (see Fig. 8.7).

Directing a narrow stream of air under a fixed sieve via a rotating slit-shaped jet is an effective means of fluidizing and deagglomerizing the feedstock so that the fine fraction can pass through the sieve apertures. In this case, transport of the particulate through the sieve openings occurs primarily through action of the applied air stream. The air stream spreads out above the sieve plate and as its velocity decreases it is suctioned into the perimeter of the sieve away from the outlet of the slit-shaped jet. Particles smaller than the aperture size are thereby transported with the gas through the sieve openings. Wet sieving is recommended for the analysis of cohesive or wet particles down to a particle size of approximately 5 μm. In this case, etched, microprecision sieves are usually employed.

8.1.4.2 **Sedimentation Analysis**

Sedimentation analysis is based on the principle that gravitational effects cause individual particles to precipitate in a stirred fluid [8.2, 8.4, 8.14]. A characteristic parameter for a particle subjected to sedimentation analysis can be calculated by measuring the stationary sedimentation velocity u_S of a particle subjected to gravitational acceleration (g) settling in a stirred fluid of infinite length with a density ρ_L and a viscosity η_L:

$$x_{ST} = \sqrt{\frac{18 \cdot \eta_L \cdot u_S}{(\rho_S - \rho_L) \cdot g}} \tag{8.21}$$

The sedimentation velocity defined in Eq. (8.21) is applicable when the particle Reynolds number, Re_p, defined as

$$Re_p = \frac{\rho_L \cdot x_S \cdot u_S}{\eta_L} \leq 0.25 \tag{8.22}$$

is ≤ 0.25. This corresponds to the Stokes resistance law region, which is valid for a sphere subjected to creeping flow at Reynolds numbers ≤ 0.25.

In order to determine the particle size distribution via sedimentation analysis, the particles must be sufficiently dispersed in the measurement liquid. Also, particle agglomeration during the measurement must be avoided. Dispersing agents are often added to the system to minimize attractive interaction forces between particles.

After a short quiescent stage, the sedimentation process is initiated and a particle concentration gradient is obtained within the suspension. During the sedimentation process, the time-dependent solids concentration is measured locally within a predetermined fixed slice across the sedimentation vessel. The real time solids concentration $C(t)$ can be determined using various methods, such as light or X-ray attenuation. The cumulative mass distribution $Q_3(t)$ is calculated as follows:

$$Q_3(t) = \frac{C(t)}{C_0} \tag{8.23}$$

where C_0 is the solids concentration at time $t = 0$.

Photosedimentometers, which function by measuring attenuation of light passing through a suspension, provide a measure of the particle area, which is of course a function of the particle cross-section. X-ray sedimentometers measure the attenuation of X-rays passing through the fluid during sedimentation, thereby providing a measure of the time-dependent particle concentration, which can be used to calculate the cumulative mass distribution. With sufficient patience, this technique can be used to measure particle sizes as small as 0.5 µm. By conducting the sedimentation in a fluid subjected to centrifugal forces, particle sizes as small as 0.1 µm can be measured and the measurement time required is reduced. When analyzing particles in the nanometer range, one must ensure that Brownian motion effects do not influence the results.

8.1.4.3 **Image Analysis**

The first step of image analysis is to obtain photographs of the sample material. Every imaged particle is then individually measured, classified according to size and the number of particles of a given size is counted. Such images usually provide a distribution of projected areas of the particles. Hence it is generally possible to draw conclusions only about two-dimensional particle size characteristics.

As with the other techniques, successful image analysis is highly dependent on effective sample preparation. Particles must be spread onto the objective surface such that they are individually separated and adventitious material must not be introduced. Generally, particles are imaged with a CCD camera and the resulting photograph is digitized. Images can be obtained using various lighting techniques (direct, transmissive, oblique or fluorescent) with either white, spectral or other types of light. Software algorithms are generally used to prepare the images so that the particle characteristics can be calculated, which is also an automated process using other specialized software routines. Images in which the object and background are difficult to distinguish (fuzzy photographs or those with poor contrast) or in which there is no sharp separation between particles are problematic to analyze.

A variety of optical instruments provide varying degrees of magnification, making it possible to use image analysis over a wide range of particle sizes. Such instruments include the optical microscope, scanning electron microscope (SEM), transmission electron microscope (TEM), scanning tunneling microscope and the scanning force microscope. Comprehensive descriptions of these instruments and an overview of image analysis methods, image preparation and signal processing are available elsewhere [8.3, 8.16]. A decisive advantage of optical techniques is that they are useful in elucidating particle morphology and structure, in addition to size and shape characteristics.

8.1.4.4 **Coulter Counter**

Coulter Counters (Fig. 8.8) use the disturbance of an electric field to determine the particle size distribution in a sample. As such, this method is clearly situated among the direct counting procedures. The particles, suspended in an electrolyte, flow one at a time through a capillary channel of known diameter in an electric field. Whenever a particle is in the channel, the resistance between the electrodes increases and hence also the voltage in the circuit (the current is held constant). The change in resistance ΔR depends on the volume of displaced electrolyte and is thus proportional to the particle volume [8.3, 8.4].

One disadvantage of this procedure is that the capillary diameter must be compatible with the particle size spectrum. If there are particles in the sample that are too large, they will block the counting channel. If the particles have a diameter < 2% of the capillary diameter, they cannot be detected [8.4]. With very small particles there is also the danger that two will enter the capillary together and be

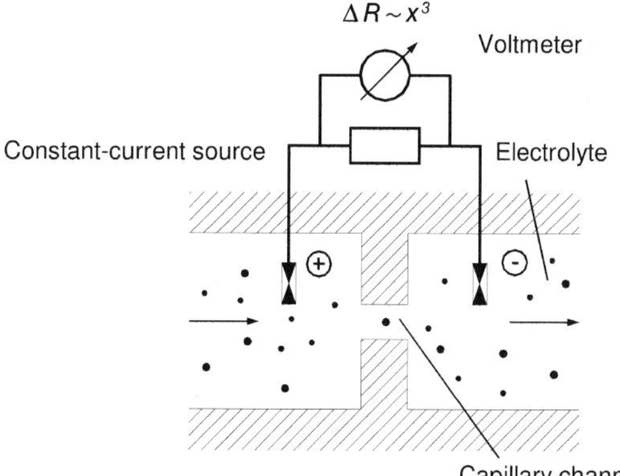

Figure 8.8. Schematic diagram of a Coulter Counter.

counted as a single, larger particle. Coulter Counters cover size ranges from 0.4 to 1200 μm. For samples containing a wide range of particle sizes, preconditioning to appropriate particle size is required.

8.1.4.5 Laser Light Diffraction Spectrometry

Fundamentals

Laser light diffraction spectrometry is a method that determines particle size distribution based on the scattered light distribution, which originates from particle-light interactions, i. e. from the scattering that occurs as light waves expand through a particulate medium. The scattered light radiation resulting from interaction between a single particle and incident light can be categorized into different ranges based on the value of the *Mie* parameter α:

$$\alpha = \frac{\pi \cdot x}{\lambda} \tag{8.24}$$

The scattering behavior is thus dependent on the light wavelength λ and the particle diameter x and is exactly described by applying the Mie theory. In general, the following ranges can be distinguished in dependence on the Mie parameter α:

$\alpha << 1$	range of *Rayleigh scattering*;
$0.5 < \alpha < 10$	range of *Mie scattering*;
$\alpha >> 1$	range of *geometric optics*.

Rayleigh scattering occurs when the particles are very small compared with the light wavelength. Rayleigh scattering is characterized by complete symmetry in the direction of the incident and reflected light, i. e. light is scattered in the forward and reverse directions in equal amounts. In this case the scattered light intensity is proportional to the sixth power of the particle diameter x and inversely proportional to the fourth power of the wavelength λ. Because the light scattering is symmetrical, the scattered light distribution cannot be used to determine the particle size distribution in the Rayleigh range.

In the Mie range the particle size can be determined from the scattered light distribution for particles with a diameter greater than ~0.1 μm. In this range there is no symmetry in the scattered light distribution. With increasing Mie parameter α, the light scattered in the forward direction gains in intensity and the light scattered in the reverse direction becomes weaker. As the particle or droplet size increases, the intensity of the scattered light maximizes at the scattering angle $\varphi = 0$. In this range the scattered light intensity is proportional to the fourth power of the particle diameter.

In the geometric optics range ($\alpha >> 1$), i. e. the relationship between the scattered light distribution and the particle size is straightforward, the narrowest range of forward scattering (scattering angle $\varphi \rightarrow 0$) is primarily determined from light diffraction. The contributions of refraction and reflection to the scattered light distribution are negligible. In this range the diffraction behavior is akin to the well-known Fraunhofer diffraction rings [8.17]. In the geometric optics region, light intensity is proportional to the second power of the particle diameter.

The interactions between spherical particles and the entering light, characterized by its wavelength λ and radiation intensity I_0, can be described using the Mie theory [8.18]. The Mie theory defines the relationship between the scattering angle distribution (or intensity) and the particle size. Comprehensive descriptions of the Mie theory are available elsewhere [8.19, 8.20]. Because the Mie range and Rayleigh range are both characterized by a strong dependence on the optical properties of the system examined, knowledge of the material's complex refraction index m is necessary:

$$m = n \cdot (1 - i\kappa) \tag{8.25}$$

where n is the refractive index and κ is the absorption index, defined as follows:

$$\kappa = \frac{k \cdot \lambda}{4 \cdot \pi \cdot n} \tag{8.26}$$

k being the absorption coefficient.

The optical material properties of a homogeneous material are described by the complex refractive index. Because optical properties such as the refractive index n and the absorption index κ are often unknown, an approximation called the Fraunhofer approximation is often used [8.21]. This approximation is valid only for scattered light in the narrow forward region (range of geometric optics), in which the scattered light intensity results predominately from diffraction of the incident light off of the particle surface.

Figure 8.9 depicts the measurement principle employed by a laser light diffraction spectrometer. The spectrometer is equipped with a light source that radiates an He-Ne laser beam with a wavelength λ = 632.8 nm and a receiver that contains imaging optics and detector electronics. Laser light diffraction spectrometry is based on the fact that the scattering of light depends on the size of individual particles in

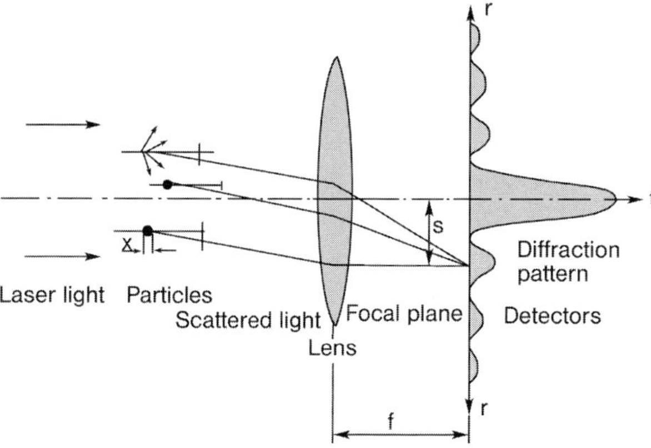

Figure 8.9. Basic design of the laser light diffraction spectrometer.

the medium and on the particle collective. The smaller the particle, the more the light is scattered. By using a Fourier lens, the light scattered from a particle of a certain size is imaged at the same spot on the detector, regardless of the particle's position in the medium. The scattered light forms a radially symmetric diffraction image on the semicircular detector. For spherical particles, the diffraction image consists of concentric dark and light rings (diffraction rings).

The scattered light intensity distribution $I(r,x)$, which depends on the particle diameter x and the radial distance to the optical axis r, is defined for a single particle as follows [8.22]:

$$I(r,x) = I_0 \cdot \left(\frac{\pi \cdot x^2}{4 \cdot f \cdot \lambda} \right)^2 \cdot \left[\frac{2 \cdot J_1 \left(\frac{\pi \cdot r \cdot x}{\lambda \cdot f} \right)}{\frac{\pi \cdot r \cdot x}{\lambda \cdot f}} \right]^2 \tag{8.27}$$

where f is the focal length and J_1 is a Bessel function of the first kind and first order.

For investigations of a particle collective, the scattered light intensities from the individual particles are superimposed. The particle size distribution of the sample can be calculated from the measured scattered light intensities [8.22].

Laser light diffraction spectrometry is a method that can determine particle sizes in the range from 0.1 to 3000 µm. A distinct advantage of laser light diffraction spectrometry is the short measurement time required, especially compared with several other non-optical methods. The procedure is particularly suitable for spherical and absorbing particles with diameters $x > 5$ µm, even when the material's optical properties are not known. In some cases the optical properties of the particles, such as the refractive index n and the absorption index κ, must be known in order to apply laser light diffraction spectrometry successfully [8.23].

Ideal transparent particles do not absorb light, instead they reflect or diffract it. They have an absorption index $\kappa = 0$. Materials that only slightly absorb light have an absorption index in the range from 10^{-2} to 10^{-3} [8.24].

Light diffraction in disperse systems is not only dependent on the material properties of the particles; the refractive index of the matrix liquid (continuous phase) n_c is also influential. The difference in the two phases can be characterized by the relative refractive index n^*:

$$n^* = \frac{n_d}{n_c} \tag{8.28}$$

which is the ratio of the refractive indices (the real component) of the disperse and continuous phases. For disperse systems with transparent particles or droplets, the two indices are often nearly identical (i.e., $n^* \rightarrow 1$). For the limiting case $n^* = 1$ the phase separation is no longer optically detectable.

Characterization of transparent particles

Particle size distribution was determined in this investigation using a Mastersizer laser light diffraction spectrometer from Malvern Instruments. The dispersed systems examined in this work consisted of demineralized water (refractive index n_c = 1.333) as the continuous phase and four types of glass spheres with different mean particle diameters (F1–F4) as the disperse phase. To produce glass sphere fractions with a narrow particle size distribution, the glass spheres were separated by size via sieving prior to determining the size characteristics of each fraction. Four fractions with different mean particle diameter ranging from $x = 3$ to $x = 115$ µm were chosen. The refractive index of the glass spheres was $n_d = 1.53$. Figure 8.10 shows representative glass spheres imaged using a scanning electron microscope.

Figures 8.11–8.15 show the cumulative distribution by volume $Q_3(x)$ and the volume density distribution $q_3(x)$ of the glass sphere fractions determined via laser light diffraction spectrometry. The size distributions were calculated using the Fraunhofer approximation and the Mie theory, using the refractive index and absorption index of the dispersed system as input parameters in the calculation based on the Mie theory.

Figure 8.11 shows the cumulative distribution by volume $Q_3(x)$ and the volume density distribution $q_3(x)$ of the narrow-distributed monomodal glass sphere fraction F1 (shown in Fig. 8.10) calculated using the Fraunhofer approximation. Using this method, the median value was $x_{50,3}$ = 115.2 µm and the Sauter diameter x_S = 113.4 µm.

In this range the calculation of the size distribution using the Fraunhofer approximation agrees exactly with that of the Mie theory, using a refractive index $n_d = 1.53$ for glass and an absorption index $\kappa = 0$ and 0.001. Even when the refractive index is chosen to be larger than that of glass and the adsorption index is varied, there is no significant change in the key parameters of the distribution. By decreasing the value of the refractive index to the value of the continuous phase (water), so that $n^* \approx 1$, the calculation predicts the existence of a fine fraction of particles that in reality does *not* exist (see Fig. 8.12).

Figures 8.13–8.15 show the results for glass sphere fraction F2. Calculations using the Fraunhofer approximation solution yield a median diameter $x_{50,3}$ = 13.7 µm and a Sauter diameter x_S = 6.9 µm. As seen in Fig. 8.13, a fine fraction is

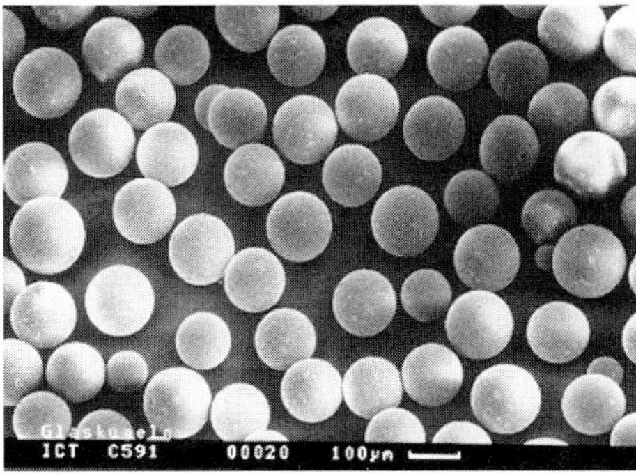

Figure 8.10. Scanning electron micrograph of glass sphere fraction F1.

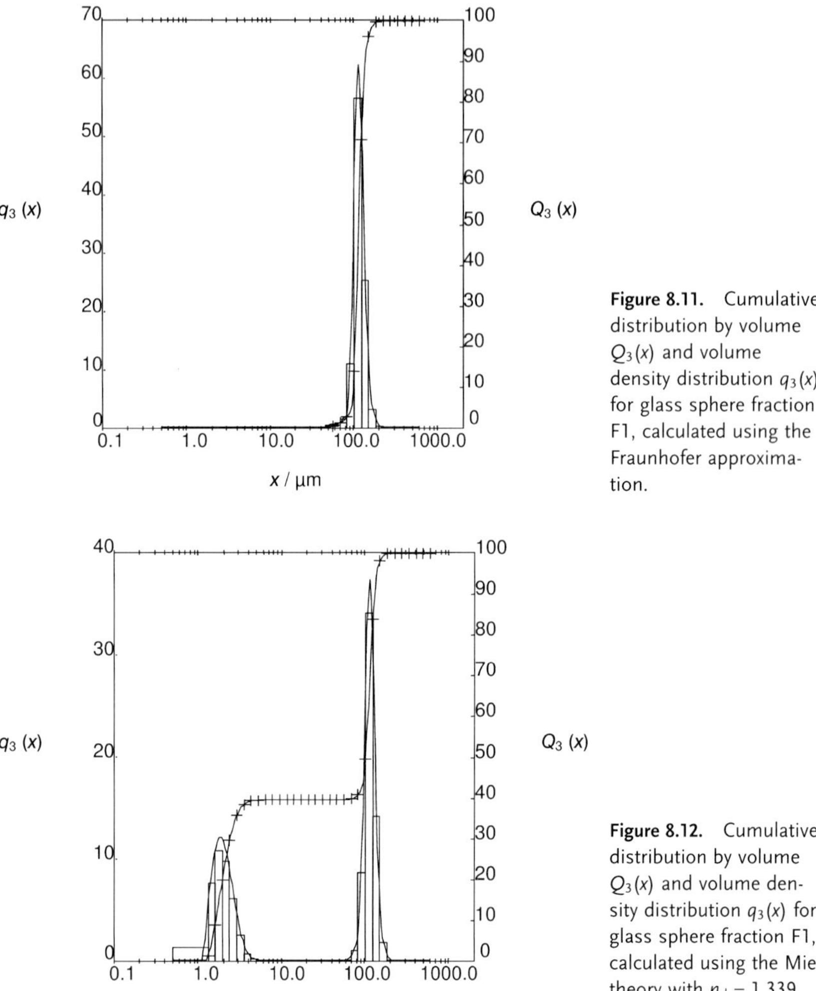

Figure 8.11. Cumulative distribution by volume $Q_3(x)$ and volume density distribution $q_3(x)$ for glass sphere fraction F1, calculated using the Fraunhofer approximation.

Figure 8.12. Cumulative distribution by volume $Q_3(x)$ and volume density distribution $q_3(x)$ for glass sphere fraction F1, calculated using the Mie theory with $n_d = 1.339$ and $\kappa = 0.001$.

predicted that in reality does not exist. This reduces the value of the Sauter diameter considerably. Light passing through the sample is scattered by the weakly absorbing (or non-absorbent) transparent particles, causing it to be deflected to the outer region of the detector. This secondary effect leads to the false prediction that a fine fraction of these glass spheres (F2) exists.

Calculations using the Mie theory with a refractive index $n_d > n_{glass}$ and an absorption index ranging from $\kappa = 10^{-3}$ to 0 also yields a prediction of a fictitious fine solids fraction. However, using the appropriate refractive index for glass, $n_d = 1.53$, and an absorption index of either $\kappa = 0$ or 10^{-3} yields the correct prediction for the particle size distribution, with no fine fraction and also yields identical values for the median diameter $x_{50,3}$ and the Sauter diameter x_S (see

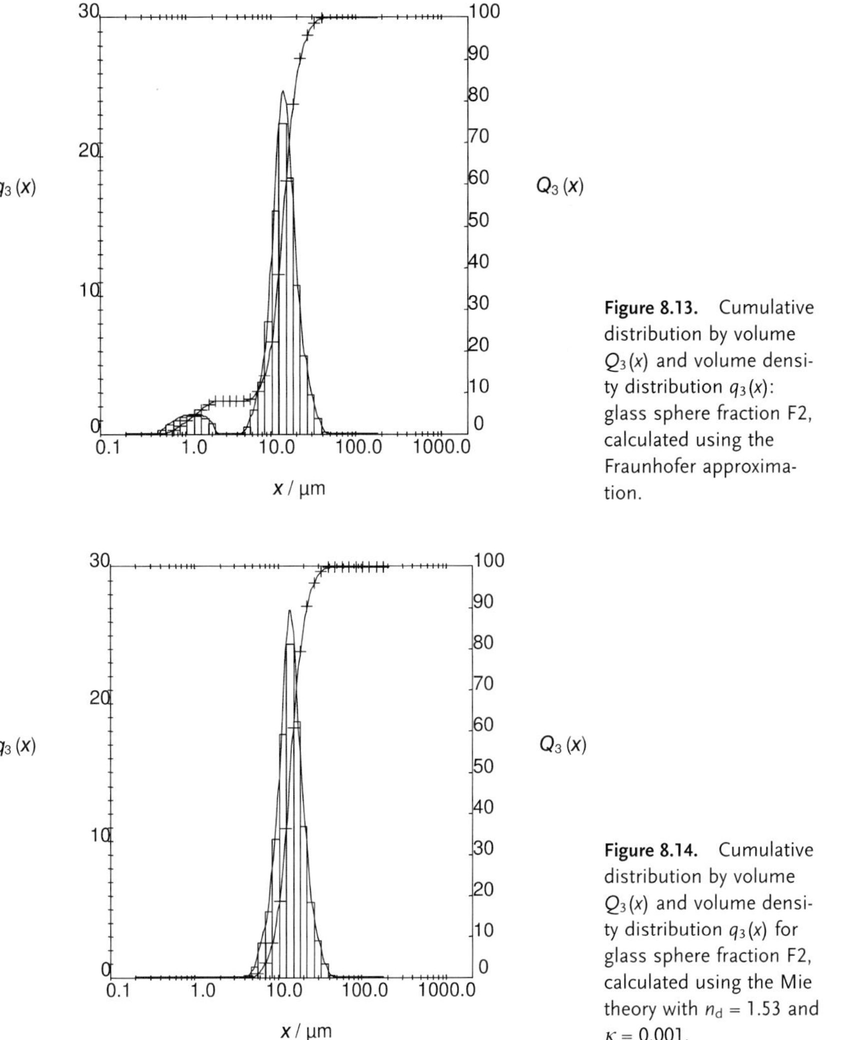

Figure 8.13. Cumulative distribution by volume $Q_3(x)$ and volume density distribution $q_3(x)$: glass sphere fraction F2, calculated using the Fraunhofer approximation.

Figure 8.14. Cumulative distribution by volume $Q_3(x)$ and volume density distribution $q_3(x)$ for glass sphere fraction F2, calculated using the Mie theory with $n_d = 1.53$ and $\kappa = 0.001$.

Fig. 8.14). If the absorption index is increased to $\kappa > 10^{-3}$ while the refractive index is held constant at $n_d = 1.53$, the Mie theory calculations predict a fine fraction similar to that predicted by the Fraunhofer approximation.

By reducing the refractive index to $n_d = 1.339$ (i.e. a relative refractive index $n^* \approx 1$), the Mie theory falsely predicts that the distribution is dominated by a fine solids fraction, with a median diameter equal to the Sauter diameter, $x_{50,3} = x_S = 0.84$ µm (see Figure 8.15).

These results show that using the Fraunhofer approximation to predict the size distribution of small ($\alpha < 10$), *transparent* particles results in the false prediction that a fine fraction of particles exists. This false bimodal particle size distribution is not

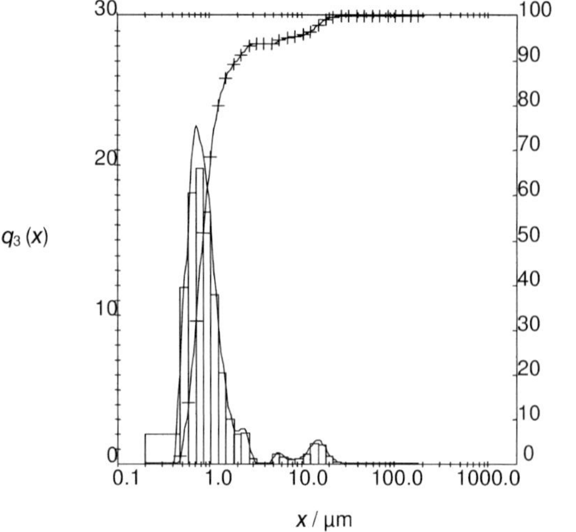

Figure 8.15. Cumulative distribution by volume $Q_3(x)$ and volume density distribution $q_3(x)$ for glass sphere fraction F2, calculated using the Mie theory with $n_d = 1.339$ and $\kappa = 0.001$.

obtained when the Mie theory is applied, provided that the refractive index and absorption index of the materials are known.

The scattering behavior of spherical, transparent particles was calculated based on the Mie theory using the simulation program *Streu* (Invent) [8.25, 8.26]. The technical parameters of the laser light diffraction spectrometry equipment were used as input parameters, including the power of the laser (P = 2 mW) and the wavelength of the linear polarized He-Ne laser λ = 632.8 nm. The polar diagram calculated from this simulation shows the scattering angle φ as a function of the intensity of the scattered light I (where φ is the angle between the direction of incident light and the light beam after scattering off a particle into the detector).

For particles smaller than the wavelength of light, a typical dipole behavior is observed, characteristic of Rayleigh scattering. With increasing particle diameter, the scattering behavior becomes increasingly evident. Figs 8.16 and 8.17 show polar diagrams of spherical particles with diameters of x = 1 and x = 20 µm, respectively. The refractive index of the particles is n_d = 1.48 and, with water as the continuous phase, the relative refractive index is n^* = 1.113. The absorption index was set at $\kappa = 10^{-3}$ for the calculation of all of the polar diagrams. These diagrams show that with increasing particle size, the scattering in the forward region becomes more dominant. In particular, the maximum in scattered light intensity at $\varphi \approx 0$ increases with increasing particle size. This clearly demonstrates that laser light diffraction spectrometry is well suited for the characterization of large particles. Of note for both particle sizes is the scattered light beam at φ = 90 and 270°. The backscattering maximum for the particles with x = 20 µm is at an angle φ = 180°, whereas for the particles with x = 1 µm the backscattering maxima are at φ = 120 and 240°. However, the backscattering intensities are several orders of magnitude lower than the intensity of the forward scattered light, which means that the Mie theory can be successfully applied to determine the particle size distribution of these systems.

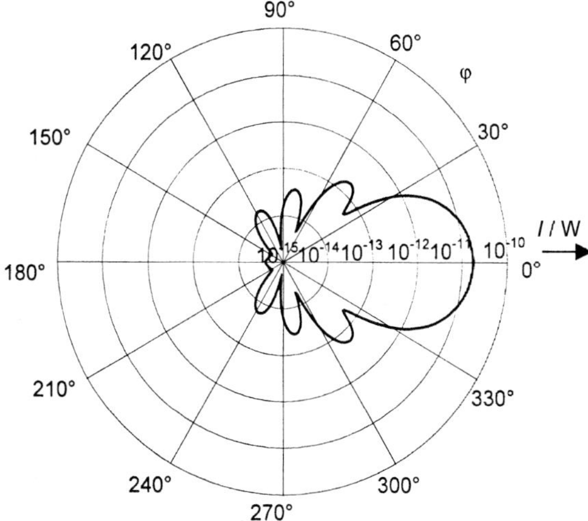

Figure 8.16. Polar diagram for a particle with $x = 1$ μm in water.

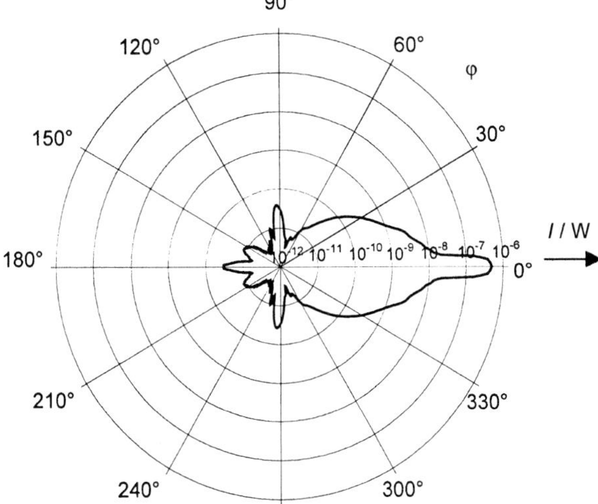

Figure 8.17. Polar diagram for a particle with $x = 20$ μm in water.

Figures 8.18 and 8.19 show polar diagrams of particles with identical diameters ($x = 5$ μm) but different refractive indices. The polar diagram in Fig. 8.18 is based on a refractive index $n_d = 1.333$. The continuous phase was water ($n_c = 1.333$), so the relative refractive index was $n^* = 1$. Noteworthy in this limiting case of relative refractive index is the large number of scattering peaks in the ranges from $\varphi = 60$ to 120° and from 240 to 300°. The intensity maximum in the forward direction is more pronounced in Fig. 8.19, which shows the polar diagram of glass spheres in water ($x = 5$ μm, $n_d = 1.53$, $n^* = 1.15$, $\kappa = 10^{-3}$). For this system, the backscattering intensity maximizes at $\varphi = 180°$ and local maxima are also observed at $\varphi \approx 110$ and 250°.

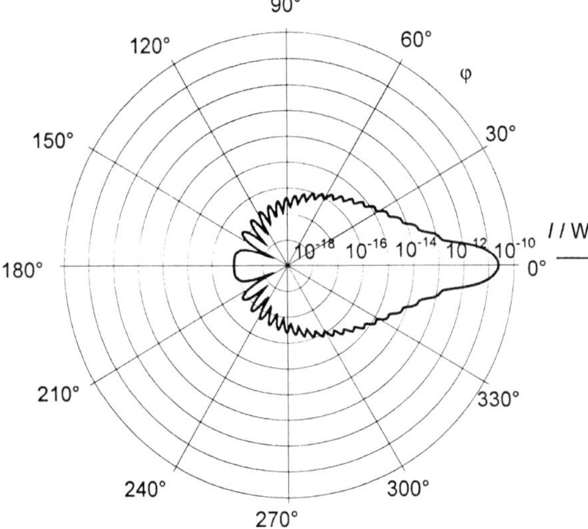

Figure 8.18. Polar diagram for a particle in water with $x = 5\ \mu m$, $n_d = 1.333$, $n^* = 1$, $\kappa = 10^{-3}$.

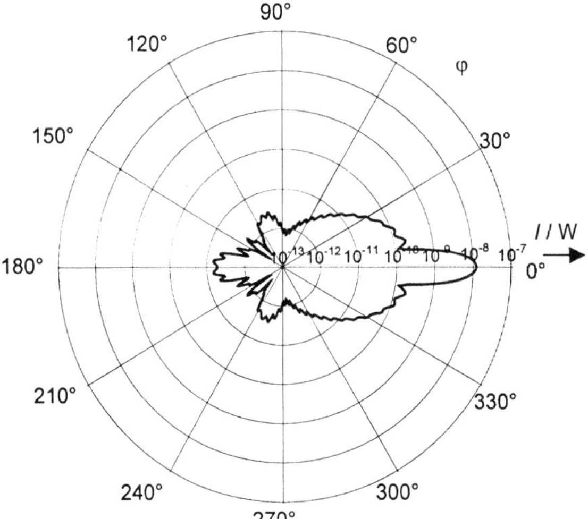

Figure 8.19. Polar diagram for a particle in water with $x = 5\ \mu m$, $n_d = 1.53$, $n^* = 1.15$, $\kappa = 10^{-3}$.

To verify the phenomenon observed in the laser light diffraction spectrometric measurements (i.e. the existence of a fictitious fine fraction of particles in the size distribution of transparent particles), comparison measurements were conducted using the Coulter Counter and the sedimentation method. The results for each method, expressed in terms of distribution curves $Q_3(x)$ of the glass sphere fractions as a function of equivalent particle diameter, are shown in Figs. 8.20–8.22.

One sees from Fig. 8.20, which shows $Q_3(x)$ for glass sphere fraction F2, that in the size range $x_{50,3} \approx 14\ \mu m$, measurements obtained via the Coulter Counter, sedimentation analysis and laser light diffraction spectrometry calculated using the

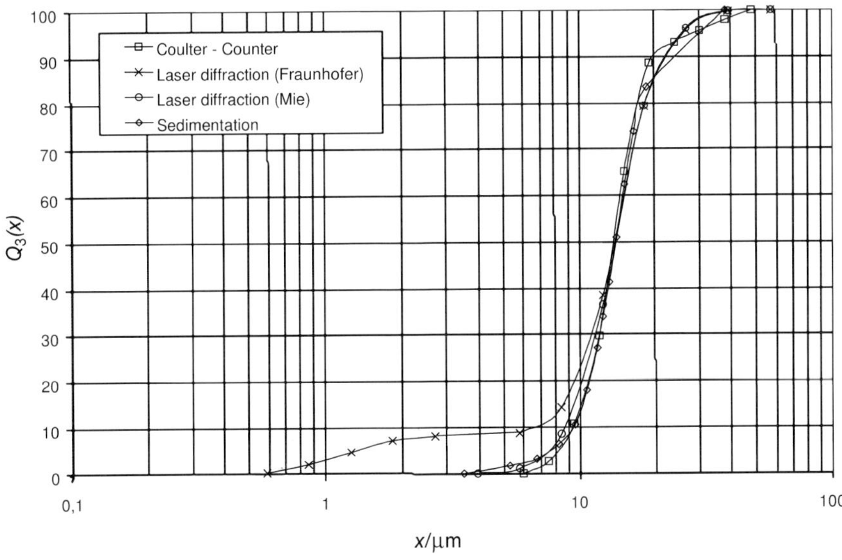

Figure 8.20. Cumulative distribution by volume $Q_3(x)$: comparison of methods for suspensions of glass sphere fraction F2.

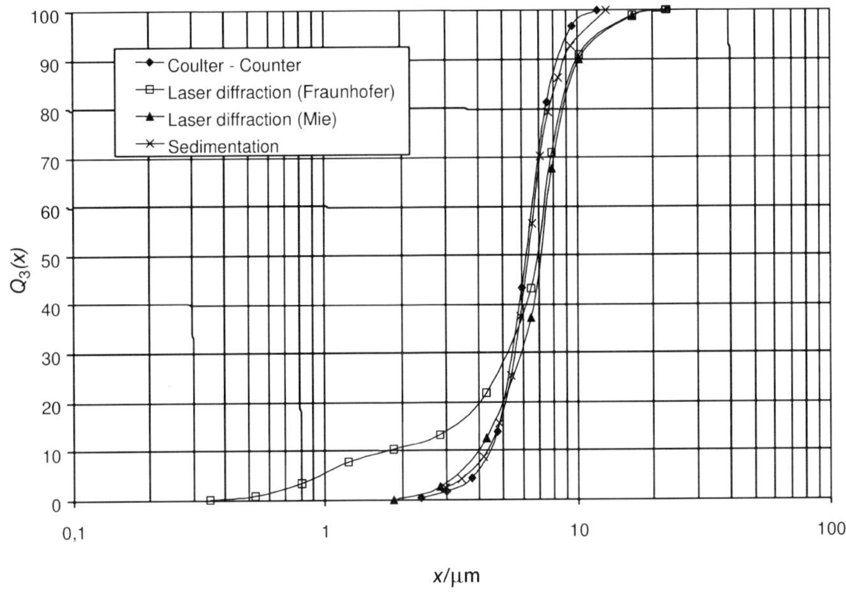

Figure 8.21. Cumulative distribution by volume $Q_3(x)$: comparison of methods for suspensions of glass sphere fraction F3.

Table 8.5. Mean particle size results (μm) (based on volume) for increased agitation time during sample preparation.

Preparation	FP-TATB	PG-TATB	UF-TATB
10 s USP/40 W	14.91	54.50	5.65
5 min USP/40 W	11.34	55.69	5.89
12 min USP/100 W	8.85	53.02	5.17

Table 8.6. Mean particle size results (μm) (based on volume) for increased sample concentration.

Preparation	High concentration, ~ 70% transmission	Low concentration, ~ 95% transmission
PG-TATB	55.12	54.50
UF-TATB	5.64	5.65
FP-TATB	17.70	14.91

ultrasonic irradiation (ultrasonication). Results from powder characterization studies on the TATBs illustrated that the arithmetic median particle diameters of FP-TATB is 15 μm whereas that of UF-TATB is 5 μm and that of PF-TATB is 55 μm.

Further studies on particle dispersion techniques indicated the need for an additional particle dispersion method, such as ultrasonic agitation and/or the use of a surfactant. The addition of a small amount of dilute Triton X-100 greatly aided sample dispersion. Although the use of the ultrasonic bath or probe alone was acceptable for sample dispersion, the addition of the surfactant made cleaning of the instrument between sample runs much easier. Excessive ultrasonic agitation appeared to have little effect on the particle sizing results of both the UF- and PF-TATB powders. However, FP-TATB showed some reduction in mean particle diameter. The use of heavily concentrated samples did not appear to affect the particle sizing results significantly. However, less concentrated samples are recommended to prevent detector overloading and powder agglomeration, in addition to conserving sample powder.

8.1.4.6 **Dynamic Light Scattering**

Dynamic light scattering, or photon correlation spectroscopy (PCS), is a technique in which laser light is focused into a suspension. Intensity fluctuations of the laser light resulting from Brownian motion of the particles are measured and used to determine the particle size [8.13, 8.32, 8.33]. After determining the translation diffusion coefficient D, the Stokes-Einstein relation can be used to determine the hydrodynamic equivalent diameter, x, of the particle:

$$x = \frac{k \cdot T}{3 \cdot \pi \cdot \eta_C \cdot D} \tag{8.29}$$

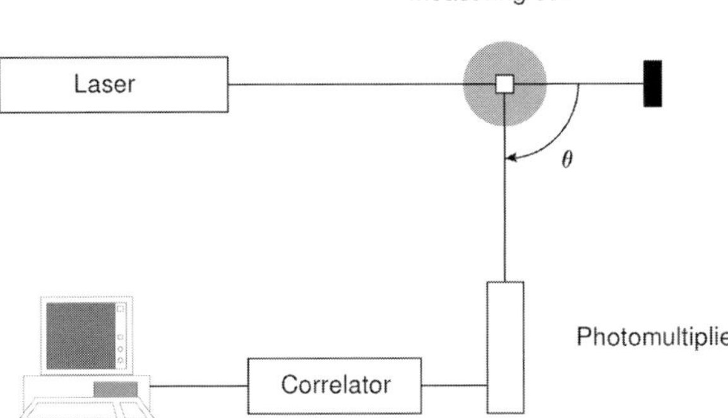

Figure 8.28. Schematic design of a dynamic light scattering apparatus.

where k is the Boltzmann constant, T is the sample temperature and η_C is the viscosity of the continuous phase.

Figure 8.28 depicts schematically the essential design of an apparatus used for dynamic light scattering experiments. Light scattered from the particle collective is captured at scattering angles of 90 and 180°. Instruments are also available that can collect light at a number of different angles and that are capable of amplifying the light signal with a photomultiplier.

Random Brownian motion of submicron particles results in constantly changing arrangements of the particles relative to each other. This leads to changes in the relative phase of the scattered light, resulting in fluctuations in the frequency of the cumulative signal. Because particles of different size change their position at different rates, the characterization time of the intensity fluctuations provides information about the particle size. The time-scale corresponding to these fluctuations is on the order of micro- to milliseconds. In order to obtain information on the diffusion coefficient from intensity fluctuations, an autocorrelation function corresponding to the time-dependent fluctuations in the scattered light intensity is calculated and compared with a known correlation function. For spherical particles the autocorrelation function can be represented by an exponential function [8.34]:

$$g(v) = \exp\left(-D \cdot q^2 \cdot v\right) \tag{8.30}$$

where D is the particle diffusion coefficient, v is the correlation time and q is the scattering vector, which can be described in terms of the refractive index n, the wave length of light λ and the scattering angle φ:

$$q = \frac{4 \cdot \pi \cdot n}{\lambda} \cdot \sin\left(\frac{\varphi}{2}\right) \tag{8.31}$$

For a polydisperse, multiphase system, the autocorrelation function is a transposition of a set of exponential functions containing different functions corresponding to each of the different particle sizes in the system. The individual functions are weighted according to the scattered light intensity contribution F of the corresponding particle fraction [8.34–8.36]:

$$g(v) = \int_0^\infty F(D \cdot q^2) \cdot \exp(-D \cdot q^2 \cdot v)\, dv \tag{8.32}$$

The photon correlation spectroscopy method is suitable for characterizing multiphase systems with a fluid continuous phase, such as suspensions, emulsions and nano-dispersions. Particle sizes ranging from approximately 5 nm to 3 µm can be detected using this technique. For photon correlation spectroscopy to be used successfully, there must be no sedimentation of the dispersed phase ($\Delta\rho \cdot g = 0$) and no structure formation among the particles when the system is subjected to external forces.

8.1.4.7 **Ultrasonic Spectrometry**

Ultrasonic spectrometry is a new method of particle size analysis based on measurement of ultrasonic extinction at multiple frequencies [8.37–8.40]. Small-amplitude ultrasound waves are used so that the effects on the energetic material and the particle collective are negligible. A key advantage of this new measurement technique is its capability of measuring the particle size in suspensions up to very high solids concentrations (from 1 to 70%) over a range of sizes from 10 to 3000 µm. Ultrasound waves can be transmitted through highly concentrated suspensions much better than light and can be detected over a wide range of dynamic frequencies up to 180 MHz.

The interaction of ultrasonic waves with particles in a dispersion is characterized by a reduction in intensity and a change in propagation velocity of the waves. The propagation rate change is strongly dependent not only on particle size but also on the physical properties of the dispersion fluid and the dispersed particles (e.g. viscosity, density, compressibility, sound velocity, attenuation). As a result, the size and size distribution of dispersed particles determined via ultrasound are a function only of the extinction behavior of ultrasonic waves of different frequency.

As in light wave extinction, ultrasonic wave extinction in a dispersion can be described by the Lambert-Beer equation:

$$-\ln\left(\frac{I}{I_0}\right) = N \cdot L \cdot \int_{x_{min}}^{x_{max}} \pi \cdot x^2 \cdot q(x) \cdot Q_{ext.}(f_i, x)\, dx \tag{8.33}$$

where:

I = reduced intensity, after passage through the measuring cell;
I_0 = initial intensity, on entry into measuring cell;
N = particle concentration (particles per unit volume);
L = pathlength of the ultrasonic waves through the dispersion;
x = particle diameter;
$q(x)$ = particle size distribution function;
$Q_{ext.}$ = extinction coefficient;
f_i = measuring frequency.

By measuring extinction at a number of different frequencies and using a discrete form of the above integral, a system of equations can be obtained that, with appropriate methods, yields particle size, size distribution and concentration.

An advantage of this technique is that particle size distributions can be determined on-line. However, little is currently known about the influence of particle-particle interactions at high solids concentrations on the measurement results.

8.2 Properties of Powders

8.2.1 Density

One of the most important parameters for energetic materials and for particle analysis is the particle density. The density is the ratio of the mass to the volume.

8.2.1.1 Particle Density

Density ρ_P can be determined by measuring the mass and volume of a sample of particles. The volume of a particle with no well-defined geometry can be determined with a pycnometer. The body displaces a liquid or a gas and its volume can be determined from the displacement either directly or, with the use of gas pycnometers, from physical laws.

Fluid pycnometers base on the displacement principle. After air is displaced from the entire empty space within the pycnometer by the fluid, one determines the volume of the test material in the pycnometer, based on the amount of the volume of the displaced fluid. With the gas pycnometers often used today, helium or another non-adsorbing inert gas is often used in place of air, in order to detect very small capillaries and pores [8.3, 8.41]. The physical principle of gas pycnometers is based on the Boyle-Mariott law:

$$p_1 \cdot V_1 = p_2 \cdot V_2 = \text{constant} \tag{8.34}$$

If pressures p_1 and p_2 and the reference volume are known, the unknown volume can be determined from Eq. (8.34).

This method makes it possible to determine the volume of solids, regardless of whether they have a regular or irregular structure. When the particulate or bulk system is porous, such as agglomerated materials, one speaks of an 'apparent density' of the dispersed system. The volume required to determine the apparent density is obtained in the presence of a non-wetting fluid, such as mercury. Such an approach ensures that, to the greatest extent possible, the inter-particulate spaces are filled with mercury.

8.2.1.2 Bulk Density

Bulk density is the density of a powder or bulk material. In contrast to the particle density, the bulk density includes the pores within individual particles in addition to inter-particle spacing between particles. Bulk density is also a measure of mass to volume, in this case the mass of a powder to its volume. The bulk density of a powder is determined by allowing a given mass of powder to fall freely and settle (without vibration) into a vessel of known volume [8.42].

The bulk density ρ_{bulk} can be calculated from the density of the solid ρ_S, the density of the fluid ρ_F that fills the pores and the porosity ε:

$$\rho_{bulk} = \rho_S (1-\varepsilon) + \rho_F \cdot \varepsilon \qquad (8.35)$$

8.2.1.3 Tap Density

Vibrating or tapping the vessel containing a bulk solid can considerably reduce the space required to hold the material, because particles subjected to vibration lose contact with their neighboring particles and the friction between particles is thus temporarily reduced. The higher bulk density associated with the denser packing achieved in this manner is referred to as the 'tap density' [8.43].

The compressibility κ is an important parameter for processing energetic materials:

$$\kappa = \frac{\text{tap density} - \text{bulk density}}{\text{tap density}} \qquad (8.36)$$

Understanding compressibility is especially important, for example, in order to successfully inject particulate material volumetrically or to achieve high packing densities. Highly compressible materials exhibit large differences between their bulk and tap densities and, likewise, exhibit poor flow and processing properties. Materials with a compressibility $\kappa \leq 0.2$ usually flow relatively well. In general, powders with a relatively low bulk density (and corresponding high friction between the particles) will exhibit a sharp increase in density when subjected to vibration. A bulk material with spherical particles has a higher tap density than a powder with irregularly shaped particles.

8.2.2 Water Content

If dry solids come into contact with moist air or steam, the particles will absorb an equilibrium concentration of moisture. The equilibrium moisture content depends on the moisture concentration of the surroundings and, even more critically, on the hygroscopic properties of the particles and/or solid material.

Materials with polar functional groups have a tendency to absorb water. The water may adsorb on the solid surface or it may be incorporated into the solid structure as hydrated or capillary water. The amount of water adsorbed is particularly dependent on the material type, and also the partial pressure of water and the relative humidity, but is less dependent on temperature.

The water content of a particle or powder is extremely significant with respect to its properties and suitability for technical applications. Crystal water, which unlike sorption water is bound within the crystalline structure of the particle, leads for instance to the formation of pseudo-modifications with varying solubilities. Sorption water, on the other hand, primarily affects the outer properties of the particle. It results in an increase in the cohesivity, which considerably reduces the flowability of the powder and increases its tendency to agglomerate.

Methods to determine the water content of powders include gravimetric treatment in a drying chamber, thermogravimetric analysis, IR spectroscopy, measurement of the dielectric constant and Karl Fischer titration.

8.2.3
Surface Area

Specific surface area is an important parameter for characterizing the degree of fineness of a particle collective. One distinguishes between a specific surface area referenced to volume S_V (the ratio of surface area to volume) and a specific surface area referenced to mass S_m. These specific surface areas are related to each other by the particle density ρ_P:

$$S_v = \rho_P \cdot S_m \tag{8.37}$$

In general, the specific surface area referenced to volume S_V is defined as the ratio of the particle surface area to the particle volume, so for spherical particles with a diameter x, S_V is calculated as follows:

$$S_v = \frac{\pi \cdot x^2}{\pi \cdot \frac{x^3}{6}} = \frac{6}{x} \tag{8.38}$$

Various physical properties can be used to characterize the specific surface area of a particle collective or dispersed system. In addition to measuring light extinction passing through a suspension (a photometric method) and determining flow resistance of a densely packed particle bed (a permeation method), the adsorption of gas on the particle surface (a sorption method) is a particularly important technique.

Whether the outer or inner surface area is measured depends on the particular technique employed. The outer surface area of the particle is primarily related to its geometric characteristics and surface roughness. The particle's inner surface area depends on its microroughness and the nature of its pores that extend to the outer surface. It therefore includes all of the information available in the outer surface area parameter, plus more. Knowledge of the inner surface area of a particulate material is particularly important because of its profound influence on the burn properties, reaction kinetics and ballistic and detonation sensitivity of particulate-based energetic materials. One seldom finds agreement between the specific surface area values calculated via indirect approximation from the measured particle size distribution [using Eq. (8.38)] and the values determined using the measurement methods described in this chapter.

8.2.3.1 Photometric Method

A light beam directed through the suspension is attenuated owing to scattering and absorption by a particle. The attenuation of the light beam can be calculated using the Lambert-Beer equation as a function of the intensity of a light beam directed through a solids-free fluid I_∞, the solids concentration C_v, the path of the light beam L and the extinction cross-section A_v (referenced to volume):

$$\frac{I_0}{I_\infty} = \exp(-A_v \cdot C_v \cdot L) \tag{8.39}$$

The specific surface area S_v can then be calculated from the extinction cross-section A_v.

8.2.3.2 Permeation method

In this method, the specific surface area is determined by pumping a fluid through the dispersed material, which has been packed as uniformly as possible into a bed. The fluid flows through the packed bed from top to bottom and the pressure loss across the bed is measured. With increasing specific surface area, the particle bed exhibits increased resistance to fluid flow. The pressure drop Δp and specific surface area S_v are related by the Carman-Kozeny equation:

$$\frac{\Delta p}{L} = k \cdot \frac{(1-\varepsilon)^2}{\varepsilon^2} \cdot S_v^{\,2} \cdot \eta_L \cdot \bar{u} \tag{8.40}$$

where L is the length of the packing, k is the Kozeny constant, which is a function of particle size and shape, η_L is the fluid viscosity, $\bar{u}$ is the fluid throughput velocity and ε is the porosity of the particle packing.

8.2.3.3 Sorption Technique

Measurement of particle surface area via sorption is based on the principle that porous or finely dispersed particles adsorb gas molecules and, under defined conditions, the amount of gas adsorbed is proportional to the particle surface area. After determining the amount of gas required to form a monolayer V_m, if the amount of space required A_m for a gas molecule is known, one can determine the solid surface area S as follows:

$$S = A_m \cdot N_A \cdot \frac{V_m}{V_0} \tag{8.41}$$

where N_A is Avogadro's number ($N_A = 6.022 \times 10^{23}$ molecules/mole) and V_0 is the molar volume of the gas.

The adsorption isotherm depicts the amount of gas adsorbed at constant temperature by a solid sample as a function of the equilibrium gas pressure. A well-known relationship describing isothermal sorption, the BET isotherm, was introduced by Brunauer, Emmett and Teller [8.3, 8.14, 8.41]:

$$\frac{p}{V(p_0 - p)} = \frac{1}{b \cdot V_m} + \frac{b-1}{V_m \cdot b} \cdot \frac{p}{p_0} \tag{8.42}$$

where V is the volume of gas adsorbed per gram of powder at a pressure p, p_0 is the saturation pressure of the gas at the measurement temperature T and b is a constant that accounts for the bond energy. Usually, the ratio $p/[V(p_0-p)]$ is plotted as a function of p/p_0. From the resulting line, one can determine the constant b and the amount of gas required to form a monolayer V_m.

An inert gas, such as nitrogen, is used with the sorption technique. Helium, which is better at permeating into the particle's very fine pores, may be employed instead of nitrogen. To purify the material following sorption measurements (e.g. to remove water, etc.) the particles should be heated, either in a vacuum or in an inert gas environment.

8.2.4
Flow Properties

In addition to characterizing individual particle properties such as mean particle size, size distribution, shape and density, it is also important to obtain information about the powder or bulk material as a whole and about the static and dynamic behavior of the particle collective. Important bulk properties include the material's flowability and how its solidification characteristics change with time, i.e. the increase in solid-like behavior of the powder with increased storage time. Schwedes and Schulze provide a comprehensive description of how to quantitatively determine this bulk parameter in references [8.44, 8.45]. Several empirical methods for determining the flow behavior of particles are described in the following sections. The flow behavior of a powder can be characterized by the time required for a specific amount of powder to flow out of a standardized funnel, such as shown in Fig. 8.29 [8.46].

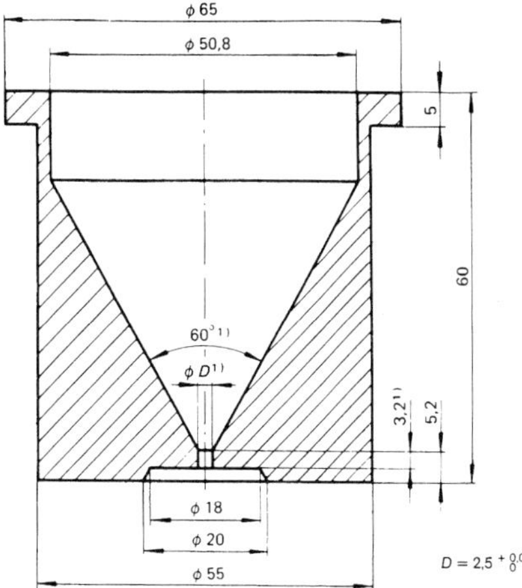

Figure 8.29. Standard test funnel to determine flow behavior [8.46].

The flow process is initiated by subjecting the powder to a shear stress proportional to the weight of the bulk material acting on a constant cross-sectional area of powder. As the particle size decreases, the interparticle interactions between particles increase, which reduces the powder's flowability. Finely sized powders also have a tendency to form particle bridges that restrict flow.

Another method to determine a powder's flow characteristics is by measuring the slope angle (Fig. 8.30). If the powder being investigated flows out of a funnel onto a flat substrate, it forms into a cone. The slope angle α can be calculated from the diameter and height of this cone.

Powders with high cohesivities exhibit a large slope angle. Bulk materials with good flow characteristics and low cohesivities tend to form a flatter cone and thus have a smaller slope angle [8.47]. In general, the slope angle can be used as a direct measure of a powder's flowability. Powders with smaller values of this angle tend to flow better. Naturally, the value of the slope angle depends strongly on the shape, size and size distribution of the particles.

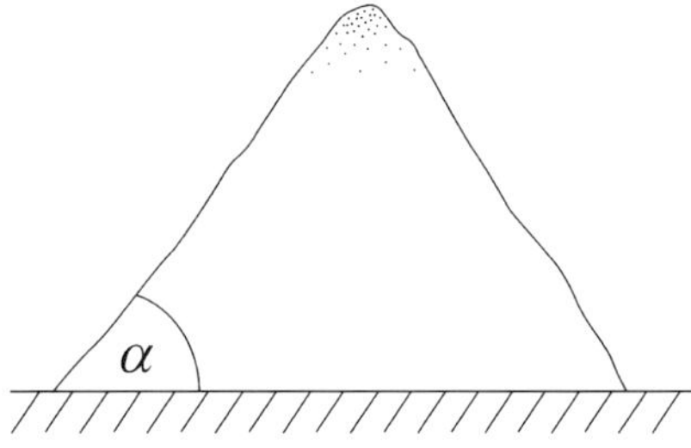

Figure 8.30. Slope angle of a powder.

8.3 References

8.1 Löffler F, Raasch J (1992) *Grundlagen der mechanischen Verfahrenstechnik*, Vieweg, Braunschweig.

8.2 Stiess M (1992) *Mechanische Verfahrenstechnik 1*, Springer, Berlin.

8.3 Allen T (1990) In: *Particle Size Measurement, Powder Technology Series*, 4th edn, Scarlett B (ed.), Chapman and Hall, London.

8.4 Leschonski K, Alex W, Koglin B (1974) Teilchengrößenanalyse, Folge von 13 Artikeln, *Chem.-Ing.-Tech.* **1**, 23–26.

8.5 VDI-Richtlinie 3491 (1980) Messen von Partikeln, Kennzeichnung von Partikeldispersionen in Gasen, Begriffe und Definitionen, *VDI-Handbuch: Reinhaltung der Luft*, Vol. 4, VDI, Düsseldorf.

8.6 Rhodes M (1998) *Introduction to Particle Technology*, Wiley, New York.

8.7 Krupp H (1967) Particle adhesion, theory and experiment, *Adv. Colloid Interface Sci.* **1**, 111–239.

8.8 Israelachvili JN (1992) *Intermolecular and Surface Forces*, Academic Press, London.

8.9 Lyklema J (1993) *Fundamentals of Interface and Colloid Science*, Academic Press, San Diego, CA.

8.10 Nelson RD (1988) *Dispersion of Powders in Liquids*, Elsevier, Amsterdam.

8.11 Xu R (2000) *Particle Characterization: Light Scattering Methods*, Kluwer, Dordrecht.

8.12 Stanley-Wood NG, Lines RW (1992) *Particle Size Analysis*, Royal Society of Chemistry, Cambridge.

8.13 Barth, HG (1984) *Modern Methods of Particle Size Analysis*, Chemical Analysis, Vol. 73, Wiley, New York.

8.14 Kaye BH (1999) *Characterization of Powders and Aerosols*, Wiley-VCH, Weinheim.

8.15 Rhewum Siebfible (1995) *Eine Übersicht über das Sieben*, Rhewum Siebfible, Remscheid.

8.16 Russ JC (1999) *The Image Processing Handbook*, CRC Press, Boca Raton, FL.

8.17 Fraunhofer J (1817) Bestimmung des Brechungs- und Farbzerstreuungsvermögens verschiedener Glasarten, *Gilberts Ann. Phys.* **56**, 193.

8.18 Mie G (1908) Beiträge zur Optik trüber Medien, *Ann. Phys.* **25**, 377–445.

8.19 Kerker M (1969) *The Scattering of Light*, Academic Press, New York.

8.20 Van de Hulst HC (1981) *Light Scattering by Small Particles*, Dover, New York.

8.21 Airy GB (1835) On the diffraction of an object glass with circular aperture, *Trans. Cambridge Philos. Soc.* **5**, 283–290.

8.22 Heuer M, Leschonski K (1985) Results obtained with a new instrument for the measurement of particle size distributions from diffraction patterns, *Part. Charact.*, 2, 7–13.

8.23 Förter-Barth U, Teipel U (2001) Characterization of particles by means of laser light diffraction and dynamic light scattering, *Eur. J. Mech. Environ. Eng.* **46**, 11–14.

8.24 Bauckhage K (1993) Nutzung unterschiedlicher Streulichtanteile zur Partikelgrößen bestimmung in dispersen Systemen, *Chem.-Ing.-Tech.* **65**, 1200–1205.

8.25 Streu (1991) *A Computational Code for the Light Scattering Properties of Spherical Particles, Instruction Manual*, Invent, Erlangen-Tennelohe.

8.26 Naqwi AA, Durst F (1991, 1992) Light scattering applied to LDA and PDA measurements, parts 1 and 2, *Part. Part. Syst. Charact.* **8**, 245–258; **9**, 66–80.

8.27 Teipel U (2002) Problems in characterizing transparent particles by laser light diffraction spectrometry, *J. Chem. Eng. Technol.* **25**, 13–21.

8.28 Dobratz BM (1995) *The Insensitive High Explosive Triaminotrinitrobenzene (TATB): Development and Characterization – 1888 to 1984*, LA-13014-H, Los Alamos National Laboratory.

8.29 Lee R, Bloom G, Holle WV, Weingart R, Erickson L, Sanders S, Slettevold C, Mc Guire R (1985) The relationship between shock sensitivity and the solid pore sizes of TATB powders pressed to various densities. In: *Proc. 8th Int. Symp. on Detonation*, NSWC MP 86–194, Albuquerque, NM, pp. 3–14.

8.30 Osborn AG, Stallings TL (1992) *Ultrafine TATB – Lot No. 91190-135M-003, PXET-92–03*, Mason & Hanger – Silas Mason, Pantex Plant, Amarillo, TX.

8.31 Horiba (1994) *Instruction Manual – Laser Scattering Particle Size Distribution Analyzer, LA-900*, Horiba, Kyoto.

8.32 Wagner J (1986) Teilchengrößen-Bestimmung mittels dynamischer Lichtstreuung, *Chem.-Ing.-Tech.* **58**, 578–583.

8.33 Pecora R (1985) *Dynamic Light Scattering, Applications of Photon Correlation Spectroscopy*, Plenum Press, New York.

8.34 Streib J (1988) Partikelgrößenbestimmung mit der Photonen-Korrelations-Spektroskopie, *Chem.-Ing.-Tech.* **60**, 138–139.

8.35 Cummins PG, Staples EJ (1987) Particle size distributions determined by a multiangle analysis of photonen correlation spectroscopie data, *Langmuir* **3**, 1109–1113.

8.36 Stock RS, Ray WH (1985) Interpretation of photon correlation spectroscopy data: a comparison of analysis methods, *J. Polym. Sci. Polym. Phys. Ed.* **23**, 1393–1447.

8.37 Riebel U, Löffler F (1989) The fundamentals of particle size analysis by means of ultrasonic spectrometry, *Part. Part. Syst. Charact.* **6**, 135–143.

8.38 Allegra JR, Hawley SA (1972) Attenuation of sound in suspensions and emulsions: theory and experiment, *J. Acoust. Soc. Am.* **1**, 1545–1564.

8.39 Babick F, Hinze F, Stintz M, Ripperger S (1998) Ultrasonic spectrometry for particle size analysis in dense submicron suspensions, *Part. Part. Syst. Charact.* **15**, 230–236.

8.40 Riebel U (1992) Ultrasonic spectrometry: on-line particle size analysis at extremly high particle concentrations. In: *Particle Size Analysis*, Stanley-Wood NG, Lines RW (eds), Royal Society of Chemistry, Cambridge, pp. 488–497.

8.41 Webb PA, Orr C (1997) *Analytical Me-*

thods in Fine Particle Technology, Micromeritics, Norcross, GA.

8.42 DIN EN 725–9, *Prüfverfahren für keramische Pulver Teil 9: Bestimmung der Schüttdichte.*

8.43 DIN ISO 787, *Allgemeine Prüfverfahren für Pigmente und Füllstoffe Bestimmung des Stampfvolumens und der Stampfdichte.*

8.44 Schwedes J, Schulze D (2003) Lagern von Schüttgütern. In: *Handbuch der Mechanischen Verfahrenstechnik*, Schubert H (ed.), Wiley-VCH, Weinheim, pp. 1137–1253.

8.45 Schwedes J, Schulze D (1990) Measurement of flow properties of bulk solids, *Powder Technol.* **61**, 59–68.

8.46 DIN ISO 4490, *Ermittlung des Fließverhaltens mit Hilfe eines kalibrierten Trichters.*

8.47 Carr RL (1965) Evaluating flow properties of solids, *Chem. Eng.* **18**, 163–168.

9
Microstructure and Morphology

L. Borne, M. Herrmann, C. B. Skidmore

9.1
Introduction

It is known that defects in an explosive formulation affect its sensitivity. Solid explosives formulations cannot be pressed to full density and therefore contain voids. These voids are mainly outside the explosive particles. These are defects of the explosive formulation and their effects have been studied in detail [9.1, 9.2]. Decreasing the density of the explosive formulation increases its sensitivity. Sensitivity changes have been recorded and correlated with density differences as small as 0.005 g/cm^3 [9.3].

Cast plastic-bonded explosives remove the residual voids located outside the explosive particle and are less sensitive that pressed compositions. Defects of the explosive particles affect the sensitivity of cast explosive formulations [3.39, 9.4–9.7]. Defects of explosive particles can be internal defects entrapped in the volume of the particle or surface defects located on the surface of the particle.

In this chapter we will focus on the tools available to look at and to characterize quantitatively the defects of explosive particles. Two kinds of defects are discussed: internal and surface defects. An accurate method to record the amount of internal defects of the explosive particles is introduced. The limited efficiency of the usual tools to measure the specific surface area applied to explosive particles is emphasized.

9.2
Defects of Explosive Particles

9.2.1
Internal Defects

Solvent inclusions, lattice imperfections (vacancies, impurities and dislocations) are internal defects of explosive particles. Different qualitative and quantitative tools are

Energetic Materials. Edited by Ulrich Teipel
Copyright © 2005 WILEY-VCH Verlag GmbH & Co. KGaA, Weinheim
ISBN: 3-527-30240-9

needed to investigate these defects, which have various sizes. Our purpose is to deal with the largest defects such as solvent inclusions. It is important to note that solvent inclusions are linked to lattice imperfections, as demonstrated with laboratory PETN single crystals [9.8]. Most of the crystals grown in solutions exhibit solvent inclusions. These are cavities entrapped in the crystal, whose sizes are in the range 1–50 μm.

9.2.1.1 Optical Microscopy with Matching Refractive Index

Optical microscopy in a liquid of matching refractive index allows in-depth particle analysis. Reducing refractive index discontinuities at particle boundaries emphasizes index gradients inside the particles. This reveals crystal internal defects. An accurate matching of the refractive index is needed to achieve the optimum performance of the method (Fig. 9.1).

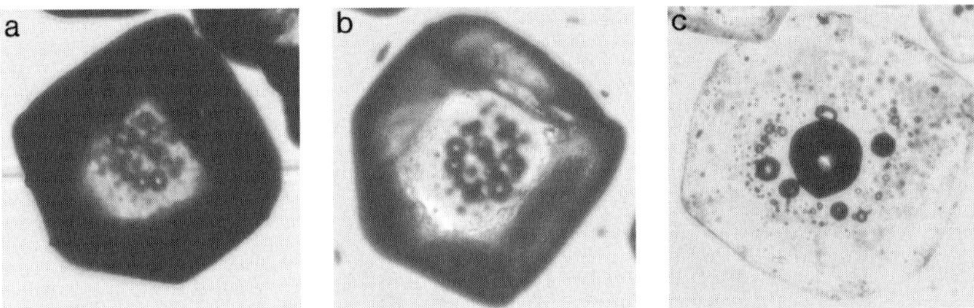

Figure 9.1. Transmitted light optical microscopy: effects of the refractive index of the surroundings. RDX 100/300 μm. (a) RDX particles in air, $n_{20} = 1.0$; (b) RDX particles in usual microscopy oil, $n_{20} = 1.5$; (c) RDX particles in a liquid mixture tuned to $n_{20} = 1.6$.

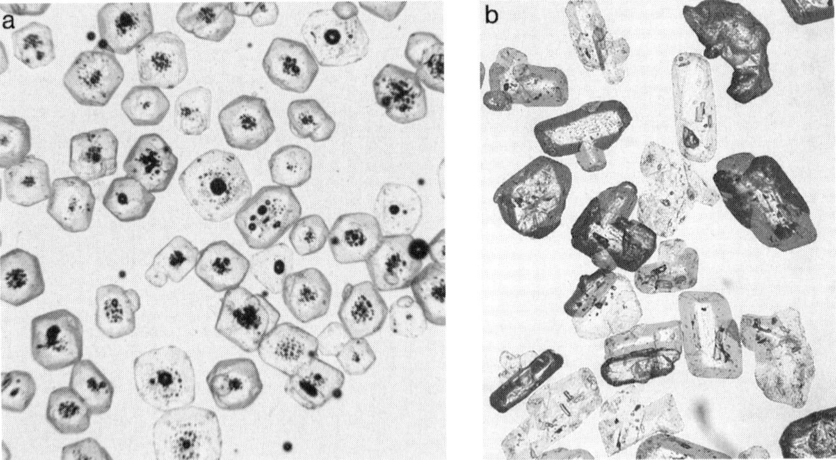

Figure 9.2. Optical anisotropy of HMX particle versus RDX particles. (a) RDX 100/300 μm, $n_{20} = 1.6$; (b) HMX 200/300 μm, $n_{20} = 1.6$.

The optical anisotropy of the crystal limits the efficiency of the optical microscopy with matching refractive index method. This is mainly sensitive with HMX particles. Depending of the orientation of the particle, the in-depth imaging of some HMX particles is reduced because of the variation of the refractive index with the crystallographic axes (Fig. 9.2).

Depending of the crystallization conditions, crystals can have different shapes, different sizes and also different populations of internal defects. These features can vary over a wide range (Fig. 9.3).

Optical microscopy with matching refractive index provides qualitative insight into the internal defects population. However, the limited statistics inhibit quantification even with the help of digital image processing.

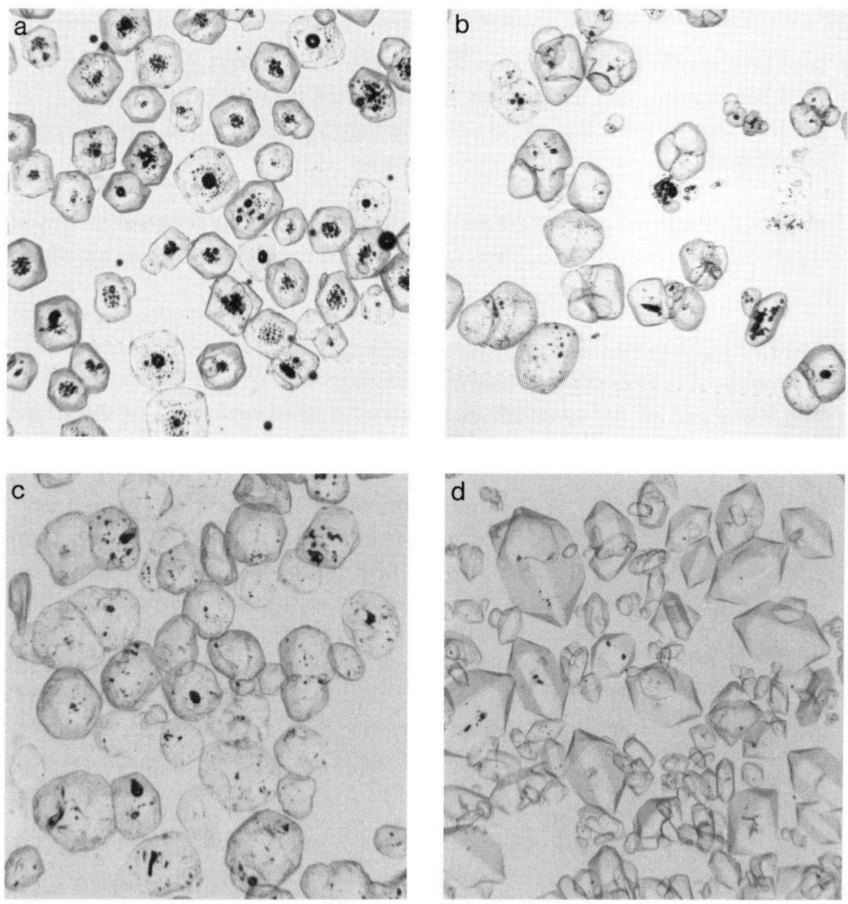

Figure 9.3. Different internal defect populations for similar RDX lots. RDX 200/315 μm. (a) Lot 1, $n_{20} = 1.6$; (b) lot 2, $n_{20} = 1.6$; (c) lot 3, $n_{20} = 1.6$; (d) lot 4, $n_{20} = 1.6$.

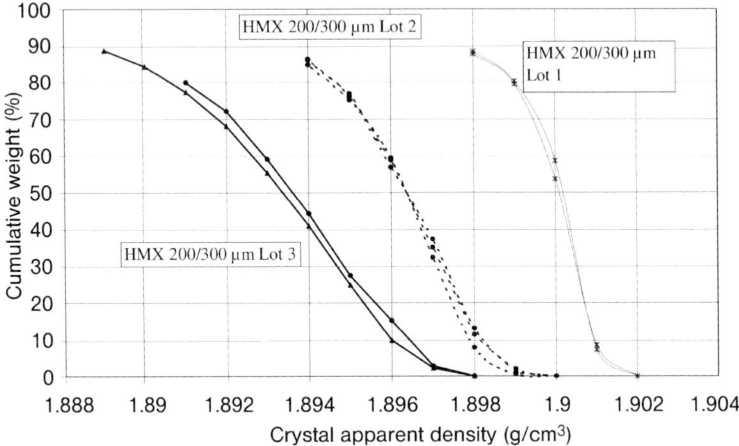

Figure 9.6. Apparent density measurements on three HMX lots.

and adsorption of chemical species. Our purpose is to deal with particle shapes and particle surface porosity [9.6, 9.11].

9.2.2.1 Microscopy

The usual scanning electron microscopy (SEM) is a simple tool to look at the explosive particle surface. Explosive particle surface properties can vary over a wide range. Figure 9.7 shows RDX particles (lot 1) with open pores in the range 1–10 µm. Lot 2 exhibits particles with smooth surfaces and marked edges. Industrial lots of explosive particles often show crystals with more complex shapes (Fig. 9.8).

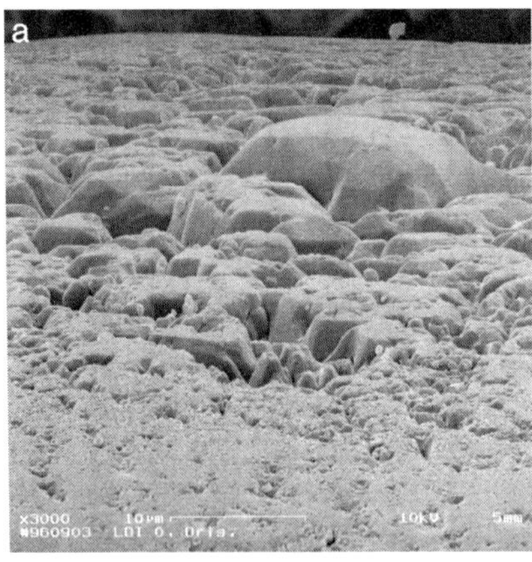

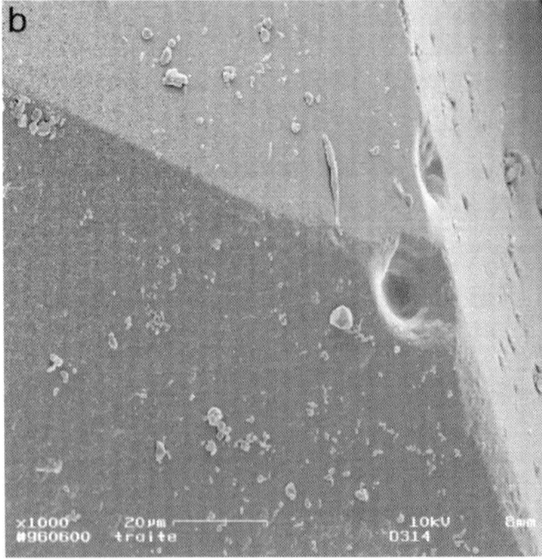

Figure 9.7. SEM of RDX particles. RDX 400/500 µm. (a) Lot 1 (raw); (b) lot 2 (processed).

Figure 9.8. SEM of HMX twin crystals.

SEM applied to energetic materials is limited by the amount of energy which can be focused on a small area without generating transformations of the material. This gives a magnification limit to SEM. An alternative method which allows finer imaging of the explosive particle surface is atomic force microscopy (AFM) [9.12].

Nevertheless the limited statistics associated with the previous imaging processes inhibit quantification even with the help of digital image processing. Two common tools are available for measurements of particle surface properties: the BET method and mercury porosimetry. These methods allow measurements of the specific surface area and the surface pore size distribution.

9.2.2.2 Gas Adsorption Method

The volumetric adsorption of a gas on a sample of particles provides a measurement of the specific surface area of the particles (BET method). The pore distribution can be computed using the BJH (Barrett, Joyner and Halenda) method based on the desorption curve. The range of measured pore sizes with the usual experimental devices is 0.001–0.02 μm.

Figure 9.9 shows the BET results obtained on the two RDX lots shown in Fig. 9.7. Experimental results are plotted as a function of the sample weight. Increasing the sample weight improves the accuracy but this is limited by the design of the experimental device and for security reasons. The specific surface areas are close to 0.01 m^2/g. This is also close to the lower limit of theusual experimental devices and so the experimental accuracy is reduced. Nevertheless, the results in Fig. 9.9 are in agreement with qualitative observations in Fig. 9.7. Particles of lot 1 give specific surface area measurements higher than particles of lot 2. The comparison of the BET measurements with the theoretical specific surface area of academic non-porous spheres of similar sizes shows that the number of pores on the particles of the two lots is very low.

The experimental correlation between the defects of explosive particles and the sensitivity of cast formulations, reported later (Chapter 13), will demonstrate the need for accurate tools to measure the defects of explosive particles. These experimental data will support the following conclusions:

- The ISL sink-float experiment is a useful, accurate and simple method to characterize the global amount of internal defects of explosive particles. Its unique drawback is that the experiment is time consuming.
- There is a need for the measurement of the size of the internal defects.
- There is a need for accurate characterization of the surface of the explosive particles. At least the accuracy of usual tools (BET method and mercury intrusion) must be improved.

Small-angle scattering (SAS) can provide new experimental data. Some of the first small-angle neutron scattering (SANS) and X-ray (SAXS) scattering measurements were performed by Mang *et al.* [9.14].

9.3
Characterization of the Microstructure by X-ray Diffraction

9.3.1 Principles

X-ray diffraction is a powerful tool for the characterization of crystalline materials, including solid energetic materials. It is based on the scattering of X-rays by the electrons associated with the atoms and the interference of the beam due to the periodic arrangement of the atoms in the crystal.

Diffraction patterns were obtained in the past by methods such as the use of the Debye-Scherrer film camera, when polycrystalline samples were measured with monochromatic radiation. Today, powder diffractometers are used, recording patterns with scintillation, position-sensitive or two-dimensional electronic detectors. The most common diffractometers are built according to the geometry of Bragg-Brentano using X-rays generated by vacuum tubes, when synchrotron radiation with its high intensity is not available. Combined with commercially available temperature chambers, temperature-resolved X-ray diffraction is a useful tool for dynamic investigations.

In addition to conventional systems, new X-ray optical components such as multilayer mirrors, channel cut monochromators and multi-capillary collimators have been developed in recent years and proposed for special applications. With the new parallel beam concept, problems of sample displacements, which are one of the major error sources in X-ray diffraction, are overcome [9.15–9.17].

The diffraction patterns include three types of information: geometric, structural and physical state. In powder patterns measured with monochromatic radiation, geometric information is contained in the angular positions of diffraction peaks and structural information in the peak intensities controlled by the Bragg equation and the powder pattern power equation, respectively [9.18, 9.19]. The profiles of diffraction peaks represent the physical state of the sample and the measuring instrument. Its use for the investigation of microcrystalline properties such as micro strain and particle size is nearly as old as powder diffractometry itself. In 1918, Scherrer reported that the widths of diffraction lines are inversely related to the sizes of crystallites and in 1925 van Arkel [9.20] found that lines are broadened by micro strains.

9.3.2 Evaluation

9.3.2.1 Phase Identification

In the field of product development, the structures of materials are generally known. The diffraction data are therefore used as fingerprints for identifying phases by comparison with reference patterns found in databases such as the powder diffraction files (PDF) of the International Centre for Diffraction Data (ICDD) or the Cambridge Crystallographic Data Centre (CCDC), providing peak intensities and positions of a variety of materials.

9.3.2.2 Crystal Structure

In a more advanced technique, diffraction patterns are evaluated based on the crystal structure. A powerful tool is a procedure named Rietveld refinement [9.21, 9.22]. Based on at least approximately known structures, patterns are calculated by Fourier synthesis and fitted numerically to the measured patterns. A successful fit delivers refined elementary cell parameters and atom coordinates. In this way, the geometric and structural information mentioned above is obtained.

9.3.2.3 Quantitative Phase Analysis

As the intensities of diffraction peaks are related to the phase concentrations of mixtures, this effect is used for quantitative phase analysis where intensities of reference or calculated patterns, e.g. by Rietveld analysis, are compared or fitted to measured patterns. Measurements with defined mixtures of HMX and quartz showed a limit of detection of the method in of ~ 1 wt% HMX. Further, the accuracy determined by measurements of defined silicon nitride mixtures is $< 0.5\%$ [9.23].

9.3.2.4 Dynamic Investigations

In the case of dynamic investigations where series of diffraction patterns are measured, phases are identified and crystal structures and phase concentrations are determined for each pattern by the methods described above. Monitoring these data versus temperature or time reveals phase transitions, solid-state reactions and thermal expansion behavior. Temperature-resolved X-ray diffraction is therefore a powerful tool for thermal analysis.

9.3.2.5 Particle Size and Micro Strain

For evaluating micro structural parameters such as particle size and micro strain, special evaluation techniques were developed by Warren and Averbach [9.24] and Williamson and Hall [9.25]. In this context, it is convenient to express the integral breadth β in terms of reciprocal units where $\beta^* = \beta\cos\theta/\lambda$ with diffraction angle θ and wavelength λ of the radiation. The method described by Williamson and Hall consists essentially in plotting this reciprocal breadth against the reciprocal lattice distances $d^* = 2\sin\theta/\lambda$. If strain broadening is negligible, the values of β^* lie on a horizontal line with an intercept at $1/t$, the inverse of the mean linear crystallite size t. If size broadening is negligible, the values of β^* lie on a straight line passing through the origin with the integral breadth ξ of the strain contribution as slope, if the strain distribution is isotropic. In the case of unisotropic micro strain, the values of β^* for different orders of reflection lie on a straight line passing through the origin with the slope of the lines varying systematically with the lattice direction. The composite peak broadening produced by simultaneous small particle size and strain depends on the broadening functions of both effects [9.19].

As measured peaks represent a convolution of instrumental and sample profiles, it is necessary to determine the pure sample profiles or width for the determination of micro strain or particle size. However, if relative values are required, the measured peak widths may be used if the parameters of the measuring system are constant.

9.3.3 Applications

9.3.3.1 Phases and Crystal Structures of CL20

CL20 is one of the most interesting new energetic materials. It consists in four phases called α-, β-, ε- and γ-CL20 with different densities and mechanical sensitivities. The crystal structures of the phases described in literature [9.26] are summarized in Table 9.2.

ε-CL20, the phase with the highest density, is commercially available. However, the qualities of purchased lots differ. Therefore, experiments were started with the aim of obtaining pure, high-quality phases, especially ε-CL20.

Table 9.2. Crystal data of CL20 phases.

Parameter	α-CL20	β-CL20	γ-CL20	ε-CL20
Space group	*Pbca*	$Pca2_1$	$P2_1/n$	$P2_1/n$
Lattice parameters				
a (Å)	9.546	9.693	14.846	8.860
b (Å)	13.232	11.641	8.155	12.581
c (Å)	23.634	12.990	13.219	13.426
β (°)			109.21	106.88
Density (g/cm^3)	1.97	1.986	1.926	2.033

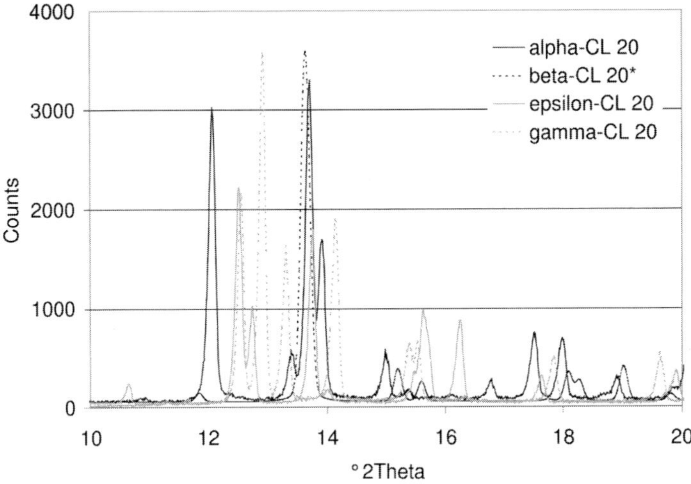

Figure 9.11. X-ray patterns of the CL20-phases at room temperature.

Rietveld analysis reveals the concentrations of phases in mixtures, which gives important feedback for optimizing the crystallization conditions.

Samples of CL20 were crystallized from solutions and characterized by applying X-ray diffraction. The measured diffraction patterns of the CL20 phases differ significantly, as shown in Fig. 9.11. Applying Rietveld analysis, the different phases are identified on the basis of the crystal data. Figure 9.12 shows a plot of calculated and measured patterns, where the agreement identifies ε-CL20.

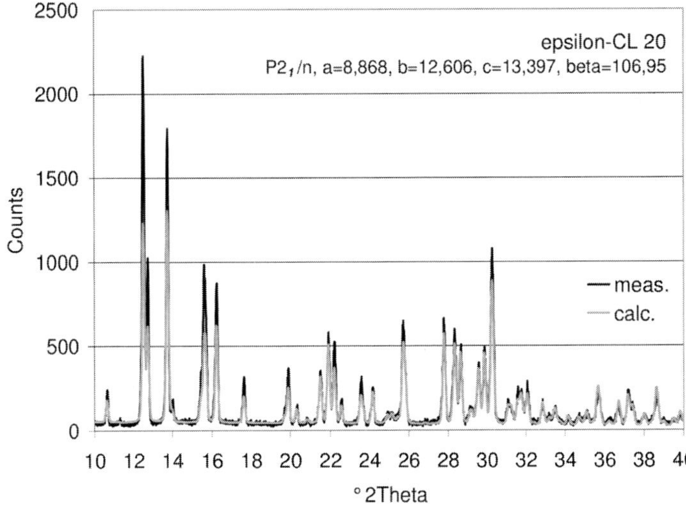

Figure 9.12. Rietveld plot for ε-CL20.

9.3.3.2 Phase Transitions and Thermal Expansion of Ammonium Nitrate

Ammonium nitrate (AN) is used as oxidizer in solid propellants, explosives and gas generator systems. Special interest fell on AN when smokeless, low-sensitivity and ecologically harmless propellants were required. Drawbacks are low performance, low burning rate and phase transitions that influence the material properties. AN crystallizes in five phases that appear at different temperatures. The crystal structures are reported in the literature [9.27–9.32].

The phase transitions were investigated by applying temperature-resolved X-ray diffraction, where samples were heated and cooled stepwise, measuring the X-ray patterns after each temperature step [9.33]. Figure 9.13 shows a series of diffraction patterns of AN measured on heating between –70 and 150 °C, indicating the phase transitions V/IV, IV/II and II/I and the thermal expansion presented by the continuously shifting peaks. Evaluating these patterns with Rietveld analysis yielded refined crystallographic data that were monitored versus temperature, as shown in Fig. 9.14 for the lattice parameters of phases II and V.

The plot reveals anisotropic expansion of phase IV with a slightly shrinking lattice parameter a and a strongly expanding parameter b on heating, which is explained by the model of Amorós *et al.* [9.34] shown in Fig. 9.15. The planar nitrate groups oscillate more strongly with increasing temperature, resulting in a strongly expanding parameter b. As the nitrate groups deviate more and more from the linear position in the direction of a, this parameter shrinks slightly. Therefore, the basis of

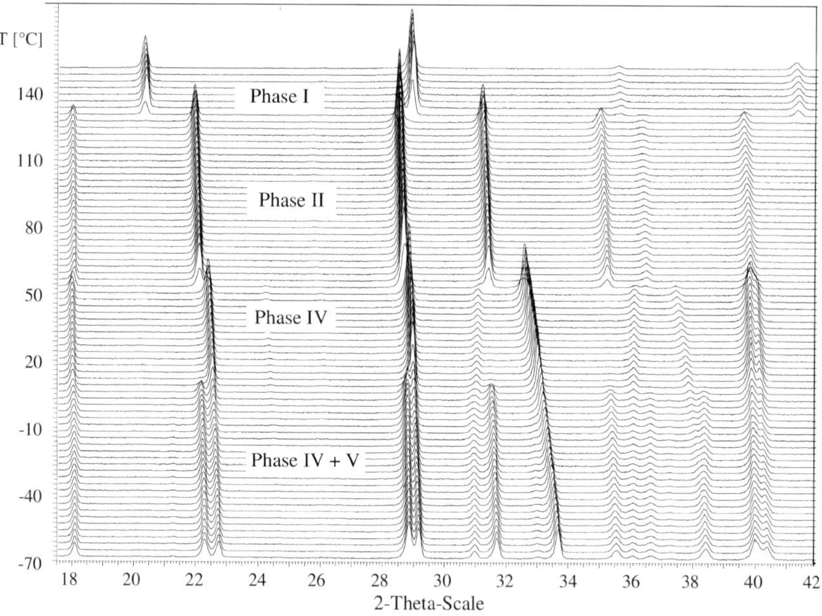

Figure 9.13. Series of diffraction patterns of AN measured on heating between –70 and 150 °C.

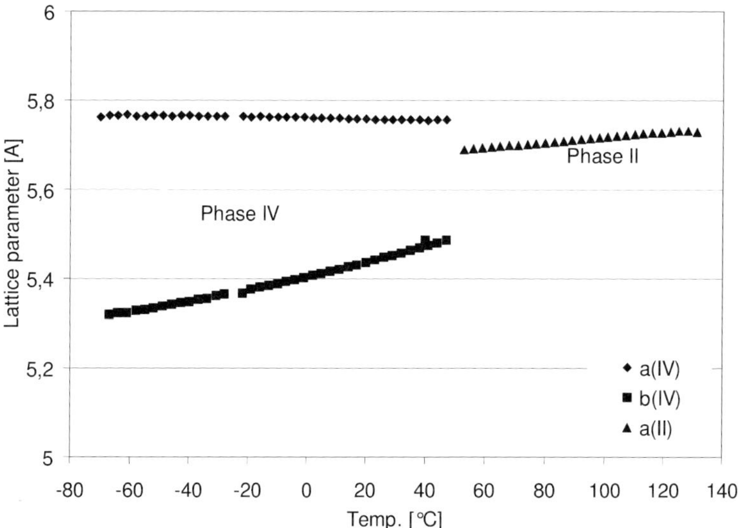

Figure 9.14. Lattice parameters of AN phases II and IV as a function of temperature.

the elementary cell approaches tetragonality, which is reached with the transition IV/II, where the nitrate groups twist into the 45° position.

Calculating the specific volume of the elementary cells of the phases using the refined cell parameters gives the thermal expansion behavior of the phases and volume changes during the phase transitions as shown in Fig. 9.16. The plot gives an idea of the problems caused by the phase transitions. The high volume changes of 3.7 and 2.0% of the phase transitions IV/III, IV/II, respectively, strain the propellant matrix when the materials are exposed to changing temperature, e.g. during storage. On the other hand, a smooth thermal expansion behavior with a volume change of 0.3% is revealed when phases III and IV are suppressed, which is part of actual stabilization concepts.

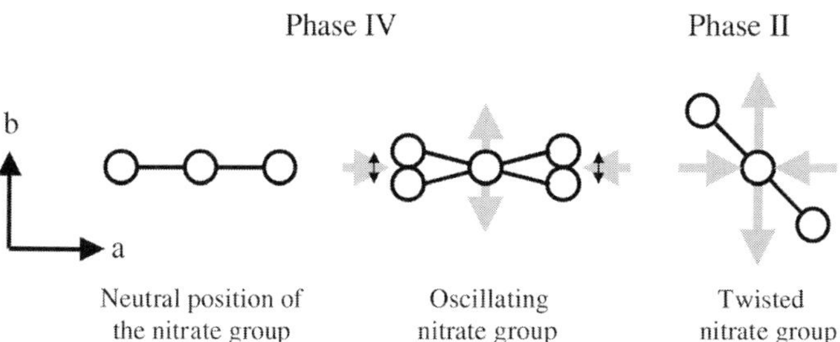

Figure 9.15. Model for anisotropic expansion behavior based on temperature-dependent oscillations of nitrate groups.

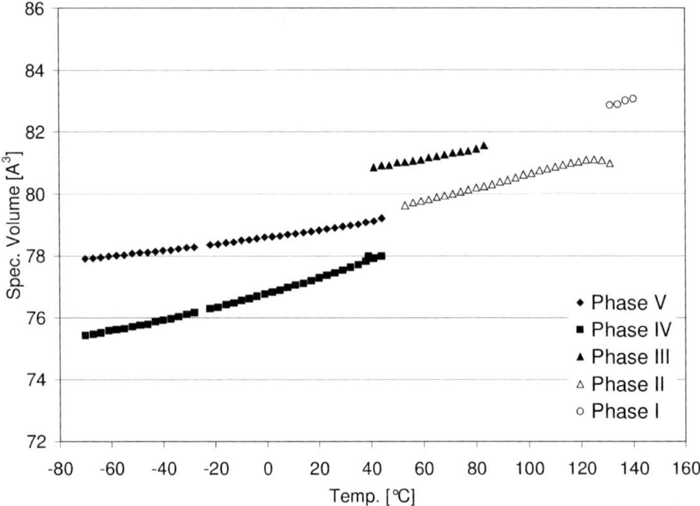

Figure 9.16. Specific volume of the AN phases as a function of temperature.

9.3.3.3 Quantitative Analysis and Interaction of ADN and AN

Ammonium dinitramide (ADN) is a new oxidizer proposed for solid propellants. Its crystal structure is described by Gidaspov *et al.* [9.35]. Drawbacks of ADN are low thermal stability and compatibility. Further, AN is created during the synthesis and decomposition of ADN, which may influence the stability of the latter. Therefore, ADN-AN mixtures were measured with temperature-resolved X-ray diffraction and evaluated by Rietveld analysis [9.36].

The investigations showed that melting of an ADN-AN mixture containing 5% AN starts at 55 °C (Fig. 9.17, squares) combined with bending of the volume curve (diamonds). At the same temperature AN phase IV disappears on heating (triangles), but no peaks of other ammonium nitrate phases (II or III) appear, as assumed. Therefore, it is suggested that the ammonium nitrate expands the ADN lattice by forming a solid solution, which enhances the formation of a eutectic melt.

9.3.3.4 Lattice Defects and Micro Strain in HMX

A new and challenging topic is the investigation of lattice defects in energetic materials. X-ray diffraction methods were tested and adjusted with a set of differently crystallized HMX samples. As the relatively large crystals create difficulties for powder measurements, modern X-ray optics and rotating sample holders had to be applied.

The X-ray diffraction patterns of the various HMX samples show significant peak broadening, which was evaluated by Williamson-Hall plots as shown in Fig. 9.18 for selected HMX samples. Relative values of micro strain were determined from the slopes of the fitted lines in the plots and compared with mechanical sensitivities.

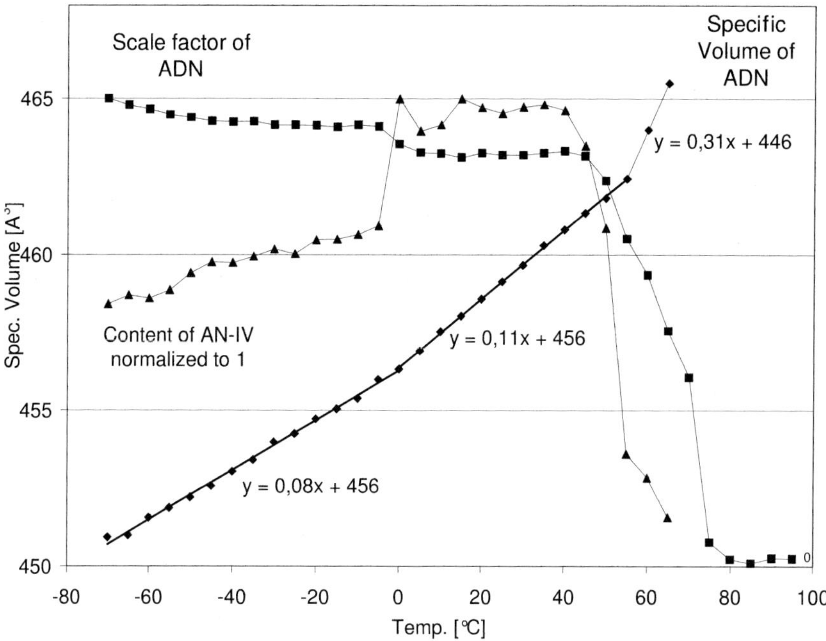

Figure 9.17. Thermal expansion and scale factor of ADN and content of AN phase IV on heating.

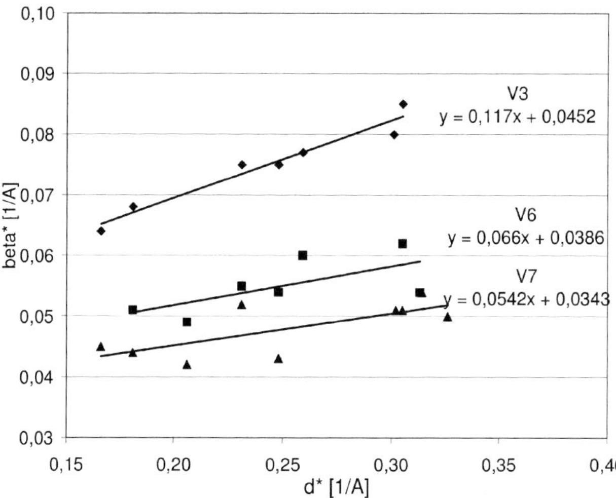

Figure 9.18. Williamson-Hall plots for three HMX samples for determining micro strain.

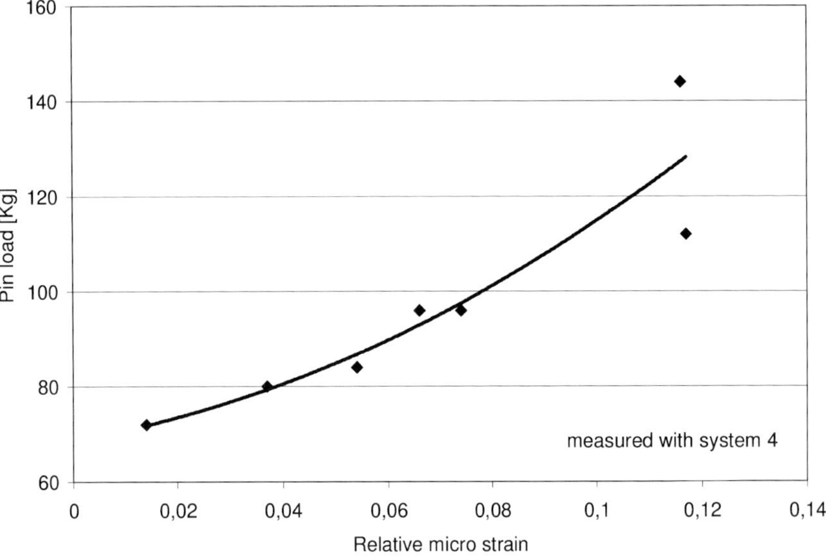

Figure 9.19. Pin load of the friction test versus micro strain. The strong correlation suggests that increasing micro strain decreases the sensitivity to friction.

Plotting the pin load of the friction tests versus relative micro strain (Fig. 9.19) shows a high correlation where the sensitivity to friction decreases significantly with increasing micro strain. The results suggest the hypothesis that the sensitivity of HMX to friction decreases with increasing concentration of defects. The effect may be explained by deformation twinning [9.37] or dislocations improving the plasticity, as shown in Fig. 9.20 for a gliding dislocation [9.38].

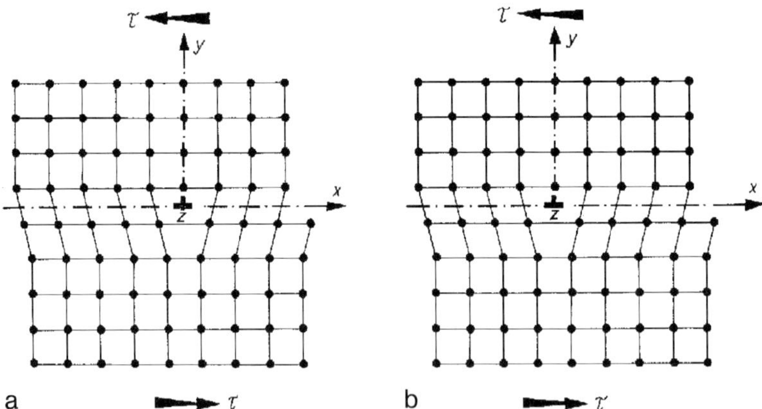

Figure 9.20. Schematic diagram of gliding of an edge dislocation through a lattice [9.38].

9.3.3.5 Simulation of Microstructures

Modern evaluation techniques such as Rietveld analysis involve the molecular and crystal structure of investigated materials. The simulation of these structures is a complementary tool, as simulated results can be checked by diffraction experiments or vice versa.

The programs WinMOPAC [9.39] and CERIUS^2 from MSI were applied for the simulation of the HMX phases starting with literature data [9.40–9.43]. After simulating the crystal structures, the crystal growth and resulting morphologies were simulated and compared with crystal habits of samples crystallized from solutions [9.44]. Figure 9.21 compares simulated and crystallized morphologies of

Figure 9.21. Comparison of simulated and crystallized habits for (a) α-, (b) β- and (c) γ-HMX.

HMX phases. The simulated habits agree excellently with the crystallized needles, compact crystals and platelets of α-, β- and γ-HMX, respectively [9.45] (see Chapter 3).

Additionally, the simulation of the interaction between energetic ingredients [9.46] or with solvents gives an idea of how stability and crystallization procedures may be improved and simulation may help in characterizing types of lattice defects comparing simulated with measured peak broadening.

9.4 Composite Explosives as Probed with Microscopy

9.4.1 Introduction

Microstructural details in bulk explosives, such as particle size, intergranular void volume and intragranular crystal defects, are known to have a significant influence on shock initiation. A similar influence is found in non-shock initiation events (as would be found in accident scenarios) such as mechanical and thermal insult. Consequently, microscopic techniques are used to probe microstructural details and characterize changes resulting from insult, in order better to understand and simulate the phenomena that lead to initiation.

The crystalline high explosive octahydro-1,3,5,7-tetranitro-1,3,5,7-tetrazocine (HMX, Octogen) has been used as an ingredient in many formulations since its discovery as a high-melting impurity in hexahydro-1,3,5-trinitro-1,3,5-triazine (RDX, Hexogen, cyclonite) around 1941 [9.47, 9.48]. Work from that era, summarized by Blomquist [9.49], included extensive material characterization by polarized, transmitted light microscopy. These studies, in the style of chemical microscopy, were chiefly directed toward phase identification by specification of indices of refraction, of dispersion and of habitual crystal morphology. The later development of electron microscopy added a complementary dimension to microscopic investigations.

Early in US practice, HMX, a very high energy density material, was mixed with molten TNT [9.50], which was then cast into molds or directly into munitions. Some years later, an aqueous slurry process was developed to coat the HMX particles with a polymer binder and agglomerate them into small granules or prills suitable for compression molding. At Los Alamos, such formulations are designated as plastic-bonded explosives or PBXs. Over the years, viscoplastic binder properties were found to affect significantly the handling safety of pressed charges. Today, the mature formulation PBX 9501 is used in weapons for the US Department of Energy. It contains ~93% (solids volume) HMX and 7% polyester-urethane/nitroplasticizer binder.

The need for insensitive munitions led to the development of the high explosive 1,3,5-triamino-2,4,6-trinitrobenzene (TATB) [9.51] and its formulations with polymer binders. The microstructural characteristics of crystalline TATB are different from those of HMX. For example, Cady reported on the prevalence of unusual void structures that he calls 'worm holes' [9.52, 9.53]. The Los Alamos formulation PBX 9502 employs ~95% (solids volume) TATB with the binder comprised of chlorotrifluoroethylene-vinylidine fluoride copolymer. This material is also compression molded to produce serviceable articles.

There is a need to study the microstructure of energetic compounds beyond the identification of phases and their crystal structures. Tools and techniques are needed to characterize the results of downstream processing. These processes include grinding to control particle size distribution, formulation with binders, pressing to specified shapes and densities and the aging of these articles in service, mechanical testing or damage to samples.

9.4.2 Methods

9.4.2.1 Reflected Parallel Polarized Light Microscopy (RPPL)

Particle sizes typically specified in composite explosives are well within the resolution of microscopic examination. However, the historical approach of dispersing powders on a glass slide and examining them in transmitted light is not well suited to consolidated samples of composite materials. Preparation of a thin slice or section of the composite for study in transmitted light requires significant additional effort and is useful for some studies involving high index materials such as TATB. We have found it more worthwhile, in general, to prepare samples for examination in reflected light.

Reasonable grain-to-grain contrast occurs naturally under reflected, plane-polarized light. This is provided by the relatively high indices of refraction of most energetic crystalline solids, compared with polymeric binders and random orientation of the birefringent grains. For most applications, etching or staining is not necessary.

In preparation for examination in reflected light, a sample of the composite explosive under study is typically vacuum-mounted in low-viscosity epoxy for viewing in cross-section. The sample is further prepared for observation by polishing with a sequence of fine abrasives and leveling the mount in clay. Details of our polishing protocol are designed to meet the individual idiosyncrasies of PBX 9501 and PBX 9502. For PBX 9501, we 'coarse grind' with silicon carbide paper with ~25-μm abrasive, 'fine grind' with 1-μm alumina and final polish with colloidal silica, all with abundant cooling water. This procedure is considerably milder than those of traditional metallography, in response to both the detonability and the low toughness of our samples. Further details are described in a previous report [9.54]. We acknowledge chemical incompatibilities between our samples and our traditional amine-cured epoxy mounts, which we accommodate by disposing of old samples. All operations are conducted remotely when practical.

The transparency of polished HMX crystals led us to the practice of matching the plane orientation of the 'polarizer' for the incident light with that of the 'analyzer' for the reflected light. This parallel arrangement minimizes the obfuscating observation of return light scattered from crystal interfaces below the plane of polish.

If accomplished perfectly, the image observed in the eyepiece would be created only by the light data corresponding to the plane of polish. We call this configuration 'reflected parallel polarized light' (RPPL).

9.4.2.2 Secondary Electron Imaging (SEI)

The effective use of RPPL requires a polished plane of observation. Hence it is well suited to a cross-sectional view of nominal or damaged material. On the other hand, scanning electron microscopy (SEM), specifically secondary electron imaging (SEI), is a more effective tool for examining surface topography in plan view, offering excellent resolution with a large depth of focused field.

The application of SEM to explosive science is non-trivial and deserves some discussion. The samples of interest are insulators, prone to local accumulation of charge, which prevents further access to the beam. They are also materials to which prudent operators endeavor to mitigate exposure to sparks. The problem is traditionally solved by (a) sputter coating the sample with a thin metallic film (~ 100 Å), especially gold or gold-palladium, prior to observation, (b) operating the SEM in low vacuum to provide some electrical conductivity through the gas or (c) operating the SEM at some low accelerating voltage where the secondary electron yield approaches the incident beam current. Sputter coating, which produces good signal-to-noise ratios and good resolution in secondary electron images as a consequence of the small interaction volumes characteristic of heavy metals, has been most useful in our experiments. However, it is possible inadvertently to anneal or thermally alter samples during coating and to decompose the energetic material under the coating during observation at excessively high voltages. The low-vacuum alternative [sometimes referred to as environmental scanning electron microscopy (ESEM)] necessarily degrades the resolution of the microscope owing to the scattering and energy loss by which the gas-phase conductivity is produced. However, Rae *et al.* [9.55] creatively used this approach to observe directly the opening of cracks in composite materials. The dynamic exposure of fresh surfaces in those experiments makes prior sputter coating pointless. The third alternative, at low voltage, limits microscope resolution owing to the large interaction volumes of low-density materials. Although none of these methods is perfect, careful application of the technique can result in new insight into microstructural aspects of HE systems.

9.4.2.3 Other Microscopic Methods

Two other microscopic techniques that have recently been applied to explosives hold promise for delivering complementary data. Confocal scanning laser microscopy has been applied to the characterization of micro-inclusions in HMX crystals by van der Heijden *et al.* [3.42] and to the HMX composite PBX 9501 by Los Alamos researchers. This method exploits the transparency of HMX to probe subsurface features using reflection. Laser light is transmitted through a microscope objective, reflected from the sample and returned through the objective following the same path as for standard optical reflectance microscopy. In the confocal configuration, both the illumination and collection are opened to allow for imaging of a narrow depth slice (sub-micron) within an optically thick sample. The thin focal plane can then be incrementally stepped through the sample thickness and images collected at every step. From these, a three-dimensional picture of the internal features can be constructed.

Application of the non-linear optical technique second harmonic generation (SHG) to detect dynamic phase changes in 'bulk' samples of HMX, reported by Henson *et al.* [9.56], has spawned work by other Los Alamos researchers probing the microscopic level. Work in progress uses this technique to measure the kinetics of phase change (including reversion) in individual HMX crystals and in the composite PBX 9501. SHG microscopy has also been used to observe (and confirm) the nucleation and growth mechanism for the beta-to-delta phase transition in HMX. As this technique matures, it may become very useful for isolating phases in thermally damaged samples.

9.4.3 Application to HMX Composites

9.4.3.1 Effects of Quasi-static Mechanical Insult

Various microscopic techniques have been employed in studies at Cambridge University on the mechanics of deformation in various explosives, including TATB- and HMX-based compositions [9.57, 9.58]. Recent work by Rae *et al.* [9.55] has shown two advanced variations on the Brazilian disk compression test (a compact, grip-free method of evaluating the tensile properties of composite explosives), one facilitating strain mapping by Moiré interferometry on polished samples and another *in situ* in an environmental SEM showing the extensive local elasticity of binder fibrils under tensile load. The real-time observation offered by these techniques provides significant insights into mechanical deformation processes.

The process of consolidation or pressing a composite explosive is a simple method for quasi-statically introducing mechanical damage. Lefrançois and Demol [9.59] and Skidmore *et al.* [9.60] have used microscopic techniques, post-test, to characterize such damage for some HMX compositions. While the pressing process is designed to increase the mechanical integrity in a bulk article, individual grains may become fractured. Figure 9.22 is a typical RPPL image of a polished sample of PBX 9501 which has been pressed at ambient temperature to a density of 1.81 g/cm^3. Individual large grains are easily distinguished from the surrounding matrix of fine particles mixed with binder. Most of the large grains exhibit fracture. This figure provides a marked contrast with thermally damaged samples of PBX 9501, to be discussed later.

Fracture in the bulk sample can be induced at quasi-static rates by a number of methods. Figure 9.23 is a secondary electron image showing typical features in a PBX 9501 sample (nominally pressed) that was notched and placed in slow, three-point bend until failure. Note the exposure of growth facets (angles typical of beta HMX) and the intragranular cleavage on the large crystal. Nearby, there is a vacancy

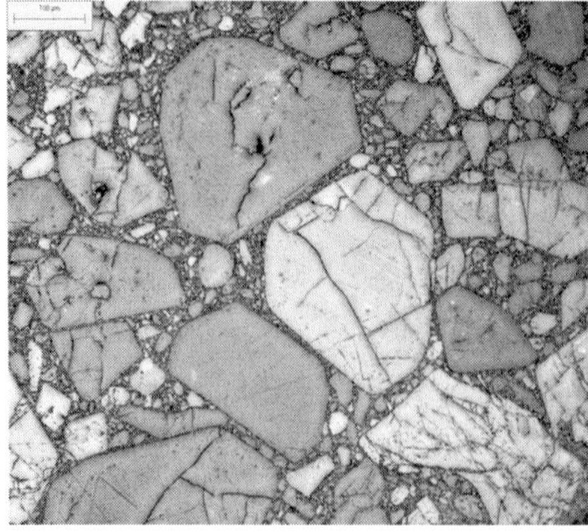

Figure 9.22. PBX 9501 pressed to 1.81 g/cm^3 (RPPL).

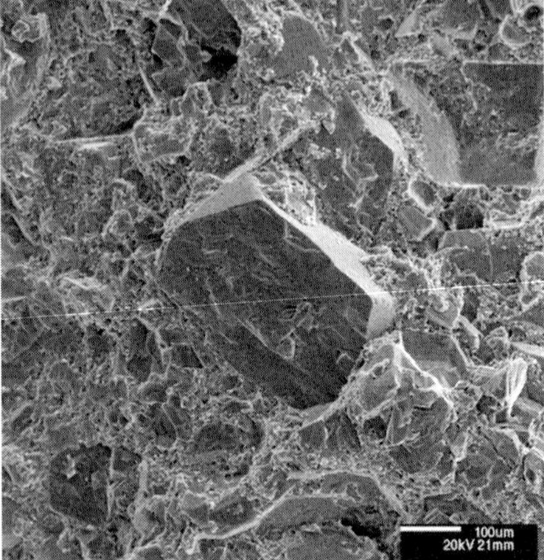

Figure 9.23. Typical surface in PBX 9501 quasi-statically fractured in bending (SEI).

from which another crystal has been removed. Crystals appear to be randomly oriented and a high density of smaller crystals is exposed. The gross cross-section of the sample retains its rectangular form and size, suggesting, along with morphological observations, that quasi-static fracture of PBX 9501 is predominantly brittle.

9.4.3.2 **Effects of Thermal Insult**

Another common damage mode for explosives is thermal insult. This may be introduced by any of the transport mechanisms, conduction, convection or radiation. The magnitude of the insult may be sufficient to produce ignition and self-sustained combustion or low enough to produce an annealing environment for affected regions of the microstructure. Microscopic techniques may be used to observe post-test microstructures and infer the dynamic conditions that created them. Although this is standard practice for materials scientists working with metals and ceramics, the much more thermally reactive nature of explosives introduces a new complexity to the interpretation. While acknowledging that some conjecture is inherent in the analysis of composite explosives treated at elevated temperatures and subsequently prepared for observation at ambient temperature, we believe that the following examples have provided useful insights into thermal damage processes.

Renlund *et al.* [9.61] used SEI to observe morphological changes in annealed pellets of various HMX and TATB compositions. For PBX 9501, a pellet heated to 190 °C then immediately cooled to ambient temperature and a pellet held at 200 °C for 1 h before cooling were chemically indistinguishable from unheated PBX 9501 (as evaluated with differential scanning calorimetry and high-performance liquid chromatography). However, the two pellets were very different morphologically, as observed with SEI. Images of the latter sample showed greater porosity, corroborating the higher mass loss and pressurization from gaseous decomposition products.

RPPL microscopy of polished epoxy mounts in cross-section can also be a useful technique for analyzing annealed samples. Consider the significantly different microstructures observed in unconfined PBX 9501 pellets (12.7 mm diameter, 12.7 mm high) placed in an oven at 180 °C for 0, 15 and 30 min. Representative micrographs are shown in Figs 9.22, 9.24 and 9.25, respectively. Figure 9.22, described previously, provides a baseline from which to assess thermally induced changes. Figures 9.24 and 9.25 are at half the resolution of Fig. 9.22 in order to show a larger field of view. The sample from the 15-min treatment had numerous pockets of apparently undisturbed structure. For example, the central portion of Fig. 9.24 is dominated by typical beta-phase microstructure for PBX 9501. The surrounding region has been significantly altered. There are individual crystals whose features suggest that portions are original beta wherease other portions have converted to another phase.

No pockets or even particles of original material were discovered in the sample from the 30-min treatment. Figure 9.25 shows the new microstructure that was

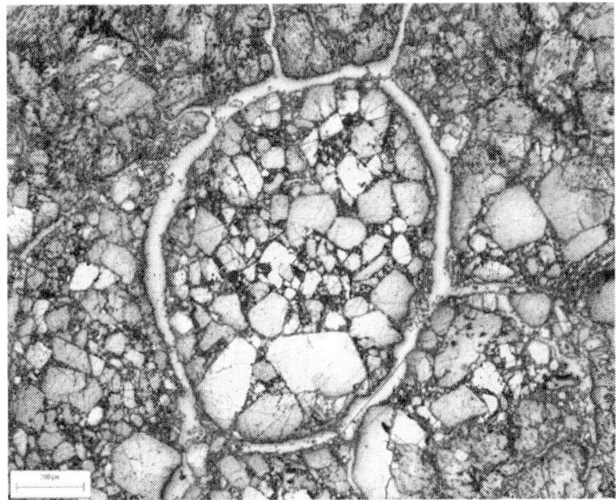

Figure 9.24. PBX 9501, 180 °C oven, 15 min (RPPL).

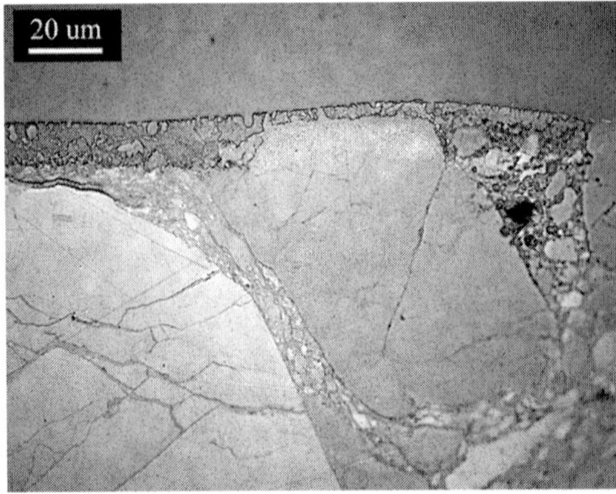

Figure 9.28. Polished cross-section of fracture surface induced by projectile impact (from the top) on PBX 9501 (RPPL).

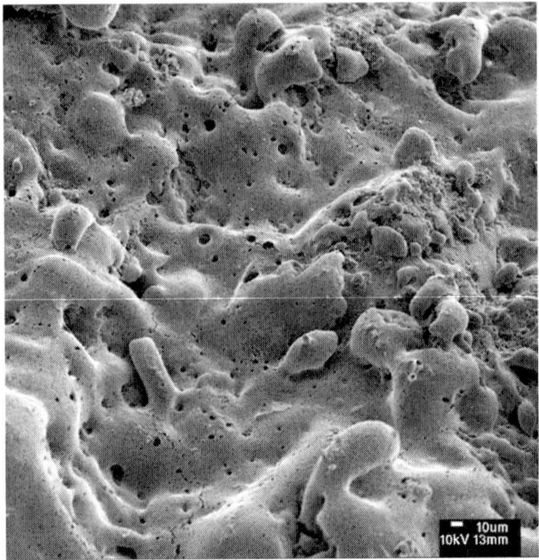

Figure 9.29. Plan view of fracture surface induced by projectile impact on PBX 9501 (SEI).

'jaggedness' shown in Fig. 9.27 may be due to a faster quench (by sudden depressurization) in the dynamic impact experiment.

9.4.4 Application to TATB Composites

9.4.4.1 Effects of Quasi-static Mechanical Insult

TATB-based materials present a different set of microstructural issues to the HMX-based materials. The anisotropy of intrinsic crystal parameters such as morphology, refractive index and coefficient of thermal expansion is much greater. Processing conditions such as modest variations in the synthesis process, consolidation (pressing) of molding powder prills using elevated temperature and pressure and the practice of reprocessing also produce unique microstructural effects.

Cady [9.52] used SEI to show the differences in crystal void structures depending on how much water was used in TATB synthesis. Skidmore *et al.* [9.54] reported contrasting microstructures observed by optical microscopy in samples of PBX 9502 with increasing levels of quasi-static mechanical damage beginning with molding powder, including a pressed piece and finally a compression-fractured piece. Blumenthal *et al.* [9.65] reported anisotropy in PBX 9502 compressive strength depending on whether specimens were machined from a billet with the axis parallel or perpendicular to the pressing direction. This effect was presumed to be due to the 'preferential alignment of the TATB crystals during densification'. In our current work, we contrast the microstructure of nominally formulated PBX 9502 with that of PBX 9502 that was formulated with 50% previously processed material.

9.4.4.2 **Effects of Reprocessing**

TATB is a relatively expensive energetic material. Nearly two decades ago, economic considerations suggested the investigation of various methods for re-using TATB. One method used 50% PBX 9502 that had already been pressed. The material to be re-used was mechanically processed until it would pass through a specified sieve, then it was added to binder solvent and mixed with fresh TATB and a fresh portion of binder-solvent lacquer. From this point the material was formulated together.

Figure 9.30 provides a representative, polished cross-sectional view from a nominally formulated (no reprocessed material) and pressed sample. The resolution was chosen to show typical grain level details. There is evidence of twinning in some grain orientations. Numerous 'worm hole' voids are present in virtually every grain. Views with a larger field suggest that the distribution of light and dark shaded grains is random.

Figure 9.31 shows typical texture from a fracture surface (SEI). The micaceous, plate-like morphology is clearly evident. Failure appears to occur by crystal cleavage. Regions dominated by binder are not obvious.

Comparative images from the reprocessed material are shown in Figs 9.32 (RPPL) and 9.33 (SEI). Figure 9.32 shows a larger field than Fig. 9.30 in order to highlight

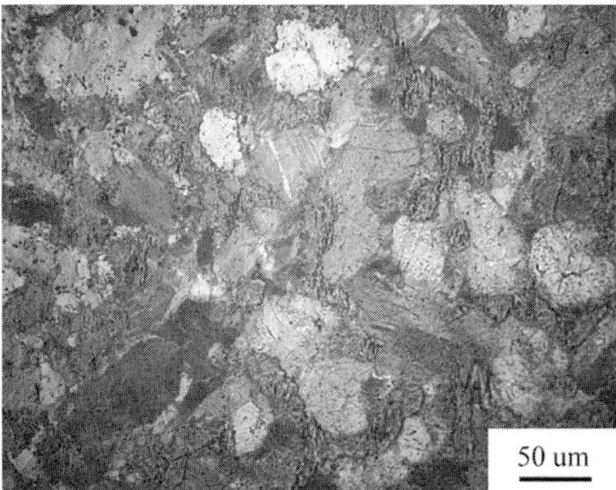

Figure 9.30. Nominal PBX 9502, polished cross-section (RPPL).

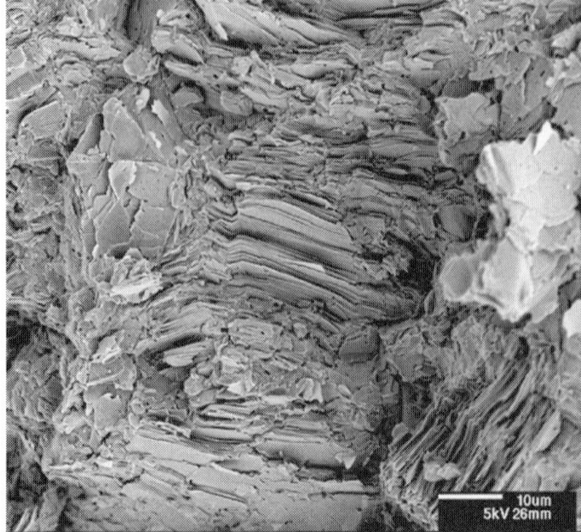

Figure 9.31. Nominal PBX 9502, fracture surface (SEI).

Figure 9.32. 50% reprocessed PBX 9502, polished cross-section (RPPL).

the more ordered arrangement of dark and light grains. The upper left quadrant of Fig. 9.32 contains a nearly complete ring of dark grains, suggesting that the reprocessed material may be more heterogeneous. Perhaps pieces of reprocessed material are not fully integrated with the fresh constituents.

Figure 9.33, which shows a fracture surface, provides corroborative evidence. The center of the field appears to shows failure through weak binder adhesion rather than crystal cleavage as shown in Fig. 9.31. This is an example of microscopic techniques providing insights into microstructural failure mechanisms.

9.4.4.3 Effects of Thermal Insults

TATB is much more thermally stable than HMX, which is one attraction of its use as an insensitive high explosive. The effects of thermal insult are not well characterized. Son *et al.* [9.66] applied optical second harmonic generation to detect dynamically structural changes that occur in TATB upon heating. Phillips *et*

Figure 9.33. 50% reprocessed PBX 9502, fracture surface (SEI).

al. [9.67] sought to define more clearly those changes using microscopy (light and electron) and powder X-ray diffraction. Very little microstructural characterization work has been reported for thermally damaged TATB composites. A notable exception is the reflected light microscopy study of Demol *et al.* [9.68].

9.5 References

9.1 Campell AW, Davis WC, Ramsay JB, Travis JR (1961) Shock initiation of solid explosives, *J. Appl. Phys.* **4**, 511–521.

9.2 Lindstrom IE (1970) Planar shock initiation of porous tetryl, *J. Appl. Phys.* **41**, 337–350.

9.3 Gustavsen RL, Sheffield SA, Alcon RR, Hill LG (1999) Shock initiation of new and aged PBX 9501 measured with embedded electromagnetic particle velocity gauges, *Los Alamos National Laboratory Report*, LA-13634-MS.

9.4 Van der Steen AC, Verbeek HJ, Meulenbrugge JJ (1989) Influence of RDX crystal shape on the shock sensitivity of PBXs. In: *Proc 9th Int. Symp. on Detonation, Portland, Oregon*, pp. 83–88.

9.5 Baillou F, Dartyge JM, Spyckerelle C, Mala J (1993) Influence of crystal defects on sensitivity of explosives. In: *Proc 10th Int. Symp. on Detonation, Boston, Massachusetts*, pp. 816–823.

9.6 Borne L, Fendeleur D, Beaucamp A (1997) Explosive crystal properties and PBX's Sensitivity, *DEA 7304: Physics of Explosives*, Berchtesgaden, Germany.

9.7 Borne L (1998) Explosive crystal microstructure and shock-sensitivity of cast formulations. In: *Proc 11th Int. Symp. on Detonation, Aspen, Colorado.*

9.8 Spitzer D, Samirant M (1993) Shock solicitation of PETN single crystals presenting defects and visualization of hot spots initiation. In: *Proc 10th Int. Symp. on Detonation, Boston, Massachusetts*, pp. 831–840.

9.9 Borne L (1995) Microstructure effects on the shock sensitivity of cast plastic-bonded explosives. In: *Proc. 6ème Congrès Int. de Pyrotechnie, EUROPYRO 95, Tours.*

9.10 Borne L, Patedoye JC, Spyckerelle C (1999) Quantitative characterization of internal defects in RDX crystals, *Propell., Explos. Pyrotech.* **24**, 255–259.

9.11 Borne L, Beaucamp A, Fendeleur D (1998) Metrology tools for the characterization of explosive crystal properties. In: *Proc 29th Int. Annual Conference of ICT, Karlsruhe.*

9.12 Sharma J, Hoover SM, Coffey CS, Tompa AS, Sandusky HW, Armstrong RW, Elban WL (1997) Structure of crystal defects in damaged RDX as revealed by an AFM. In: *Proc. Conference of the American Phys. Soc. Topical Group on Shock Compression of Condensed Matter, Amherst, Massachusetts*, pp. 563–566.

9.13 Cherin H, Bournisien D (1992) Characterization of the porous structure of explosive powders. Correlation with their combustion rate. In: *Proc 23rd Annual International Conference of ICT, Karlsruhe.*

9.14 Mang JT, Skidmore CB, Howe PM, Hjelm RP, Rieker TP (1999) Structural characterization of energetic materials by small angle scattering. In: *Am. Phys. Soc. Topical Conference on Shock Compression of Condensed Matter, Snowbird, Utah.*

9.15 Kumachov MA, Komarov FA (1990) Channeling of photons and new X-ray optics, *Nucl. Instrum. Methods B* **48**, 283–286.

9.16 Scardi P, Setti S, Leoni M (2000) Multicapillary optics for materials science studies, *Mater. Sci. Forum*, **321–324**, 162–167.

9.17 Schuster M, Göbel H (1995) *J. Phys. D* **28**, A270.

9.18 Klug HP, Alexander LE (1974) *X-ray Diffraction Procedures for Polycrystalline and Amorphous Materials*, Wiley, New York.

9.19 Chung FH, Smith DK (2000) *Industrial Applications of X-Ray Diffraction*, Marcel Dekker, New York.

9.20 Van Arkel AE (1925) *Physica* **5**, 208–212.

9.21 Rietveld HM (1969) A profile refinement method for nuclear and magnetic structures, *J. Appl. Crystallogr.* **2**, 65–71.

9.22 Young RA (1995) *The Rietveld Method.* International Union of Crystallography, Oxford Science, Oxford.

9.23 BAM (2000) *ICT Contribution to the α-/β-*Silicon Nitride Round Robin Organized by the Bundesanstalt für Materialsforschung, BAM, Berlin.

9.24 Warren BE, Averbach BL (1950) The effect of cold-worked distortion on X-ray patterns, *J. Appl. Phys.* **21**, 595–599.

9.25 Williamson GK, Hall WH (1953) X-ray line broadening from filed aluminum and wolfram, *Acta Metall.* **1**, 22–31.

9.26 Jacob G, Toupet L, Ricard L, Cagnon G (1997) *CCDC 124947–12450.* Cambridge Crystallographic Data Centre, Cambridge.

9.27 Athee M, Smolander KL, Lucas BW, Hewat AW (1983) The structure of the low temperature phase V of ammonium nitrate, *Acta Crystallogr., Sect. C* **39**, 651–655.

9.28 Choi CS, Prask HJ (1983) The structure of ND_4NO_3 phase V by neutron powder diffraction, *Acta Crystallogr., Sect. B* **39**, 414–420.

9.29 Choi CS, Mapes JE, Prince E (1971) The structure of ammonium nitrate (IV), *Acta Crystallogr., Sect. B* **28**, 1357–1361.

9.30 Lucas BW, Ahtee M, Hewat A (1980) The structure of phase III ammonium nitrate, *Acta Crystallogr., Sect. B* **36**, 2005–2008.

9.31 Lucas BW, Ahtee M, Hewat A (1979) The crystal structure of phase II ammonium nitrate, *Acta Crystallogr., Sect. B* **35**, 1038–1041.

9.32 Yamamoto S, Shinnaka Y (1974) X-ray study of polyoriential disorder in cubic NH_4NO_3, *J. Phys. Soc. Jpn.* **37**, 724–732.

9.33 Herrmann M (1998) Temperaturverhalten von Ammoniumnitrat, *Wiss. Schriftenreihe des Fraunhofer ICT*, 15.

9.34 Amorós JL, Alonso P, Canut ML (1958) Transformaciones polimorfas en monocristales. Transición IV/II del nitrato amónico y forma metaestable II, *Bol. R. Soc. Esp. Hist. Nat. (G)* **56**, 77.

9.35 Gidaspov BV, Tselinskii IV, Mel'nikov VV (1995) Crystal and molecular structure of dinitramide salts and acid-base properties of dinitramide, *Russ. J. Gen. Chem.* **65**, Part 2, 906–913.

9.36 Herrmann M, Engel W (1999) Thermal expansion of ADN measured with X-ray diffraction. In: *Proc 30th Int. Annual Conf. of ICT*, p. 118.

9.37 Armstrong AW, Ammo HL, Du ZY, Elban WL, Zhang XJ (1993) Energetic crystal-lattice-dependent response, *Mater. Res. Soc. Symp. Proc.* **296**, 227–232.

9.38 Bohm J (1995) Realstruktur von Kristallen, Schweizerbart, Stuttgart.

9.39 Stewart JJP (1990) MOPAC: a general molecular orbital package, *Quantum Chem. Prog. Exch.* **10**, 86.

9.40 Cady HH, Larson AC, Cromer DT (1963) The crystal structure of α-HMX and a refinement of the structure of β-HMX, *Acta Crystallogr.* **16**, 617–623.

9.41 Choi CS, Boutin HP (1970) A study of the crystal structure of β-HMX by neutron diffraction, *Acta Crystallogr., Sect. B* **26**, 1235–1240.

9.42 Main P, Cobbledick RE, Small RWH (1985) Structure of the fourth form of HMX (γ-HMX), *Acta Crystallogr., Sect. C* **41**, 1351–1354.

9.43 Cobbledick RE, Small RWH (1974) Crystal structure of the δ-form of HMX, *Acta Crystallogr., Sect. B* **30**, 1918–1922.

9.44 Herrmann M, Engel W, Eisenreich N (1993) Thermal analysis of the phases of HMX using X-ray diffraction, *Z. Kristallogr.* **204**, 121–128.

9.45 Thome V, Kempa PB, Herrmann M, Teipel U, Engel W (2000) Molecular simulation of the morphology of energetic materials. In: *Proc 31st Int. Annual Conf. of ICT*, p. 64.

9.46 Thome V, Kempa PB, Bohn MA (2000) Erkennen von Wechselwirkungen der Nitramine HMX und CL20 mit Formulierungskomponenten durch Computer Simulation. In: *Proc 31st Int. Annual Conf. of ICT*, p. 63.

9.47 McCrone WC (1959) Explosives crystallization, *Ordnance* 506–507.

9.48 McCrone WC (1999) Playing with HMX, *Chem. Eng. News* **77**, 2.

9.49 Blomquist AT (1944) *Microscopic Examination of High Explosives and Boosters*, National Defense Research Committee of the Office of Scientific Research and Development, Washington, DC.

9.50 Wilson, EK (1999) Science/technology, HMX: by any name, a powerful explosive, *Chem. Eng. News* **77**, 26.

9.51 Benziger TM (1981) Manufacture of triaminotrinitrobenzene. In: *Proc 12th Int. Annual Conf. of ICT, Karlsruhe.*

9.52 Cady HH (1986) Microstructural differences in TATB that result from manufacturing techniques. In: *Proc 17th Int. Annual Conf. of ICT, Karlsruhe*, p. 53.

9.53 Cady HH (1992) Growth and defects of explosives crystals. In: *Proc. Structure and Properties of Energetic Materials, Boston, Massachusetts*, pp. 243–254.

9.54 Skidmore CB, Phillips DS, Crane NB (1997) Microscopic examination of plastic-bonded explosives, *Microscope* **45**, 127–136.

9.55 Rae PJ, Goldrein HT, Palmer SJP (1982) Studies of the failure mechanisms of polymer-bonded explosives by high resolution Moire interferometry and environmental scanning electron microscopy. In: *Proc. 11th Int. Detonation Symp., Aspen, Colorado.*

9.56 Henson BF, Asay BW, Sander RK (1999) Dynamic measurement of the HMX beta-delta phase transition by second harmonic generation, *Phys. Rev. Lett.* **82**, 1213–1216.

9.57 Palmer SJP, Field JE (1982) The deformation and fracture of beta-HMX. *Proc. R. Soc. London, Ser. A* **383**, 399–407.

9.58 Palmer SJP, Field JE, Huntley JM (1993) Deformation, strengths and strains to failure in polymer bonded explosives. *Proc. R. Soc. London, Ser. A* **440**, 399–419.

9.59 Lefrançois A, Demol G (1998) Increase of sensitivity of HMX-based pressed explosives resulting of the damage induced by hydrostatic compression. In: *Proc. 29th Int. Annual Conf. ICT*, p. 33.

9.60 Skidmore CB, Phillips DS, Howe PM (1998) The evolution of microstructural changes in pressed HMX explosives. In: *Proc 11th Int. Symp. on Detonation, Aspen, Colorado*, pp. 537–545.

9.61 Renlund AM, Miller JC, Trott WM (1998) Characterization of thermally degraded energetic materials. In: *Proc 11th Int. Symp. on Detonation, Aspen, Colorado*, p. 341.

9.62 Skidmore CB, Phillips DS, Idar DJ (1999) Characterizing the microstructure of selected high explosives. In: *Proc. Europyro 99, Brest*, pp. 2–10.

9.63 Skidmore CB, Phillips DS, Asay BW (1999) Microstructural effects in PBX 9501 damaged by shear impact. In: *Proc. American Phys. Soc. Topical Conf. on Shock Compression of Condensed Matter, Snowbird, Utah*, pp. 659–662.

9.64 Idar DJ, Lucht RA, Straight JW (1998) Low amplitude insult project: PBX 9501 high explosive violent reaction experiments. In: *Proc 11th Int. Detonation Symp., Aspen, Colorado.*

9.65 Blumenthal WR, Gray III GT, Idar DJ (1999) Influence of temperature and strain rate on the mechanical behavior of PBX 9502 and Kel-F 800. In: *Proc. American Phys. Soc. Topical Conf. on Shock Compression of Condensed Matter, Snowbird, Utah,* pp. 671–674.

9.66 Son SF, Asay BW, Henson BF (1999) Dynamic observation of a thermally activated structure change in 1,3,5-triamino-2,4,6-trinitrobenzene (TATB) by second harmonic generation, *J Phys. Chem. B* **103**, 5434–5440.

9.67 Phillips DS, Schwarz RB, Skidmore CB (1999) Some observations on the structure of TATB. In: *Proc. American Phys. Soc. Topical Conf. on Shock Compression of Condensed Matter, Snowbird, Utah,* pp. 707–710.

9.68 Demol G, Lambert P, Trumel H (1998) A study of the microstructure of pressed TATB and its evolution after several kinds of solicitations. In: *Proc 11th Int. Symp. on Detonation, Aspen, Colorado.*

10
Thermal and Chemical Analysis

S. Löbbecke, M. Kaiser, G. A. Chiganova

10.1
Characterization of Energetic Materials by Thermal Analysis

10.1.1
Introduction

The combined use of different thermoanalytical methods allows the extensive characterization of energetic materials and their fundamental properties. In addition to the determination of characteristic physical data such as temperature or enthalpy values, thermal analysis gives access to a detailed understanding of the decomposition and pyrolysis behavior of energetic materials. Especially in combination with appropiate spectroscopic and/or spectrometric evolved gas analysis (EGA) techniques, products, intermediates and residues of pyrolysis reactions can be identified and thus chemical pathways and mechanisms of thermal decompositions can be revealed. Moreover, thermal analysis allows also a qualitative and quantitative description of the decomposition exothermicity and kinetics and gives reliable information about thermal stability and compatibility.

In addition to the analysis of the decomposition behavior of energetic materials, thermal analysis is a very helpful tool to characterize their phase behavior. The latter is of significant importance regarding performance, stability and safety and also processing of energetic materials. Therefore, thermal analysis can be used to investigate the polymorphism of crystalline energetic materials, their melting, sublimation and evaporation behavior or – in the case of energetic polymers such as binders and plastizisers – to analyze their glass transition, crystallinity or other mechanical properties.

The application of thermoanalytical techniques provides a characteristic 'fingerprint' of energetic materials by acquiring a large number of physical and chemical data. Such data are used not only as reference values for the identification of specific energetic materials but also for purity analysis and to permit a qualitative and quantitative description of potential modifications of samples. Therefore, the influence of chemical, thermal or mechanical processing and the influence of storage and aging on samples can be easily controlled.

Energetic Materials. Edited by Ulrich Teipel
Copyright © 2005 WILEY-VCH Verlag GmbH & Co. KGaA, Weinheim
ISBN: 3-527-30240-9

In the following, the capability of thermoanalytical techniques for the characterization of energetic materials is described with respect the analysis of two relatively new energetic materials, ammonium dinitramide (ADN) and 2,4,6,8,10,12-hexanitrohexaazaisowurtzitane (HNIW or CL 20) (a detailed discussion of their structural properties is given in Sect. 10.2). The exemplary results given in the following two sections are based on a combined application of the following thermoanalytical methods:.

- (modulated) differential scanning calorimetry [(M)DSC].
- (high-resolution) thermogravimetric analysis [(HiRes) TGA].
- evolved gas analysis (EGA) – either by mass spectrometry (MS) adapted to TGA or by Fourier transform infrared (FTIR) spectroscopy based on a self-constructed heatable optical cell.
- thermomicroscopy (TM).

The thermal analysis of energetic materials requires particularly exact experimental procedures. Owing to the strong exothermic decomposition reactions, small sample sizes of < 1.0 mg, slow heating rates and open sample crucibles (e. g. DSC pans with pierced lids) are required. The sample itself has to be well characterized in advance with respect to its chemical and polymorphic purity, its particle size and its particle size distribution. To ensure statistical certainty of measured thermoanalytical data, a sufficient number (> 6) of repeated measurements is needed.

10.1.2
Thermal Analysis of Ammonium Dinitramide (ADN)

Ammonium dinitramide, $NH_4N(NO_2)_2$ is a crystalline substance with high oxygen and nitrogen contents considered to be of interest as a new explosive and in particular as a new oxidizer for solid-fuel rocket propellants [10.1–10.3]. Therefore, ADN could be a potential candidate to replace the widely used oxidizers ammonium perchlorate (AP) and ammonium nitrate (AN). In contrast to the combustion of AP-containing propellants, ADN formulations show a significantly reduced plume signature and atmospheric ozone destruction due to the absence of emitted hydrochloric acid. In comparison with AN, a higher energy content in combination with strong decomposition exothermicities under low pressures and the absence of solid-phase transitions are the most important benefits of ADN.

Figure 10.1 shows a characteristic DSC curve of ADN consisting of three main thermal effects. After melting at 90.37 ± 0.49 °C, ADN decomposes quantitatively at an extrapolated onset temperature of 120.65 ± 0.77 °C, releasing a heat of decomposition of 234.06 ± 5.55 kJ/mol in an open system. One of the decomposition products formed is ammonium nitrate. Its endothermic sublimation, overlaid by further decomposition steps, can be clearly observed when the exothermic decomposition of ADN is completed.

Moreover, DSC measurements also show that AN has a significant influence on the melting behavior of ADN. Figure 10.2 shows the DSC curve of an ADN sample

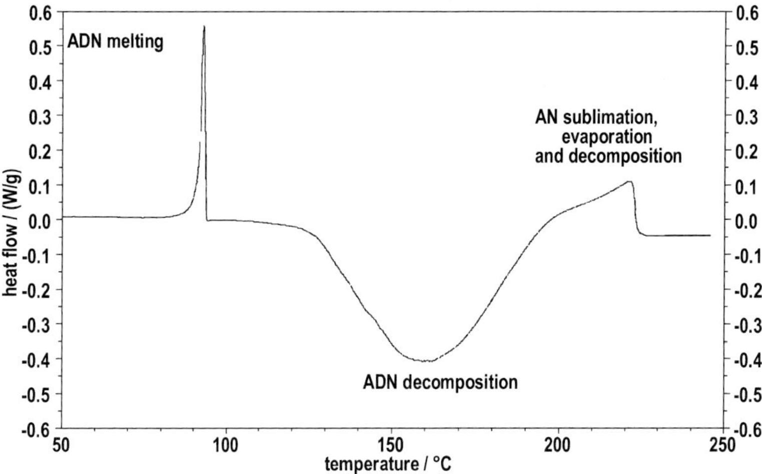

Figure 10.1. Characteristic DSC curve of ADN (heating rate, 0.5 K/min; sample weight, 2.15 mg; Ar atmosphere; Al pan with pierced lid). Exothermic processes are shown with negative heat flows.

contaminated with 3.8 wt% AN. In addition to lowering of the melting-point of ADN, an additional endothermic peak can be observed at lower temperatures. Further DSC experiments with a systematic variation of the AN content have shown that the additional endothermic effect is caused by an eutectic ADN/AN phase which occurs quantitatively at a mass ratio of 70 : 30 (ADN : AN). The AN content has an important impact on the application of ADN in energetic formulations. Melting-point depression and formation of a eutectic phase could, for example, result in an uncontrolled burning behavior of ADN propellants or could lead to segregation effects within formulations.

Moreover, a higher AN content reduces significantly the thermal stability of ADN samples, as could be shown by isothermal TGA experiments, for instance. Owing to these negative impacts, the AN content of ADN samples must be below 0.5 wt%. Since AN is not only a decomposition product but also a by-product of ADN synthesis, DSC measurements are very suitable to prove easily the purity of ADN samples, both qualitatively and quantitatively. DSC measurements can therefore also be used routinely to control the AN content of processed ADN samples, for example to check the quality of ADN prills that are formed in a process based on thermal treatment of the crude material (see Sect. 3.2.7).

Thermogravimetric data for ADN (Fig. 10.3) show a residue-free decomposition. By applying the HiRes TGA method, two decomposition steps can be clearly observed and quantitatively described. The two-step decomposition is also indicated by a weak low-temperature shoulder of the exothermic DSC peak.

With FTIR spectroscopic EGA, the gaseous decomposition products of ADN can be identified. Figure 10.4 shows the IR gas-phase spectra which were detected on-line during linear heating of ADN.

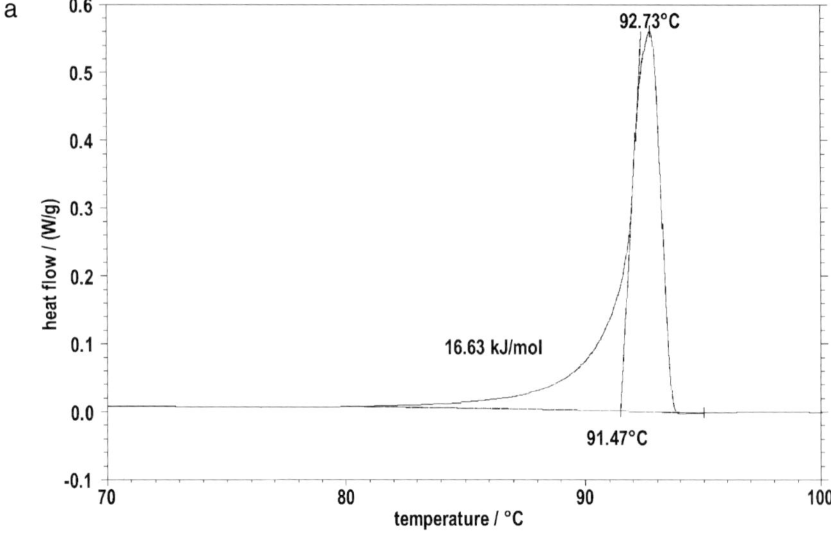

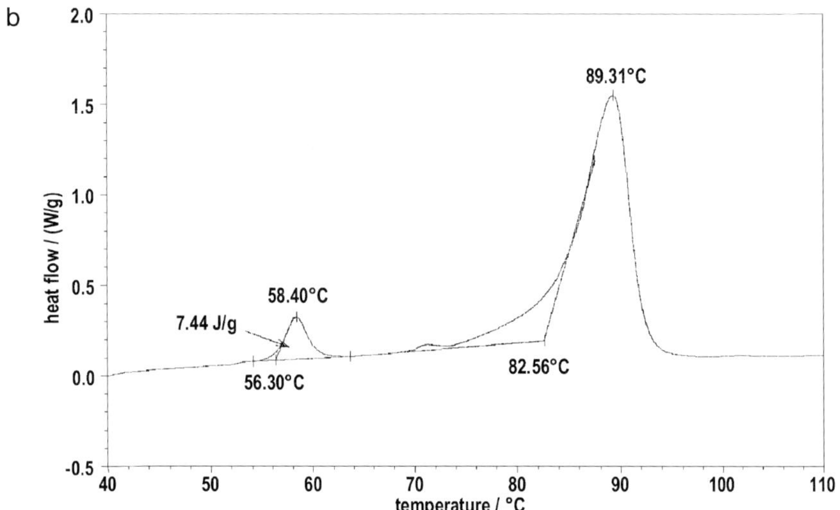

Figure 10.2. DSC curves of (a) pure and (b) AN-containing ADN samples comparing their melting behaviors (heating rate, 2.0 K/min; sample weight, 2.25 mg; AN content, 3.8 wt%; Ar atmosphere; Al pan with pierced lid). Endothermic processes are shown with positive heat flows.

As the main decomposition products N_2O, NH_4NO_3 and H_2O and also NO_2, NH_3 and NO were detected. The evaporation profiles of N_2O, NH_4NO_3 and NO_2 calculated from the EGA data (Fig.10.5) and the 30% mass loss which occurs during

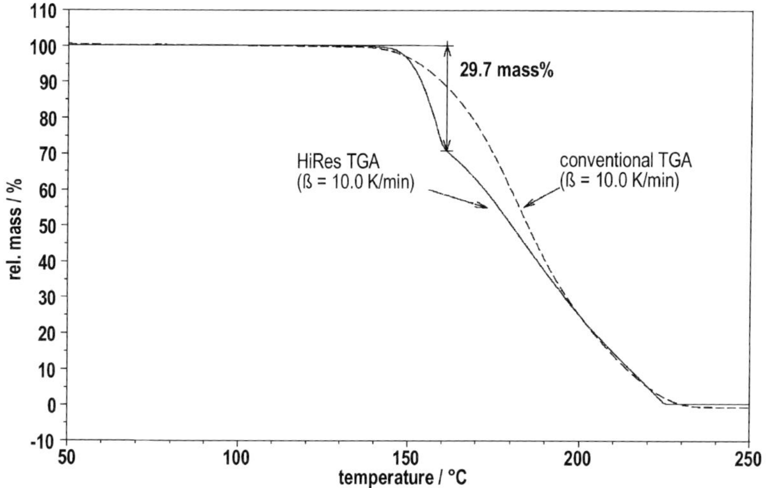

Figure 10.3. HiRes TGA measurement of ADN in comparison with conventional TGA showing a two-step decomposition (heating rate, 10.0 K/min; resolution, 2.0; sensitivity, 8.0; sample weight, 2.45 mg; Ar atmosphere; Pt pan).

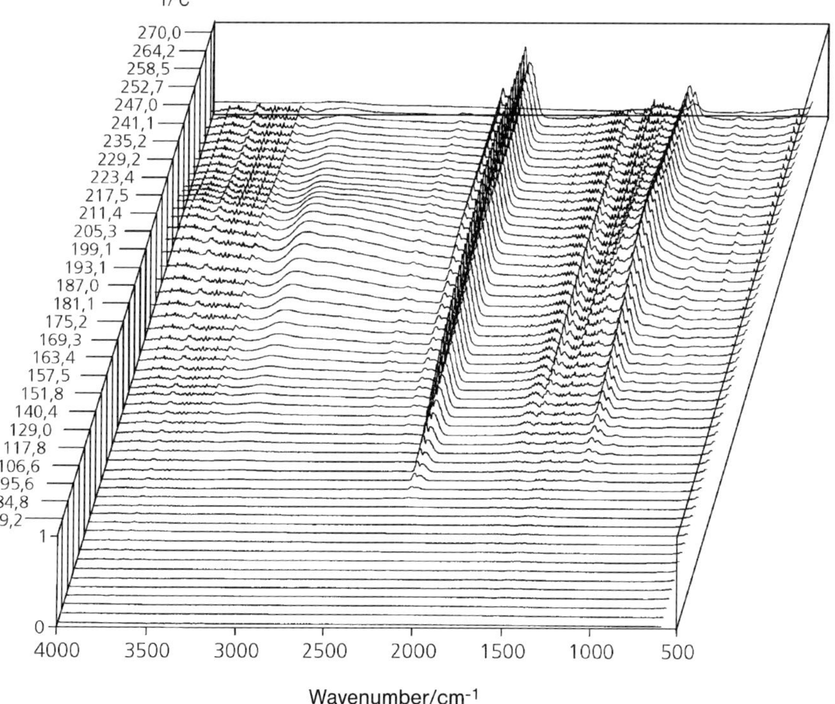

Figure 10.4. IR spectroscopic EGA measurement of ADN pyrolysis (heating rate, 5.0 K/min; Ar atmosphere; Al pan).

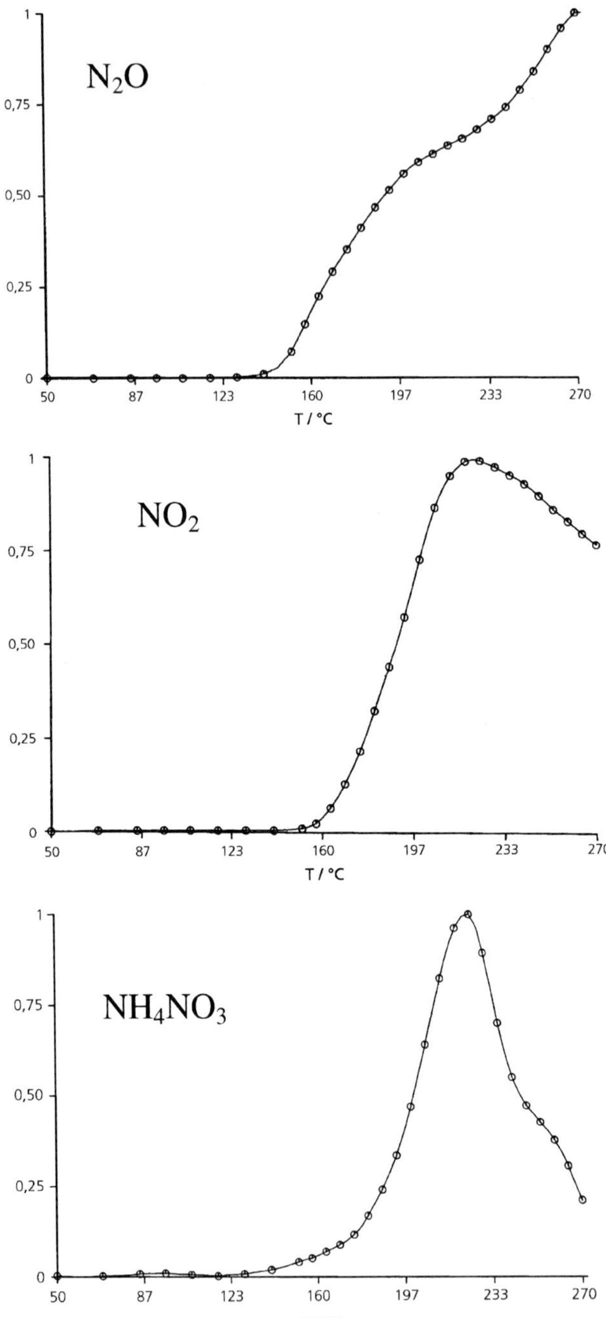

Figure 10.5. EGA profiles of the ADN decomposition gases N_2O (2240–2234 cm^{-1}), NO_2 (1635–1631 cm^{-1}) and NH_4NO_3 (3281–3176 cm^{-1}) (relative absorbances).

the first decomposition step of ADN (measured by HiRes TGA) allow one to describe the main decomposition pathways of ADN pyrolysis.

In a first decomposition step, ADN collapses by forming N_2O and NH_4NO_3 (theoretical mass loss: 35 %). NH_4NO_3 decomposes in a consecutive reaction step by forming H_2O and N_2O as indicated by the maximum in the EGA profile of NH_4NO_3 and the release of further amounts of N_2O in its EGA profile. NO_2 is released by the intermediates $HN(NO_2)_2$ and HNO_3 reacting consecutively with ammonia (a dissociation product of ADN) by forming N_2O, H_2O, NO and elemental nitrogen (proved by TGA/MS experiments). The EGA profile of NO_2 confirms its degradation due to consecutive gas-phase reactions. A detailed description of the decomposition mechanism of ADN is given elsewhere [10.4–10.6]. The thermoanalytical data discussed there show also that the decomposition mechanism is initiated by an acid-catalysed mechanism releasing N_2O as an initial decomposition product instead of a homolytic cleavage of the N-NO_2 nitramine bond as is conventionally observed for organic nitramines.

In addition to investigating the quantitative decomposition of ADN, its slow thermal decomposition was also considered by applying TM, isothermal TGA and isothermal EGA and also MDSC experiments [10.7]. All these experiments show in good accordance that already during melting of ADN a slow decomposition is initiated. Gas bubbles, for example, released from the ADN melt can be clearly observed by TM. Moreover, in MDSC experiments slow decomposition of molten ADN is detected by a constant reversible heat flow after endothermic melting in conjunction with a decreasing total heat flow (Figure 10.6). A continual mass loss and gas release measured by isothermal TGA and EGA experiments at 100 °C confirm the thermal instability of ADN as soon as it is molten. This restricted thermal stability has again an important impact on the practical handling and processing of ADN.

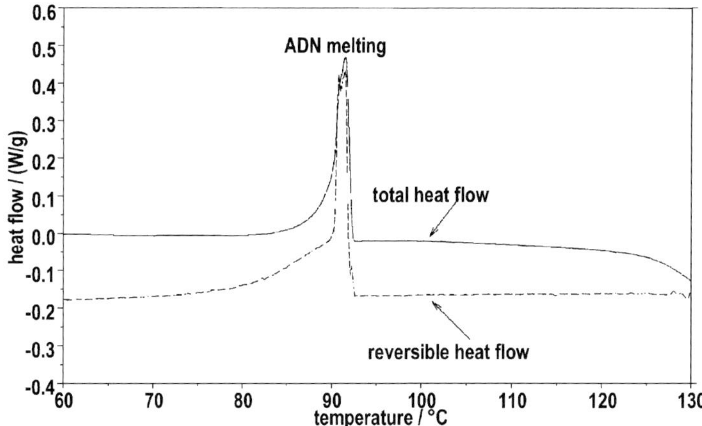

Figure 10.6. MDSC measurement of ADN indicating slow thermal decomposition (slope of total heat flow) already after melting of ADN (heating rate, 0.5 K/min; modulation: period, 40 s; amplitude, 0.053 K).

10.1.3 Thermal Analysis of Hexanitrohexaazaisowurtzitane (CL 20)

A further crystalline substance considered to be of interest as a new high-performance energetic material is the polycyclic nitramine 2,4,6,8,10,12-hexanitro-2,4,6,8,10,12-hexaazatetracyclo[5.5.0.$0^{5,9}$.$0^{3,11}$]dodecane, better known as hexanitrohexaazaisowurtzitane (HNIW) or CL 20 (see Chapter 1).

Owing to certain structural similarities with monocyclic energetic nitramines CL 20 is discussed as a substitutent for HMX and RDX. The cage structure of the CL 20 molecule ensures a high enthalpy of formation and one of the highest molecular densities known for organic substances, 2.04 g/cm^3 (ε-modification). Therefore, the use of ε-CL 20 in energetic formulations is expected to improve significantly performance properties, such as an increase in specific impulse and burning rate concerning propellants or higher detonation velocities and pressures concerning explosives [10.8–10.10].

Figure 10.7 shows a characteristic DSC curve of ε-CL 20 consisting of two main thermal effects. At an extrapolated onset temperature of 165.17 ± 0.87 °C a solid-phase transition from the most dense ε-modification to the γ-modification with the lowest molecular density can be observed. Further heating of the condensed γ-phase leads to a strong exothermic decomposition of CL 20 starting at an extrapolated onset temperature of 200.65 ± 0.14 °C. The released heat of decomposition in an open system is extremely high (1129.5 ± 19.5 kJ/mol) and is spread over a relatively narrow temperature range. These data explain very pragmatically why this polycyclic nitramine is considered as a promising candidate for energetic applications.

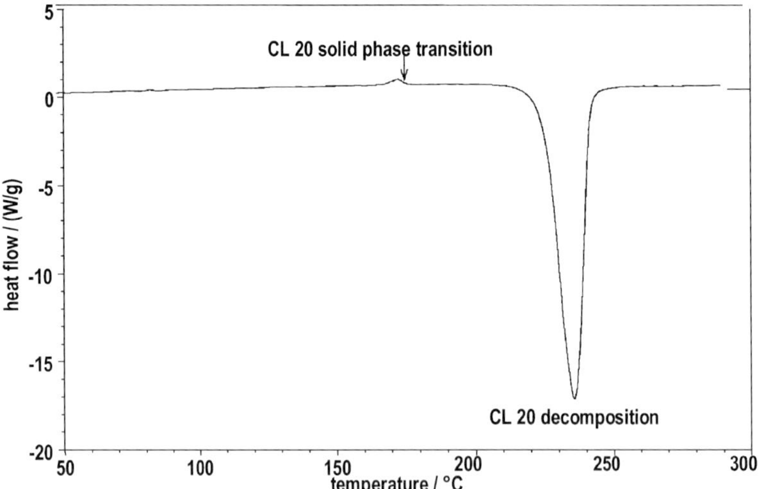

Figure 10.7. Characteristic DSC curve of CL 20 (heating rate, 5.0 K/min; sample weight, 0.82 mg; Ar atmosphere; Al pan with pierced lid). Exothermic processes are shown with negative heat flows.

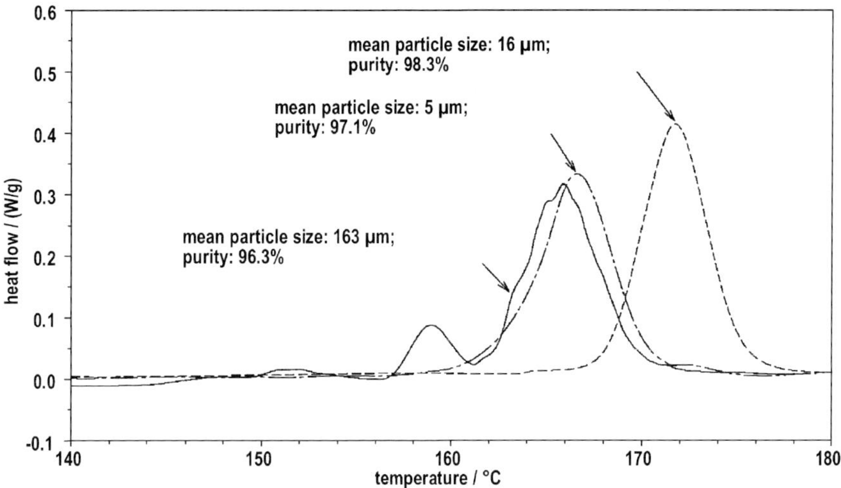

Figure 10.8. DSC measurements of the $\varepsilon \rightarrow \gamma$ solid-phase transition of CL 20 samples with different purities and mean particle sizes. Endothermic processes are shown with positive heat flows.

Moreover, DSC data show also that the $\varepsilon \rightarrow \gamma$ phase transition of CL 20 is not reversible and is shifted to lower temperatures with higher amounts of impurities (Fig. 10.8).

The 6% increase in volume during this phase transition is accompanied by an intensive crystal strain which leads to uncontrolled microscopic fragmentations of CL 20 crystals. Such 'micro-cracking' effects can be clearly observed under a thermomicroscope and also by artifacts in DSC, TGA and EGA experiments, especially for large CL 20 crystals. However, such artifacts are also very valuable information, indicating potential instabilities or increased hazardous potential of certain CL 20 samples.

The exothermic decomposition of CL 20 consists of two very close overlapping decomposition steps forming a residue of ~ 11–18 wt% at 300 °C with polyazine-like structures (Figs 10.9 and 10.10).

FTIR spectroscopic EGA measurements (Fig. 10.11) show that NO_2, N_2O, CO_2 and HCN and also traces of H_2O, CO and NO can be detected as the main thermal decomposition gases, as was also confirmed by TGA/MS experiments.

The calculated EGA profiles (Fig. 10.12) show clearly that, in contrast to ADN, the thermal decomposition of γ-CL 20 is initiated by the homolytic cleavage of the N-NO_2 nitramine bond releasing NO_2 as an initial gaseous decomposition product. The formed radical centers in the CL 20 backbone cause an immediate collapse of the polycyclic cage structure by releasing the thermodynamically stable products N_2O, HCN and CO_2.

By back-attack reactions of the NO_2 radicals, the degradation of the hydrocarbon backbone is autocatalytically accelerated. Further amounts of HCN, CO_2 and N_2

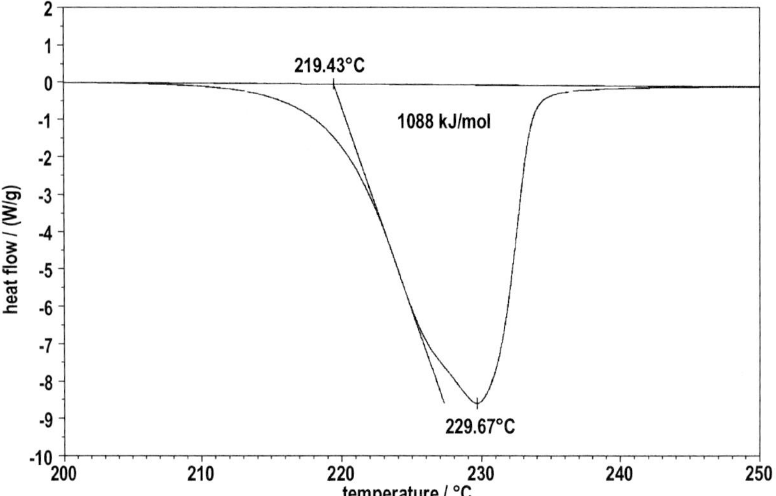

Figure 10.9. DSC curve of the exothermic decomposition of CL 20 (heating rate, 2.0 K/min; sample weight, 0.74 mg; Ar atmosphere; Al pan with pierced lid). Exothermic processes are shown with negative heat flows.

(proved by mass spectrometry) are released, as indicated by the maximum in the EGA profile of NO_2 and increasing concentration profiles of HCN and CO_2. Further amounts of N_2O could not be released because they are solely formed from the peripheric nitramine groups. Consequently, the EGA profile of N_2O shows a

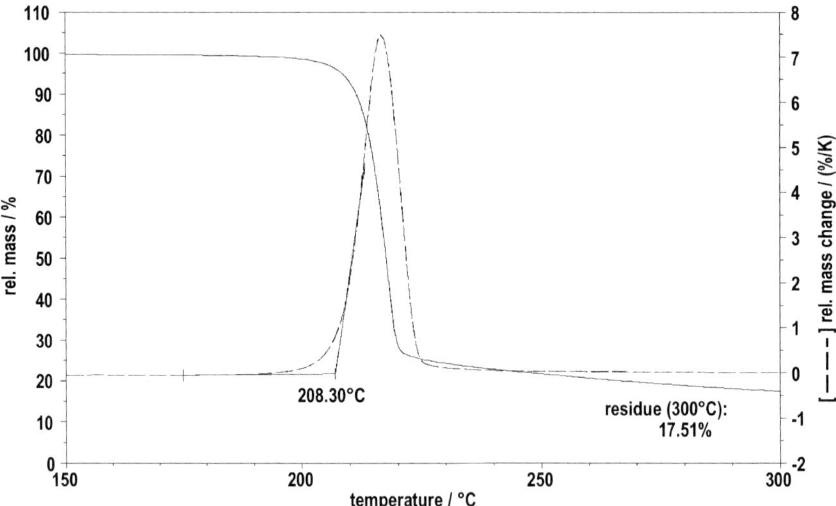

Figure 10.10. TGA measurement of CL 20 with corresponding DTG curve (heating rate, 1.0 K/min; sample weight, 0.98 mg; Ar atmosphere; Pt pan).

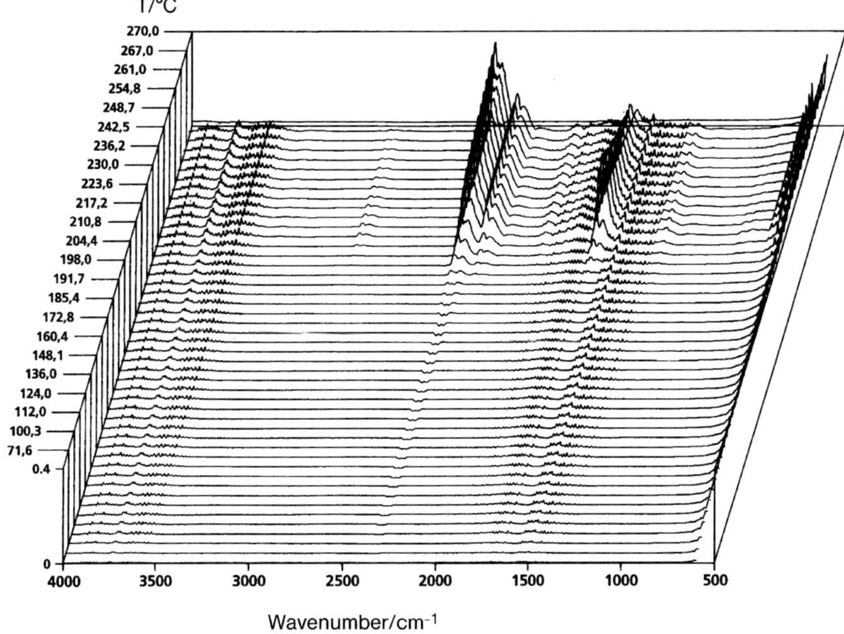

Figure 10.11. IR spectroscopic EGA measurement of CL 20 pyrolysis (heating rate, 5.0 K/min; Ar atmosphere; Al pan).

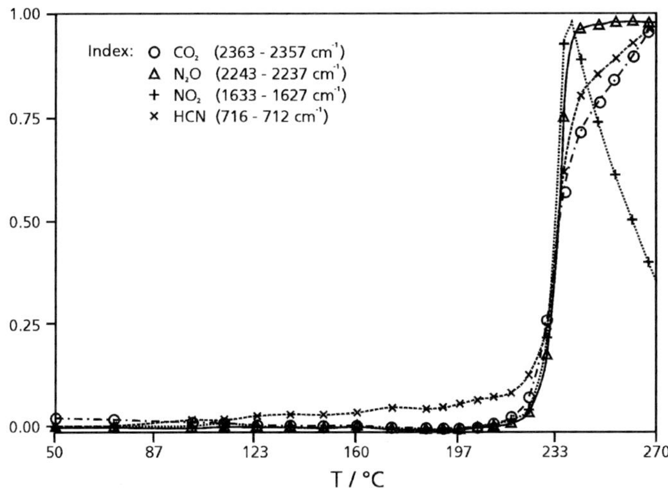

Figure 10.12. EGA profiles of the CL 20 decomposition gases NO_2, N_2O, CO_2 and HCN (relative absorbances).

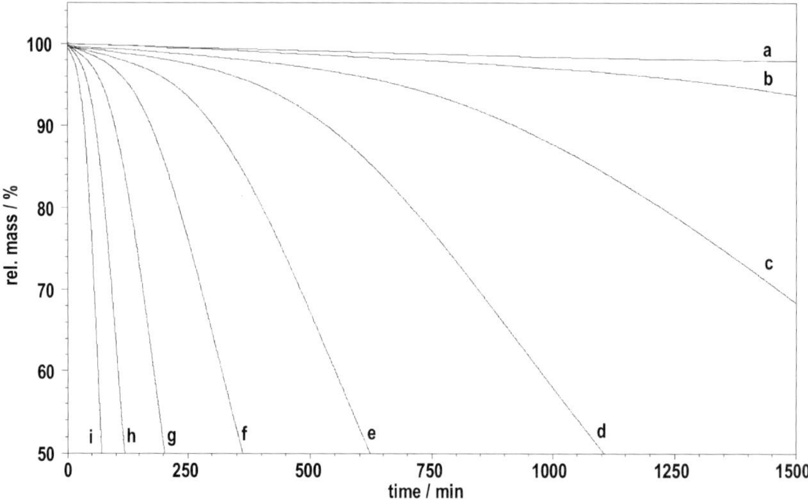

Figure 10.13. Isothermal mass loss of CL 20 at different temperatures (a, 150; b, 155; c, 160; d, 165; e, 170; f, 175; g, 180; h, 185; i, 190 °C).

maximum concentration which, in contrast to NO_2, does not decrease during further pyrolysis. A detailed description of the HNIW decomposition pathway and mechanisms is given elsewhere [10.6, 10.11].

The autocatalytic acceleration of the thermal decomposition of CL 20 is confirmed by isothermal TGA measurements (Fig. 10.13). They show also that besides a quantitative decomposition of CL 20 at 200.65 ± 0.14 °C, slow decomposition can be observed even at lower temperatures. More detailed TGA and MDSC measurements have revealed that the thermal stability of CL 20 is significantly weakened after it is transformed into the γ-phase.

On basis of the isothermal TGA data, kinetic parameters of the CL 20 decomposition were calculated by applying a 'first-order plus autocatalysis' model, giving an activation energy of 183.2 ± 0.5 kJ/mol and a pre-exponential factor of 17.76 ± 0.05 [10.11].

10.2 Characterization of Energetic Materials by NMR Spectroscopy

10.2.1 Introduction

Although the phenomenon of nuclear magnetic resonance was discovered only in 1946 by Purcell *et al.* [10.12] and Bloch *et al.* [10.13, 10.14], the method has developed very rapidly. A reason is the introduction of Fourier transform spectrometers and the application of high magnetic field strengths with superconducting magnets [10.15]. The major interest of analysts began with the discovery of the chemical shift, since different groups of a molecule could thus be distinguished [10.16]. A further tool for the structure determination is the magnetic interaction of kernels within a molecule. This interaction is called coupling of kernels and is observed as a splitting in the spectrum (coupling constant). NMR spectroscopy nowadays is a very powerful method for the analysis of organic substances. The method is not only very suitable for the examination of pure, solid and liquid substances, but is also used for the analysis of mixtures and the examination of the purity of substances.

Many NMR-active kernels are known, which can be examined and virtually each element has an NMR-active isotope. For the examination of explosives, particularly the NMR-active nuclei ^{1}H, ^{13}C, ^{14}N, ^{15}N and ^{17}O are of interest. The isotopes have different natural abundances and different physical properties (e.g. gyromagnetic ratio, quadrupole moment), and therefore they differ strongly in their sensitivity and signal width in the spectrum. The frequency ranges of the individual NMR-active kernels are so different that they do not overlap. The NMR method is an absolute method, i.e. the signal size is strictly proportional to the number of measured NMR nuclei, if certain additional conditions such as relaxation times and the nuclear Overhauser effect (NOE) are considered with the selection of the experiment.

Solid-state NMR spectroscopy is distinguished from high-resolution NMR spectroscopy. In solid-state NMR spectroscopy, the substance is supplied to a rotor of ceramics under a magic angle of 54° 44' relative to the outer magnetic field with a rotation frequency of 2000–5000 Hz and a ^{13}C spectrum is measured under high-power decoupling and cross-polarization/magic angle spinning (CP/MAS).

In high-resolution NMR spectroscopy, the substance is dissolved in a deuterated solvent for the measurement. The deuterated solvent is necessary in order to use the deuterium frequency for stabilization of all applied frequencies in the spectrometer. The resolution of these spectra is clearly larger than in solid-state NMR spectroscopy. In the following, examples of high-resolution NMR spectroscopy are presented.

As an example of the use of NMR spectroscopy, the new explosives ammonium dinitramide (ADN) and 2,4,6,8,10,12-hexanitrohexaazaisowurtzitane (CL 20) were examined in a solvent under high-resolution conditions [10.11, 10.17–10.19]. Structure determination using NMR spectroscopy is shown with the example of an isolated impurity in CL 20.

10.2.2
Theory of the NMR Method

The nucleus can be considered to be a small sphere in which a positive charge is distributed symmetrically or asymmetrically. If the charge is distributed asymmetrically and a rotation around the diameter is assumed, the charge is consequently rotating in a circular orbit around the rotation axis. A rotating charge induces an electrical current. We know from electricity that current produces a magnetic field, which follows the three-finger rule. The rotating nucleus produces a small magnetic field, which means the nucleus behaves like a small magnet.

A nucleus having an asymmetric distribution of charge, which is rotating around the diameter, has an intrinsic angular momentum. This intrinsic angular momentum is also called nuclear spin. The magnetism of a nucleus is quantitatively described by the magnetic momentum.

If a nucleus possesses a nuclear spin, the nucleus also has a magnetic momentum. The magnetic momentum and the nuclear spin are proportional. The proportionality constant is the gyromagnetic ratio. Not all nuclei possess a magnetic momentum, namely nuclei with an even number of protons and an even number of neutrons. Therefore, nuclei such as ^{13}C and ^{17}O are NMR active whereas ^{12}C and ^{16}O are not.

Because the nucleus with a magnetic spin behaves like a small magnet, it is oriented in a magnetic field. The number of possible orientations of a nucleus is different between different nuclei and depends on the magnetic quantum number. For simplification, we look at a nucleus with a magnetic quantum number $I = \frac{1}{2}$ such as ^{1}H, ^{19}F, ^{13}C and ^{15}N. In this case two adjustments relative to the magnetic field are possible. These are orientations in the direction of the magnetic field and opposite to it. In the former case the energy is lower than that in the latter. From this reason, orientation in the field direction is preferred. The transition between the two energy levels is the change in orientation of the nucleus. The energy which is necessary for reorientation is given by

$$\nu = \frac{1}{2 \cdot \pi} \cdot \gamma \cdot H \tag{10.1}$$

where ν = frequency, γ = gyromagnetic ratio and H = field strength of the magnetic field.

This reorientation means absorption of energy, which gives the NMR signal. The gyromagnetic ratio γ is a specific constant for each type of nucleus. The field strength of the magnetic field should be as high as possible, because the energy difference of the two orientations and sensitivity of the method are proportional to the field strength. Nowadays superconducting magnets are used. The resonance frequency for the ^{1}H nucleus lies between 100 and 800 MHz, depending on the magnet system used. Because of the different gyromagnetic ratio, the resonance frequency for each type of nucleus is different, i. e. if the resonance frequency for ^{1}H is 400.13 MHz, the resonance frequency for ^{13}C is 100.62 MHz. In a sample of an organic substance containing protons, a huge number of ^{1}H nuclei are present. The numbers of the nuclei in the two orientations can be calculated by the Boltzmann distribution.

A sample outside a magnetic field is not oriented and the direction of the magnetic spins is pointed in various directions. Placing the sample in a high magnetic field leads to an orientation of the spins of the nuclei. This orientation towards the two preferred directions is not spontaneous and takes some time. The time is called the spin-lattice relaxation time T_1. After orientation of the spins, a high-frequency pulse is applied to the sample, followed by recording of a spectrum. Nuclei with a magnetic quantum number $I > \frac{1}{2}$ have not only a magnetic moment but also a quadrupole moment, which causes a quick transfer of energy. This leads to broad lines in high-resolution spectra, which is an unwanted effect. Long relaxation times (e. g. ^{15}N nuclei) also give problems in observing signals, because saturation (equilibration between the two spin orientations) occurs. In this case, long waiting periods have to be inserted between each NMR stimulation.

We have seen that each type of nucleus (^{1}H, ^{13}C, ^{15}N) has a characteristic resonance frequency, because of different values of the gyromagnetic ratio. From Eq. (10.1) we should expect one sharp line for all protons. In practice, however, we see a spectrum of different resonances for, let us say, protons in an organic molecule. The reasons for this difference are first the electrons which surround a nucleus in a molecule and second the neighborhood of this nucleus. The electrons which surround a nucleus are rotating around it. A rotating charge induces a current and a small magnetic field, which is oriented opposite to the outer magnetic field. Hence the effective magnetic field at the position of the nucleus is weakened. The effect increases with the number of electrons near the nucleus. This is called the shielding effect of the electrons and leads to an NMR signal which is different for each group of equivalent nuclei in a molecule. The position of the NMR signal relative to a standard substance in the spectrum is called the chemical shift. Different types of protons (aliphatic groups, olefinic protons and aromatic protons) in an NMR spectrum have typical chemical shifts. Most of the chemical shifts in a proton spectrum lie in a range of 0–10 ppm.

In high-resolution proton spectra, signals of different groups were seen which are split into different lines. This splitting is not caused by different chemical shifts of proton groups, but from indirect nuclear coupling. Different position of lines occur because a proton is influenced by the orientation of the neighboring protons. The effect is transferred through the electronic bonds when measurement is made in a solvent, because direct coupling of the protons through space is averaged by molecular motion. More detailed information on NMR theory can be found elsewhere [10.20–10.26].

10.2.3
Instruments and Methods

The investigations were accomplished with a Bruker DMX 400 NMR spectrometer with a proton resonance frequency of 400.13 MHz. The resonance frequency for the ^{13}C nucleus lies at 100.62, that for ^{14}N at 28.91 and that for ^{15}N at 40.56 MHz. The ^{17}O NMR spectra were measured at a resonance frequency of 54.24 MHz.

For the measurement, the substances were dissolved in D_2O (ADN) or acetone-d_6 (CL 20). A 5- and a 10-mm multinuclear high-resolution probe were used. The number of the data points totalled characteristically 32K and/or 64K. All measurements were executed at ambient temperature. The spectra were calibrated taking ^{1}H and ^{13}C spectra in relation to tetramethylsilane (TMS), where the proton signal of the methyl group in acetone was set to 2.04 and the ^{13}C signal was set to 29.8. The chemical shifts in the ^{14}N and ^{15}N spectra refer relative to external nitromethane, which was measured before the exposure of the substances and was calibrated to the value of $\delta = 0$. Before the measurement of ^{17}O spectra, deionized water was measured and its chemical shift was fixed at $\delta = 0$. The CL 20 samples were technical products of Thiocol (USA) and SNPE (France). ADN was synthesized by ICT (Germany).

Different one-dimensional spectra such as ^{13}C with broad band decoupling or ^{13}C gated spectra [10.15, 10.27–10.31] were recorded. One-dimensional proton NOE experiments were also measured. As 2D NMR techniques, ^1H–^1H COSY [10.24, 10.32–10.34], ^1H–^{13}C HSQC [10.35–10.39], ^1H–^{13}C HMBC [10.37, 10.40], ^1H–^{13}C RELAY [10.41, 10.42] and ^1H–^1H NOESY [10.43–10.46] were used. Usually 256 spectra with 4K data points were recorded for 2D NMR experiments. The heteronuclear 2D correlations were recorded with use of pulse programs with z-gradient selection to suppress t_1 noise. If not specifically indicated, all measurements were executed at 298 K.

The analytical separation of the impurities was accomplished by means of HPLC employing a Gynkotek instrument with an RP-18 column (250 × 4 mm i.d.) and a 20-mm pre-column of LiChrospher Si 100 (Merck). The eluent was methanol-water (50:50). The preparative separation was carried out with the help of a medium-pressure column from Merck, which was filled with silica gel Si 60. CL 20 technical products from Thiocol (USA) and SNPE (France) were used.

10.2.4 Characterization of ADN and CL 20 by NMR Spectroscopy

10.2.4.1 ^1H NMR Spectroscopy

CL 20 shows two signals at $\delta = 8.34$ and 8.20 in the ^1H NMR spectrum. The assignments can be accomplished through integration of the signals. The first signal corresponds to four protons and the second to two protons [Fig. 10.14(a)]. The determination of the purity of the substance is also shown in Fig. 10.14(b) and (c), which show the spectra of products from two different manufacturers at higher amplification of the spectrum. Additional impurities of the products are visible.

The measurement of the ^1H NMR spectrum of ADN was omitted, since an exchange between the NH_4^+ signal and the signal of the HDO appears in the spectrum. Thus a mean value of the signal appears in the spectrum, whose chemical shift depends on the ADN concentration and temperature.

10.2.4.2 ^{13}C NMR Spectroscopy

In contrast to ^1H NMR spectroscopy, ^{13}C NMR spectroscopy is less sensitive because of the natural abundance of 1.1% of the ^{13}C nucleus.

The ^{13}C NMR spectrum of CL 20 shows two signals at $\delta = 72.1$ and 75.1 [Fig. 10.15(a)]. An assignment can be effected by quantitative inverse gated NMR spectroscopy [10.15, 10.20–10.24]. In this case one signal corresponds to two and the other to four carbons.

The determination of the purity of the substance is also shown in the ^{13}C NMR spectrum [Fig. 10.15(b) and (c)]. The determination of the purity of the substance is also shown in Fig. 10.15(b) and (c), which show the spectra of products from two

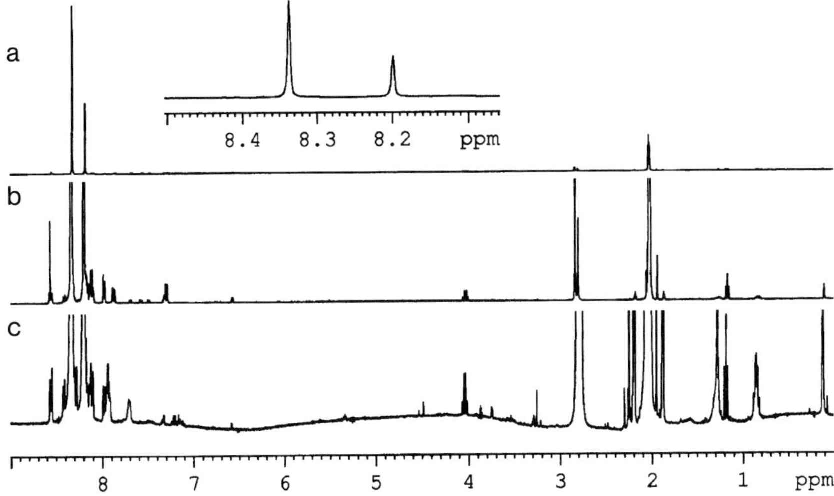

Figure 10.14. (a) ^{1}H NMR spectrum of CL 20 in acetone-d_6; (b) amplified spectrum of sample from manufacturer A with additional impurities; (c) amplified spectrum of sample from manufacturer B showing different impurities to that from manufacturer A.

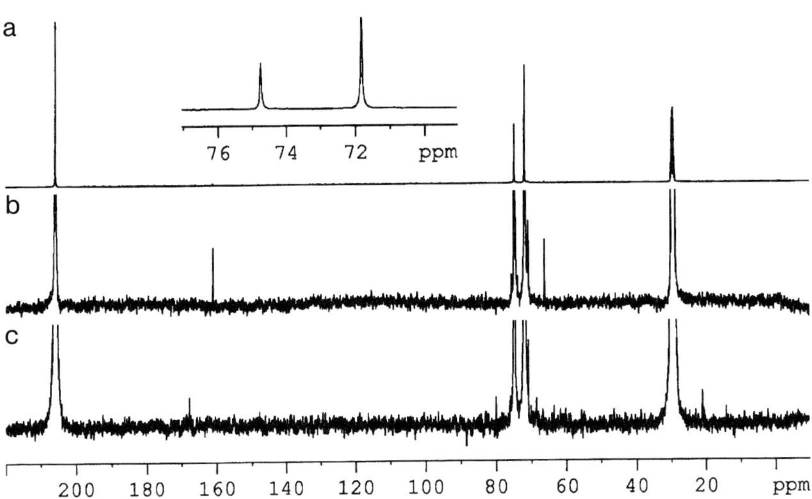

Figure 10.15. (a) ^{13}C NMR spectrum of CL 20 in acetone-d_6 with broad band decoupling of protons; (b) amplified spectrum of sample from manufacturer A with additional impurities; (c) amplified spectrum of sample from manufacturer B showing different impurities to that from manufacturer A.

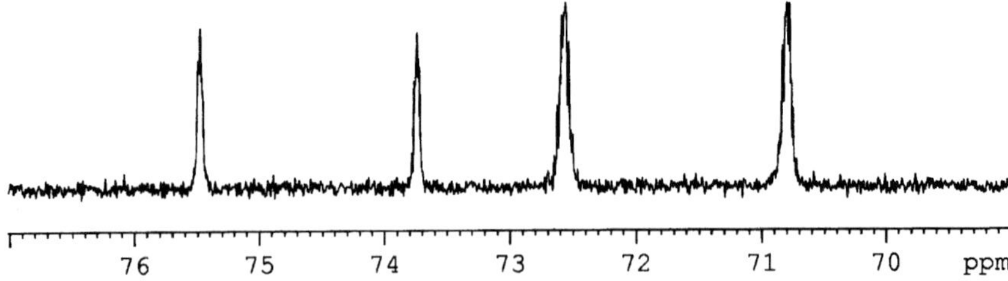

Figure 10.16. ^{13}C gated NMR spectrum of CL 20 in acetone-d_6.

different manufacturers at higher amplification of the spectrum. Additional impurities of the products are visible.

The $^{1}J_{CH}$ coupling constants were determined by means of a gated NMR spectrum [10.15, 10.20–10.23] (Figure 10.16). The ^{13}C NMR signal at $\delta = 75.1$ shows a coupling constant of $^{1}J_{CH} = 175.9$ Hz and at $\delta = 72.1$ a $^{1}J_{CH} = 176.6$ Hz. The two coupling constants hardly differ, since the chemical environments of the carbon atoms are very similar.

In the case of ADN, no ^{13}C spectrum was acquired, since no carbon is available in the substance.

10.2.4.3 ^{14}N NMR Spectroscopy

Although the ^{14}N nucleus has a large natural abundance of 99.63%, the measurement of ^{14}N nuclei is not simple, because the nuclear spin is $I = 1$, hence the nucleus shows a quadrupole moment [10.47, 10.48]. This leads to broadening of the signals and a degradation of the signal-to-noise ratio.

The ^{14}N NMR spectrum of ADN (Fig. 10.17) shows three signals at $\delta = -12.0$, -60.2 and -360.1. The signal at $\delta = -12.0$ can be associated with the nitro group in the dinitramide. At $\delta = -60.2$, the central nitrogen atom of the dinitramide appears and at $\delta = -360.1$ the nitrogen atom of the ammonium ion is visible. A small signal at $\delta = -3.9$ is a nitrate impurity of the ADN. The linewidths of the signals of ADN are very different; the signal of the ammonium ion shows the smallest linewidth,

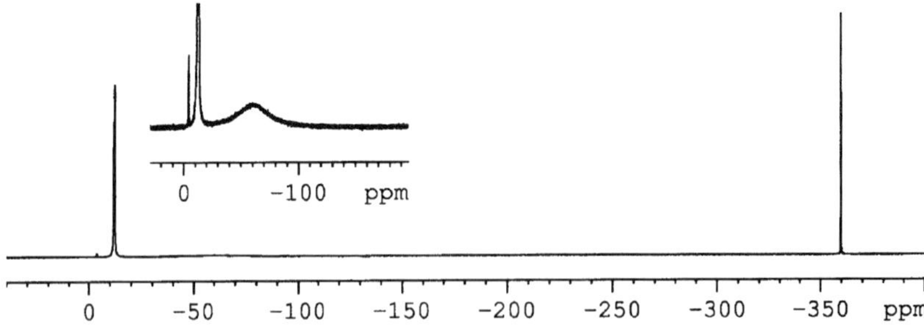

Figure 10.17. ^{14}N NMR spectrum of ADN in D_2O.

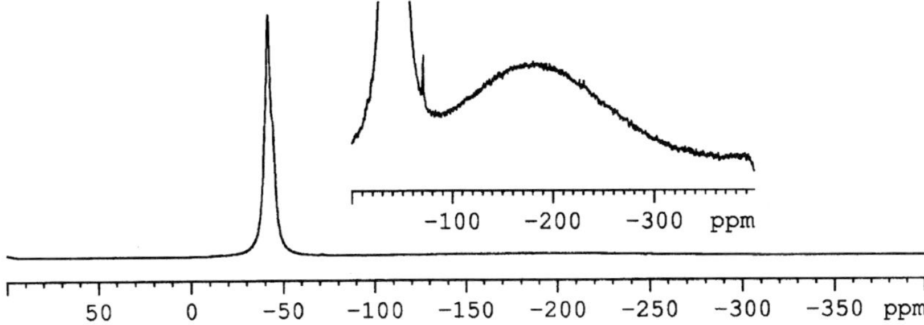

Figure 10.18. ^{14}N NMR spectrum of CL 20 in acetone-d_6.

followed by the nitro groups and the central nitrogen atom of the dinitramide. The linewidths correlate with the quadrupole moment of the ^{14}N nucleus. Sharp lines will only be obtained in a symmetrical environment of the ^{14}N nucleus and with very fast rotating groups around the nitrogen nucleus.

The ^{14}N NMR spectrum of CL 20 (Fig. 10.18) shows two signals at $\delta = -41.6$ and -180.6. The signal at $\delta = -41.6$ is assigned to nitro groups and thatat $\delta = -180.6$ to the nitramine nitrogens. Both lines are fairly broad; the signal at $\delta = -180.6$ is so broad that it almost disappears into the baseline. So again the quadrupole moment of the ^{14}N is responsible for the broad lines and for the missing splitting of the signals.

10.2.4.4 ^{15}N NMR Spectroscopy

In contrast to ^{14}N NMR spectroscopy, ^{15}N NMR spectroscopy has the advantage of having very sharp lines in the spectrum [10.49–10.51]. The reason lies in the missing quadrupole moment, since the nuclear spin quantum number is $I = \frac{1}{2}$. However, the measurement of ^{15}N spectra is very time consuming, because the natural abundance of this isotope is only 0.36%. Hence a measuring time of 1 day with a sample amount of 1 g was needed for the spectrum shown in Fig. 10.19.

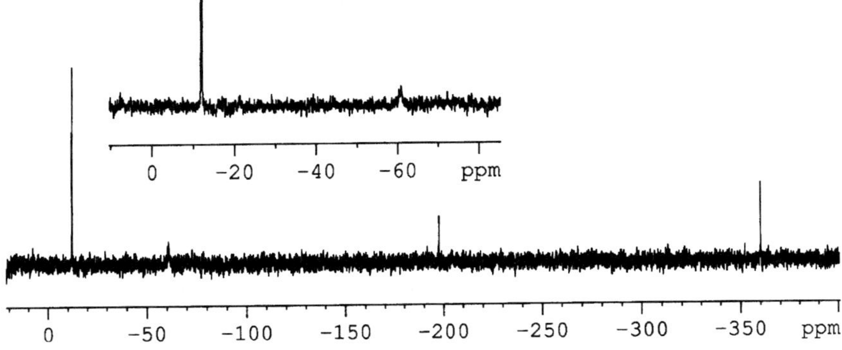

Figure 10.19. ^{15}N NMR spectrum of ADN in D_2O.

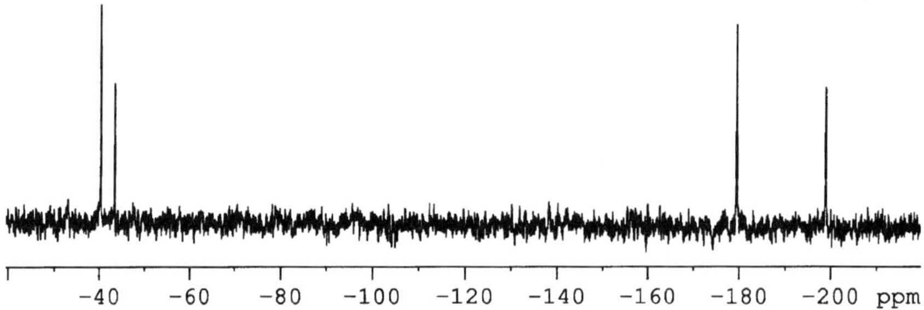

Figure 10.20. ^{15}N NMR spectrum of CL 20 in acetone-d_6.

Additional reasons are the long relaxation times and the negative NOE because of the negative gyromagnetic ratio.

ADN shows three signals at $\delta = -12.2$, -60.8 and -360.1 in the ^{15}N spectrum (Fig. 10.19). The signal at $\delta = -12.2$ corresponds to the nitro group and that at $\delta = -360.1$ to the ammonium ion. The signal of the central nitrogen atom of the dinitramide at $\delta = -60.8$ is broadened in the spectrum because of chemical exchange.

The ^{15}N spectrum of the CL 20 is much more resolved than the ^{14}N spectrum. Four lines can be seen at $\delta = -40.3$, -43.4, -179.5 and -199.0. The two signals at $\delta = -40.3$ and -43.4 are assigned to nitro groups. The signals at $\delta = -179.5$ and -199.0 correspond to the nitramine nitrogen atoms. An assignment according to the intensity of the signals is presented in Fig. 10.19.

10.2.4.5 ^{17}O NMR Spectroscopy

The NMR-active ^{17}O nucleus has some unfavorable characteristics in NMR spectroscopy [10.52], e.g. its low natural abundance of 0.037% and its nuclear spin quantum number of $I = 5/2$. This leads to a quadrupole moment, which has a negative effect on the linewidths.

The oxygen signal of the nitro groups in ADN appears at $\delta = 469.6$ (Fig. 10.21). In the spectrum a water signal is visible at $\delta = -3.0$. Calibration was performed by

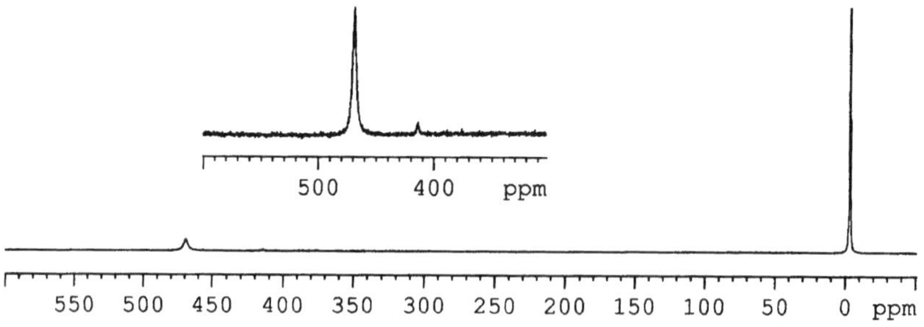

Figure 10.21. ^{17}O NMR spectrum of ADN in D_2O.

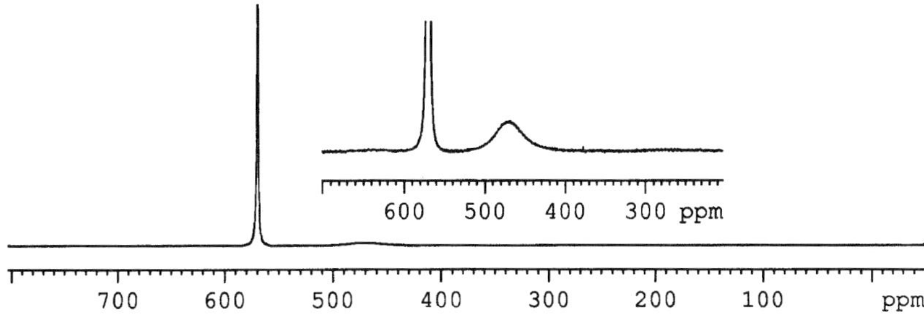

Figure 10.22. ^{17}O NMR spectrum of CL 20 in acetone-d_6.

measuring external H_2O. Since the sample is dissolved in water the chemical shift of the water differs from the external calibration. The small signal at $\delta = 414.2$ is a nitrate impurity of ammonium dinitramide. The ^{17}O signal of the nitro groups of CL 20 is visible at $\delta = 468.7$ (Fig. 10.22). The large signal in the spectrum is caused by the carbonyl group of acetone, which was used as solvent.

All the NMR signals of ADN and CL 20 are presented in Table 10.1.

Table 10.1. All NMR signals of ADN and CL 20.

Nucleus	MHz	ADN	CL 20
^{1}H	400.13	–	8.20 (2H) (H-1, H-7) 8.34 (4H) (H-3, H-5, H-9, H-11)
^{13}C	100.62	–	72.1 ($^{1}J_{CH}$ = 175.9 Hz) (C-1, C-7) 75.1 ($^{1}J_{CH}$ = 176.6 Hz) (C-3, C-5, C-9, C-11)
^{14}N	28.91	−12.0 ($N\underline{N}O_2^-$) $\Delta 1/2$ = 12.4 Hz −60.2 ($\underline{N}NO_2^-$) $\Delta 1/2$ = 940 Hz −360.1 ($\underline{N}H_4^+$) $\Delta 1/2$ = 4.5 Hz	−41.6 ($N\underline{N}O_2$) $\Delta 1/2$ = 140 Hz −180.6 ($\underline{N}NO_2$) $\Delta 1/2$ = 4.3 kHz
^{15}N	40.56	−12.2 ($N\underline{N}O_2^-$) −60.8 ($\underline{N}NO_2^-$) −360.1 ($\underline{N}H_4^+$)	−40.3 ($N\underline{N}O_2$) −43.4 ($N\underline{N}O_2$) −179.5 ($\underline{N}NO_2$) −199.0 ($\underline{N}NO_2$)
^{17}O	54.24	469.6 ($NN\underline{O_2}^-$), $\Delta 1/2$ = 204 Hz	468.7 ($NN\underline{O_2}$), $\Delta 1/2$ = 2.1 kHz

10.2.5 Structure Determination of 4-FPNIW by NMR Spectroscopy

For the characterization of the quality of an explosive, it is necessary to determine the purity accurately. This can be done by chromatographic and/or spectroscopic methods. In an earlier investigation, impurities could be examined in HNIW (CL 20) by means of HPLC. At that time the structures of the main impurities were not known, because the substances were present as small impurities in the main product [10.53]. NMR investigations were also carried out on HNIW [10.54]. Two

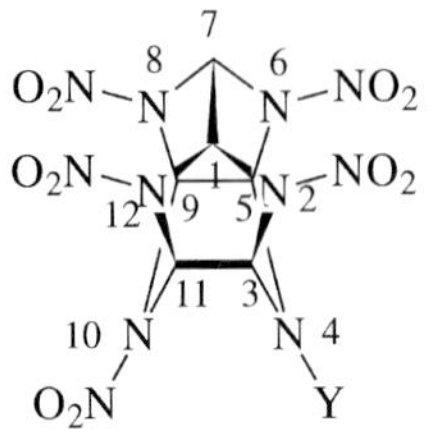

Figure 10.23. Structure and numbering of 4-FPNIW (Y = CHO).

publications have appeared that deal with impurities containing an acetyl group [10.55, 10.56].

In the present investigation, the main impurities were isolated and investigated by NMR spectroscopy. NMR spectroscopy is very useful because it can give exact statements towards the structure and can also explore mixtures of substances. In the following, the isolated impurity was explored with the help of different 1D and 2D NMR techniques and with the aid of different NMR nuclei.

The structural formula of the compound studied, 4-formyl-2,6,8,10,12-pentanitro-2,4,6,8,10,12-hexaazaisowurzitane (4-FPNIW) is shown in Fig. 10.23. The position of the substituent was determined by means of NMR spectroscopy and is explained later.

10.2.5.1 ^{1}H NMR Spectroscopic Investigations

Part of the proton spectrum of -FPNIW is shown in Fig. 10.24(a). One can recognize six protons of the isowurzitane frame. The sharp signal at $\delta = 8.57$ of the formyl proton can be seen in Fig. 10.26.

The problem exists in the assignment of the individual protons in the rings. The assignment of the protons is done by the determination of the coupling constants [Fig. 10.24(b)]. It is seen that the protons H-1 and H-7 display a coupling constant of 6.32 Hz. The two other couplings over three bonds have distinctively larger coupling constants and were determined as 7.86 and 8.1 Hz. The protons H-3/H-5 and H-9/H-11 show a W coupling over four bonds of the order of 2.3–2.7 Hz.

For the determination of the adjacent protons H-1/H-7, H-3/H-11 and H-5/H-9 a 2D ^{1}H-^{1}H COSY45 spectrum (Fig. 10.25) and a 2D ^{1}H-^{1}H NOESY spectrum (Fig. 10.26) are very helpful. After one has determined the three pairs with the ^{1}H-^{1}H COSY45 and the NOESY spectra, an assignment of the position of the pairs must be done. This is accomplished with the ^{1}H-^{1}H NOESY spectrum. The formyl proton can only be adjacent to the protons H-3 and H-5. This was confirmed once more by a 1D NOE experiment (Fig. 10.27), whereby H-3 has an increase in intensity of 16.2% and H-5 an increase of 6.0% if the formyl proton is irradiated.

The appearance of six protons in this molecule with substitution in the 4-position is unusual, because one would expect only three signals on the basis of symmetry conclusions. This applies likewise in the ^{13}C spectrum. Owing to the ^{1}H coupling constants, the given assignment must result. The molecule is twisted or hindrance

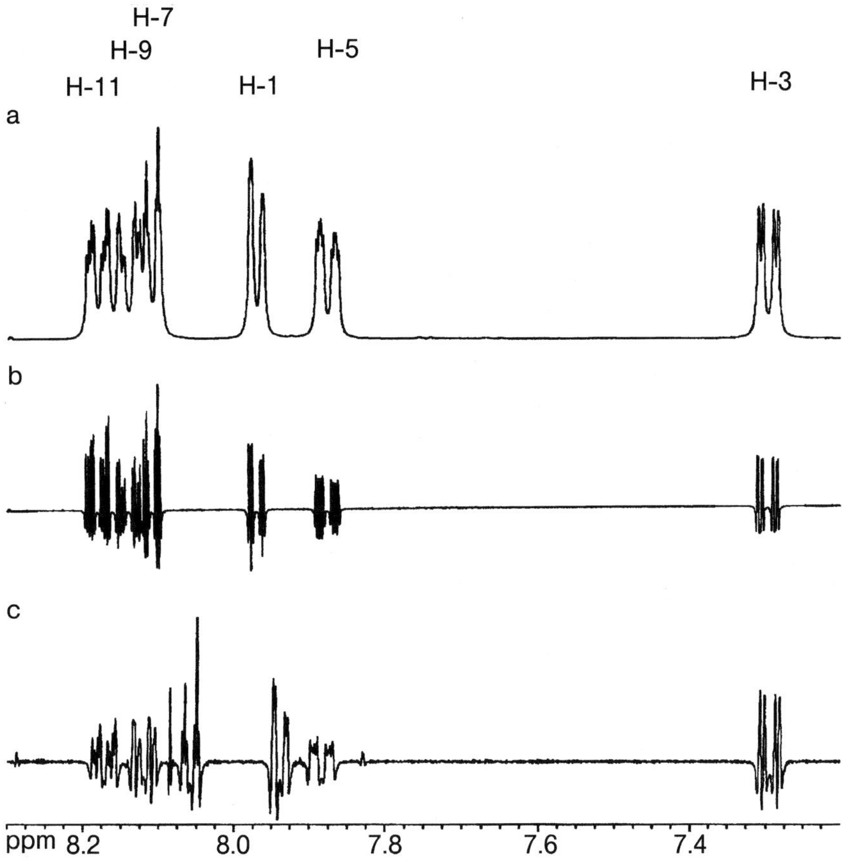

Figure 10.24. ^{1}H NMR spectrum of (a) 4-FPNIW in acetone-d_6 at 295 K; (b) resolution enhanced with square sine multiplication of the FID at 298 K; (c) resolution enhanced with square sine multiplication of the FID at 318 K.

of the rotation of the formyl group exists, so that no mirror plane is available in the molecule.

The spectrum of the substance was taken at 318 K. Thereby a change in the splitting of the protons takes place. Higher temperatures could not be set owing to the boiling-point of the solvent. The change of the spectrum suggests that a coalescence appears in the spectrum.

Assignments of the ^{1}H NMR signals for 4-FPNIW are given in Tables 10.2 and 10.3.

Figure 10.28 shows the ^{13}C NMR spectrum of 4-FPNIW. The above-mentioned symmetry conclusions also apply to the ^{13}C NMR spectrum. The assignment was done by 2D HSQC, HMBC and RELAY experiments (Figs 10.29–10.31). The ^{13}C-^{1}H

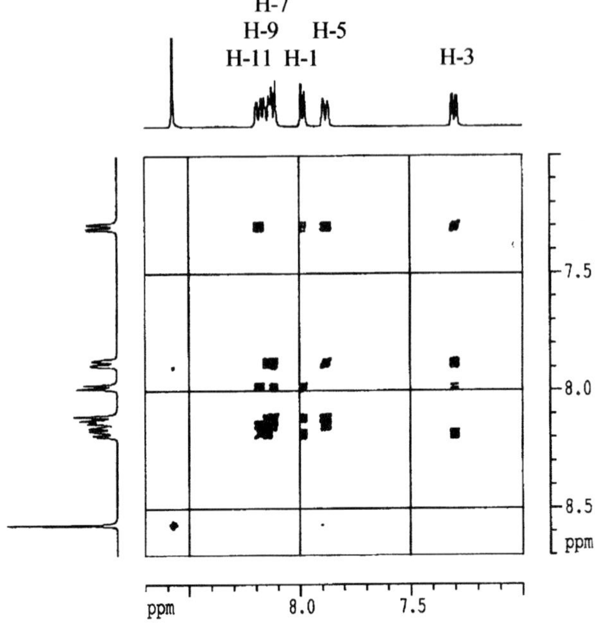

Figure 10.25. ^1H–^1H correlation of 4-FPNIW (COSY45) in acetone-d_6.

Table 10.2. Assignments of the ^1H NMR signals for 4-FPNIW at 295 K.

Proton No.	Proton chemical shift δ at 295 K	Proton-proton coupling constant $^3J_{CH}$ (Hz) (± 0.03 Hz)	Carbon-proton coupling constant $^4J_{CH}$ (Hz) (± 0.03 Hz)
1	7.98	6.32 (H-1/H-7)	0.58 (H-1/H-3); 1.35 (H-1/H-11)
3	7.30	7.86 (H-3/H-11)	0.58 (H-1/H-3); 2.36 (H-3/H-5)
5	7.88	8.11 (H-5/H-9)	2.36 (H-3/H-5); 1.27 (H-5/H-7)
7	8.10	6.32 (H-7/H-1)	1.27 (H-5/H-7); 1.09 (H-7/H-9)
9	8.14	8.05 (H-9/H-5)	1.07 (H-7/H-9); 2.75 (H-9/H-11)
11	8.18	7.89 (H-11/H-3)	2.72 (H-9/H-11); 1.32 (H-1/H-11)
13	8.57	–	–

Table 10.3. Assignments of the ^1H NMR signals for 4-FPNIW at 318 K.

Proton No.	Proton chemical shift δ at 318 K	Proton-proton coupling constant $^3J_{CH}$ (Hz) (± 0.03 Hz)	Carbon-proton coupling constant $^4J_{CH}$ (Hz) (± 0.03 H)
1	7.94	6.33 (H-1/H-7)	– (H-1/H-3); 1.27 (H-1/H-11)
3	7.29	7.93 (H-3/H-11)	– (H-1/H-3); 2.44 (H-3/H-5)
5	7.88	8.11 (H-5/H-9)	2.44 (H-3/H-5); 1.37 (H-5/H-7)
7	8.07	6.32 (H-7/H-1)	1.37 (H-5/H-7); 1.09 (H-7/H-9)
9	8.12	8.06 (H-9/H-5)	0.95 (H-7/H-9); 2.69 (H-9/H-11)
11	8.17	7.89 (H-11/H-3)	2.76 (H-9/H-11); 1.27 (H-1/H-11)
13	8.56	–	–

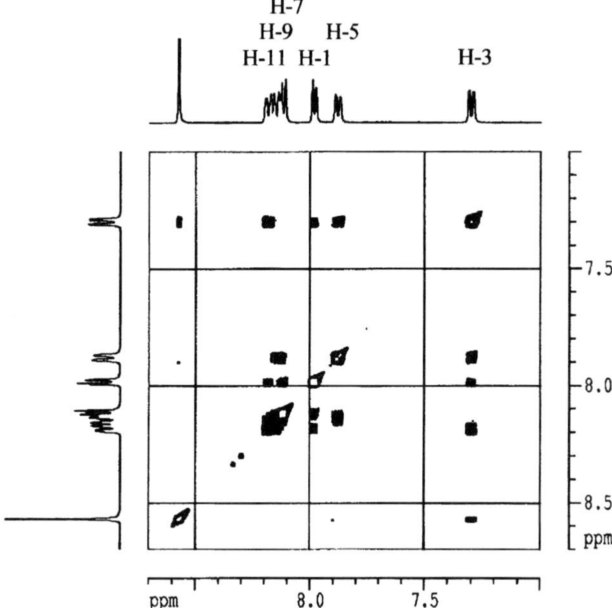

Figure 10.26. ^1H–^1H correlation of 4-FPNIW with NOE (NOESY) showing protons in the vicinity. The vicinity of H-3 and the formyl proton is clearely seen.

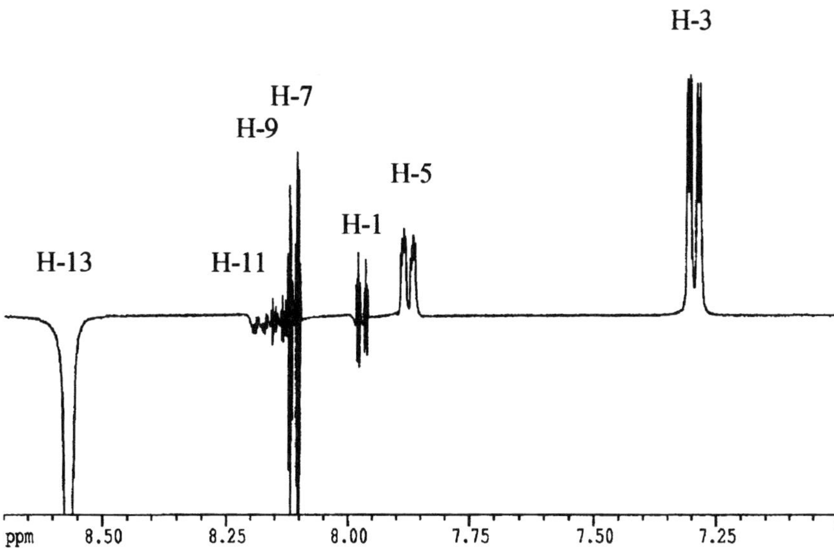

Figure 10.27. ^1H NMR NOE difference spectrum of 4-FPNIW with irradiation of the formyl proton at 298 K.

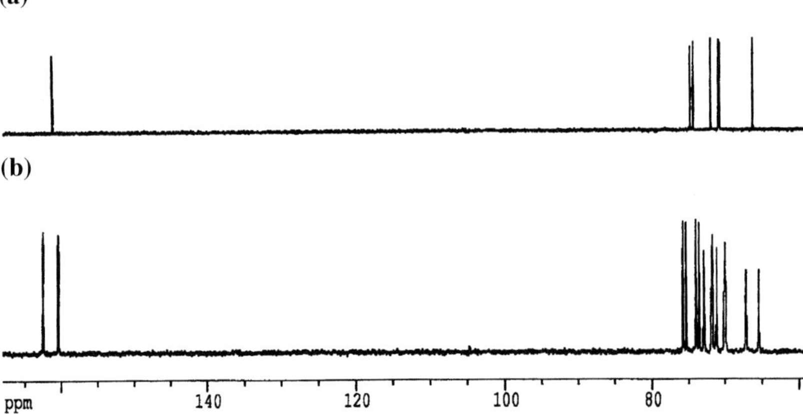

Figure 10.28. (a) ^{13}C NMR spectrum of 4-FPNIW in acetone-d_6; (b) ^{13}C gated NMR spectrum of 4-FPNIW in acetone-d_6.

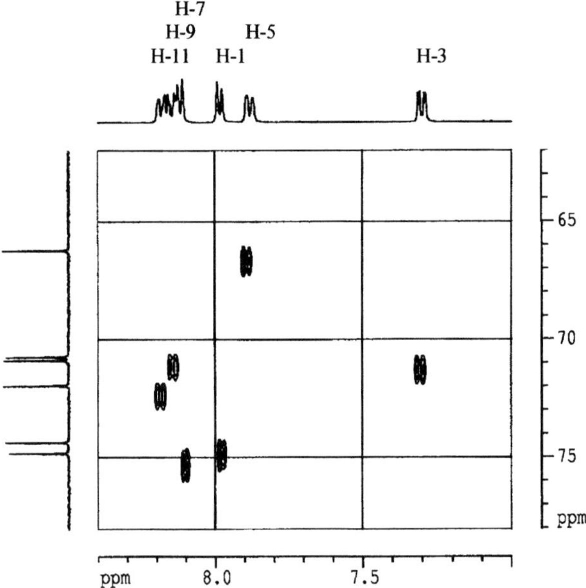

Figure 10.29. ^{1}H–^{13}C correlation of 4-FPNIW (HSQC) in acetone-d_6.

coupling constants were determined by a ^{13}C gated experiment (Fig. 10.27). The resulting data are given in Table 10.4.

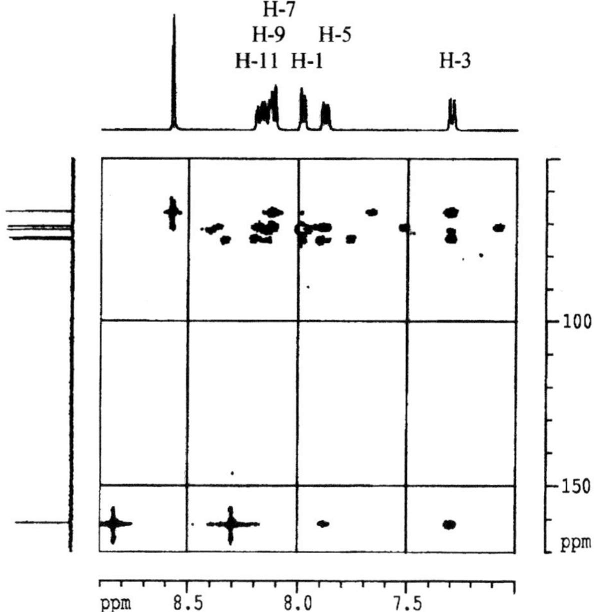

Figure 10.30. Long-range $^1H-^{13}C$ correlation of 4-FPNIW (HMBC) in acetone-d_6. The experiment is without decoupling in F_2. The long-range coupling between the carbonyl carbon and protons H–3 and –5 is seen.

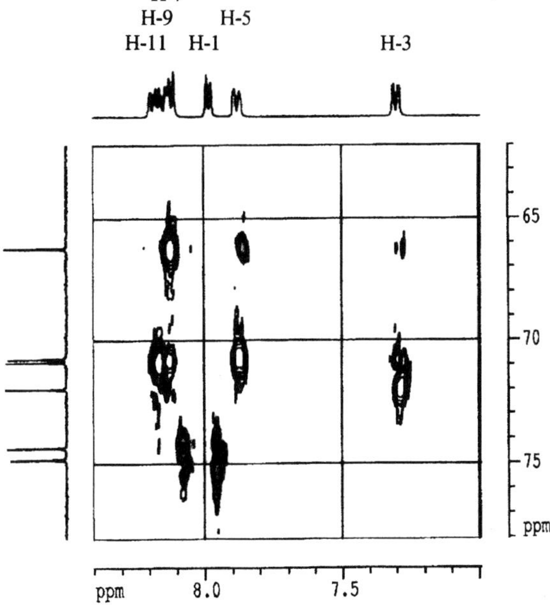

Figure 10.31. $^{13}C-^1H$ heteronuclear RELAY experiment of 4-FPNIW in acetone-d_6.

Table 10.4. Assignments of the ^{13}C NMR signals for 4-FPNIW at 298 K.

Carbon No.	Carbon chemical shift δ at 298 K	Carbon-proton coupling constant $^1J_{CH}$ (Hz) (± 0.18 Hz)	Carbon-proton coupling constant $^2J_{CH}$ and $^3J_{CH}$ (Hz) (± 0.18 Hz)
1	74.7	176.78	3.0/3.0
3	71.2	175.37	5.30/5.30/2.65
5	66.5	172.01	5.12/2.12/5.12
7	75.1	175.02	4.24/2.65
9	71.0	176.61	5.65/4.24
11	72.3	175.19	5.47/4.24/1.77
13	161.6	214.22	3.71/2.12

10.3 Chemical Decomposition in Analysis of Shock Wave Synthesis Materials

10.3.1 Introduction

Of special interest, owing to their peculiar physical and chemical properties, are the ultrafine powders obtained with the use of explosion energy. The particles' small size is the factor accounting for the major contribution of the surface energy to the total energy of the particles; under conditions of high gradients of temperature and pressure this results in additional bulb and surface defects. The relaxation of such energy-saturated particles follows in a most favorable way, including the change in characteristics of the crystal structure and also through the adsorption of environmental components. As a consequence, it is fairly common that ultrafine materials obtained under non-equilibrium conditions exhibit heterogeneity of phase and impurity composition. The conventional methods of analysis of phase and surface impurity composition in ultrafine materials provide mainly a semi-quantitative assessment. This is due to the amorphous phases and the distortions in the crystal structure and also because of the low sensitivity and the locality of the probe methods. A possible approach best suited to expand the possibilities of these methods is their combination with the controlled chemical decomposition of the ultrafine materials. The reactivity, different in polymorphous modifications of a substance, will allow the selection of different conditions of chemical decomposition and the quantitative determination of the ratio of different phases. Layer-wise etching in combination with highly sensitive methods of elemental analysis permits the determination of the amount of impurities and their distribution in the material.

Here we consider the results of studies of the use of chemical decomposition in combination with conventional methods in the analysis of ultrafine powders obtained in special-purpose explosion chambers by means of the explosion energy.

10.3.2
Experimental

10.3.2.1 Samples

Ultrafine powders of aluminum oxide were obtained at a shock-wave loading of an aluminum powder layer with a contact high explosive (HE) charge in a gaseous atmosphere [10.57], the volume of the chamber being 0.3 m^3. X-ray phase analysis of the powders according to the ASTM catalogue revealed α-Al_2O_3 (corundum), δ-Al_2O_3 and the nitride-containing modification of the oxide of composition $Al_{(8/3+x/3)}O_{4-x}N_x$ ($0.22 < x < 0.5$). The simultaneous presence of the last two phases hinders quantitative X-ray phase analysis considerably. Depending on the loading parameters, the size of the particles varied from 50 to 300 nm. The authors [10.57] suggested a multiphase structure of the largest particles of the oxide: the nucleus consists of the α-phase and the surface layer of the δ- and the nitride-containing phases.

Detonation carbon was obtained in the process of detonation of TNT/RDX alloy in a conserving medium [10.58], the chamber volume being 2 m^3. In accord with electron microscopy data, the powders obtained included a diamond phase, a graphite-like phase and amorphous carbon of a low degree of ordering [10.59]. The specific surface area of carbon powders determined with the BET method is ~400 m^2/g.

Ultrafine diamonds were extracted from the products of detonation of TNT/RDX alloy through thermal oxidation of carbon non-diamond forms with boiling in a mixture of nitric and sulfuric acids. The diamond powders are characterized by a cubic structure of the lattice and by X-ray purity. The specific surface area is ~280 m^2/g, the average size of the coherence scattering regions is 4 nm and the average particle size in the low-concentrated hydrosols is ~40 nm. Qualitative analysis of the chemical composition of the diamond surface by secondary ion mass spectrometry (IMS-3F) and also with the X-ray photoelectron spectroscopy (ESCA-3) showed the surface to contain calcium, copper, potassium and aluminum.

All the reagents used were of pure grade. The data presented are the averaged results for 3–6 powders for each method.

10.3.2.2 Chemical Decomposition in a Phase Analysis

Aluminum Oxide

For ultrafine powders of the aluminum oxide, the conditions for the selective dissolution of the δ- and nitride-containing phases were chosen in an empirical way: boiling in 70% sulfuric acid, the ratio of the powder mass to the volume of reagent being 1:100. The data for the X-ray phase analysis showed the powders to contain only the α-phase after boiling for 25 min. Corundum powder is insoluble under the given conditions. This provided a gravimetric method of determining the content of the α-phase. The analysis of the solutions obtained for the content of nitrogen using the Kjeldahl method and the corresponding calculations of the stoichiometric composition made it possible to determine an average ratio between the δ- and N-containing aluminum oxide forms. The determination procedure was tested for powders and mixtures of corundum and aluminum nitride in plasma-chemical synthesis.

Detonation Carbon

The closeness of the values of the activation energy in the oxidation reactions of various forms of carbon (164–182 kJ/mol) and the exothermicity of these reactions required the introduction of a selective inhibitor of the diamond oxidation reaction, namely boric anhydride. The oxidation of ultrafine carbon with heating in the presence of boric anhydride made possible a choice of temperature (750 K) providing, in essence, removal of the amorphous carbon only. A rise in temperature to 800 K results in gasification of a less disordered graphite-like carbon also. Consequently, the quantitative estimation of the composition of detonation carbon powders becomes possible. The determination procedure has been tested for ultrafine diamonds and detonation carbon powders synthesized under conditions unfavorable for diamond formation.

10.3.2.3 Chemical Decomposition in the Analysis of Impurity Distribution

Ultrafine Diamonds

The layer-wise analysis of powders under gradual decomposition is possible if the rate of the reaction front movement is uniform, i.e. if the particles conserve their shape. In this case the dependence of the degree of impurity dissolution on the degree of the basic substance decomposition reflects the manner of admixture distribution within the particles' volume. Simulation models to calculate the dissolution curves in accord with the character of theimpurity distribution have been described [10.60]. In the analysis of ultrafine diamonds extracted from the products of detonation synthesis, this method was used in combination with atomic absorption spectrometry at AAS-1N (acetylene + N_2O and acetylene + air flames). The conditions for the analysis of diamond powders in layers were defined: slow oxidation at boiling in a mixture of nitric and sulfuric acids (2:1). The elemental analysis of powders obtained in the process of oxidation and the construction of dependence curves of impurity dissolution versus oxidizing decomposition of ultrafine diamonds allowed the elucidation of the kind and the amount of elements composing the inclusion impurities, the distinct poorly soluble phases or the surface compounds.

10.3.3 Results and Discussion

10.3.3.1 Phase Composition of Aluminum Oxide Powders

The content of the α-phase in powders varied from 18 to 41%, the δ-phase 43–71% and the nitride-containing phase of aluminum oxide 3–22% on average, depending

Table 10.5. Phase composition of the aluminum oxide powders.

No.	HE	Medium	α-Al_2O_3 (%)	δ-Al_2O_3 (%)	$Al_2O_{2.6}N_{0.26}$ (%)
1	Ammonite	Air	36.5	47.5	17.0
2	RDX	Air	34.5	43.7	21.8
3	RDX	O_2	41.0	55.1	3.9
4	RDX	CO_2 + air	31.7	61.6	6.7
5	RDX	N_2 + air	18.5	70.2	11.4

on the type of high explosive (HE), the parameters of loading and the atmosphere in the chamber. The simultaneous determination of the specific surface area changes under chemical decomposition revealed a prevailing mono-phase of the aluminum oxide particles, the largest ones containing mainly α-Al_2O_3. Under certain synthesis conditions the powders contained α- and δ-phases (20 and 75%), and also α-Al_2O_3 and AlN (81 and 14%, respectively) for the most part. The results obtained for the same ratio between the masses of aluminum powder and the HE and the initial pressure in the chamber (1 atm) are given in Table 10.5.

The correlation of the tabulated data shows that the phase composition of the powders obtained is conditioned for the most part at the expansion stage during the reaction with environmental components. The effect of the HE type is considerably weaker. Nevertheless, on transition from ammonite to RDX (experiments 1 and 2), as the content of nitrogen increases and that of oxygen decreases in charges of the same mass, a decrease in the content of α-Al_2O_3 and the oxide δ-phase along with an increase in the aluminum N-containing oxide is observed. This may suggest that some part of the aluminum undergoes oxidation in the vapor phase burning regime with interaction with the gaseous products of detonation in the course of loading of the porous metal layer.

The amount of N-containing phase of the aluminum oxide is mainly determined by nitrogen in the explosion chamber (experiments 2 and 3). The deficiency of oxidizer in experiments 4 and 5 (products of synthesis contain aluminum: No. 4, 6%; No. 5, 15%) results in a decrease in the corundum content It seems likely that the phase transition of the oxide into a high-temperature modification is mainly conditioned by the total heat effect of the highly exothermic reaction of aluminum oxidation during the interaction with the components of the explosion chamber medium. The decrease in the N-containing aluminum oxide in experiment 5 (as compared with experiment 2) may be a consequence of the decrease in the local heat-up temperature due to the reduction in mass of the oxidized aluminum [10.61]. The content of a metastable δ-phase derives mainly from the rate of cooling of the product under the expansion.

10.3.3.2 **Phase Composition of Detonation Carbon**

Depending on the experimental conditions the samples studied contained 14–75% of diamond phase, 12–26% of graphite-like carbon and 3–64% of a more disordered amorphous carbon. Table 10.6 displays the data on detonation carbon obtained with

Table 10.6. Composition of detonation carbon.

No.	TNT:RDX	Medium	Total yield from HE (%)	Diamond (%)	Graphite-like carbon (%)	Amorphous carbon (%)
1	60:40	H_2O	7.5	56	26	18
2	60:40	$H_2O + CO_2$	6.0	44	23	33
3	60:40	CO_2	3.6	36	20	44
4	50:50	CO_2	3.4	36	16	48
5	40:60	CO_2	3.2	37	12	50

the same chamber and mass of charge. The comparison of the results of the analysis of powders obtained under similar conditions of detonation and different chamber media (experiments 1–3) allows one to calculate the changes in the yield of each phase as the conserving conditions become worse, i. e. as the residual temperatures and the effect of the oxidizer (carbon dioxide) rise. Taking into account these data and the difference in the reactivities of various forms of carbon, one may conclude that the main losses of diamonds at the stage of product expansion can be attributed, for the greater part, to their amorphization.

The comparison of the data on the phase composition of the powders obtained with content-varying charges and the same conserving medium (experiments 3–5) shows the diamond content in the condensed carbon to be practically unchanged with increase in the content of RDX from 40 to 60%. The observed decrease in the graphite-like carbon content suggests an effect of the structure of the HE molecules not only on the formation of the diamond phase [10.62] but also on the amount of graphitized carbon. Then, in the case of insufficient power of the detonation wave and incomplete destruction of the HE molecules, the prevailing sp^3 hybridization of carbon atoms in aromatic compounds may slow the transition into the diamond form.

10.3.3.3 Distribution of Impurities in Ultrafine Diamonds

The basic impurities in ultrafine diamonds were shown experimentally to be at the surface. By way of example, a high degree of dissolution of copper, potassium and aluminum even during the initial etching is indicative of surface contamination, conceivably with specifically absorbed ions. The character of the dependence of the iron content in diamond powders on the degree of their decomposition showed that approximately two-thirds of the total amount of iron in powders is dissolved in proportion to the oxidative decomposition of powders. This is evidence of a uniform distribution of iron in diamonds, i. e. this iron is an inclusion impurity. The enrichment with iron that takes place on more prolonged etching suggests that the rest of the impurity is in the form of a separate poorly soluble phase. The degree of calcium dissolution depends on the oxidative decomposition of the diamonds and is initially of a linear character and practically does not undergo further changes. This testifies to the presence of calcium in two forms – in a form that contaminates the product at the surface and in a form that is present in a separate poorly soluble form. Hence the studied powders contain calcium in the form of a surface compound ~0.24%, Ca in the form of a separate phase ~0.16%, copper, potassium and aluminum relating to the surface contamination ~0.02% of each, iron as an impurity of intrusion ~0.12% and iron as a separate phase ~0.06%. Apart from the specific data to characterize the materials, the effect of the synthesis conditions may also be considered. For instance, the use of metals in the region of diamond formation (the iron in the composition of detonators) results in the emergence of impurity phases of intrusion and the application of hard water in the processes is responsible for irreversible contamination with calcium compounds.

10.4 Conclusion

Consideration of the thermal analysis of the two crystalline substances ADN and CL 20 has shown that thermoanalytical techniques are very helpful tools for characterizing energetic materials and to obtain various information about their physicochemical properties and thermal behavior. The combination of different thermoanalytical methods allows one to obtain not only reliable reference data on physical and chemical transformations but also a well-founded understanding of decomposition pathways and mechanisms. Since thermoanalytical measurements require only small amounts of samples and can be carried out within a relatively short period, they are very helpful for both routine analysis (e.g. to check purity or influences of processing, storage, etc.) and detailed investigations (e.g. in addition to conventional performance tests) for a more in-depth understanding of chemical and physical processes during thermal treatment.

ADN and CL 20 were explored by NMR spectroscopy under high-resolution conditions. The measurements were performed with an NMR spectrometer at 400.13 MHz proton resonance frequency. The ^{1}H, ^{13}C, ^{14}N, ^{15}N and ^{17}O NMR spectra were measured and the signals were assigned. With the help of ^{13}C NMR spectroscopy the $^{1}J_{CH}$ coupling constants could be determined.

A CL 20 technical product was studied by liquid chromatography and the main impurity was isolated by medium-pressure liquid chromatography. The main impurity, 4-FPNIW, was investigated by NMR spectroscopy under high-resolution conditions. One-dimensional ^{1}H and ^{13}C and two-dimensional NMR spectra were recorded and the signals were evaluated. From proton spectra with resolution enhancement by special filter functions, proton coupling constants were assigned. The spectra were recorded at ambient temperature.

The use of chemical decomposition in the analysis of the phase and impurity composition of ultrafine materials is adding extensively to the possibilities of conventional methods. It allows the necessary quantitative assessment to characterize these materials and provides useful insights into the mechanism of synthesis and obtaining materials with pre-set properties.

10.5 References

10.1 Luk'yanov OA, Gorelik VP, Tartakovsky VA (1994) Dinitramide and its salts, *Russ. Chem. Bull.* **43**, 89.

10.2 Schmitt RJ, Bottaro JC, Penwell PE, Bomberger DC (1991) Manufacture of ammonium dinitramide salt for rocket propellant, *US Patent* 5 316 749.

10.3 Christe KO, Wilson W, Petrie MA, Michels HH, Bottaro JC, Gilardi R (1996) The dinitramide anion $N(NO_2)_2^-$, *Inorg. Chem.* **35**, 5068.

10.4 Löbbecke S, Keicher T, Krause H, Pfeil A (1997) The new energetic material ammonium dinitramide and its thermal decomposition behavior, *Solid State Ionics* **101**, 945.

10.5 Löbbecke S, Krause H, Pfeil A (1997) Thermal decomposition and stabilization of ammonium dinitramide (ADN), In: *Proc. 28th Int. Annual Conference of ICT, Karlsruhe*, p. 112.

10.6 Löbbecke S (1999) *Einsatz thermischer Analysenmethoden zur Charakterisierung neuer energetischer Materialien am Beispiel von Ammoniumdinitramid (ADN) und Hexanitrohexaazaisowurtzitan (HNIW)*, Fraunhofer IRB Verlag, Stuttgart.

10.7 Reading M, Luget A, Wilson R (1994) Modulated differential scanning calorimetry, *Thermochim. Acta* **238**, 295.

10.8 Nielsen AT (1988) Caged polynitramine compound, *US Patent* 88–253106.

10.9 Wardle RB, Hinshaw JC, Braithwaite P, Rose M, Johnston G, Jones R, Poush K (1996) Synthesis of the caged nitramine HNIW (CL-20), In: *Proc. 27th Int. Annual Conf. of ICT, Karlsruhe*, p. 27.

10.10 Golfier M, Graindorge H, Longevialle Y, Mace H (1998) New energetic molecules and their applications in energetic materials, In: *Proc. 29th Int. Annual Conf. of ICT, Karlsruhe*, p. 3.

10.11 Löbbecke S, Bohn MA, Pfeil A, Krause H (1998) Thermal behavior and stability of HNIW (CL 20), In: *Proc. 29th Int. Annual Conf. of ICT, Karlsruhe*, p. 145.

10.12 Purcell EM, Torrey HC, Pound RV (1946) *Phys. Rev.* **69**, 37.

10.13 Bloch F, Hansen WW, Packard M (1946) *Phys. Rev.* **69**, 127.

10.14 Bloch F, Hansen WW, Packard M (1946) *Phys. Rev.* **70**, 474.

10.15 Kalinowski H-O, Berger S, Braun S (1984) *^{13}C-NMR-Spektroskopie*, Georg Thieme, Stuttgart.

10.16 Knight WD (1949) *Phys. Rev.* **76**, 1259.

10.17 Nielsen AT (1988) Caged polynitramine compound, *US Patent* 88–253106.

10.18 Wardle RB, Hinshaw JC, Braithwaite P, Rose M, Johnston G, Jones R, Poush K (1996) Synthesis of the caged nitramine HNIW (CL-20), In: *Proc. 27th Int. Annual Conf. of ICT, Karlsruhe*, p. 27.

10.19 Golfier M, Graindorge H, Longevialle Y, Mace H (1998) New energetic molecules and their applications in energetic materials, In: *Proc. 29th Int. Annual Conf. of ICT, Karlsruhe*, p. 3.

10.20 Günther H (1983) *NMR-Spektroskopie*, Georg Thieme, Stuttgart.

10.21 Friebolin H (1978) *NMR-Spektroskopie*, Verlag Chemie, Weinheim/Bergstrasse.

10.22 Ault A, Dudek GO (1978) *Protonen-Kernresonanz-Spektroskopie*, Dietrich Steinkopf, Darmstadt.

10.23 Ernst L (1980) *^{13}C-NMR-Spektroskopie*, Dietrich Steinkopf, Darmstadt.

10.24 Gruber U, Klein W (1995) *NMR-Spektroskopie für Anwender*, Verlag Chemie.

10.25 Zschunke (1971) *Kernmagnetische Resonanzspektroskopie in der Organischen Chemie*, Akademie Verlag, Berlin; Pergamon Press, Oxford; Vieweg + Sohn, Braunschweig.

10.26 Michel D (1981) *Grundlagen und Methoden der Kernmagnetischen Resonanz*, Akademie Verlag, Berlin.

10.27 Derome AE (1987) *Modern NMR Techniques for Chemistry Research*, Pergamon Press, Oxford.

10.28 Sanders KKM, Hunter BK (1993) *Modern NMR Spectroscopy*, 2nd edn, Oxford University Press, Oxford.

10.29 Friebolin, HP (1993) *Basic One- and Two-Dimensional NMR Spectroscopy*, 2nd edn, VCH, Weinheim.

10.30 Günther H (1995) *NMR Spectroscopy*, 2nd edn, Wiley, Chichester.

10.31 Kalinowski H-O, Berger S, Braun S (1988) *Carbon 13-Spectroscopy*, Wiley, Chichester.

10.32 Jeener J (1971) Presented at the Ampère International Summer School, Basko Polje.

10.33 Aue WP, Bartholdi E, Ernst RR (1975) *J. Chem. Phys.* **64**, 2229–2246.

10.34 Nagayama K (1980) *J. Magn. Reson.* **40**, 321.

10.35 Müller L (1979) *J. Am. Chem. Soc.* **101**, 4481–4484.

10.36 Bax A, Griffey RH, Hawkins BL (1983) *J. Magn. Reson.* **55**, 301–315.

10.37 Martin GE, Zektzer AS (1988) *Two Dimensional NMR Methods for Establishing Molecular Connectivity*, VCH, Weinheim.

10.38 Hurd RE, John BK (1991) *J. Magn. Reson.* **91**, 648–653.

10.39 Ruiz-Cabello J, Vuister GW, Moonen CTW, van Gelderen P, Cohen JS, van Zijl PCM (1992) *J. Magn. Reson.* **100**, 282–303.

10.40 Bax A, Summers MF (1986) *J. Am. Chem. Soc.* **108**, 2093–2094.

10.41 Lerner L, Bax A (1986) *J. Magn. Reson.* 69, 375–380.

10.42 Willker W, Leibfritz D, Kerssebaum R, Bermel W (1993) *Magn. Reson. Chem.* **31**, 287–292.

10.43 Jeener J, Meier BH, Bachmann P, Ernst RR (1979) *J. Chem. Phys.* **71**, 4546–4563.

10.44 States DJ, Haberkorn RA, Ruben DJ (1982) *J. Magn. Reson.* **48**, 286–292.

10.45 Bodenhausen G, Kogler H, Ernst RR (1984) *J. Magn. Reson.* **58**, 370–388.

10.46 Neuhaus D, Williamson M (1989) *The Nuclear Overhauser Effect in Structural and Conformational Analysis*, VCH, Weinheim.

10.47 Harris RK, Mann BE (1978) *NMR and the Periodic Table*, Academic Press, London.

10.48 Yoder CH, Schaeffer CD Jr (1987) *Introduction to Multinuclear NMR*, Benjamin/Cummings, Menlo Park, CA.

10.49 Martin GJ, Martin ML, Gouesnard J-P (1981) *^{15}N NMR Spectroscopy*, Springer Verlag, Berlin,

10.50 Witanowski M, Webb GA (1973) *Nitrogen NMR*, Plenum Press, London.

10.51 Berger S, Braun S, Kalinowski H-O (1992) *NMR-Spektroskopie von Nichtmetallen – ^{15}N-NMR-Spektroskopie*, Band 2, Georg Thieme, Stuttgart.

10.52 Berger S, Braun S, Kalinowski H-O (1992) *NMR-Spektroskopie von Nichtmetallen – Grundlagen, ^{17}O-, ^{33}S- und ^{129}Xe-NMR-Spektroskopie*, Band 1, Georg Thieme, Stuttgart.

10.53 Bunte G, Pontius H, Kaiser M (1998) Charaterization of impurities in new energetic materials, In: *Proc. 29th Int. Annual Conf. of ICT, Karlsruhe*, p. 148.

10.54 Kaiser M (1998) Characterization of ADN and CL20 by NMR spectroscopy, In: *Proc. 29th Int. Annual Conf. of ICT, Karlsruhe*, p. 130.

10.55 Zhao X, Liu J (1996) *Henneng Cailiao* **4**, 145–149.

10.56 Liu J (1997) *Huozhayao* **20**, 26–28.

10.57 Beloshapko AG, Bukaemski AA, Staver AM (1990) *Fiz. Gorenia Vzryva* **26**, 93–98.

10.58 Lyamkin AI, Petrov EA, Ershov AP (1988) *Dokl. Akad. Nauk SSSR* **302**, 611–613.

10.59 Mal'kov IY (1996) Ultrafine powders, materials and nanostructures, In: *Proc. Interregional Conf. Krasnoyarsk*, Tech. Univ. Publ., Krasnoyorsk, pp. 17–18.

10.60 Vaivads YK, Smilshkalne GL, Miller TN, *Activation Analysis. Methodology and Application*, Tashkent, pp. 107–112.

10.61 Il'in AP, Proskurovskaya LT (1990) *Fiz. Gorenia Vzryva* **26**, 71–72.

10.62 Kozyrev NV, Brylyakov PM, Sen CC (1990) *Dokl. Akad. Nauk SSSR* **314**, 889–891.

11
Wettability Analysis

U. Teipel, I. Mikonsaari, S. Torry

11.1
Introduction

In addition to satisfying the requirements for explosive performance and energetic content, plastic bonded explosives and propellants must have good mechanical properties. Propellants must withstand stresses and strains induced during storage and deployment. In most cases, the mechanical properties of energetic formulations depend on the interaction of the binder with the surface of the filler. Good interactions of the binder with the surface of the filler are required for high-performance propellant motors.

The interfacial layer between the filler and the binder can be modified by the addition of a bonding agent. This has been discussed in some detail [11.1]. A number of chemicals have been tested as candidates for bonding agents. However, interactions between these bonding agents and the filler surface have been poorly characterized. Indeed, there is some debate as to whether some materials act as interfacial bonding agents or as additional crosslinking chemicals. Little work has been done on characterizing filler surfaces. Hence the aim of this work was to characterize the surfaces of energetic materials, such as PSAN, RDX and HMX, using different methods, e.g. contact angle measurements of polymer plates and bulk energetic materials by the capillary penetration method and chromatographic techniques, and to test the applicability of these methods for the characterization of bulk material. Additionally, the work with chromatographic techniques on RDX and HMX has identified molecular species that have a good affinity for nitramine surfaces. This information can be used in the longer term to design potential bonding agents.

Energetic Materials. Edited by Ulrich Teipel
Copyright © 2005 WILEY-VCH Verlag GmbH & Co. KGaA, Weinheim
ISBN: 3-527-30240-9

11.2 Determination of Surface Energy

Surface tension forces are observed in materials consisting of two or more phases. Phases are macroscopic regions of constant density and chemical composition separated from other phases by an interfacial boundary. Contact between phases occurs only at this interfacial boundary, giving rise to an interfacial surface with specific characteristics. According to Gibbs, interfacial surfaces can be defined as intermediate or two-dimensional macroscopic phases. The thickness of interfacial surfaces ranges from one to several molecular diameters.

The properties of interfacial surfaces are significantly influenced by molecular interaction forces. Five different types of interfacial surfaces exist: liquid/gas, liquid/liquid, solid/liquid, solid/solid and solid/gas.

Energetic effects that occur within the solid/liquid interface are especially significant when processing energetic materials. These effects have a particular influence on the mechanical stability of particulate solid/binder systems. Improved wetting of the liquid (polymeric binder) on a particulate solid leads to smaller contact angles between the two phases and results in higher mechanical stability of the entire system. Very few data on the surface tension of energetic materials are available in the literature, e. g. one source suggests that the surface tension of HMX is 41.6 mN/m [11.2].

While liquid surface tensions can be measured directly, surface tensions between solid phases are difficult to measure directly because the surfaces cannot be deformed reversibly. To determine surface energies of solids, the contact angle is measured as a characteristic parameter. Young's equation (Eq. (11.5)) is then used to determine quantitatively the surface energy from contact angle measurements.

11.2.1 Theory of Surface Tension

The extensive free enthalpy g of a closed system is the energy of the system available as work. It can be represented as a function of temperature T, pressure p, amount of material n and surface area A for a one-component system as follows:

$$g = g(T, p, n, A) \tag{11.1}$$

A differential change in the free enthalpy can be expressed in terms of a sum of differential changes in the individual variables:

$$g = -s \cdot dT + v \cdot dp + \mu \cdot dn + \sigma \cdot dA \tag{11.2}$$

where s is the entropy, v the volume, μ the chemical potential and σ the surface tension.

Each term of the total differential represents a specific variable differential contribution of energy. When the temperature, pressure and amount of material are held constant, the surface specific term can be simplified as follows:

$$(dg)_{T,p,n} = \sigma \cdot dA = dW_{rev} \tag{11.3}$$

Under these conditions, the change in free enthalpy is equivalent to the reversible work done by the system, W_{rev}. The surface tension is then defined as

$$\left(\frac{\partial g}{\partial a}\right)_{T,p,n} = \frac{\mathrm{d}W_{rev}}{\mathrm{d}A} = \sigma \tag{11.4}$$

Thus, for isothermal, isobaric systems in which the amount of material is constant, the surface tension is equal to the change in free enthalpy. Creation of any part of a surface is related to the specific expenditure of work [11.3].

While the surface tension of liquids σ_{lv} can be measured directly without much difficulty, the measurement of the surface tension of solids is much more difficult. One possibility is to use Young's equation in combination with contact angle measurements:

$$\sigma_{sl} = \sigma_{sv} - \sigma_{lv} \cdot \cos\Theta \tag{11.5}$$

where Θ is the contact angle, σ_{sl} the surface tension (solid/liquid), σ_{sv} the solid surface tension (solid/vapor) and σ_{lv} the liquid surface tension (liquid/vapor) (Fig. 11.1).

11.2.2
Models for Determining the Free Interfacial Energy

A number of models have been proposed in the literature for determining the free interfacial energy at the phase boundary between solids and vapors, because this quantity cannot be measured directly.

The model proposed by Owens, Wendt, Rabel and Kaelble [11.4] postulates that the free interfacial energy is the sum of disperse and polar contributions:

$$\sigma = \sigma^{\text{disperse}} + \sigma^{\text{polar}} \tag{11.6}$$

Assuming that the two polar contributions and two non-polar contributions of the interfacial pair interact reciprocally, the relationship for the geometric mean of the free interfacial energy can be written as follows:

$$\sigma_{12} = \sigma_1 + \sigma_2 - 2\left(\sqrt{\sigma_1^{d} \cdot \sigma_2^{d}} + \sqrt{\sigma_1^{p} \cdot \sigma_2^{p}}\right) \tag{11.7}$$

where the subscripts 1 and 2 represent phases 1 and 2, respectively, and superscripts d and p represent the disperse and polar contribution, respectively.

If the harmonic or geometric-harmonic mean is used for the calculation rather than the geometric mean, then the resulting equations, as presented by Wu [11.5], are

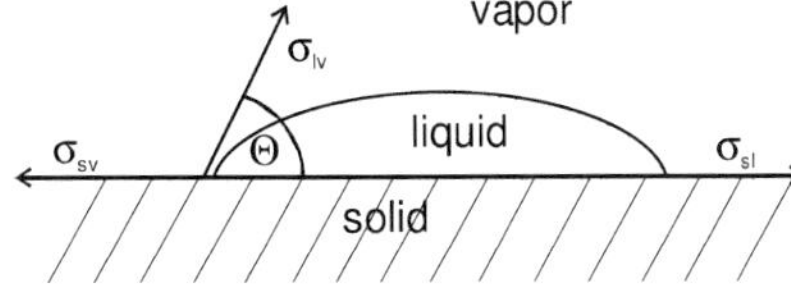

Figure 11.1. Wetting of a solid with a liquid.

$$\sigma_{12} = \sigma_1 + \sigma_2 - \frac{4 \cdot \sigma_1^d \cdot \sigma_2^d}{\sigma_1^d + \sigma_2^d} - \frac{4 \cdot \sigma_1^p \sigma_2^p}{\sigma_1^p + \sigma_2^p} \tag{11.8}$$

and

$$\sigma_{12} = \sigma_1 + \sigma_2 - 2\sqrt{\sigma_1^d \cdot \sigma_2^d} - \frac{4 \cdot \sigma_1^p \sigma_2^p}{\sigma_1^p + \sigma_2^p} \tag{11.9}$$

The harmonic mean (Eq. (11.8)) provides good results for low-energy systems such as organic solutions, water, polymers and organic pigments, whereas the geometric-harmonic mean (Eq. (11.9)) is more successful for high-energy systems such as glass, mercury, metal oxides and graphite. The equation based on the geometric mean proposed by Owens, Wendt, Rabel and Kaelble (Eq. (11.7)) yields good results for non-polar systems.

The model by Owens, Wendt, Rabel and Kaelble provides another approach for determining the surface tension of a solid. For experimental studies one must know both the disperse and polar contributions of the measurement fluid and the contact angle in order to calculate both parts of the surface tension of the solid from the model equation. The result is a linear equation whose intercept is equal to the disperse component and slope is the polar component of the free interfacial energy.

Another method for determining the free interfacial energy was developed by Zisman [11.6]. The contact angles for a series of liquids (e.g. hydrocarbons) on a single type of solid surface are determined. The cosines of these angles are then plotted as a function of the liquid surface tension. Such a plot should yield an approximately linear relationship as shown in Fig. 11.2.

By extrapolating the line to an angle $\Theta = 0°$, one obtains an interfacial energy value for the theoretical case of perfect wetting, i.e. a contact angle of zero. This value is considered as the free surface energy of the solid. However, the model is valid only for non-polar liquids.

11.2.3 Contact Angle Determination on Flat Surfaces

The Wilhelmy plate method (Fig. 11.3) uses force measurements to determine the surface or interfacial tension [11.7].

The measurement fixture is a rectangular platinum or platinum-iridium plate with an exactly defined geometry and a roughened surface [11.8]. By treating the

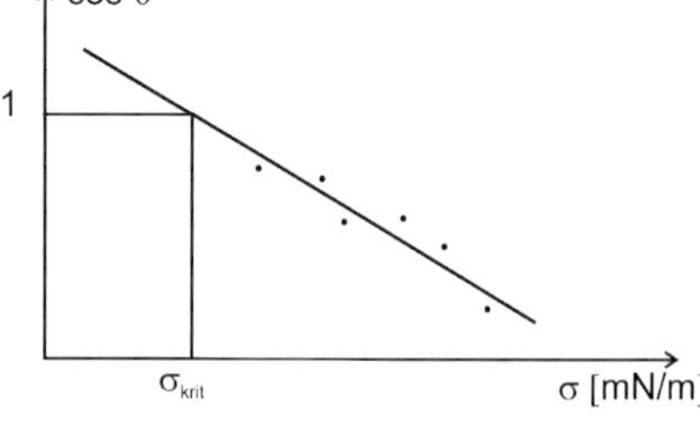

Figure 11.2. Calculation proposed by Zisman.

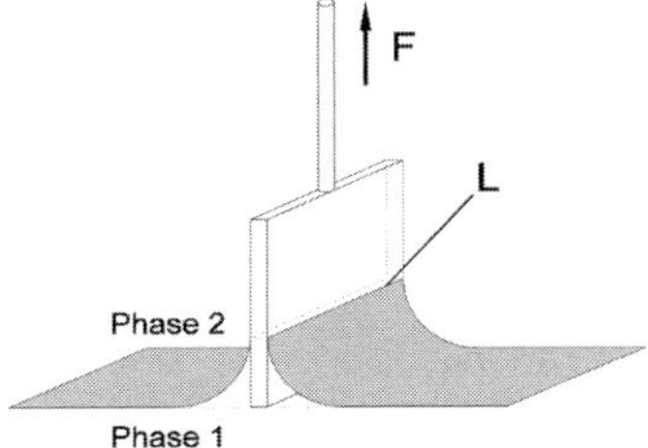

Figure 11.3. Vertical plate method proposed by Wilhelmy.

surface one obtains a contact angle between the liquid and plate of $\Theta = 0°$. The plate is brought into contact vertically against the surface or interfacial area so that the plate is completely wetted and the underside of the plate is exactly even with the level of the interface. One measures the resistance force resulting from wetting of the plate in this manner. The interfacial or surface tension is determined via the following equation:

$$\sigma = \frac{F}{L_b \cdot \cos\Theta} \tag{11.10}$$

where F = resistance force, L_b = wetted length and Θ = contact angle.

In the Wilhelmy plate method, no new interfacial surface is formed during the measurement and the plate does not move relative to the interfacial surface:

$$\cos\Theta = \frac{F_W}{\sigma_{lv} \cdot L} \tag{11.11}$$

This result should be interpreted as the main contact angle among the many different contact angles formed along the wetted surface. Hence it is important that the sample area along the meniscus be as homogeneous as possible.

For the case in which the test plate is dipped into the liquid and then withdrawn, one can determine a dynamic contact angle in addition to forward and reverse angles. The force, which also depends on how deep the plate is dipped into the solution, is composed of a buoyancy component and a wetting component:

$$\vec{F}_{total} = \vec{F}_{buoyancy} + \vec{F}_{Wilhelmy} \tag{11.12}$$

The Wilhelmy force acts only at the three-phase interface, i.e. on the fluid surface area, and remains constant during the entire dipping and pull-out process (provided that the geometry of the plate is constant). The buoyancy force is then a linear function of the dip depth. The contact angle can be calculated from the following equation:

$$\cos\Theta = \frac{1}{\sigma_{lv} \cdot L} \cdot (F_{Wilhelmy} + F_{buoyancy}) \tag{11.13}$$

11.2.4
Contact Angle Measurements on Bulk Powders using the Capillary Penetration Method

For porous materials and powders, the classical methods for determining the contact angle at the solid/liquid interface cannot be used. The problem of determining wettability of powders and fiber strands has been addressed by applying the capillary penetration method and using the Washburn equation to calculate the resulting data.

Washburn [11.9] combined the Hagen-Poiseuille equation (Eq. (11.14)) with the equation for the capillary curvature pressure of liquids (the Laplace equation, Eq. (11.15)) to obtain an expression that describes fluid flow in a cylindrical capillary (Eq. (11.16)):

$$\frac{\mathrm{d}V}{\mathrm{d}t} = \frac{\pi \cdot \Delta p \cdot r^4}{8 \cdot \eta \cdot h}, \tag{11.14}$$

where $\mathrm{d}V = \pi \cdot r^2 \cdot \mathrm{d}h$

$$\Delta p = 2 \cdot \sigma_{\mathrm{lv}} \cdot \cos\Theta \cdot \frac{1}{r} \tag{11.15}$$

$$h^2 = \frac{t \cdot r \cdot \sigma_{\mathrm{lv}} \cdot \cos\Theta}{2 \cdot \eta} \tag{11.16}$$

In these equations, h is the length travelled by the fluid of volume $\mathrm{d}V$ in time t, Δp is termed the Laplace pressure, r is the capillary radius, η is the viscosity of the fluid, t is the flow time, σ_{lv} is the surface tension of the liquid and Θ is the advancing angle.

The dependence of the length traveled by the fluid, h, on $t^{1/2}$ as calculated from the Washburn equation, was confirmed in single capillaries and also experiments with a bundle of parallel capillaries with equal diameters [11.10–11.13]. Results were calculated based on the following modified Washburn equation:

$$h^2 = \frac{t \cdot (c \cdot \bar{r}) \cdot \sigma_{\mathrm{lv}} \cdot \cos\Theta}{2 \cdot \eta} \tag{11.17}$$

where c is a constant that accounts for the orientation of the capillaries. The quantity $c\,\bar{r}$ is a constant that depends on the packing density of the powder material.

When using the Washburn equation to calculate results from capillary elevation experiments, the increase in mass is measured as a function of time. From Eq. (11.17), and substituting Eq. (11.18):

$$h = \frac{m}{\rho \cdot A} \tag{11.18}$$

where A is the area through which the fluid flows, one obtains

$$\frac{m^2}{t} = \frac{\rho^2 \cdot A^2 \cdot (c \cdot \bar{r}) \cdot \sigma_{\mathrm{lv}} \cdot \cos\Theta}{2\eta} \tag{11.19}$$

Bartell [11.14] determined that the contact angle can be determined solely from wetting kinetic data and capillary pressure measurements of liquids whose surface tension is close to that of the solid surface energy. When the contact angle influence is larger, a form factor related to porosity is also apparently determined. If the boundary layer is assumed to have a thickness much greater than a single molecular dimension, then the advancing angle can be calculated from the Navier-Stokes equation and depends only on the flow velocity, viscosity and surface tension of the fluid. If such a boundary layer is not present, then the contact angle can only be calculated by considering all interactions, including the anomalies in density and viscosity that may exist in the fluid.

In contrast to the literature cited above, Schindler [11.15] examined the significance of the form of the porosity on the rate of wetting. In this context, it was necessary to assume two equivalent pore radii, based on comparisons with values of radii from geometric models, in order to describe physically the suction and flow experiments. The porosity of the bulk material can be described based on a model consisting of cylindrical capillary bundles with a distribution of capillary radii.

11.2.4.1 Measurement Principles for the Capillary Penetration Method

The measurement methods are based on three important assumptions. One assumes that only laminar flow exists in the bulk material being studied and that the packing constant and structure of the bulk material remain constant during the measurement. The effect of gravity is also neglected [11.16].

The powder being studied is placed in a closed tube containing a frit on the bottom end. It is treated like a capillary bundle with a mean capillary radius r. For a given powder packing, the term $Ac\bar{r}$ is constant. The value of this packing constant can be calculated from the governing equation when a fluid with a contact angle $\Theta = 0°$ is used (i. e. perfect wetting). The tube is brought into contact with this fluid (which perfectly wets the powder) at the frit end (Fig. 11.4).

The fluid is imbibed through the glass frit into the powder. Measuring the

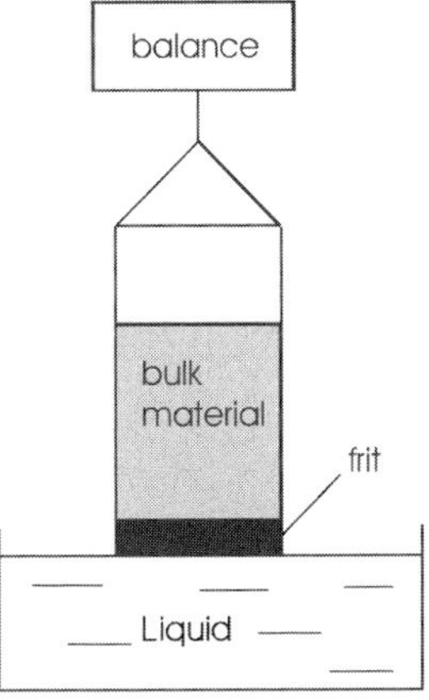

Figure 11.4. Schematic diagram of the capillary penetration method.

increase in weight as a function of time and applying the modified Washburn equation allows one to characterize the wetting behavior as follows:

$$\frac{m^2}{t} = \frac{[(c \cdot \bar{r}) \cdot \varepsilon^2 \cdot (\pi \cdot R)^2] \cdot \rho^2 \cdot \sigma_{lv} \cdot \cos\Theta}{2 \cdot \eta} \tag{11.20}$$

where m is the weight of fluid imbibed into the tube, ρ the density of the measurement fluid, ε the relative porosity and R the inner diameter of the measurement tube. The terms in the square brackets determine the geometric factor C (cm^5), which, as described above, is determined under conditions of perfect wetting. If this factor is known, one can conduct a second experiment using the specific materials of interest to determine the contact angle.

11.2.5 Experimental Results

11.2.5.1 Contact Angle Determination by the Plate Method

Various systems composed of different types of the energetic binder HTPB (hydroxy-terminated polybutadiene) with different types and amounts of catalyst were examined by Teipel *et al.* [11.17]. Table 11.1 shows these different systems, which were studied using a modified Wilhelmy plate method. Irganox is an antioxidant based on substituted *tert*-butylphenol. The catalysts employed included FeAA [iron-(III) acetylacetonate], D22 (dibutyltin dilaurate) and TPB (triphenylbismuth).

The contact angle of these systems was determined using distilled water and lauric alcohol. In order to obtain flat test samples of HTPB binder, the liquid HTPB was applied to glass carrier plates and then cured.

Table 11.2 shows the dynamic advancing contact angles measured for the HTPB systems listed in Table 11.1.

It can be seen that the contact angles with water and lauric alcohol increase with increasing iron(III) acetylacetonate content. The poorest wetting behavior of the fluids examined was exhibited by sample 3 with the D22 catalyst.

The interfacial surface energy of these systems was also determined using the method proposed by Owens, Wendt, Rabel and Kaelble. The free interfacial energy of the HTPBs ranged from 27 to 29 mN/m.

Table 11.1. Systems examined using the Wilhelmy flat plate method [11.17].

Composition (wt%)	HTPB 1	HTPB 2	HTPB 3	HTPB 4	HTPB 5
HTPB	91.42	91.42	91.42	91.42	91.42
Irganox	1	1	1	1	1
Isophorone diisocyanate	7.58	7.58	7.58	7.58	7.58
Catalyst/amount	FeAA/ 40 ppm	FeAA/ 400 ppm	D22/ 4 Drops/100 g	TPB/ 200 ppm	FeAA/ 200 ppm

Table 11.2. Contact angles (°) of the systems studied.

System	Distilled water	Lauric alcohol
HTPB 1	94.0	24.8
HTPB 2	91.2	10.0
HTPB 3	90.6	19.5
HTPB 4	93.2	19.4
HTPB 5	92.7	21.0

11.2.5.2 Contact Angle Measurements of Bulk Powders using the Capillary Penetration Method

The capillary penetration method was used at the Fraunhofer Institute (ICT) to examine the wettability of various bulk powders using different solvents as the wetting agents (see Table 11.3).

Data reduction was accomplished using the Washburn equation. For the calculation the mass increase is measured as a function of time and the slope $\partial m^2/\partial t$ in Eq. (11.20) is determined. Figure 11.5 shows a significant plot of such a measurement. Spray-crystallized ammonium nitrate (SCAN) and phase-stabilized ammonium nitrate (PSAN) with different mean particle sizes were wetted with n-hexane. Depending on the packing density, porosity and other factors, different penetration velocities are obtained. The different bulk characteristics influence the solvent flow characteristics considerably. The value of the upper level depends on the filled mass of powder and the porosity of the bulk.

In order to obtain the best possible reproducibility, efforts were made to maintain constant density of the bulk solid during the measurements. This was accomplished using a stamping volumeter, which is typically used to determine the tap density of solids according to DIN ISO 787. A setting of 500 was used during the tapping procedure. This allowed the scatter in the measured values of capillary flow velocity to be reduced by a factor of four. Figure 11.6 shows an example of the reproducibility obtained (SCAN/n-hexane).

Table 11.3. Material systems examined.

Bulk powder	Solvent
Microcrystalline cellulose	n-Hexane
(three different size fractions)	DMSO
	DMF
	Water
	Methanol
Phase-stabilized ammonium nitrate (PSAN)	n-Hexane
(two different size fractions)	DMF
	DMSO
Spray-crystallized ammonium nitrate (SCAN)	n-Hexane
(two different size fractions)	Ethanol

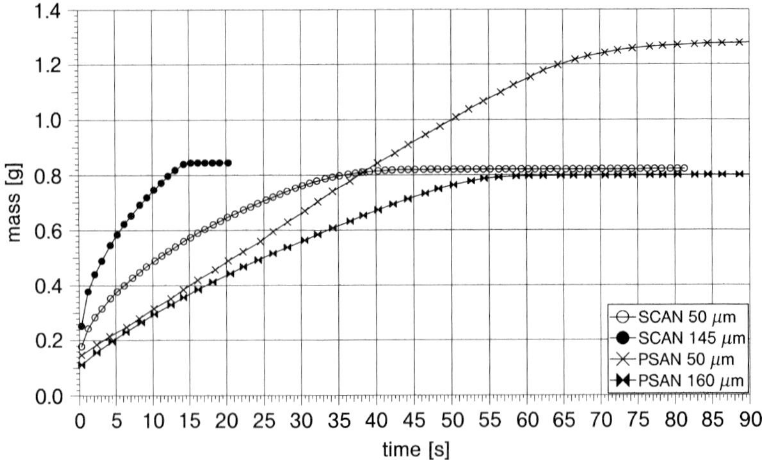

Figure 11.5. Wetting of SCAN and PSAN with *n*-hexane.

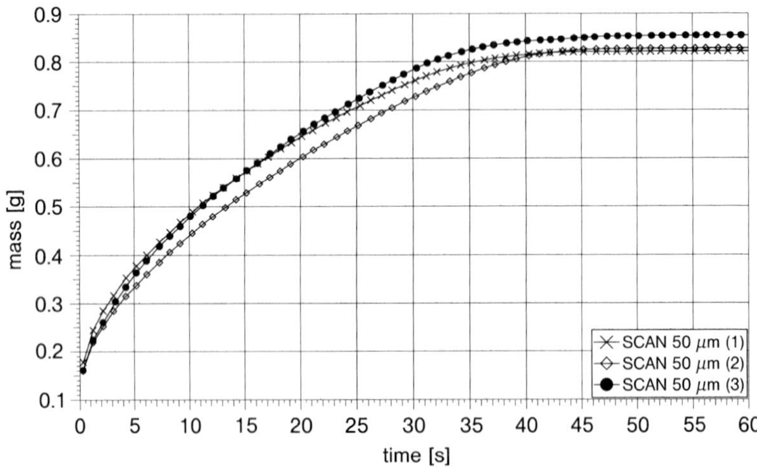

Figure 11.6. Reproducibility obtained in the sampling preparation.

As mentioned before, the results of the capillary penetration method depend on the sample preparation. The extent to which it matters if sampling devices of different radii are used was examined. Figure 11.7 shows the results for the wetting of SCAN by *n*-hexane in six different sampling devices. The sampling devices differ in their diameter. The commercial application commonly uses a sampling device of 10 mm. As expected, the penetrated mass per unit time increases with increasing diameter. However, the differences are not proportional to the wetted circle area.

Table 11.4 summarizes contact angle determinations on the bulk powder using DMF (dimethylformamide), ethanol and DMSO (dimethyl sulfoxide) as wetting

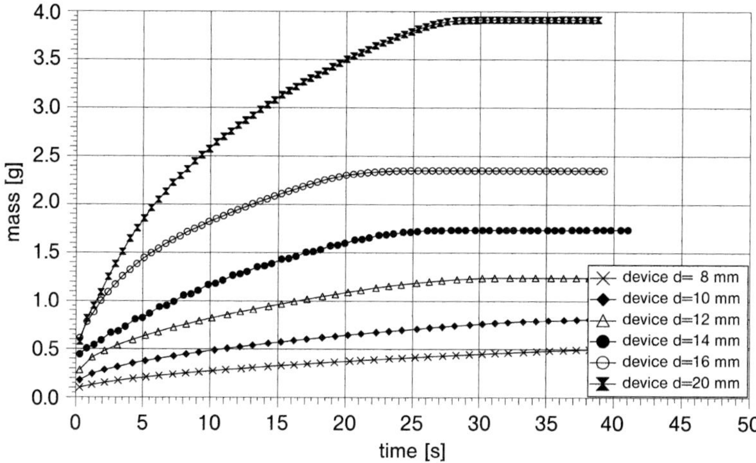

Figure 11.7. Wetting of SCAN with *n*-hexane in different sampling devices.

Table 11.4. Results of the wetting experiments.

Bulk powder	Fluid	Θ (°)
Microcrystalline cellulose; $x_{50,3}$ = 67 μm	DMF	42
Microcrystalline cellulose; $x_{50,3}$ = 110 μm	DMF	57
Microcrystalline cellulose; $x_{50,3}$ = 170 μm	DMF	60
Microcrystalline cellulose; $x_{50,3}$ = 67 μm	DMSO	58
Microcrystalline cellulose; $x_{50,3}$ = 110 μm	DMSO	80
Microcrystalline cellulose; $x_{50,3}$ = 170 μm	DMSO	75
PSAN; $x_{50,3}$ = 55 μm	DMF	80
PSAN; $x_{50,3}$ = 160 μm	DMF	75
PSAN; $x_{50,3}$ = 55 μm	DMSO	70
PSAN; $x_{50,3}$ = 160 μm	DMSO	75
SCAN; $x_{50,3}$ = 150 μm	Ethanol	38
SCAN; $x_{50,3}$ = 50 μm	Ethanol	37

agents. For the determination of the geometric factor *C*, *n*-hexane was used. Figure 11.8 shows exemplarily a SCAN/ethanol wetting experiment. For comparison, the wetting kinetics with *n*-hexane are also shown.

PSAN exhibited good wetting with both solvents, with a contact angle of ~ 75°. For PSAN and SCAN the wetting was not particle size dependent as it is for many other material types reported in the literature [11.18]. The cellulose exhibited different wetting properties with the different solvents. Unlike PSAN, the wetting behavior of cellulose was dependent on particle size. With both solvents, small particles (67 μm) exhibited better wetting characteristics than larger particles. Also, DMF appeared to be a superior wetting agent to the other solvents.

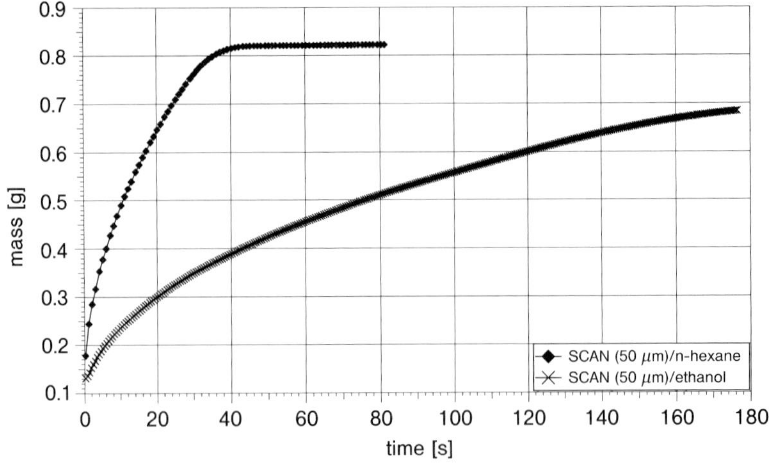

Figure 11.8. Wetting of SCAN with *n*-hexane and ethanol.

11.3
Surface Characterization using Chromatographic Techniques

The surfaces of HMX and RDX using inverse gas chromatography (IGC) and, to a lesser extent, inverse high-performance liquid chromatography (IHPLC) were characterized by DERA (Defense Evaluation and Research Agency). In normal chromatographic experiments, the column is a tool used to separate known/unknown species in a sample. The injected sample is the prime subject of study, i.e. the chromatogram is analyzed to identify unknown species or concentrations of species in the analyte. In the case of inverse chromatography, the material in the column is the prime subject of study. The injected sample is a probe molecule of known characteristics. The interaction of the probe molecule with the column material, more specifically the probe molecule's retention volume, is used to characterize the column's surface.

For inverse gas chromatography, there are essentially two phases of the probe molecule: the probe on the solid surface and the probe in the gaseous state. It is assumed that the helium carrier gas does not interfere with any adsorption processes between the probe molecule and the solid surface. The adsorption process is carried out at infinite dilution. In inverse gas chromatography, it is assumed that there are no nearest neighbors during the adsorption process. Only the most energetically favorable sites are filled initially.

As the name suggests, the inverse liquid chromatography (ILC) experiment is similar in concept to the IGC experiment. However, unlike the IGC experiment, where it is assumed that there are no nearest neighbors during the adsorption process, the ILC probe molecule has to compete for surface sites with the more highly concentrated solvent species. Consequently, the partitioning of the probe in the solvent and on the surface depends very much on the nature of the solvent. This complicates the analysis process. Indeed, attempts by other workers to characterize the surface of solid materials by applying IGC methods to the analysis of IHPLC data have failed. In this work, the natures of both HMX and RDX surfaces were characterized in a qualitative and semiquantitative fashion. An advantage of using the ILC technique is that the explosives are kept in solvent at all times and are less hazardous than in dry IGC columns.

11.3.1 Inverse Gas Chromatography

IGC can be used to study adsorption isotherms [11.19–11.21], specific surface areas [11.22, 11.23], isosteric heats of adsorption at low coverage [11.24], polymer characteristics [11.25], acidic properties of oxides [11.26], solute polarizability and hydrogen bond acidity [11.27], acid-base properties [11.28, 11.29] and surface energy [11.30]. IGC has been used to characterize a number of materials such as carbon fibers, glass, coal powder and polymers [11.31]. The thermodynamics of adsorption of probe molecules on substrates and the acid-base characteristics of the substrates have been correlated with the modulus and other mechanical properties of polymer/solid composites.

The acid-base properties of RDX [11.32] and TATB [11.33] have been characterized previously by IGC. Bailey *et al.* [11.33] calculated the specific heat of adsorption of probe molecules on TATB and sorted the molecules in terms of increasing interaction with the surface. Botija *et al.* [11.32] correlated the specific heat of adsorption of the probe molecules with the Gutmann donor number scale [11.34], the Mayer acceptor number scale [11.35] and the Drago electrostatic and covalent EC scale [11.36].

The surface properties of RDX and HMX were studied at DERA via IGC using probe molecules at infinite dilution. The specific surface areas of RDX and HMX were measured and the specific free energies of adsorption of a variety of probe molecules were calculated. The dispersive surface energy of the explosives at different temperatures was measured and a ranking system of probe species suitable as potential bonding agents was made.

11.3.2 Typical IGC Experimental Conditions

It is advantageous for the IGC column fillers to have a surface area that is as high as possible as this allows the probe vapor adsorption behavior to equilibrate rapidly. However, if the particle size is too small, the column pressure increases beyond the capacity of the GC instrumentation. A compromise has to be reached. In this work, the column fillers were coarse, grade 1 RDX (sieve fraction 100–212 μm) and spheridized HMX (sieve fraction 106–150 μm). The surface areas of RDX and HMX were found to be 5.9×10^{-2} and 6.8×10^{-2} m^2/g, respectively.

The columns were conditioned overnight by raising the temperature of the GC oven at a rate of 1 °C/min from room temperature to 75 °C and then holding the temperature constant for ~13 h. High-purity helium gas (BOC) was dried and deoxygenated using Chrompack Gas Clean Oxygen and Gas Clean Moisture filters. The gas was passed through the column at a rate of 5 cm^3/min during the temperature ramp. This ensured that the surface of the explosive was dry and free from volatile contaminants. The column surfaces were deactivated with dimethyldichlorosilane to minimize the effect of the glass on the probe's retention volume.

An essential requirement of this IGC method is to introduce probe vapor into the GC column approaching infinite dilution. Reliable thermodynamic data from IGC measurements require the probe to be at infinite dilution relative to the surface of the test solid (i.e. to fit the regime of Henry's law). This was achieved using a saturator [11.37] – a thermally isolated bed of inert material soaked with the probe liquid. Helium gas, spiked with methane (948 volumes per million), was passed through the saturator to form a gas stream saturated with the probe vapor. The vapor was injected into the GC column by a Valco gas-sampling valve with an internal volume of 0.1 μl. Varying the temperature of the saturator controlled the concentration/vapor pressure of the probe molecule. The vapor pressure was calculated using Antoine's equation:

$$\log P = A - \frac{B}{C + T} \tag{11.21}$$

where P is the pressure in mmHg, T is the temperature in °C and A, B and C are solvent-specific constants. As the saturator was kept below room temperature, there was no need to heat the pipes between the saturator and the column oven. The probe vapor did not condense out during the transfer to the GC column.

The partial pressure of the vapor in helium was typically 1–0.05 kPa. Owing to the small volume and low concentration of the vapor, the typical amount of probe injected on to the column was 10^{-13}–10^{-11} mol. This is of the order of 10^6-fold less than that obtained by injection of liquid samples.

11.3.3 IGC Theory

Thermodynamic properties of the interaction of a vapor with a GC column can be calculated from the corrected net retention volume, V_n, which is defined as the amount of carrier gas required to elute a probe from a column. Consider the chromatographic plot in Fig. 11.9. The corrected net retention volume is calculated according to

$$V_n = (T_p - T_m) \cdot F \cdot J \cdot C \cdot \frac{T_C}{T_{RT}} \tag{11.22}$$

where T_m and T_p are the retention times of the methane marker peak and probe peak, respectively. All columns have dead volumes, which are the sum of the volumes associated with the injector ports, detector ports, etc.. Hence the actual

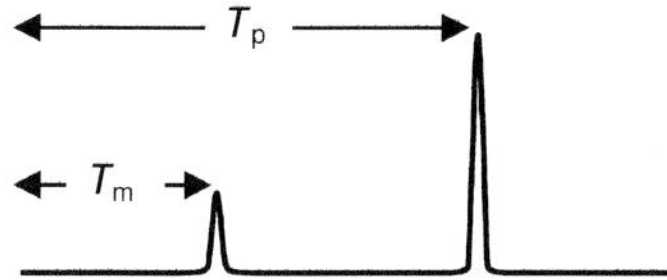

Figure 11.9. A simplified chromatographic plot. T_m and T_p are the retention times of the methane marker peak and the probe molecule, respectively.

time to elute the probe molecule is $T_p - T_m$. F is the flow-rate of the helium carrier gas in cm^3/min, J is the correction for the pressure drop across the column:

$$J = \frac{3}{2} \cdot \frac{\left(\frac{P_i}{P_0}\right)^2 - 1}{\left(\frac{P_i}{P_0}\right)^3 - 1} \tag{11.23}$$

where P_i is the pressure at the inlet of the column and P_0 is atmospheric pressure, and T_C and T_{RT} are the temperatures of the column and the room respectively.

The helium rates were measured using a bubble flowmeter; corrections due to the partial pressure of the detergent/water mixture were made using the equation

$$C = 1 - \frac{P_{H_2O}}{P_0} \tag{11.24}$$

where P_{H_2O} is the partial pressure of water at room temperature.

The Gibbs free energy of adsorption, $\Delta G^0{}_A$, of the probe sample on the column material is related to the retention volume according to the equation [11.38]

$$\Delta G_D^0 = -\Delta G_A^0 = RT \ln\left(\frac{V_n P_0}{Sg\pi_0}\right) \tag{11.25}$$

where $\Delta G_D{}^0$ is defined as the 'free energy of desorption of 1 mol of solute from a reference adsorption state defined by the bidimensional spreading pressure π_0 of the adsorbed film to a reference gas phase state defined by the partial pressure P_0 of the solute. The surface area of the particles is defined as S and g is the mass of column material.

De Boer defined the reference states as $P_0 = 1.013 \times 10^5$ Pa and $\pi_0 = 3.38 \times 10^{-4}$ N/m [11.39] whereas Kimball and Rideal [11.40] defined P_0 as 1.013×10^5 Pa and π_0 as 6.08×10^{-5} N/m. It is not necessary to know the column filler's surface area and standard states for IGC measurements as the values drop out during the IGC calculations. For the sake of consistency and labeling of diagram axes, we used the Kimball and Rideal's convention. The surface pressure π_0 is analogous to P_0. It is the pressure exerted by molecules absorbed on the surface separated by an average distance similar to that observed by the same molecules in the gaseous state.

Equation (11.25) assumes that Henry's law is obeyed and that the column material is not saturated with probe molecules. This can be tested in two ways. The first of these is to increase the vapor concentration by raising the saturator temperature. If the retention volume remains constant in the concentration range studied than the probe behaves as if it were at infinite dilution. Another measure of the infinite dilution regime is to plot the natural logarithm of the retention volume versus the inverse of temperature in kelvin. The plot should be linear.

The Gibbs free energy of adsorption, ΔG^0_A [Eq. (11.25)], can be related to the work of adhesion, W_A, according to the equation [11.41]

$$\Delta G^0 \approx NaW_A \tag{11.26}$$

where N is Avogadro's number and a is the cross-sectional area of the molecule. The work of adhesion can be split into contributions due to dispersion, acid-base and dipole interactions:

$$W_A = W_A^D + W_A^{AB} + W_A^{dipole} \tag{11.27}$$

The dispersive surface energy of the solid is calculated by measuring the adsorption behavior of non-polar alkane probe molecules in the IGC column. Two methods for calculating the dispersive surface energy of a substrate by IGC were developed by Schultz and Lavielle *et al.* [11.41] ('Schultz method') and Anhang and Gray [11.42] ('Gray method'). According to the Shultz method, the total work of interaction is calculated according to the equation

$$RT \ln\left(\frac{V_n P_0}{Sg\pi_0}\right) = 2N\sqrt{\gamma_S^D} \cdot a\sqrt{\gamma_L^D} \tag{11.28}$$

where $\gamma_S{}^D$ and $\gamma_L{}^D$ are the dispersive surface energies of the GC column's filler and probe liquid, respectively, and a is the cross-sectional area. The Gray method is similar: the solid's dispersive surface energy is proportional to ΔG_{CH_2}, the square of the free energy of interaction of the $(n + 1)$th alkane minus the free energy of the nth alkane:

$$\gamma_S^D = \frac{1}{?_{CH_2}}\left(\frac{\Delta G_{CH_2}}{2Na_{CH_2}}\right)^2 \tag{11.29}$$

where γ_{CH_2} is the estimated surface energy of a single CH_2 group (35.6 mJ/m^2) and a_{CH_2} is the cross-sectional area of a CH_2 group (6 Å^2).

The Schultz method assumes that both the surface free energy of the probe molecules and the cross-sectional area of the probe molecule do not change with temperature. Moreover, the method assumes that the surface free energy of the probe molecule at infinite dilution is the same as the surface free energy of the molecule in the liquid state. The Gray method assumes that dispersive free energy of the CH_2 group is the same as the surface free energy of a CH_2 group in a closely packed reference solid analogous to polyethylene. Likewise, the cross-sectional area of the CH_2 group is assumed not to vary with temperature.

In addition to dispersive interactions, an important contribution to the work of adhesion is made by specific interactions, the most important being acid-base interactions. A number of methods have been developed to estimate the specific free energy of interaction of a probe molecule with a surface using data from GC measurements. They all rely on calculating the difference in the free energy of the

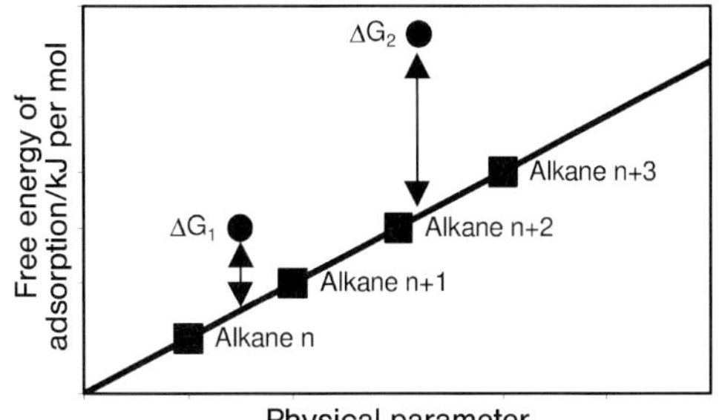

Figure 11.10. Determination of the specific free energy of adsorption between a probe molecule and the surface.

polar probe molecule (e.g. ΔG_1 and ΔG_2) relative to a series of linear alkane molecules for the same comparative physical parameter. This is summarized in Fig. 11.10.

The specific heat of adsorption, ΔH_{sp}, is calculated from the equation

$$\Delta G_{sp} = \Delta H_{sp} - T\Delta S_{sp} \tag{11.30}$$

This parameter has been correlated with the acid-base properties of the surface [11.43] according to

$$\Delta H_{spc} = K_A DN + K_D AN^* \tag{11.31}$$

where DN and AN^* are the Gutmann-Mayer donor number and modified acceptor number [11.44], respectively. The donor number is a measure of the molecule's ability to donate electron pairs (i.e. Lewis base). It is defined as the molar enthalpy of reaction between the sample and antimony pentachloride in a dilute solution of 1,2-dichloroethane. The modified acceptor number is a measure of the material's electron pair acceptor properties (i.e. Lewis acid) and is derived, prior to modifications, from the NMR shift of triethylphosphine oxide in the sample of interest.

11.3.4 Typical IGC Results for HMX and RDX Surfaces

Both HMX and RDX have a number of different surfaces, the proportion of which depends very much on the morphology of the crystal. Therefore, the surface energy of different RDX crystals should differ in proportion to the area of different crystallographic surfaces. However, contact angle studies [11.45] have shown that the dispersive surface energies of the different HMX surfaces are similar. Hence it is reasonable to expect the IGC experiments to be consistent despite the surface energy being measured over an average of surfaces. The dispersive energy calculations compare well with those measured using contact angle studies. The great advantage of IGC measurements is that the surface energy is measured over a range of temperatures – from the mixing temperatures at which propellants and PBXs are formulated down to service temperatures.

The dispersive surface energies of HMX and RDX as measured by the Schultz method is illustrated in Fig. 11.11. At 25 °C, the dispersive surface energy of RDX as

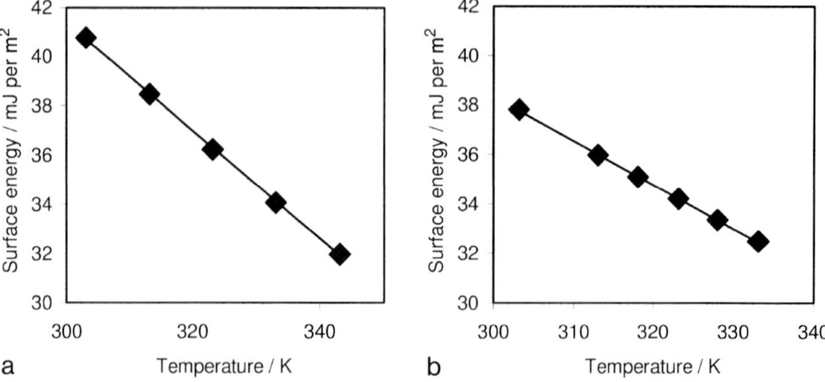

Figure 11.11. The dispersive surface energy of (a) RDX and (b) HMX as measured by the Shultz method.

calculated by the Schultz method (41.8 mJ/m^2) compared well with the value calculated by the Gray method (41.7 mJ/m^2). In the case of HMX, there was a slight difference between the dispersive surface energy calculated by the two methods. At 25 °C, the dispersive free energy of HMX was 42.1 mJ/m^2 (Gray) and 38.4 mJ/m^2 (Schultz).

Unlike the dispersive surface energy calculations, a number of methods have been developed to study the specific acid-base characteristics of the solid's surface. Early empirical methods used the boiling-point of the probe molecules [11.46] and the vapor pressure of the probe molecules [11.43] as a basis for comparing polar probes against the alkane probes. Work at DERA found that these methods could be unreliable – some apparently polar molecules appeared to exhibit lower energies of interaction than that measured for the non-polar probes. Moreover, the two methods appeared to yield contradictory results. Schultz and Lavielle [11.41] and Donnet *et al.* [11.47] derived more rigorous methods for measuring the specific energy of interaction.

The Schultz method [11.41] is based on measuring the difference between the energy of adsorption of polar probes and non-polar probes using a function of surface energy as a comparative property:

$$\theta_L = K \cdot \sqrt{(h\nu_S)} \cdot \alpha_{0 \cdot S} \sqrt{(h\nu_L)} \cdot \alpha_{0 \cdot L} \tag{11.32}$$

the abscissa function being $a\ (\gamma^D{}_L)$. This method assumes that the probe's surface area of interaction, a, is the same for all surfaces and is independent of temperature. Likewise, it is assumed that the dispersive surface energy of the probe molecule, $\gamma^D{}_L$, is also independent of temperature. This method of analysis does not take into account any specific orientation induced by the nitramines surfaces. Likewise, it is assumed the bulk properties of the probe molecule in the liquid state are the same as in the vapor state.

According to Table 11.5, both butanol and acetonitrile appear to interact readily with the surfaces of RDX and HMX. Note also that ethyl acetate, benzene and

Table 11.5. Summary of the specific free energy of adsorption of various non-polar probes on RDX and HMX using both the Schultz and Donnet analysis methods.

Probe molecule	Schultz method				Donnet method			
	RDX		HMX		RDX		HMX	
	ΔH_{sp} (kJ/mol)	ΔS_{sp} (J/K/mol)	ΔH_{sp} (kJ/mol)	ΔS_{sp} (J/K/mol)	ΔH_{sp} (kJ/mol)	ΔS_{sp} (J/K/mol)	ΔH_{sp} (kJ/mol)	ΔS_{sp} (J/K/mol)
Ethyl acetate	−0.4	10	18.7	−38	5.9	−0.1	22.2	−38
Benzene	−3.3	11	7.2	−20	4.7	−2.0	12.3	−22
Acetone	4.8	−0.8	10.3	−7	16.6	−19.3	19.2	−14
THF	6.2	−12	11.0	−20	18.6	−31.2	20.1	−27
CCl_4	3.1	−14	2.5	−10.5	8.5	−22.3	5.0	−8
Nitromethane	5.1	−13	6.1	−4.9	30.7	−52.4	27.3	−29
n-Butanol	27	−46	20.0	−19.3	31.2	−52.3	21.7	−16
Acetonitrile	29.4	−45	29.8	−35.7	36.7	−56.2	34.8	−37
Toluene	−9.5	−21	–	–	3.6	1.1	–	–
Chloroform	–	–	3.8	−7.7	–	–	9.4	−9
Diethyl ether	–	–	13.7	−37.4	–	–	12.0	−29
Nitroethane					2.8	20.2	–	–
Dioxane					11.7	−13.2	18.7	−23
DMF					35.7	−41.5	–	–

toluene appear to have negative heats of adsorption on RDX. This suggests that either the molecules are repulsed from the surface of RDX or the adsorption is such that the area of interaction is not the same as the area of interaction as measured on a neutral solid such as polyethylene as used by Schultz and Lavielle.

Another way of calculating the specific free energy of adsorption was developed by Donnet *et al.* [11.47] using a simplified version of the London equation. The potential energy of interaction, Θ_L, of two species which exchange dispersion forces only, is described by the equation

$$\theta_L = K \cdot \sqrt{(h\nu_S)} \cdot \alpha_{0 \cdot S} \sqrt{(h\nu_L)} \cdot \alpha_{0 \cdot L} \tag{11.33}$$

where

$$K = \frac{3}{4} \cdot \frac{N_A}{(4\pi\varepsilon_0)^2} \cdot \left(\frac{1}{r_{S,L}}\right)^6$$

ν_L is the characteristic electronic frequency of the probe molecule, ν_S is the characteristic electronic frequency of the solid, $\alpha_{0 \cdot S}$ is the deformation polarizability of the solid, $\alpha_{0 \cdot L}$ is the deformation polarizability of the probe, $r_{S,L}$ is the distance between the probe and the surface, N_A is Avogadro's number and ε_0 is the permittivity in a vacuum.

Donnet *et al.* have shown that for interactions involving non-polar molecules, e.g. alkanes, the potential energy of interaction given by the London equation is proportional to the dispersive surface energy of the surface. Hence the specific free energy of adsorption of polar probes can be calculated using the function $(h\nu_S)^{1/2} \cdot \alpha_{0 \cdot L}$ as the abscissa in Fig. 11.10.

Table 11.5 summarizes the specific heat of interaction of polar molecules with RDX and HMX using the Donnet method calculations. The absolute magnitude of the specific free energy of adsorption is larger for most of the probes compared with the previous methods of analysis. However, there appear to be some similarities between the Schultz and the Donnet methods. Acetonitrile appears to interact readily with the surface of RDX and HMX. Likewise, butanol and nitromethane have hight specific free energies of adsorption on HMX and RDX. The nitromethane data contrast with the data calculated by the Schultz surface energy method. According to the Schultz method, nitromethane has a smaller interaction energy than that calculated by the Donnet method.

Both analysis methods indicated that the specific surface energy of adsorption was smaller than the dispersive component, suggesting that the dispersive surface energy of RDX and HMX is greater than the specific surface energy. This agrees qualitatively with contact angle measurements on RDX [11.48] and HMX [11.45].

A number of workers have correlated the specific heat of adsorption with the acid-base properties of solid fillers in rubber composites. The product of the acid and base constant was found to correlate strongly with the physical properties of solid-filled composites. We therefore attempted to measure acid-base properties of RDX and HMX. However, with both the Schultz and Donnet methods for calculating ΔH_{sp}, the correlation between the specific heats of adsorption and the acid-base donor numbers is poor (Figs 11.12 and 11.13). This may be due to limitations of the donor/acceptor number scales – it may be unreasonable to correlate bulk effects in solutions with diluted IGC interactions. Alternatively, the poor correlation may be due to the explosive's heterogeneous surfaces. The different nitramine crystal faces may exhibit similar dispersive surface energies but different specific surface energies. This problem has recently been addressed in the literature for non-energetic materials [11.49–11.51].

Additionally, the two analytical methods yield different acid-base numbers (Table 11.6). This may be due in part to the empirical nature of the acid-base scales. The molecular interactions of the solvents in the bulk may deviate substantially from that in the infinitely dilute gas phase. Certainly some points appear to deviate consistently from the trend. Benzene appears to interact with the nitramines surfaces more than one would expect based on the classical acid-base schemes. Nevertheless, general trends can be discerned from the data. Both HMX and RDX have relatively high K_D (0.8 and 0.4, respectively) and lower K_{A^*} numbers (0.7 and 0.2, respectively). Hence, according to the acid-base model both explosives acts as weak Lewis bases.

Table 11.6. Acid-base characteristics of RDX and HMX.

Explosive and experiment type	K_{A^*}	K_D
RDX (surface energy term)	0.7	0.2
RDX (polarization term)	0.2	0.4
HMX (surface energy term)	0.1	0.6
HMX (polarization term)	0.7	0.8

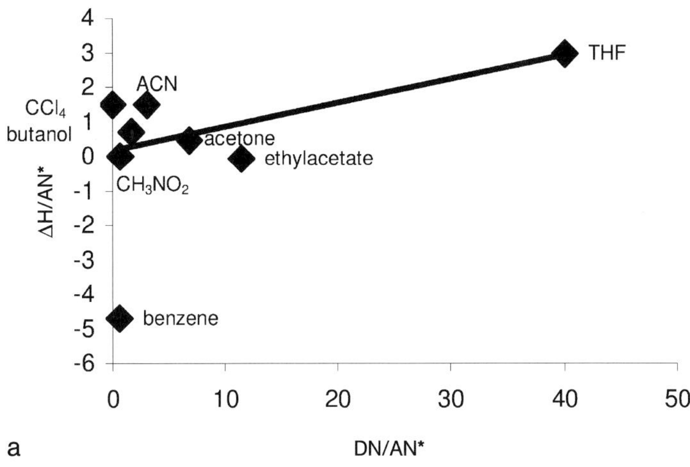

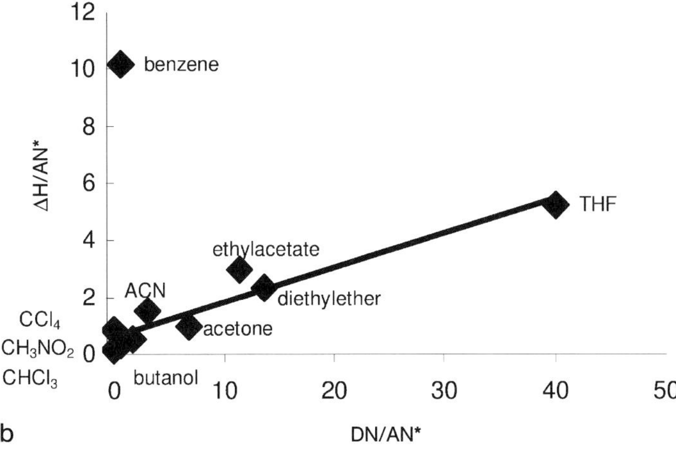

Figure 11.12. Relationship between ΔH_{sp} (Shultz method) on (a) RDX and (b) HMX and the *DN* and AN^* scale.

With regard to selecting bonding agents for nitramine surfaces, it may be simpler to choose materials based on specific heat of adsorption. For both nitramines, acetonitrile and butanol readily interact with the nitramine surfaces, thereby suggesting that nitrile and alcohol functionalized moieties may exhibit good bonding agent properties.

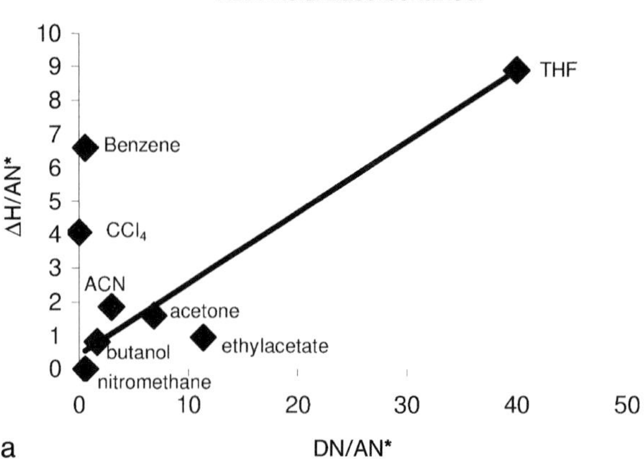

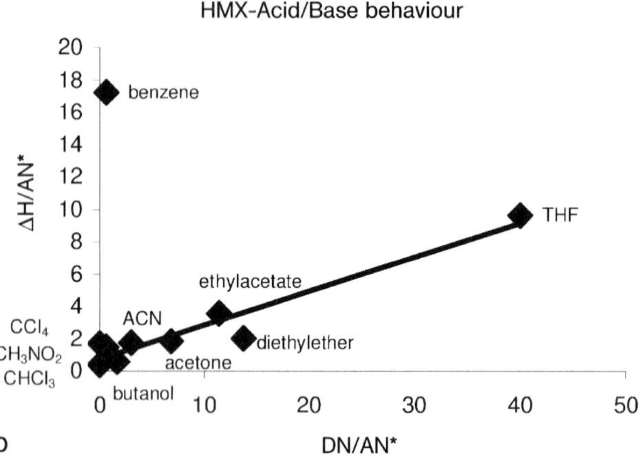

Figure 11.13. Relationship between ΔH_{sp} (Donnet method) on (a) RDX and (b) HMX and the *DN* and AN^* scale.

11.3.5 Inverse Liquid Chromatography

As the name suggests, the inverse liquid chromatography experiment is similar to inverse gas chromatography. The solid phase is characterized by the retention volume of probe molecules of known properties. In IGC, there are essentially two phases of the probe molecule: the probe on the solid surface and the probe in the gaseous state. It is assumed that the helium carrier gas does not interfere with any adsorption processes between the probe molecule and the solid surface. In the case of ILC, the probe molecule has to compete for surface sites with the more highly concentrated solvent species. In IGC, it is assumed that there are no nearest neighbors during the adsorption process. This is obviously not the case for ILC. Consequently, the partitioning of the probe in the solvent and on the surface depends very much on the nature of the solvent. This complicates the analysis process. ILC methods have been developed to study the surface properties of coals [11.52–11.55].

ILC experiments need materials of high surface area, hence the technique is suitable to study micronized explosives – this would not be possible using IGC columns owing to pressure constraints. In this work, micronized explosive surfaces were characterized in a qualitative and semiquantitative fashion. Explosive slurry was packed into glass LC columns (25 mm × 6.6 mm i. d.). The average particle sizes and surface areas of RDX and HMX were 39 μm and 0.21 m^2/cm^3 and 13 μm and 0.92 m^2/cm^3, respectively. Despite the particle size of HMX being ~13 μm, the column could not be described as having good resolving capability – the number of theoretical plates was calculated to be 55! The column was cooled to 1 °C during use by means of water circulating around an outer jacket.

An essential requirement of ILC is that the column filler should not dissolve in the eluting solvent. In this work, UV spectroscopic measurements suggested that over a period of 24 h, the RDX and HMX columns lost < 0.2 mg of material during elution with hexane. This assumed that the explosives exhibited extinction coefficients of around 10 000 $dm^3/mol/cm$ per nitro group at ~210 nm. An additional ILC requirement is that the solvents must be dry – water would compete with the probe molecule at the explosive's surfaces. In this work, we distilled HPLC-grade hexane from calcium hydride into the solvent reservoir under a dry, inert atmosphere. Prior to elution, the hexane was degassed by the application of ultrasound and vacuum. Thereafter, solvent was sparged with dry helium to ensure that no air was admitted into the solvent reservoir. Owing to the absence of oxygen, the UV cut-off point of the solvent was < 200 nm. Air adsorbed in the injected sample could be seen in the chromatogram – this was used as a non-adsorbing marker peak for the calculation of the column's dead volume.

An attempt to quantify the surface of the explosives was made using activity coefficients at the surface and in the solution. The partitioning of the probe molecule from solution on to the surface is represented by the equation

$$K = \frac{\text{weight of probe per cm}^2 \text{of solid}}{\text{weigh of probe per cm}^3 \text{ of eluent}} = \frac{V_n}{Sg} \tag{11.34}$$

where V_n is the retention volume of the probe (corrected for the dead time of the column), S is the surface area and g is the mass of the column material. The partition constant can be calculated from the activity coefficients according to the followinf equations:

$$\mu_{sol} = \mu_{sol}^0 + RT\ln(x_{sol}\gamma_{sol})\,;\; \mu_{sur} = \mu_{sur}^0 + RT\ln(x_{sur}\gamma_{sur}) \tag{11.35}$$

$$\Delta G_a = -RT\ln\left(\frac{V_n\mu_{sol}^0}{Sg\mu_{sur}^0}\right) \tag{11.36}$$

This assumes that the diluted probe molecule obeys Henry's law and that the probe molecule in hexane behaves as an ideal solution.

The activity coefficient of the probe molecule is a measure of the non-ideal behavior of the probe in solution and at the surface. In this work, it has been assumed that there are no specific interactions between the hexane eluent and the probe molecules, i. e. no acid-base, polar and hydrogen bonding interactions take place between the probe and the solvent. It has also been assumed that the dispersive interaction between the hexane and the different probe molecules is approximately the same. This condition would not be valid for solvents with an ability to undergo specific interactions such as polar, hydrogen bonding or acid-base interactions. It has also been assumed that there is little change in the entropy of the probe molecule when it undergoes adsorption on to the surface of the solid from the solution. This contrasts with IGC analysis. The change of entropy from a gaseous state to a liquid/solid state on the surface of the GC column is relatively large.

The enthalpy of mixing two solvents is related to the cohesive energy of the components [11.56] according to the equation

$$\Delta H_{\mathrm{a}} = V \cdot \Phi_1 \cdot \Phi_2 \cdot (\partial_1 - \partial_2)^2 \tag{11.37}$$

where V is the volume of the mix, Φ_n the volume fractions of the nth components and ∂_n is the cohesion parameter of the nth component.

The cohesion parameter is a measure of the cohesive energy of a non-polar solvent. It can predict the mutual solubility of two components and can be related to a number of other physical measurements such as activity coefficients, polymer solvent interaction parameters, van der Waals gas constant, critical pressure, surface tension and dipole moments [11.57]. The cohesion parameter is also known as the solubility parameter. It gives a measure of the solubility behavior of the material – it measures the bulk properties of the material rather than just surface behavior. In this work, we used the cohesion parameter scale to give a measure of the surface properties of HMX and RDX. Consequently, the value of the surface cohesion parameter will be different to that calculated by group contribution methods for the bulk cohesion parameter.

In this work, we related the retention volume to the cohesive parameter according to the equation

$$RT \ln\left(V_n k\right) \approx K\left(\partial_{\mathrm{solid}} - \partial_{\mathrm{probe}}\right)^2 \tag{11.38}$$

We do not know the values of k and K, but by employing several molecular probes, one can use simultaneous equations to solve for $\partial_{\mathrm{solid}}$.

In an analogous way to surface energy, the cohesive parameter can be divided into dispersive, hydrogen bonding and polar components according to the equation

$$\partial^2 = \partial^2_{\mathrm{dispersive}} + \partial^2_{\mathrm{polar}} + \partial^2_{\mathrm{hydrogen\ bonding}} \tag{11.39}$$

Assuming that only dispersive interactions take place with the solvated probe and the solid surface, probe molecules with large dispersive cohesion parameters (relative to the polar and hydrogen bond contributions) can be used to study dispersive characteristics of the explosive surface. Note that probe molecules do not necessarily require zero ∂_{polar} and $\partial_{hydrogen\ bonding}$ contributions to the cohesion parameter. The total cohesion parameter is the square root of the sum of the squares of the cohesion contributions. Hence, in the case of chloroform, ∂_{polar} and $\partial_{hydrogen\ bonding}$ are 3.1 and 5.7, respectively, and $\partial_{dispersive}$ is 17.8 $MPa^{1/2}$. The total cohesion parameter of chloroform is 19.0, which is similar enough to the dispersive component to class chloroform as a molecule which engages in dispersive interactions only.

We can qualitatively rank the extent of interaction of the probe molecules with HMX surfaces (Fig. 11.14) based on their retention volumes. Generally, the alcohols appear to interact with the surface of HMX better than most of the other probe molecules. Both nitro- and nitrile-containing species also appear to interact well with the surface. Amines, DMF and pyridine absorb readily on to the surface. The halogen-substituted materials and isopropyl nitrate were eluted through the column relatively quickly. That is, the energy of interaction of these materials with the HMX column was low. Although the peaks are broad, they appeared to be symmetrical and hence the retention volume should reflect the thermodynamic interaction of the probe molecules with the surface of HMX.

The total cohesion constant of HMX was measured by solving all permutations of the simultaneous equations for the adsorption of the following probes on to the surface of the chromatographic column: nitrohexane, *N*-methylformamide, DMF, propionitrile, nitropropane, butyronitrile, nitrobutane, acetone, hexanenitrile, heptanenitrile, nitromethane, nitroethane and octanenitrile. The average cohesion parameter of HMX was found to be 22 ± 6.5 $MPa^{1/2}$. The large error is probably due to the inability to discriminate between the varying degrees of the polar, dispersive and hydrogen bonding interactions of the probe molecules with HMX.

An average dispersive cohesion parameter of HMX was obtained by solving simultaneous equations for the following probes: benzene, chloroform, carbon tetrachloride, dichloromethane, 1,2-dichloroethane and 1,1,2-trichloro-trifluoroethane. $\partial_{dispersive}$ was found to be 17.1 ± 0.8 $MPa^{1/2}$. The polar and hydrogen bonding component of the HMX cohesion parameter was found to be 14 $MPa^{1/2}$. Note that this is less than that observed for the dispersive component of the HMX surface.

Generally, DMF, DMSO, *N*-methylformamide, ethyl sulfone and diethyl sulfate appear to interact readily with the surface of RDX. Nitriles, amines, alcohols and species with nitro groups also interact with the explosive surface reasonably well. Isopropyl nitrate, diiodomethane and bromoform do not interact with the surface of RDX. The correlation of the remaining peaks is relatively poor in comparison with that observed for the HMX column. It was not possible to calculate the dispersive component of the cohesion parameter for the present RDX column owing to poor data quality. Simultaneous equations were solved using the retention volumes of the following probes: nitromethane, nitroethane, nitrobutane, nitrobenzene and nitropentane. The cohesion parameter of RDX was found to be 23 ± 8 $MPa^{1/2}$.

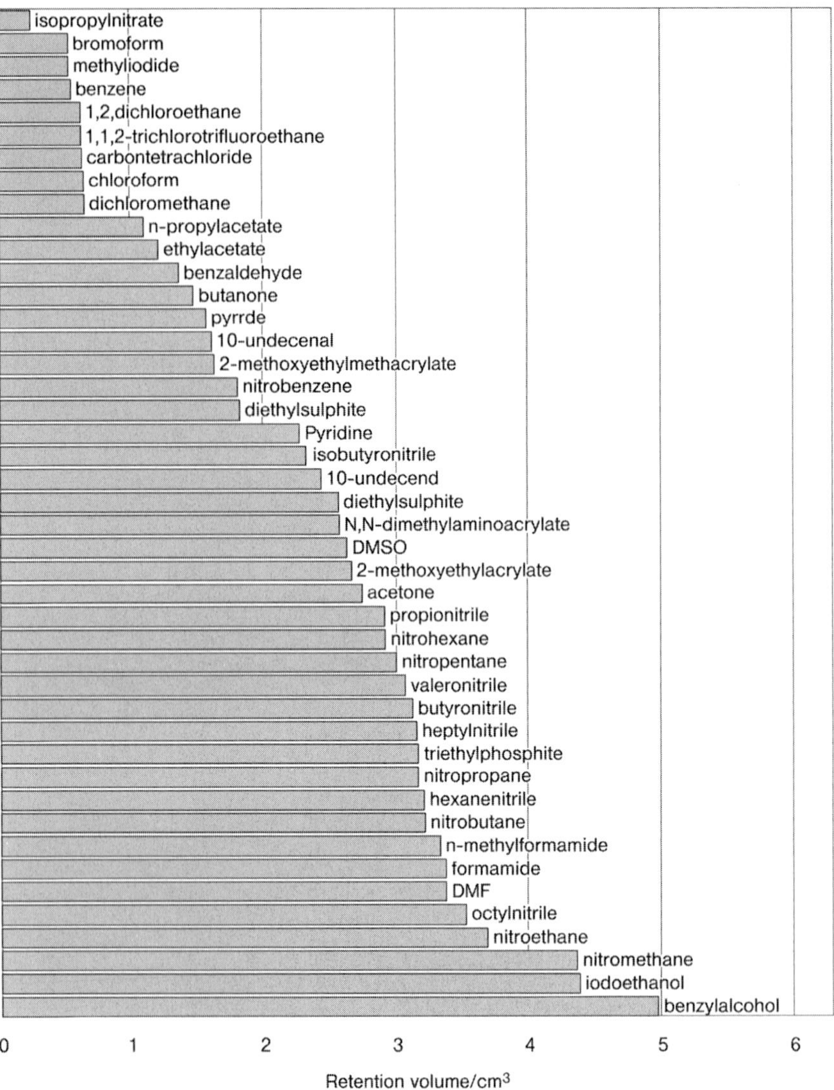

Figure 11.14. Retention volume of probe molecules in HMX column.

11.3.6 Conclusions

The dispersive surface energies of RDX and HMX are similar. At 25 °C, the dispersive surface energies of RDX and HMX are 42.1 and 41.7 mJ/m^2, respectively, as calculated by the Shultz method. The specific heat of adsorption of polar molecules on to the surfaces of RDX and HMX does not appear to correlate well with the AN^* and DN parameters.

Although, in this work, RDX and HMX surfaces did not show a strong acid-base correlation, the results do indicate the type of species that interact well with the surfaces of explosives. Both RDX and HMX readily interact with acetonitrile, which suggests that polymers with nitrile groups could be used as potential bonding agents. Butanol has a high specific heat of interaction with both types of explosive. Hence polymeric alcohols may be suitable bonding agents. These materials could only be used if the rate of reaction of the curing agent with the energetic binder is greater than that with the alcohol groups on the bonding agents. According to the Donnet method, nitromethane interacts strongly with both HMX and RDX.

The ILC technique, although less well developed, also qualitatively indicates that species with nitro groups, nitrile groups, amines, DMF, DMSO and some alcohols interact reasonably well with the surface of both HMX and RDX. Materials containing these groups may have potential as bonding agents.

The semiquantitative analysis of the nitramines surfaces by ILC indicates that the dispersive component of the cohesion parameter is greater than the polar component. This agrees qualitatively with the IGC data and RDX contact angle measurements performed by other workers.

It is recognized that both RDX and HMX have a number of different surfaces. Hence the measurements in this work represent the interaction of the dilute probe molecule with the most energetically favorable sites. These sites are unlikely to be the same for each polar probe molecule. Therefore, it may not be meaningful to assign average donor and acceptor numbers to nitramine surfaces using infinitely dilute IGC. New methods for measuring the distribution of surface energies by IGC techniques have recently been developed and could be applied to the study of nitramine surfaces [11.49–11.51].

11.4 References

11.1 Oberth E (1974) *Bonding Agent Review*, CPIA Publ. No. 260, pp. 337–368.

11.2 Kaully T, Kimmel T (1998) Failure mechanism in PBX, In: *Proc. 29th Int. Annual Conference of ICT, Karlsruhe.*

11.3 Döefler H-D (1994) *Grenzflächen- und Kolloidchemie*, VCH, Weinheim.

11.4 Owens DK, Wendt RC (1969) Estimation of the surface free energy of polymers, *J. Appl. Polym. Sci.* **3**, 1741–1747.

11.5 Wu S (1973) Polar and non-polar interaction in adhesion, *J.Adhesion* **5**, 39–55.

11.6 Zisman WA (1964) Relation of the equilibrium contact angle to liquid and solid constitution, *Adv. Chem. Ser.* **43**, 1–51.

11.7 Defay R, Pétré G (1971) Dynamic surface tension, *Surf. Colloid Sci.* **3**, 27–79.

11.8 Miller R, Kretzschmar G (1991) Adsorption kinetics of surfactants at fluid interfaces, *Adv. Colloid Interface Sci.* **37**, 97–121.

11.9 Washburn EW (1921) The dynamics of capillary flow, *Phys. Rev.* **17**, 273.

11.10 Schubert H (1982) *Kapillarität in Porösen Feststoffsystemen*, Springer, Berlin.

11.11 Siebold A, Walliser A, Nardin M, Oppliger M, Schultz J (1997) Capillary rise for thermodynamic characterization of solid particle surface, *J. Colloid Interface Sci.* **186**, 60–70.

11.12 Kossen NWF, Heertjes PM (1965) The determination of the contact angle for systems with a powder, *Chem. Eng. Sci.* **20**, 593–599.

11.13 Grundke K, Bogumi, T, Gietzelt T, Jacobasch H-J, Kwok D-Y, Neumann AW (1996) Wetting measurements on smooth, rough and porous solid surfaces, *Prog. Colloid Polym. Sci.* **101**, 58–68.

11.14 Bartell F (1927) Determination of the wettability of a solid by a liquid, *Ind. Chem. Phys.* **19**, 1277.

11.15 Schindler B (1973) Über einen Zusammenhang zwischen Benetzungskinetik und Oberflächenenergie pulverförmiger Stoffe, *Dissertation*, Universität Stuttgart.

11.16 Grundke K, Augsburg A (2000) On the determination of the surface energetics of porous polymer materials, *J. Adhesion Sci. Technol.* **14**, 765–775.

11.17 Teipel U, Marioth E, Heintz T, Mikonsaari I (1998) Bestimmung der Oberflächenenergie von Polymerbindern und Explosivstoffpartikeln, In: *Proc. 29th Int. Annual Conference of ICT, Karlsruhe.*

11.18 Palzer S, Sommer K (1998) Benetzbarkeit von Pulvern und Pulvergemischen – Einflüsse auf die Netzungseigenschaften von Pulvern und deren verfahrenstechnische Bedeutung, *Chem. Ing. Tech.* **9**, 70.

11.19 Meng-Jiao Wang, Wolff S (1992) Filler-elastomer interactions. Part VI. Characterization of carbon blacks by inverse gas chromatography at finite concentration, *Rubb. Chem. Technol.* **65**, 890–907.

11.20 Sa M M, Sereno A M (1992) Effect of column material on sorption isotherms obtained by inverse gas chromatography, *J. Chromatogr.* **600**, 341–343.

11.21 Roles J, Guiochon G (1992) Experimental determination of adsorption isotherm data for the study of the surface energy distribution of various solid surfaces by inverse gas-solid chromatography, *J. Chromatogr.* **591**, 233–243.

11.22 Jagiello J, Papirer E (1991) A new method of evaluation of specific surface area of solids using inverse gas chromatography at infinite dilution, *J. Colloid Interface Sci.* **142**, 232–235.

11.23 Song H, Parcher JF (1990) Simultaneous determination of Brunauer-Emmett-Teller (BET) and inverse gas chromatography surface areas of solids, *Anal. Chem.* **62**, 2313–2317.

11.24 Okonkwo JO, Colenutt BA, Theocharis CR (1992) Inverse gas chromatography of uncoated and stearate coated calcium carbonate, In: Mottola HA, Steinmetz JR (eds), *Chemically Modified Surfaces*, Elsevier, Amsterdam, pp. 119–129.

11.25 Munk P (1991) Polymer characterization using inverse gas chromatography, In: Barth HG, Mays JW (eds), *Modern Methods of Polymer Characterization*, p. 151.

11.26 Contescu CR, Jagiello J, Schwarz JA (1991) Study of the acidic properties of pure and composite oxides by inverse gas chromatography at infinite dilution, *J. Catal.* **131**, 433–444.

11.27 Li J-J, Zhang Y-K, Dallas AJ, Carr PW (1991) Measurement of solute dipolarity/polarizability and hydrogen bond acidity by inverse gas chromatography, *J. Chromatogr.* 550, 101–134.

11.28 Panzer U (1991) Characterization of solid surfaces by inverse gas chromatography, *Colloids Surf.* **57**, 369–374.

11.29 Tiburcio AC, Manson JA (1991) Acid-base interactions in filler characterization by inverse gas chromatography, *J. Appl. Polym. Sci.* **42**, 427–438.

11.30 Nardin M, Papirer E (1990) Relationship between vapor pressure and surface energy of liquids: application of inverse gas chromatography, *J. Colloid Interface Sci.* **137**, 534–545.

11.31 Lloyd DR, Ward TC, Schreiber HP, Pizana CC (eds) (1989) *Inverse Gas Chromatography, Characterization of Polymers and Other Materials, ACS Symposium Series*, American Chemical Society, Washington, DC.

11.32 Botija JM, Rego JM (1992) Determination of the acid-base characteristics of RDX by means of inverse gas chromatography, In: *Proc. 23rd Int. Annual Conference of ICT, Karlsruhe*, p. 60/1.

11.33 Bailey A, Bellerby JM, Kinloch SA (1992) The identification of bonding agents for TATB/HTPB polymer bonded explosives, *Philos. Trans. R. Soc. Lond. A* **339**, 321–333.

11.34 Gutmann V, Resch G (1982) The unifying impact of the donor acceptor approach, *Stud. Phys. Theor. Chem.* 203–218.

11.35 Mayer U, Gutmann V, Gerger W (1977)

NMR-spectroscopic studies on solvent electrophilic properties. Part II. Binary aqueous-non aqueous solvent systems, *Monatsh. Chem.* **108**, 489–498.

11.36 Drago RS, Parr LB, Chamberlain CS (1977) Solvent effects and their relationship to the E and C equation, *J. Am. Chem. Soc.* **99**, 3203–3209.

11.37 Torry SA, Cunliffe AV, Tod D (1999) Surface characteristics of HMX and RDX as studied by inverse gas chromatography, In: *Proc. 30th Int. Annual Conference of ICT, Karlsruhe*, p. 86.

11.38 Schultz J, Lavielle L, Martin C (1987) The role of the interface in carbon fiber-epoxy composites, *J. Adhesion* **23**, 45–60.

11.39 de Boer JH (1953) *The Dynamic Character of Adsorption*, Clarendon Press, Oxford.

11.40 Kimball C, Rideal EK (1946), *Proc. R. Soc. Lond. A* **147**, 53.

11.41 Schultz J, Lavielle L (1989) Interfacial properties of carbon fiber-epoxy matrix composites, *ACS Symp. Ser.* 391.

11.42 Anhang J, Gray DG (1982) Surface characterization of polyethylene terephthalate film by inverse gas chromatography, *J. Appl. Polym. Sci.* **27**, 71–78.

11.43 Saint Flour C, Papirer E (1983) Gas-solid chromatography: a quick method of estimating surface free energy variations induced by the treatment of short glass fibers, *J. Colloid Interface Sci.* **91**, 69–75.

11.44 Riddle FL, Fowkes FM (1990) Spectral shifts in acid base chemistry 1. Van der Waals contributions to acceptor numbers, *J. Am. Chem. Soc.* **112**, 3259–3264.

11.45 Yee RY, Adicoff A, Dibble EJ (1980) Surface properties of HMX crystal, In: *JANNAF Combustion Meeting*, pp. 461–468.

11.46 Sawyer DT, Brookman DJ (1986) Thermodynamically based gas chromatographic retention index for organic molecules using self modified aluminas and porous silica beads, *Anal. Chem.* **40**, 1847–1853.

11.47 Donnet JB, Park SJ, Balard H (1991) Evaluation of specific interactions of solid surfaces by inverse gas chromatography. A new approach based on polarizability of the probes, *Chromatographia* **31**, 434–440.

11.48 Wylie P, Tod D (1996) QinetiQ, Fort Halstead, Sevenoaks, Kent, UK, unpublished data.

11.49 Papirer E, Brendle E (1998) Recent progress in the application of inverse gas chromatography for the determination of the acid-base properties of solid surfaces, *J. Chim. Phys. Phys.-Chem. Biol.* **95**, 122–149.

11.50 Katsanos NA, Iliopoulou E, Roubani-Kalantzopoulou F, Kalogirou E (1999) Probability density function for adsorption energies over time on heterogeneous surfaces by inverse gas chromatography, *J. Phys. Chem. B* **103**, 10228–10233.

11.51 Bogillo VI, Shkilev VP, Voelkel A (1998) Determination of free surface energy components of heterogeneous solids by means of inverse gas chromatography at finite concentrations, *J. Mater. Chem.* **8**, 1953–1961.

11.52 Glass AS, Wenger EK (1998) Surface thermodynamics for polar adsorbates on Wyodak coals, *Energy Fuels* **12**, 152–158.

11.53 Kaneko H, Morino M, Takanohashi T, Iino M (1996) Inverse size exclusion chromatography using extraction residues or extracts of coals as a stationary phase, *Energy Fuels* **10**, 1017–1021.

11.54 Hayashi J, Amamoto S, Kusakabe K, Morooka S (1995) Evaluation of interaction between aromatic penetrants and acidic OH groups of solvent-swollen coals by inverse liquid-chromatography, *Energy Fuels* **9**, 1023–1027.

11.55 Hayashi J, Amamoto S, Kusakabe K, Morooka S (1993) Characterization of structural and interfacial properties of solvent-swollen coals by inverse liquid-chromatography technique, *Energy Fuels* **7**, 1112–1117.

11.56 Hildebrand JH, Scott RL (1959) *The Solubility of Non-Electrolytes*, 3rd edn, Reinhold, New York.

11.57 Brandrup J, Immergut EH (eds) (1989) *Polymer Handbook*, 3rd edn, p. 519.

12
Rheology

U. Teipel, A. C. Hordijk, U. Förter-Barth, D. M. Hoffman, C. Hübner, V. Valtsifer, K. E. Newman

12.1
Stationary Shear Flow

Figure 12.1 shows uniaxial stationary shear flow between two long parallel flat plates and the deformation of a volume element subjected to Couette flow.

The lower plate remains stationary while the upper plate moves at a constant velocity u due to the force F applied tangentially. Assuming there is no slip at the wall, the velocity increases linearly from $u = 0$ at the fixed plate to the value u at the moving plate. The shear stress τ and deformation (strain) γ are defined from the parameters in Fig. 12.1 as follows:
shear stress:

$$\tau = \frac{F}{A} \tag{12.1}$$

strain:

$$\gamma = \frac{dx}{dy} = \tan\alpha \tag{12.2}$$

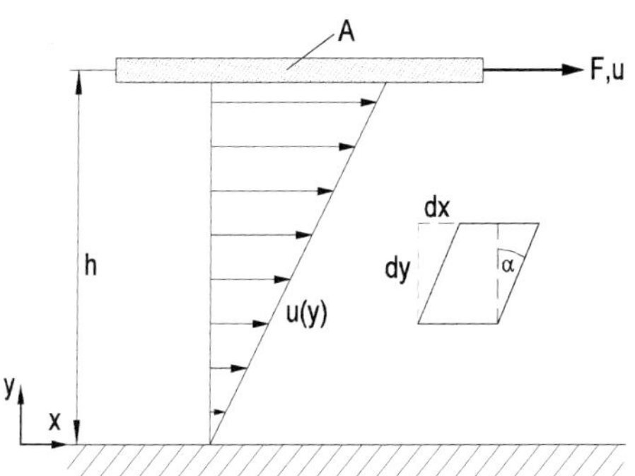

Figure 12.1. Uniaxial stationary shear flow.

Energetic Materials. Edited by Ulrich Teipel
Copyright © 2005 WILEY-VCH Verlag GmbH & Co. KGaA, Weinheim
ISBN: 3-527-30240-9

The shear rate $\dot{\gamma}$ is constant across the entire gap between the two plates and can be defined as follows:

$$\dot{\gamma} = \frac{d\gamma}{dt} = \frac{u}{h} \tag{12.3}$$

By subjecting the fluid to stationary shear flow, material functions such as the shear stress function $\tau(\dot{\gamma})$ and viscosity function $\eta(\dot{\gamma})$ can be determined. The relationship between these functions is given by the following equation for the shear stress function:

$$\tau(\dot{\gamma}) = \eta(\dot{\gamma}) \cdot \dot{\gamma} \tag{12.4}$$

where $\eta(\dot{\gamma})$ is a characteristic material function of a fluid that describes its flow properties when the fluid is subjected to a rheometric flow.

12.2 Flow Behavior of Fluids

When the viscosity η of a fluid is independent of the shear rate $\dot{\gamma}$, its flow behavior is called Newtonian. In this case, the shear stress τ increases linearly with $\dot{\gamma}$:

$$\tau = \eta \cdot \dot{\gamma} \tag{12.5}$$

When the relationship between the shear stress τ and shear rate $\dot{\gamma}$ is non-linear, the fluid is called non-Newtonian. Figure 12.2 shows the shear stress and viscosity functions for a Newtonian fluid and for a typical example of a non-Newtonian fluid.

When the viscosity increases with increasing shear rate, the fluid is said to exhibit dilatant or shear thickening behavior. In contrast, the viscosity of shear thinning fluids decreases with increasing shear rate or shear stress. In this case, the fluid undergoes a shear-induced structural change that causes the fluid to flow more readily. At sufficiently small shear rates, the viscosity function often becomes nearly horizontal, meaning that in this region the viscosity is (nearly) independent of the

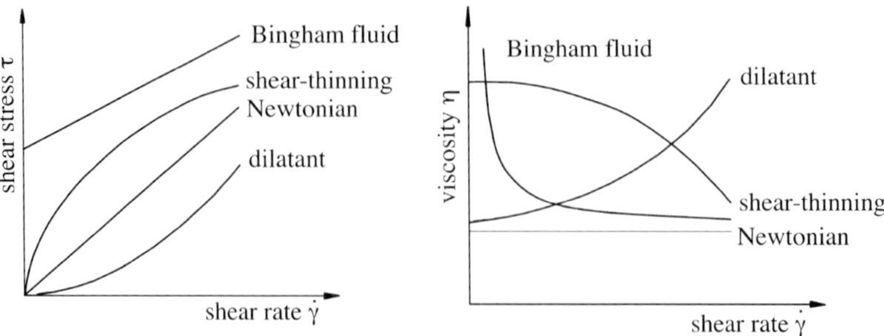

Figure 12.2. Shape of typical flow and viscosity functions for fluids exhibiting various types of flow behavior.

shear rate. In this shear rate range, called the 'lower Newtonian limiting region', the fluid exhibits approximately Newtonian behavior when subjected to stationary shear flow and the shear viscosity in this region is called the limiting viscosity at zero shear rate η_0, defined as follows:

$$\eta_0 = \lim_{\dot{\gamma}\to 0} \eta(\dot{\gamma}) \tag{12.6}$$

The power law equation of Ostwald and de Waele is suitable for describing shear thinning and shear thickening behavior:

$$\tau = K_1 \cdot \dot{\gamma}^n \tag{12.7}$$

where K_1 is the consistency coefficient and n is the flow index. When $0 < n < 1$ the fluid is shear thinning and when $n > 1$ the fluid is shear thickening. In terms of the dynamic viscosity, the power law equation is written in the following form:

$$\eta(\dot{\gamma}) = K_1 \cdot \dot{\gamma}^{n-1} \tag{12.8}$$

Many empirical equations have been proposed in the literature to describe shear thinning flow behavior. Several examples of such equations are reviewed below. Complete descriptions of these and others can be found elsewhere [12.1–12.7].

Equation (12.8) was expanded by Sisko to cover the higher shear rate range:

$$\eta(\dot{\gamma}) = \eta_\infty + K_1 \cdot \dot{\gamma}^{n-1} \tag{12.9}$$

Other models for describing shear thinning flow behavior are those of Ellis (Eq. (12.10)) and Carreau (Eq. (12.11)):

$$\eta(\tau) = \frac{\eta_0}{1 + K_2 \cdot \eta_0 \cdot \tau^{m'}} \tag{12.10}$$

$$\frac{\eta(\dot{\gamma}) - \eta_\infty}{\eta_0 - \eta_\infty} = \frac{1}{\left[1 + (K_3 \cdot \dot{\gamma})\right]^{\frac{m}{2}}} \tag{12.11}$$

where K_2, K_3, m' and m are empirically determined fitting parameters.

When the shear stress applied to a fluid must exceed a threshold value (the so-called yield stress, τ_0) before deformation begins, it is said to exhibit plastic behavior. Below this yield stress, such a fluid system deforms elastically, i. e. reversibly. Very highly concentrated suspensions of solid particles, in which the particles tend to aggregate, exhibit such viscoplastic behavior. Materials that exhibit linear dependence between the shear stress and shear rate above the yield stress τ_0 are referred to as Bingham fluids and can be described by the following equation:

$$\tau = \tau_0 + \eta_B \cdot \dot{\gamma} \tag{12.12}$$

If the material exhibits a non-linear dependence between the shear stress and shear rate above the yield stress, the Herschel-Bulkley equation can be used to model the flow behavior:

$$\tau = \tau_0 + K \cdot \dot{\gamma}^n \tag{12.13}$$

The Hersche-Bulkley equation describes viscoplastic flow behavior well when the fluid exhibits shear thinning (or thickening) behavior at stresses above the yield stress. A more complicated two-parameter model was presented by Casson to describe viscoplastic flow behavior:

$$\sqrt{\tau} = \sqrt{\tau_{0,C}} + \sqrt{\eta_C \cdot \dot{\gamma}} \tag{12.14}$$

In addition to shear rate-dependent non-Newtonian behavior, changes in the flow behavior can also occur that are a function of shear time. One speaks, for instance, of thixotropic flow behavior when the viscosity of a fluid decreases with time when subjected to a constant shear rate.

12.3 Non-stationary Shear Flow

In addition to the material properties that can be determined in stationary shear flow experiments, one can use oscillatory shear flows to characterize certain material properties, in particular the viscoelastic material functions. Compared with stationary shear flow, dynamic loading of a material using oscillatory strain (or stress) is advantageous because it causes little change to the material's quiescent inner structure. Such experiments are thus especially advantageous when testing systems with a deformable dispersed phase. A further advantage of oscillatory experiments is that periodic signals can be measured with higher accuracy, because non-periodic adventitious signals can easily be eliminated by filtering.

During oscillatory shear loading, the fluid is subjected to a periodic (e. g. sinusoidal) deformation $\gamma(t)$ with an amplitude $\hat{\gamma}$ and a radial frequency $\omega = 2\pi f$:

$$\gamma(t) = \hat{\gamma}\sin(\omega t) \tag{12.15}$$

The shear rate $\dot{\gamma}(t)$, which is the time derivative of the deformation $\gamma(t)$, oscillates at an identical frequency as the strain amplitude, but is shifted in phase by an angle $+\pi/2$:

$$\dot{\gamma}(t) = \omega\hat{\gamma}\cos(\omega t) = \omega\hat{\gamma}\sin\left(\omega t + \frac{\pi}{2}\right) \tag{12.16}$$

At sufficiently small strain amplitudes, i.e. in the linear viscoelastic region, the response to an oscillatory (sinusoidal) shear stress input is a sinusoidal shear stress output $\tau(t)$. If the material exhibits viscoelastic behavior, a phase shift δ is observed between the shear stress function $\tau(t)$ and the strain function $\gamma(t)$:

$$\tau(t) = \hat{\tau} \cdot \sin(\omega t + \delta) \tag{12.17}$$

A purely elastic solid has a phase shift $\delta = 0°$ and a purely viscous fluid has a phase shift $\delta = \pi/2$, whereas the phase shift of a viscoelastic fluid ranges from 0° to $\pi/2$.

The shear stress behavior can be described in terms of the frequency-dependent complex modulus $G^*(\omega)$ as follows:

$$\tau(t) = \hat{\gamma}\left|G^*(\omega)\right| \cdot \sin\left[\omega t + \delta(\omega)\right] \tag{12.18}$$

The complex shear modulus is expressed as

$$\left|G^*(\omega)\right| = \frac{\hat{\tau}(\omega)}{\hat{\gamma}} \tag{12.19}$$

The complex shear modulus $G^*(\omega)$ of a viscoelastic material consists of a real and an imaginary term and can be separated accordingly into two material functions, the storage modulus $G'(\omega)$ and the loss modulus $G''(\omega)$. The storage modulus $G'(\omega)$ is a measure of the elastic (reversible) deformation energy stored by the material, whereas the loss modulus $G''(\omega)$ (or viscous component) is a measure of the energy dissipated by the material:

$$\left|G^*(\omega)\right| = \sqrt{G'(\omega)^2 + G''(\omega)^2} \tag{12.20}$$

The ratio of dissipated energy to stored energy is called the loss factor or $\tan\delta$:

$$\tan\delta = \frac{G''(\omega)}{G'(\omega)} \tag{12.21}$$

Analogous to the Newtonian equation for the case of stationary flow, one can define a frequency-dependent complex viscosity $\eta^*(\omega)$ as the ratio of the complex shear modulus $G^*(\omega)$ and the oscillatory frequency ω:

$$\left|\eta^*(\omega)\right| = \frac{\left|G^*(\omega)\right|}{\omega} \tag{12.22}$$

$$\left|\eta^*(\omega)\right| = \sqrt{\eta'(\omega)^2 + \eta''(\omega)^2} \tag{12.23}$$

The elastic and viscous components of the complex viscosity can be expressed as

$$\eta'(\omega) = \frac{G''(\omega)}{\omega} \tag{12.24}$$

$$\eta''(\omega) = \frac{G'(\omega)}{\omega} \tag{12.25}$$

Oscillatory shear experiments must be conducted in the linear viscoelastic range [12.8, 12.9]. In this region the strain amplitude $\hat{\gamma}$ and the resulting shear stress amplitude $\hat{\tau}$ are proportional ($\hat{\tau} \sim \hat{\gamma}$) at an applied frequency ω. This is true only at sufficiently small strain amplitudes. A material is said to exhibit linear viscoelastic behavior when the moduli $G'(\omega)$, $G''(\omega)$ and $G^*(\omega)$ are independent of strain amplitude at a constant applied frequency (see Fig. 12.3).

When experiments are not conducted in the linear viscoelastic region, one should examine the asymptotic transition region between the linear viscoelastic and non-linear viscoelastic region to determine the limiting strain amplitude $\hat{\gamma}_{\text{lim}}$ that characterizes this transition, as shown in Fig. 12.3. This amplitude is the maximum strain at which the deviation ΔG^* of the complex modulus from its maximum (plateau) value $G^*(\hat{\gamma} \rightarrow 0)$ is negligible. For a non-linear substance the dependence of the shear modulus on the strain amplitude should be determined at various frequencies.

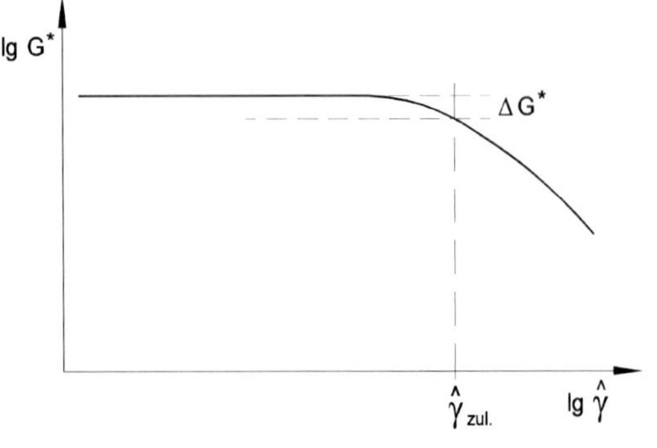

Figure 12.3. The complex storage modulus $G^*(\omega)$ as a function of $\hat{\gamma}$ at a constant frequency ω.

12.4 Rheometers

12.4.1 Rotational Rheometers

Typical rotational rheometers include coaxial cylinders, cone and plate and parallel plate rheometers. Within the measurement gap a smooth, stationary shear flow is effected (also called a 'rheometric' flow). If the moving surface is a coaxial cylinder that rotates at a defined angular speed, it is called Couette flow. If the measurement gap between the cylinders is sufficiently small, then the shear rate $\dot{\gamma}$ across the gap is nearly constant. This is also true for cone and plate flow when the cone is truncated. Because the shear stress τ depends only on the shear rate $\dot{\gamma}$, the shear stress is also constant across the gap. The shear stress effected within the gap is proportional to the torque M, which is measurable. By measuring the torque and using the known geometry of the measurement device, one can determine the shear stress and subsequently calculate the dynamic viscosity (see Eq. (12.4)).

Rheometers in which the strain γ or shear rate $\dot{\gamma}$ are given and the torque M is measured are called shear rate (or strain rate) controlled rheometers. If instead a torque (or shear stress) is applied, the device is called a shear stress controlled rotational rheometer.

12.4.1.1 Coaxial Rotational Rheometers

Coaxial cylinders are divided into two types. If the inner cylinder rotates and the outer remains stationary, the device is termed a 'Searle' system, whereas the converse arrangement is called a 'Couette' rheometer. The Searle system has the disadvantage that the centrifugal forces can lead to flow instabilities (e.g. Taylor vortices), especially at high shear rates. Figure 12.4 shows an example of a Couette rotational rheometer.

From the measured torque M one can determine the shear stress τ in the gap as a function of the radius of the cylinder r:

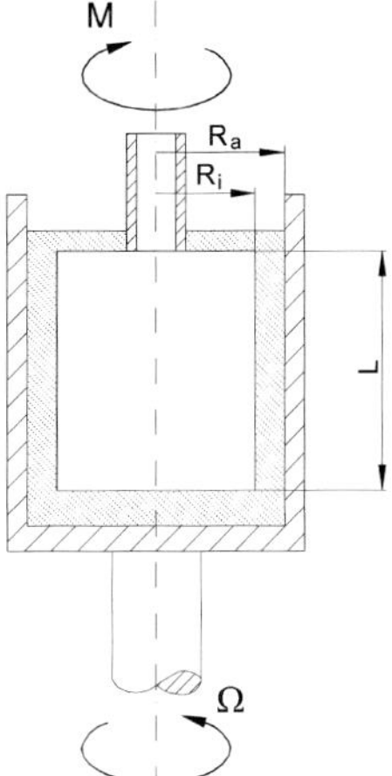

Figure 12.4. A Couette rotational rheometer.

$$\tau = \frac{M}{2\pi \cdot r^2 \cdot L} \tag{12.26}$$

where L is the wetted cylinder length.

For coaxial cylinder systems, the shear rate in the gap is dependent on the material behavior of the fluid system being measured. For Newtonian fluids measured in a Couette device, $\Omega = 2\pi n$:

$$\dot{\gamma} = 2 \cdot \Omega \cdot \frac{R_i \cdot R_a}{R_a^2 - R_i^2} \tag{12.27}$$

The viscosity can then be calculated using the Margules equation [12.7]:

$$\eta = \frac{M}{4\pi \cdot L \cdot R_i^2 \cdot \Omega}\left[1 - \left(\frac{R_i}{R_a}\right)^2\right] \tag{12.28}$$

For non-Newtonian flow behavior, the flow constitutive equation must be considered when taking the derivative of the equations written above. For fluids with unknown material behavior, the shear rate can be calculated using the series expansion method presented by Krieger and Elrod. Complete details are provided elsewhere [12.4, 12.5, 12.7].

12.4.1.2 **Cone and Plate Rheometers**

Using cone and plate fixtures, the fluid is sheared in a narrow gap between a truncated cone and a flat plate. Figure 12.5 shows the basic design of this device.

When small cone angles are used ($\beta < 6°$) the shear rate $\dot{\gamma}$ in the entire gap is constant, allowing the viscosity of the material to be calculated without employing a correction for varying shear rate, even for non-Newtonian fluids. The shear stress τ and shear rate $\dot{\gamma}$ can be determined from the following equations:

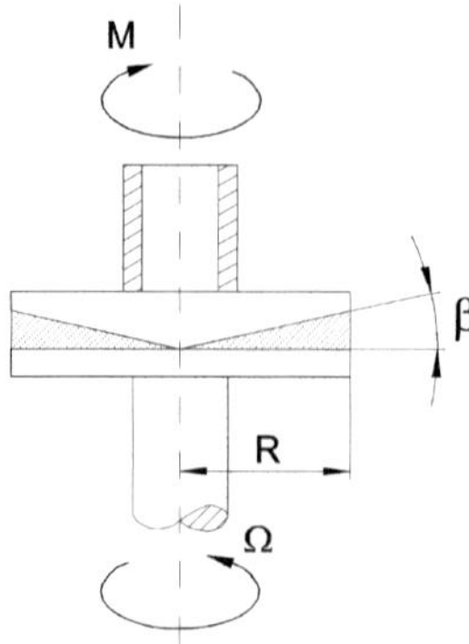

Figure 12.5. Basic design of the cone and plate rheometer.

$$\tau = \frac{3 \cdot M}{2\pi \cdot R^3} \tag{12.29}$$

$$\dot{\gamma} = \frac{\Omega}{\beta} \tag{12.30}$$

By determining the torque M and the angular rotation speed Ω one can calculate the viscosity of both Newtonian and non-Newtonian fluids directly from the equation

$$\eta = \frac{3 \cdot \beta \cdot M}{2\pi \cdot R^3 \cdot \Omega} \tag{12.31}$$

12.4.2 Capillary Rheometer

In contrast to rotational rheometers, the fluid measured in a capillary rheometer flows through a capillary with a cross-section that is either circular, toroidal or slotted. Depending on the properties of the material being investigated, low- or high-pressure capillary rheometers can be employed. Low-pressure capillary rheometers use the hydrostatic pressure of the fluid column as the driving force. The Ubbelohde viscometer is commonly used as a low-pressure capillary device to measure the properties of low-viscosity fluids. High-pressure capillary rheometers are used to determine the viscosity of polymer melts and other polymer-based materials. A particular advantage of a capillary rheometer is that the heat generated by the fluid during shearing is at least in part dissipated to the surrounding environment while the fluid flows through the capillary. The shear rate range accessible with most conventional high-pressure capillary rheometers ranges from 1 to $10^5\ s^{-1}$.

Assuming a shear stress distribution analogous to that found in pipe flow, the shear stress at the wall τ_w is determined for flow through a capillary rheometer as follows:

$$\tau_w = \frac{\Delta p}{L} \cdot \frac{R}{2} \tag{12.32}$$

where L is the capillary length, R is the capillary radius and Δp is the measured pressure loss. For Newtonian fluids, the viscosity can then be determined:

$$\eta = \frac{\pi}{8} \cdot \frac{R^4}{\dot{V}} \cdot \frac{\Delta p}{L} \tag{12.33}$$

where $\dot{V}$ is the volumetric flow determined during the experiment.

When testing non-Newtonian fluids via capillary rheometry, it is only possible to determine an apparent viscosity η_S via Eq. (12.33), because the fluid's viscosity depends on the shear rate, which is a function of the capillary radius and therefore not constant across the capillary cross-section. A correction proposed by Weissenberg and Rabinowitsch can be used to account for the changing shear rate [12.7]. In addition, one must also correct for frictional effects at the entrance to the capillary. For highly viscoelastic fluids, a further correction is necessary to account for the effect of additional pressure losses resulting from elastic deformation and viscous extension as the fluid enters the capillary. This effect can be accounted for using the well-known Bagley correction [12.7].

Highly viscoelastic fluids sometimes retain their elastic deformation during the transport process through the capillary and relax only when the fluid exits the capillary. This phenomenon is called die swell. Because all fluids can be subjected to only a limited amount of elastic deformation before undergoing physical or chemical changes, under extremely high capillary flow-rates and/or pressures a phenomenon called melt fracture may occur. This problem is identifiable by the increasing surface roughness observed on the extrudate surface as the shear rate of the process is increased above a threshold value.

12.5 Rheology of Suspensions

As mentioned in Chapter 6, it is common to measure and follow the rheological properties of the mixture of binder and filler materials in order to characterize (end-of-mix viscosity) and to determine the end time of mixing. Dispersions or suspensions of low molecular weight polymers highly filled with energetic materials and additives show a complex flow behavior, which will be elucidated in the following sections.

12.5.1 Relative Viscosity of Dispersed Systems

The relative viscosity η_{rel} is commonly defined as the ratio of the viscosity of the dispersion to the viscosity of the matrix fluid. This defintion is only applicable, however, when the matrix fluid is Newtonian. For non-Newtonian fluids one must determine which reference viscosity will be used to calculate the relative viscosity.

The relative viscosity η_{rel} can be defined as the ratio of the viscosity of the dispersion η_{Disp} to that of the continuous phase η_c at a constant shear rate $\dot{\gamma}$:

$$\eta_{\text{rel}} = \frac{\eta_{\text{Disp}}\big|_{\dot{\gamma}}}{\eta_{\text{c}}\big|_{\dot{\gamma}}} \qquad (12.34)$$

or at a constant shear stresss τ:

$$\eta_{\text{rel}} = \frac{\eta_{\text{Disp}}\big|_{\tau}}{\eta_{\text{c}}\big|_{\tau}} \qquad (12.35)$$

When comparing data from the literature, one often encounters the problem that various authors do not describe the reference viscosity used to determine the relative viscosity, making it difficult to compare different relative viscosity values. The relative viscosity of a dispersion depends not only on the concentration of the dispersed phase but also on the properties of the continuous and dispersed phases, interparticle and particle-matrix interactions, in addition to the specific loading applied to the system. The literature contains a number of equations that allow the relative viscosity to be determined solely as a function of the volume concentration of the dispersed phase. Some of these relationships have been derived theoretically, whereas others were determined empirically or semi-empirically. Kamal and Mutel [12.10] provide an overview of various equations for determining the relative viscosity of dispersed systems. Table 12.1 shows examples of several of these relationships, which express the relative viscosity of dispersions of spherical particles in terms of the volume fraction solids [12.11–12.20].

12.5.2 Matrix Fluid

The curable compositions described here are based on HTPB (hydroxy-terminated polybutadiene) or PPG (polypropylene glycol). An energetic binder, such as GAP, contains an energetic group in the monomer, such as an azide group in GAP or a nitro group as in PolyGLYN or PolyNIMMO. These prepolymers behave as Newtonian fluids under processing conditions.

These prepolymers are cured to a polymer matrix by the reaction between the hydroxyl groups and an isocyanate which is added in the last stage of the mixing process. During this reaction the prepolymer is extended and cross-linked forming a network in which the solid filler is secured by physical and physicochemical forces. The mechanical properties are a measure of the extent of these forces under various conditions.

In Table 12.2 some data are presented of the above-mentioned curable binders – it is clear that owing to their relative high melting-points, PPG compositions have to be processed at temperatures of 50 °C and higher.

The viscosity of a fluid is strongly influenced by temperature and may be described by a kind of Arrhenius equation:

$$\eta_T = \eta_0 \cdot \exp(BT) \qquad (12.36)$$

for a small temperature region and by

$$\eta_T = \eta_{T_1} \cdot \exp(-E_\eta/RT) \qquad (12.37)$$

Table 12.1. Relative viscosity η_{rel} of dispersions with spherical particles.

Author	Relative viscosity	Parameters
Einstein [12.11]	$\eta_{rel} = 1 + 2.5 \cdot c_v$	c_v = volume concentration; $c_v < 0.02$
Eilers [12.12]	$\eta_{rel} = \left(1 + \frac{1.25 \cdot c_v}{1 - \frac{c_v}{c_{v,max}}}\right)^2$	
Mooney [12.13]	$\eta_{rel} = \exp\left(\frac{2.5 \cdot c_v}{1 - s \cdot c_v}\right)$	s = displacement factor
Maron, Pierce [12.14]	$\eta_{rel} = \left(1 - \frac{c_v}{c_0}\right)^{-2}$	c_0 = geometric packing parameter
Dougherty, Krieger [12.15]	$\eta_{rel} = \left(1 - \frac{c_v}{c_{v,max}}\right)^{-[\eta] c_{v,max}}$	$c_{v,max}$ = maximum particle concentration; $[\eta]$ = power law factor
Thomas [12.16]	$\eta_{rel} = 1 + 2.5 \cdot c_v + 10.05 \cdot c_v^2 + A \cdot \exp(B \cdot c_v)$	$0 \le c_v \le 0.6$; $A = 2.73 \times 10^{-3}$; $B = 16.6$
Frankel, Acrivos [12.17]	$\eta_{rel} = \frac{9}{8}\left(\frac{\left(\frac{c_v}{c_{v,max}}\right)^{\frac{1}{3}}}{1 - \left(\frac{c_v}{c_{v,max}}\right)^{\frac{1}{3}}}\right)$	
Chong, Christiansen, Baer [12.18]	$\eta_{rel} = \left(1 + 0.75 \cdot \frac{\frac{c_v}{c_{v,max}}}{1 - \frac{c_v}{c_{v,max}}}\right)^2$	$c_{v,max}$ = maximum particle concentration
Quemada [12.19]	$\eta_{rel} = \left(1 - \frac{c_v}{c_{v,max}}\right)^{-2}$	$c_{v,max}$ = maximum particle concentration
Batchelor [12.20]	$\eta_{rel} = 1 + 2.5 \cdot c_v + 6.2 \cdot c_v^2$	

for a larger temperature region. For prepolymers such as HTPB, the value of E_η is ~ 60–80 kJ/mol [12.4].

As will be discussed in the next section on filled compositions, the ultimate viscosity of the mixture is also determined by the binder viscosity. The temperature dependence is very interesting from the viewpoint of the castability; an increase in

Table 12.2. Some characteristic properties of curable binders.

	Molecular weight	State at 20 °C	η_1 (Pa s)	T_g (°C)	T_s (°C)
HTPB	2700	Fluid	1	−65	–
PPG	2000	Solid	–	50	50
GAP	2200	Fluid	~1	−30	–

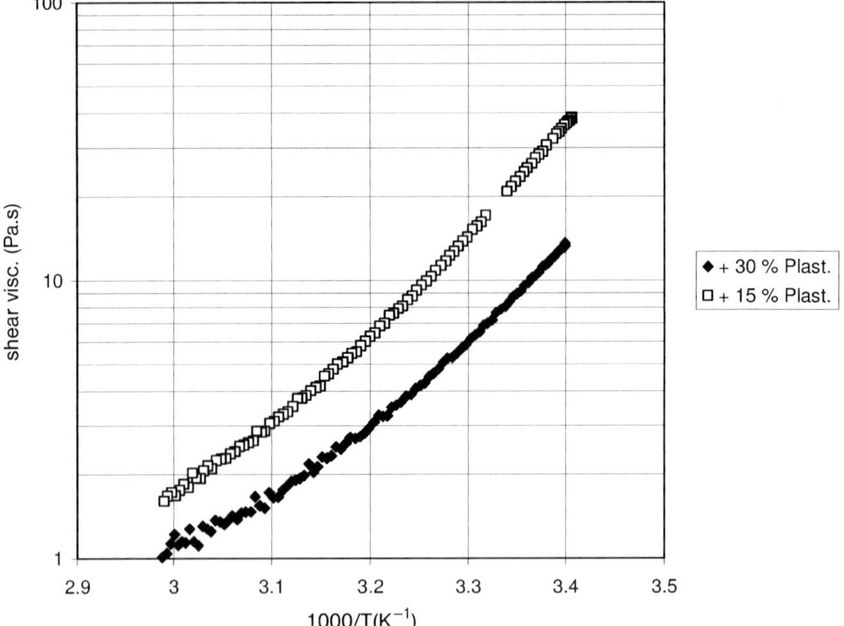

Figure 12.6. Effect of temperature and plasticizer content (wt%) on shear viscosity of GAP.

temperature is one of the available means to improve the castability of a composition without any change in the composition itself.

Another way to reduce the viscosity is to introduce a suitable plasticizer. Suitable plasticizers are those which form a homogeneous mixture with the prepolymer, which do not exudate and which have an optimal decreasing effect on the viscosity of the mixture. Suitable plasticizers for HTPB are IDP, DOA and DOS.

In Fig. 12.6, the effect of temperature on an energetic binder (GAP) with 15 and 30% plasticizer (GAP-A) on shear viscosity is presented; it shows that an increase of 15–30% of plasticizer reduces the viscosity by a factor of 2. The strong effect of temperature on viscosity is also obvious from Fig. 12.6.

It is noted that even in a relatively small temperature region of 20–60 °C, no straight line is obtained in a semilogarithmic plot of viscosity versus $1/T$; the energy of viscosity (E_η) changes from about 75 to 50 kJ/mol with increasing temperature.

12.5.3
Disperse Phase

In filling a prepolymer with solid particles, a number of aspects of the disperse phase are very important:

- mean particle size
- particle size distribution
- particle shape, morphology
- the modality of the solids – monomodal or polymodal particle sizes
- the packing density of the solids
- the volume fraction of particles to be applied
- the chemical nature of the particles with regard to a possible physico-chemical interaction between particles and binder and/or other binder ingredients (curing agent, plasticiser, bonding agent, etc.).

In the literature, much attention has been given to these aspects [12.21, 12.22]. In view of the requirements set for propellants and PBXs, the maximum volume fraction ($c_{v,max}$) is of high importance. An acceptable estimate of the value of $c_{v,max}$ is obtained by the determination of the maximum packing density of the particles.

When filling a prepolymer, the shear viscosity will often at first show a linear increase with the volume fraction up to about 0.02 (a fill of 1–2 vol.%). This was predicted theoretically by Einstein (see Table 12.1) [12.11]. For $c_v > 0.02$ the viscosity increase is no longer linear. This non-linear behavior is explained by particle interactions. Thus the increase in solids content will change the rheological behavior from Newtonian to non-Newtonian. It is noted that most mixtures are 'shear thinning', which means that shear viscosity decreases with increasing shear rate. Further filling leads to time-dependent behavior (thixotropy) and elastic behavior [12.21, 12.23]. The latter appears as a yield stress; a relatively large stress has to be applied before the mixture will flow.

The choice of the solid filler as a bimodal or trimodal mixture and the mean particle diameter ratio determines the maximum volume fraction to be obtained. For a bimodal filler mixture, both sizes with a very small particle size distribution, a model has been developed which predicts the porosity of the particle packing [12.24], which is the volume to be occupied by fluid to reduce particle to particle friction. The larger the minimum porosity, the more binder has to be used and the smaller the volume fraction (solid load) that can be obtained. Therefore, it is worthwhile to select the filler carefully.

However, this model takes only the mean diameter ratio into account and assumes spherical particles. For a large number of RDX lots this model has been validated (see Fig. 12.7). The vertical axis shows the minimum porosity (%) which equals $(1-c_{v,max}) \times 100\%$.

It is noted that for fine particles with a mean diameter smaller than about 20 μm, particle-particle interaction play a more important role, resulting in a less reliable and higher value for the minimum porosity. From Fig. 12.7, the maximum filler volume fraction $c_{v,max}$ may be determined; for a diameter ratio of 0.1 (a commonly used ratio), a minimum porosity of 0.22 is found, which means a maximum volume fraction of the solids ($c_{v,max}$) of 0.78. The maximum mass fraction may now be calculated:

$$c_{m,max} = [1 + (\rho_b/\rho_c)(c_{v,max}^{-1} - 1)]^{-1} \quad (12.38)$$

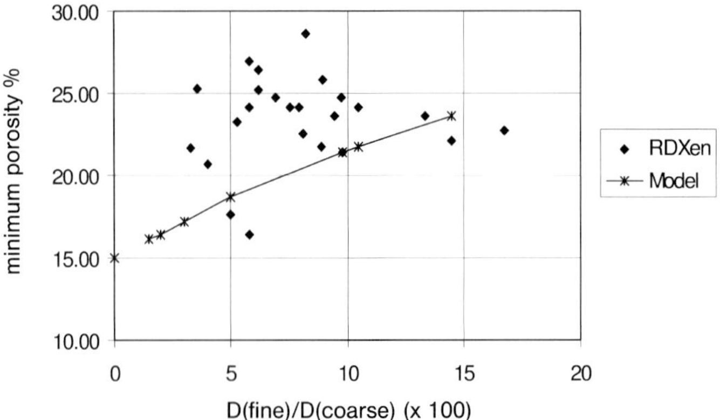

Figure 12.7. Effect of diameter ratios in bimodal mixes of RDXs on porosity: tap density of RDXs (bimodal) – comparison of model with tap density results.

which is about 0.87 (87 wt.%) for a PBX consisting of HTPB /RDX.

In practice, the maximum packing density is determined and the maximum solid loading is calculated using

$$c_{m,\,max} = [1 + \rho_b\,(\rho_t^{-1} - \rho_c^{-1})]^{-1} \tag{12.39}$$

where ρ_b is the density of the binder, ρ_c is the crystal density and ρ_t is the maximum packing density (tap density). The actually obtainable maximum solid loading has to be determined by mixing the complete formulation including the additives.

At TNO-PML, an investigation has been carried out [12.23, 12.25] with respect to the influence of the volume fraction of RDX and AP on the rheological data of monomodal and bimodal compositions. The slurry consisted of HTPB and AP or RDX without any other compound or additive normally used in a propellant or explosive. In Fig. 12.8, the results are given including a fit using $c_{v,max} = 0.62$. It is noted that in Ref. [12.8] the maximum packing fraction of various arrangements of monodisperse spheres ranges from 0.52 for simple cubic to 0.64 for random close packing and 0.74 for face-centered cubic, indicating that 0.62 is a realistic value.

From the results in Fig. 12.8, it can be concluded that a strong dependence exists between the volume fraction and reduced (shear) viscosity. In the absence of processing aids such as lecithin and a bonding agent, the measured values agree reasonably well with the equation from Dougherty and Krieger (see Table 12.1) [12.25], provided that the particles are not too small (> 30 μm), since otherwise cohesive forces start to play a role.

It is noted that the deviation of the measured data from the fit (calculated value) in Fig. 12.8 is most probably caused by interaction of AP with the HTPB binder.

The same experiment was performed using a bimodal mix of RDX with a 67:33 coarse: fine ratio, mean diameters of 230 and 5 μm, respectively, (mean diameter ratio = 0.022) and the fines showing cohesive behavior. The binder consists of

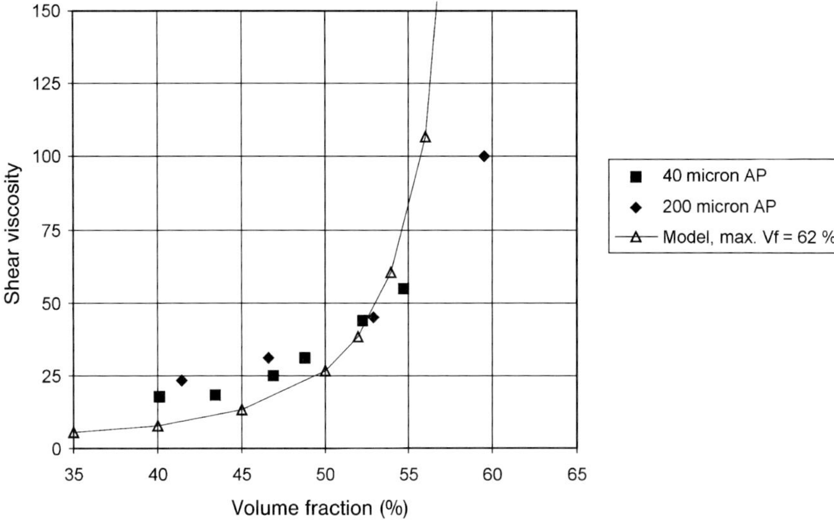

Figure 12.8. Shear viscosity of monomodal AP in HTPB at a constant shear rate of 5 s^{-1}.

HTPB, plasticized with IDP (33%), Flexzone 6H (antioxidant), Dantocol DHE (bonding agent, 0.72% in binder), lecithin (flow improver, 0.29% in mixture) and a small amount of zinc stearate to reduce cohesive particle interaction. The experiment started at a volume fraction of 45 vol.% with increments of about 3 vol.%. After half an hour mixing a sample was taken and the viscosity measured (shear rate of 1.25 s^{-1}). The results are presented in Fig. 12.9, in which the calculated values are also

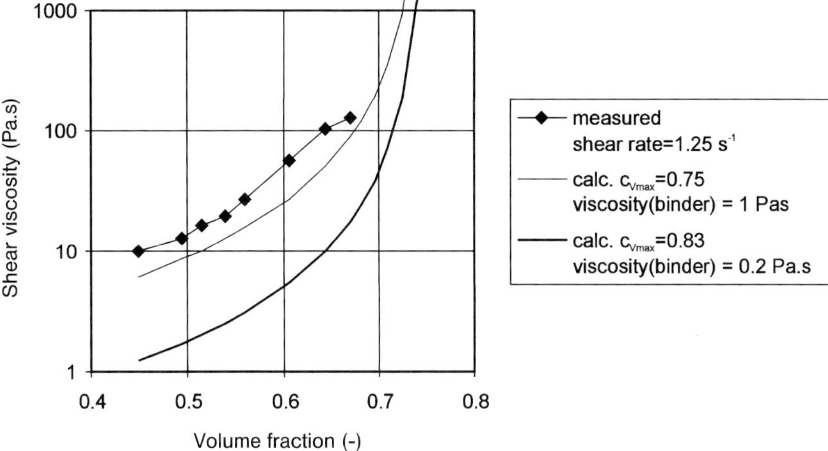

Figure 12.9. Comparison of measured and calculated values ($\theta = 60$ °C); HTPB-based bimodal RDX composition.

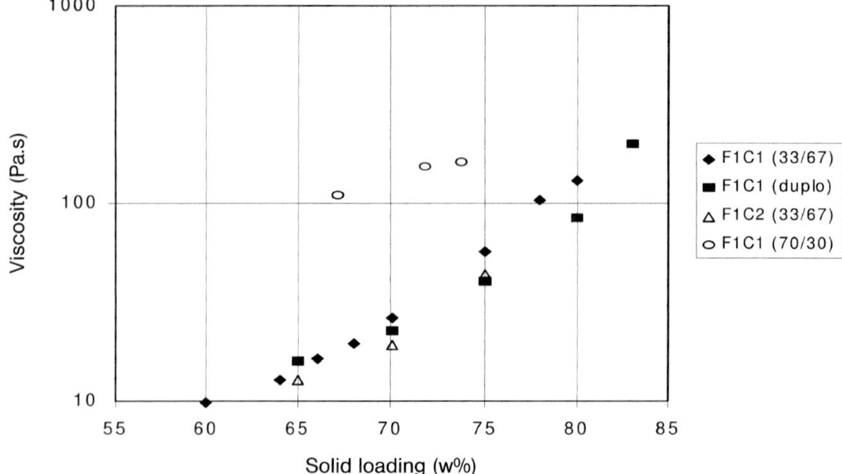

Figure 12.10. The effect of fines content on shear viscosity.

given – the lowest fit values were obtained using for $c_{v,max}$ the value from Fig. 12.7 for the diameter ratio of 0.022, $c_{v,max} = 0.83$, and with a binder viscosity of 0.2 Pa s.

One may conclude that the bimodal filler distribution does not behave ideally – $c_{v,max}$ must be smaller than 0.83. Second, the complete binder in combination with RDX shows a much higher viscosity at lower volume fractions, indicating that other factors such as binder-solid interaction and possibly H-bonding also play a role.

The introduction of fines, for instance 6 μm AP, to enhance the burning rate may result in a decrease in the maximum attainable solid load. Because of the increase in total surface area to be wetted by the binder, the solid load has to decrease in order to be sure of a castable mixture. This is demonstrated in Fig. 12.10 [12.24], in which the shear viscosity at a common shear rate is shown for various mixtures of a fine particle size ($F_1 \approx 5$ μm) and two coarse particle sizes ($C_1 \sim 230$ and $C_2 \sim 500$ μm) at different fine:coarse ratios (3:67 and 7:30). From Fig. 12.10, it may be concluded that changes in the mean coarse particle size (from C_1 to C_2) do not influence the viscosity very much, but the concentration change of F_1 (from 33 to 70) has a tremendous effect on viscosity.

12.5.4 Castability

For a number of propellant and explosive mixtures, the rheological data have been determined using rotational and oscillatory measurements. From these measurements the end-of-mix (EoM) and end-of-cast (EoC) viscosity have been determined, which are the viscosity as determined after homogenization of the complete mixture, including the curing agent and the viscosity when the cast is finished, respectively. Depending on the number and scale of the items to be cast, the time needed for the cast will be between 0.5 and several hours.

The pot life is defined as the time during which the mixture may be cast; hence the time needed for the cast should be smaller than the pot life. This time span will influence the desired curing rate and the EoM viscosity.

As yet it is unknown with which parameters the sensitivity to vibration may be described. Many times it has been observed that the mixture did not flow or only very slowly, but that a vibrational treatment forced the mixture to flow.

In case castability problems arise, a number of measures may be taken. Without being exhaustive, a number of measures are listed below, which of course may affect other properties, such as the mechanical and/or ballistic properties:

- use a higher casting temperature
- use a higher plasticizer content
- use or optimize the flow improver content
- optimize the bimodal filler distribution
- use less fine material.

12.5.5
Curing and the Effect of Time

The cure reaction the prepolymer is extended and cross-linked, forming a network in which the solid filler is secured by physical and physicochemical forces. Hence the viscoelastic behavior will change continuously during the cure and after completion of the cure the propellant shows a far more solid (elastic) behavior.

The curing reaction as a chemical reaction is temperature dependent and shows a kind of activation energy, presumably related to an overall reaction:

$$\eta_c = \eta_0 \cdot \exp(-E_c/RT) \qquad (12.40)$$

Hence a temperature increase lowers the initial viscosity but increases the curing rate, so the pot life, as an overall effect, may be shortened.

Following the curing process by oscillatory measurements, one sees an increase in complex viscosity, while at first the ratio $\tan\delta$ between the viscous and elastic behavior (Eq. (12.21)) shifts in favor of viscous behavior although both may increase. As a result, δ increases first to about 80–90° and later, when the elastic behavior increases rapidly, strongly decreases to about 10° (see Fig. 12.11).

The cure rate is also influenced by the type of isocyanate. For the cure of HTPB mostly IPDI is used, which is a safe isocyanate (low vapor pressure) with a slow reaction rate with HTPB, giving a long pot life. Aromatic isocyanates such as TDI will react much faster, which in some cases may be an advantage.

The activation energy (E_c) is of the order of 30–50 kJ/mol. Hence the viscosity may be extrapolated by using the following equation:

$$\eta(T_1)/\eta(T_2) = (E_c/R)(1/T_2 - 1/T_1) \qquad (T_2 > T_1) \qquad (12.41)$$

Hence a temperature increase of 10 °C results in a decrease in viscosity by a factor of about 1.5.

In Fig. 12.11, the effect of the concentration of cure catalyst (FeAA) and of cure

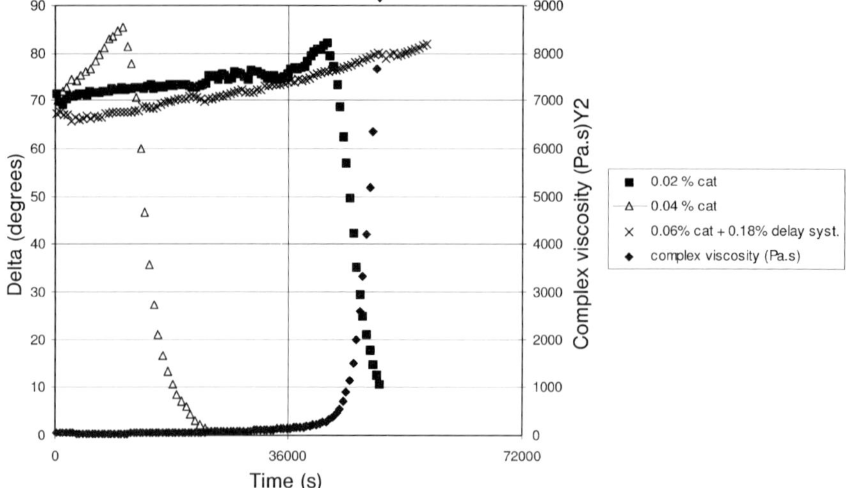

Figure 12.11. Influence of composition and concentration of a cure cat on pot life.

catalyst inhibitor (HAA + ZnO) on the cure of a PPG-based propellant is presented, showing the increase and subsequently the strong decrease in δ. The latter effect is accompanied by a strong increase in complex viscosity. This is interpreted by chain extension first taking place whereas in a later stage cross-linking is the dominant reaction accompanied by a strong increase in elastic behavior.

12.5.6 Nano-scale Suspensions

In the following, studies with nano-scale aluminum are presented. Aluminum particles are well known as an ingredient in energetic materials. The typical diameter of aluminum used in explosives and propellants is of the order of ~ 30 μm [12.21]. To enhance aluminum's reactivity, for instance during combustion of solid rocket propellants, it is advantageous to use particles with the largest possible specific surface area, i. e. particles with a smaller mean particle size are desirable. By vaporizing and subsequently condensing aluminum in argon or by electric explosion of an aluminum wire, it is possible to produce aluminum particles in the nanometer size range [12.26, 12.27]. Particles in this size range exhibit very different physical properties to those in the micrometer range. At such small sizes, inter-particle interactions become significantly more influential and, as a result, nano-scale particles have a higher tendency to agglomerate [12.28]. In addition, the changed material behavior of nano-scale particles can lead to processing difficulties when they are mixed with a fluid polymer matrix.

The suspensions investigated consisted of nano-scale aluminum particles dispersed in paraffin oil or hydroxy-terminated polybutadiene (HTPB). The paraffin oil

Figure 12.12. Nano-scale aluminum powder.

exhibited Newtonian flow behavior with a dynamic viscosity of $\eta(20\ °C)$ = 198 mPa s. It had a density $\rho = 874.7\ kg/m^3$ and a surface tension $\sigma = 30.5\ mN/m$. The HTPB (designated HTPB R 45-M) also exhibited Newtonian flow behavior with a dynamic viscosity of $\eta(20\ °C) = 9300$ mPa s.

The nano-scale aluminum (ALEX) was obtained from Argonide (Stanford, FL, USA). The density of the aluminum particles was determined via gas pycnometry as $\rho = 2.4\ g/m^3$ and the specific surface area determined via gas adsorption was $S = 11.2\ m^2/g$. An SEM image of the aluminum powder is shown in Fig. 12.12.

12.5.6.1 **Flow Behavior of Paraffin Oil/Aluminum Suspensions**

Prior to rheological characterization, the paraffin oil/aluminum suspensions were stirred for several hours and the aluminum was well dispersed using an ultrasound homogenizer. This process ensured that aluminum agglomerates were broken down and the suspension was adequately homogenized. After mixing, the flow behavior was characterized under steady shear flow. The relative viscosity as a function of shear rate is shown in Fig. 12.13 for the suspensions, which ranged in solids concentration from 2 to 45 vol.%, and also for the pure paraffin oil [12.29].

With increasing aluminum concentration, the suspensions exhibit ever-increasing shear thinning behavior. This non-Newtonian response is attributable to particle-particle interactions and the changed hydrodynamics of the suspension compared with the single-phase fluid. At small shear rates, the viscosity increase as a function of concentration is particularly pronounced. In this shear rate range, the interparticulate forces dominate over the relatively weak hydrodynamic forces, so that the rheological response of the suspension is very dependent on the solids concentration and the resulting structural interactions. As the shear rate is increased, the hydrodynamic forces also increase, leading to flow-induced structuring of the nano-scale particles and a corresponding decrease in the viscosity at a given solids concentration. The effect of solids concentration on the suspension viscosity is much less pronounced at higher shear rates because of the flow-induced structuring of the system.

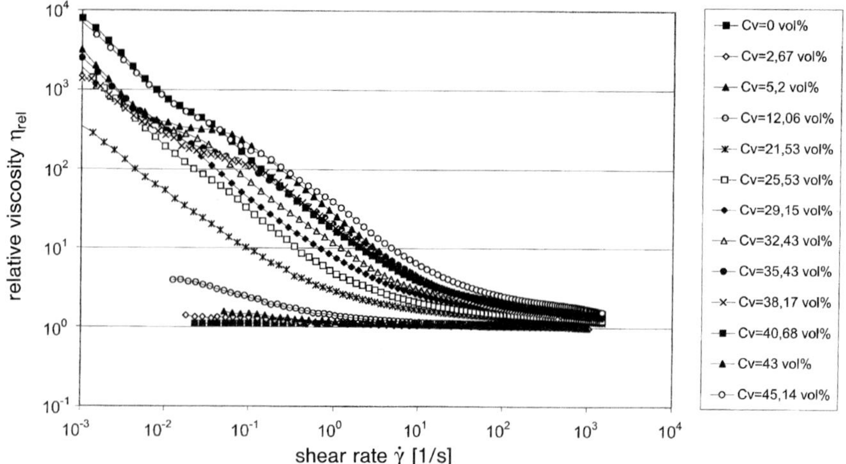

Figure 12.13. Relative viscosity of the paraffin oil/aluminum suspensions as a function of shear rate; $\vartheta = 20$ °C.

Figure 12.14 shows the relative viscosity of the suspension as a function of solids concentration for the limiting viscosity at zero shear rate ($\dot{\gamma} \to 0$) and at a relatively high shear rate ($\dot{\gamma} = 1000\ s^{-1}$).

At the highest shear rate examined, a linear increase in the relative viscosity as a function of solids concentration is observed, up to a concentration $c_V \approx 2$ vol.%. At very low concentrations, there is almost no difference in the relative viscosity

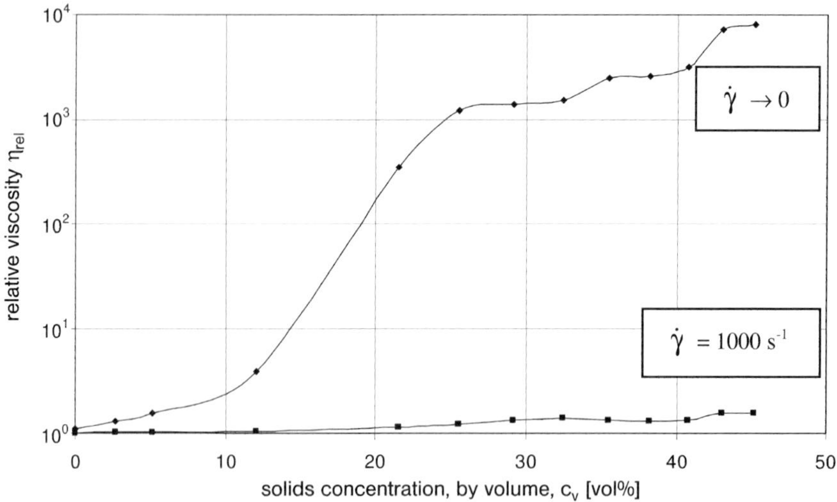

Figure 12.14. Relative viscosity of the paraffin oil/aluminum suspensions as a function of solids concentration.

determined at the minimum and maximum shear rates, because in this range of concentration there is sufficient distance between the particles that the contribution from particle-particle interactions is small and, likewise, flow-induced orientation of the particles has a relatively minor effect on the viscosity. Increasing the concentration leads to an increased contribution to the viscosity from particle-particle interactions. The quiescent particle structure that forms with increasing concentration at low shear rates is one of the reasons for the strong concentration dependence of the limiting viscosity at zero shear rate, as shown in Fig. 12.14. At higher shear rates a flow-induced structure is formed, leading to a reduction in the relative viscosity at a given solids concentration. The difference in the viscosity functions at the two shear rates $\dot{\gamma} \to 0$ and $\dot{\gamma} = 1000\ s^{-1}$, which increases with increasing solids concentration, is attributable primarily to the behavior of the particles in the Couette flow. At a concentration of 45 vol.%, the viscosity difference is of the order of $\sim 10^4$.

12.5.6.2 Flow Behavior of HTPB/Aluminum Suspensions

Figure 12.15 shows the relative viscosity of the HTPB/aluminum suspensions as a function of shear rate for solids concentrations of $0 \le c_V \le 47$ vol.%. HTPB R 45-M without additives exhibits Newtonian flow behavior (see Fig. 12.15 and Refs [12.23, 12.30]). In contrast to the paraffin oil/aluminum suspensions, the HTPB suspensions exhibited Newtonian behavior over a wide shear rate range up to a solids concentration of ~ 50 vol.%. With increasing concentration, the relative viscosity of the suspensions increased; however, the behavior remained linear. Figure 12.16

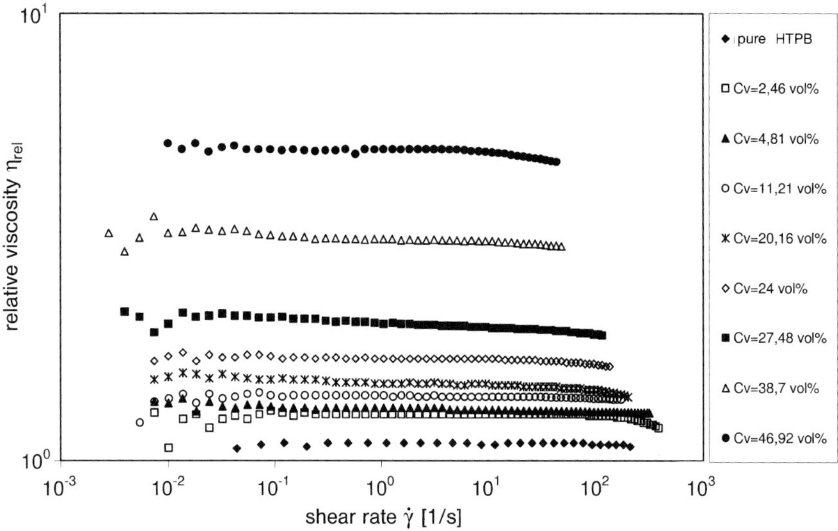

Figure 12.15. Relative viscosity of the HTPB/aluminum suspensions as a function of shear rate; $\vartheta = 20$ °C.

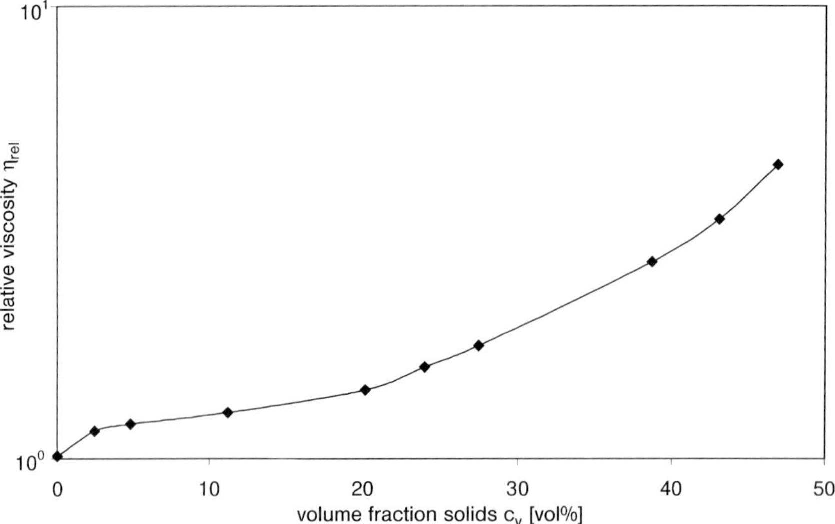

Figure 12.16. Relative viscosity of the HTPB/aluminum suspensions as a function of solids concentration.

shows the relative viscosity of the suspensions as a function of solids concentration.

The rheological characterization of the HTPB-based suspensions filled with nanoscale aluminum yielded the following relationship for the relative viscosity as a function of solid volume concentration:

$$\eta_{\text{rel,Alex-S}} = \frac{\eta_{\text{Suspension}}}{\eta_{\text{HTPB}}} = 1 + 5.5 \cdot C_v - 31.4 \cdot C_v^2 + 74.5 \cdot C_v^3 \tag{12.42}$$

Equation (12.42) is valid for solid volume concentrations c_v up to 50 vol.%.

12.5.6.3 Viscoelastic Properties of the Suspensions

The storage and loss modulus functions of the paraffin oil/aluminum suspensions are shown in Fig. 12.17 for various solids concentrations. At low frequencies the storage modulus is smaller than the loss modulus, meaning that the viscous properties are dominant in this frequency range. Both functions increase steadily with frequency; however, the slope of the storage modulus function is greater than that of the loss modulus and, as a result, the two functions intersect at a characteristic frequency ω_i, which differs depending on the solid volume concentration. Above this frequency, the elastic properties are dominant. The structural relaxation time λ is equal to the reciprocal of the frequency at which the storage and loss moduli intersect:

$$\omega_i \cdot \lambda = 1 \tag{12.43}$$

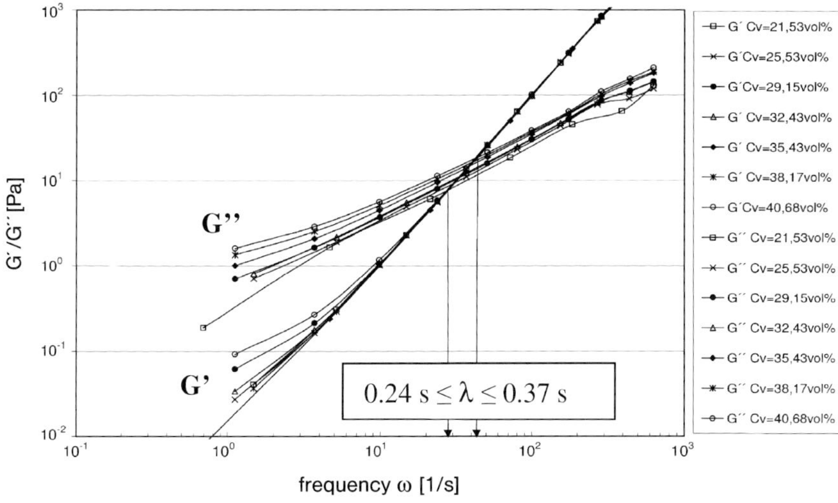

Figure 12.17. Storage and loss modulus functions of the paraffin oil/ aluminum suspensions.

For the paraffin oil/aluminum suspensions up to solids concentrations $c_V \leq$ 40 vol.%, the structural relaxation times λ ranged from 0.24 to 0.37 s. It was also observed that the storage modulus $G'(\omega)$ was essentially independent of the aluminum concentration. One concludes that for these suspensions, filled with nano-scale particles, the stored (elastic) deformation energy is independent of the particle concentration.

The storage and loss modulus functions of the HTPB/aluminum suspensions are shown in Fig. 12.18 for various solids concentrations. As in the previous case, the storage modulus $G'(\omega)$ is independent of solids concentration. However, the structural relaxation times λ for the HTPB-based suspensions are significantly smaller than those for the paraffin oil-based suspensions, falling in the millisecond range (0.0021–0.0062 s).

12.6 Gel Propellants

In recent years, the challenge of achieving high performance while also improving the safety characteristics of rocket propellants has become an important goal within the industry. Gel propellants offer the potential to satisfy such demands because they combine certain advantages of liquid propellants with other attractive properties typical of solid propellants. Gel propellants can be designed as mono- or bi-propellants. When used as a bi-propellant, both the fuel and oxidizer can be prepared as gels. The safety of the system is improved by separating the fuel and oxidizer. In general, gel propellants exhibit a specific impulse comparable to that of

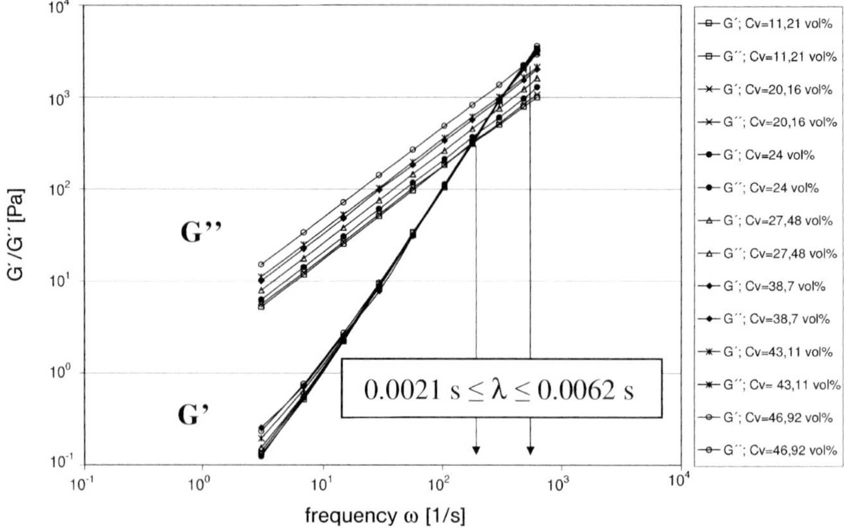

Figure 12.18. Storage and loss modulus functions of the HTPB/aluminum suspensions.

liquid propellants, but their performance can be increased even further through the addition of additives such as metal particles. A significant advantage of gel propellants over solid rocket propellants is the ability to control the thrust by controlling the mass flow of propellant into the combustion chamber. The rocket motor can even be turned on and off or be pulse driven as required. Furthermore, gel propellants are less sensitive than liquid propellants and can be handled, stored and transported more securely because of their solid-like properties. This is especially important when, for instance, fissures or leakage sites develop within the combustion chamber of a gel propellant-driven rocket motor. The viscoelasticity of the gel propellant (i.e. its unique rheological properties) significantly reduces the risk that the propellant will leak from the motor and inadvertently ignite.

The rheological properties of a gel propellant significantly affect a number of key operational and production requirements, including the propellant material behavior, casting and spraying operations and combustion within the rocket motor. Rheological characterization of the gel provides basic information critical to the production and storage of gel propellants, rocket motor casting and the design of the entire rocket motor system.

The rheological properties of gel propellants enhanced with metal particulate additives are described in the literature. Gupta *et al.* [12.31] characterized virgin and metallized unsymmetrical dimethylhydrazine (UDMH) gel fuels with methylcellulose as the gelling agent. The flow properties of this gel were determined as a function of time (in experiments conducted at various constant shear rates) and as a function of shear rate and temperature. At a given constant shear rate, the gel viscosity increased with time. As the shear loading was increased, the gels exhibited pronounced shear thinning behavior. At constant metal concentrations, shear thinning became less pronounced as the temperature increased, but at constant temperatures the shear thinning behavior became more pronounced with increasing metal concentration. Likewise, the yield stress of the gel increased with increasing metal concentration and decreased with increasing temperature. Rapp and Zurawski [12.32] examined the rheological behavior of gelled aluminum/kerosene fuels. They found that the yield point increased with increasing concentration of metal particles in the gel. During long-term storage the yield stress of the gels decreased, a phenomenon that was attributed to physical, thermal and chemical instability. Varghese *et al.* [12.33, 12.34] investigated UDMH and kerosene gels with different gellants and metal ingredients. The material was subjected to a range of different shear stresses and temperatures. The gel fuels exhibited shear thinning and thixotropic behavior, which became more pronounced with increasing metal concentration and less pronounced at higher temperatures. Compared with the kerosene gels, the flow behavior of the UDMH gels was more strongly dependent on the shear loading applied. The particle size of the metal additive and characteristics of the other ingredients, such as the gellant, stabilizer and wetting agent, all have an important influence on the rheological properties of the gel. Rahimi *et al.* [12.35] examined fuels such as hydrazine (N_2H_4), monomethylhydrazine (MMH) and kerosene, which were gelled using various cellulose compounds. They also examined the oxidizers inhibited red fuming nitric acid (IRFNA) and hydrogen peroxide (H_2O_2), which were gelled using silica particles. From these rheological studies, gel propellants can be divided into three categories depending on their degree of shear thinning behavior, their yield stress characteristics and their viscoelastic and thixotropic properties.

This study examined the rheological behavior of nitromethane gelled with nanometer-sized silicon dioxide. In combination with suitable oxidizers and additives, nitromethane exhibits a specific impulse $I_s > 2400$ N s/kg^{-1} and is much less toxic than hydrazine derivatives, thus providing environmental and handling advantages over such compounds.

12.6.1
Materials and Methods

The gel propellants examined consisted of nitromethane as the continuous phase and nanometer-sized silicon dioxide particles as the dispersed phase. Nitromethane exhibits Newtonian flow behavior with a dynamic viscosity of $\eta(25\ ^\circ\mathrm{C}) = 0.61$ mPa s. Its density is $\rho = 1.139$ kg/m^3. The silicon dioxide particles were obtained from Degussa (Frankfurt, Germany) and had a density $\rho = 1.51$ g/cm^3 (determined by gas pycnometry) and a specific surface area $S_v = 260$ m^2/g (determined by gas adsorption). The mean size of the primary particles was $\bar{x} = 7$ nm.

The rheological behavior of the gels prepared was examined in steady-state and oscillatory shear flow using a UDS 200 rotational rheometer manufactured by Physica Messtechnik. The measurement fixtures included cone and plate.

12.6.2
Steady-state Shear Flow Behavior of Nitromethane/Silicon Dioxide Gels

Prior to rheological characterization, the nitromethane/silicon dioxide gels were stirred for several hours to deagglomerate the particles and homogenize the gel. The rheological properties were then determined under steady-state shear flow. Figure 12.19 shows the relative viscosity of the gel as a function of shear rate. The concentration of dispersed particles was varied from 4 to 8 vol.%. Figure 12.19 also shows the viscosity function of pure nitromethane.

The relative viscosity, η_{rel}, is defined as the ratio of the gel viscosity to that of the matrix fluid at a constant shear rate $\dot{\gamma}$:

$$\eta_{rel} = \frac{\eta_{gel}\big|\dot{\gamma}}{\eta_{nitromethane}} \tag{12.44}$$

With increasing silicon dioxide concentration, more pronounced shear thinning flow behavior is observed. This non-linear material behavior of the gel can be attributed to particle-particle interactions and to the changed hydrodynamics of the multiphase system compared with the single-phase fluid. This viscosity increase as a function of concentration is especially pronounced at low shear rates. In this shear rate region the interparticulate interactions dominate compared with the relatively small hydrodynamic forces, so that the rheological properties of the suspension depend very strongly on the solids concentration and structural interactions within the suspension. Increasing the shear rate leads to an increase in the hydrodynamic forces, which in turn results in a shear-induced structuring of the nanometer-sized particles and a corresponding decrease in the viscosity at a given concentration. The viscosity difference as a function of concentration is therefore much smaller in the

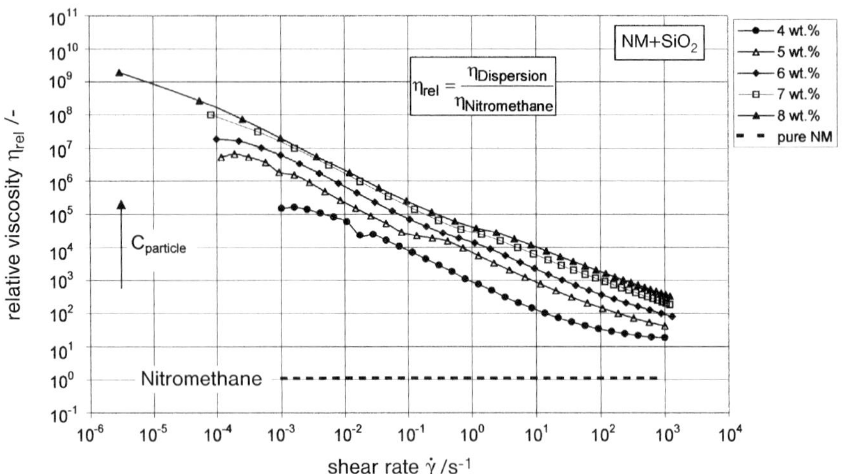

Figure 12.19. Relative viscosity of nitromethane/silicon dioxide gels as a function of shear rate.

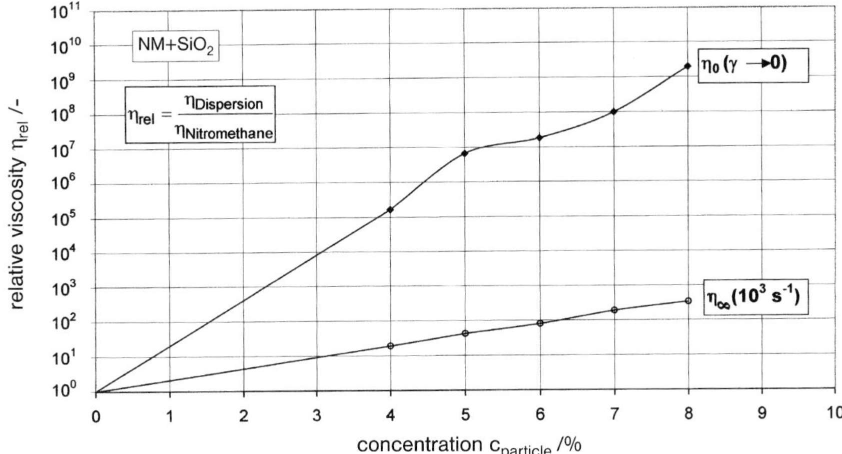

Figure 12.20. Relative viscosity of the nitromethane/silicon dioxide gels as a function of particle concentration.

high than in the low shear rate region, owing to the hydrodynamic structuring that occurs in the system at higher shear rates.

Figure 12.20 shows how the viscosity of the gel depends on solids concentration at the limiting viscosity at zero shear rate ($\dot{\gamma} \rightarrow 0$) and at a shear rate $\dot{\gamma} = 1000\ s^{-1}$.

As the silicon dioxide concentration increases, the inner particulate structure of the system becomes ever more pronounced. This inner quiescent structure in the nitromethane/silicon dioxide gel leads to the limiting viscosity at zero shear rate behavior shown in Fig. 12.20. The viscosity difference between the suspension and the pure fluid is $\sim 10^6$ at a solids concentration $c_{particle} = 8$ wt%. In contrast to the behavior of the limiting viscosity at zero shear rate, the slope of the relative viscosity function at the maximum shear rate is much lower. This relatively small increase in the viscosity of the nitromethane/silicon dioxide gel within this shear rate range arises because hydrodynamic effects are dominant and lead to development of a shear-induced structure within the silicon dioxide particles. The viscosity difference may also indicate that the gel's inner structure undergoes reversible breakdown at such high shear rates [12.9].

Figure 12.21 shows the viscosity of the nitromethane/silicon dioxide gels as a function of shear stress. The yield stress of these suspensions was determined using the tangent method [12.7]. As seen in Fig. 12.21, all of the nitromethane/silicon dioxide gels examined exhibited a yield stress. For irreversible flow to occur, the stress applied must be above the yield stress value; at stresses below the yield stress the gel exhibits pure elastic deformation like a solid.

The flow behavior of the nitromethane/silicone dioxide gel can be described using the following equation for the shear stress:

$$\tau = \tau_0 - \eta_\infty \cdot \dot{\gamma} + \eta^* \cdot \dot{\gamma}^\alpha \qquad (12.45)$$

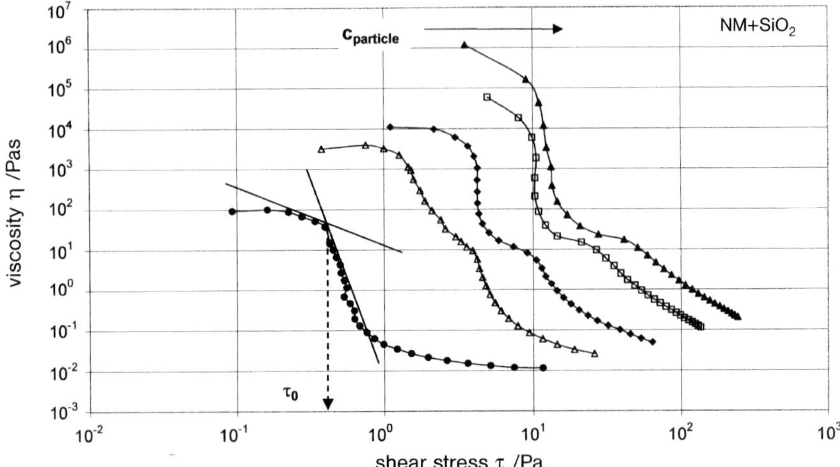

Figure 12.21. Determination of the yield stress of nitromethane/silicon dioxide gels.

where τ_0 is the yield stress of the gel, η_∞ is the viscosity at a shear rate $\dot{\gamma} \to \infty$, η^* is the viscosity that characterizes structuring within the disperse system and α is the exponent that characterizes structural changes within the system.

Figure 12.22 shows the measured shear stress values for the nitromethane/

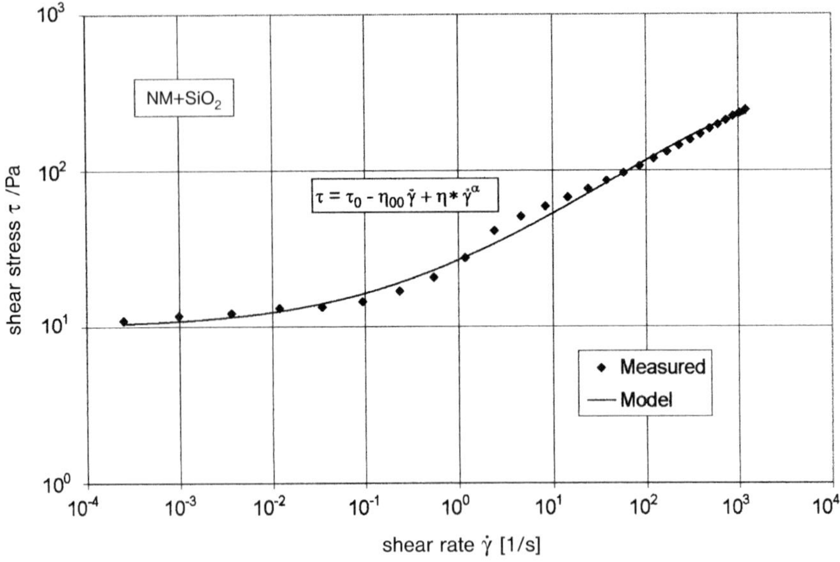

Figure 12.22. Measured values and ICT model for the shear stress of a nitromethane/silicon dioxide gel.

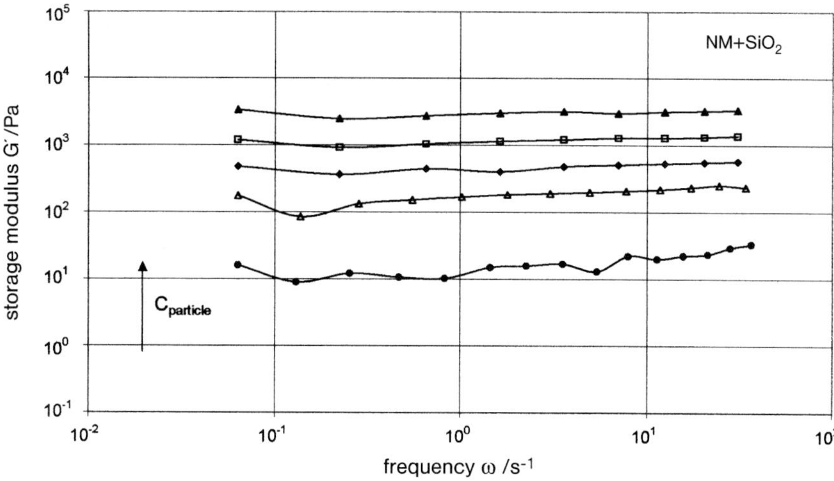

Figure 12.23. Storage modulus as function of frequency of nitromethane/silicon dioxide gels.

silicone dioxide gel with a concentration $c_{\text{particle}} = 8$ wt.% compared with the shear stress function calculated from Eq. (12.14). There is good agreement between the calculated and measured values.

12.6.3
Viscoelastic Properties of Nitromethane/Silicon Dioxide Gels

Viscoelastic properties can be determined via oscillatory shear experiments. The complex shear modulus determined via dynamic experiments in the linear viscoelastic region can be separated into two material functions as shown in Eq. (12.20), the storage modulus $G'(\omega)$ and the loss modulus $G''(\omega)$. Figure 12.23 shows the storage modulus of nitromethane/silicon dioxide gel at various solids concentrations.

In the concentration region examined, the storage modulus was independent of the radial frequency, indicating the existence of a compact inner structure in the nitromethane/silicone dioxide gel. Figure 12.24 shows example results of the storage and loss moduli as a function of frequency for the gel with a solids concentration $c_{\text{particle}} = 8$ wt.%. The moduli $G'(\omega)$ and $G''(\omega)$ are independent of frequency, meaning that within this frequency range the nitromethane/silicone dioxide gel exhibits elastic, solid-like behavior.

From this rheological study, the following hypothesis concerning the material behavior of gel propellants is proposed:

- A gel propellant must exhibit shear thinning behavior when subjected to stationary shear flow.

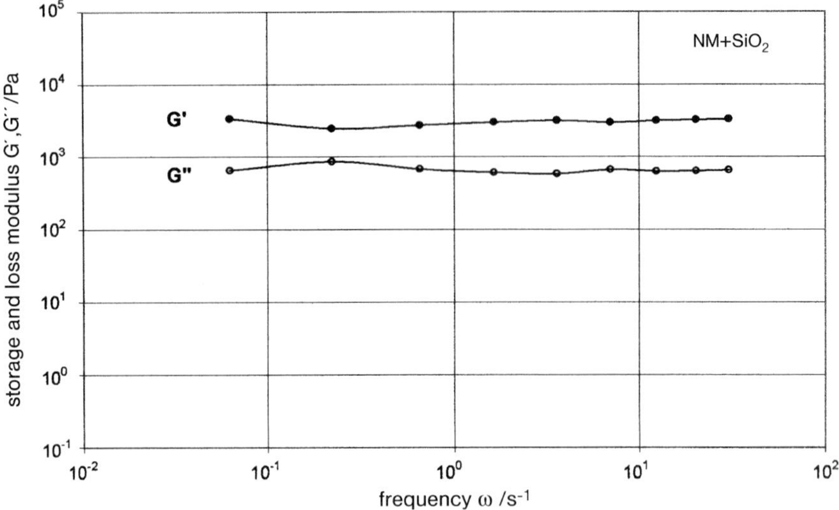

Figure 12.24. Storage and loss modulus as function of frequency of a nitromethane/silicon dioxide gel ($c_{\text{particle}} = 8$ wt%).

- The limiting viscosity at zero shear rate (η_0) of a gel propellant should be as high as possible.
- The shear viscosity at high shear rate (η_∞) of a gel propellant should be sufficiently small.
- A gel propellant should exhibit a yield stress τ_0.
- The elastic component of the complex modulus must always be higher than the viscous component.
- The moduli $G'(\omega)$ and $G''(\omega)$ should be independent (or nearly independent) of the oscillatory frequency ω.

12.7 Rheology as a Development Tool for Injection Moldable Explosives

Lawrence Livermore Laboratory (LLNL) has been interested in explosives that could be injected into components for a variety of reasons [12.36–12.42]. In the DOE nuclear weapons effort, safety is of paramount importance. One proposed method of avoiding dispersal of radioactive material in various accident scenarios is to separate the high explosive and the radioactive core of the weapon. One means of implementing this concept would be to use paste extrudable explosives (PEX), which can be stored in or near the weapon and then extruded into place when needed [12.40]. Further safety enhancement and reduced vulnerability are possible with high solids loadings of insensitive explosives such as triaminotrinitrobenzene (TATB). A unique aspect of injection-moldable explosives is that they can be prepared nearly void free. Under impact or shock loading, voids in the explosive cause 'hot spots' which act as ignition sources for detonation. Void-free injection-moldable explosives (IMX) show reduced vulnerability to accident scenarios [12.43–12.46]. Consistent rheology of the transferable insensitive explosive (TIE) is essential for void-free fills when the system is deployed.

Another area of interest in the DOE and DOD is shaped charges [12.47]. If the explosive and liner are in intimate contact, shaped charge performance improves. Machined or press-fit explosive tends to have gaps between the liner and explosive because exact tolerances are very difficult to maintain. Explosives that can be degassed and injected into the shaped charge without voids can substantially improve the quality of the jet formed when the explosive detonates. Adhesion between the liner and the explosive as the formulation cures insures intimate contact with the liner. Again, rheology of the injection moldable explosive plays a critical role in void-free loading of these shaped charges.

The preparation of injection-moldable and paste explosives generally requires three principal ingredients [12.48]: (i) a solid, crystalline high explosive; (ii) an energetic liquid carrier; and (iii) viscosity modifiers to prevent settling and/or allow the explosive to cure in place. For uniform flow in small channels and capillaries where the flow-rate varies with the fourth power of the radius, low viscosity is a critical requirement. To minimize viscosity, bimodal mixtures of HMX or TATB explosive crystals were formulated based on the theoretical work of Farris [12.39], which predicts reduced viscosities for solids mixtures when the particle size of coarse solids is an order of magnitude greater than that of fine solids. The two grades of TATB currently in production have about an order of magnitude size difference. A wide variety of particle size distributions are available as different classes of HMX.

Since it will constitute from 25 to 40% of the explosive, the liquid carrier should be an energy contributor. For DOE applications the liquid must not freeze over the stockpile-to-target-sequence (STS) temperature range (−54 to 74 °C) [12.49, 12.50], it must be a compatible, non-solvent for the crystalline explosive, it should wet the surface of the solids and it should have good thermal stability. Other desirable properties include low viscosity, low vapor pressure and no toxicity. No single energetic liquid meets all these requirements, but several candidates are given in Table 12.3.

To date, efforts to modify the liquid carriers to prevent settling have followed two approaches: (i) colloidal particles and (ii) polymer viscosity modifiers. Colloidal particles tend to require hydrogen bonding to thicken whereas polymers must be soluble in the energetic liquid. Colloidal particles are often extremely thixotropic and explosive pastes made from them tend to have yield stresses. Explosive pastes based on low molecular weight oligomers tended not to have yield stresses until the solids loading exceeds 65%. As the polymer molecular weight increases, a solution with energetic liquid tends to shear thin. This makes the rheology of the paste or injection-molded explosive very complex.

A Rheometrics Mechanical Spectrometer (Model 800) was used to collect all the data given in this section. Flow characteristics of suspensions of explosive solids in

Table 12.3. Explosives used in or proposed for TIE and IMX formulations.

Explosive common name	Chemical name, formula	Density (g/cm^3)	Mol. wt (g/mol)	Heat of formation (kJ/mol)
BDNPF	Bis(2,2-dinitropropyl)formal, $C_7H_{12}N_4O_{10}$	1.39	312.2	−597.1
DFF	Bis(2-fluoro-2,2-dinitroethyl)-difluoroformal, $C_5H_4N_4O_{10}F_4$	1.67	356	−1152.2
FEFO	Bis(2-fluoro-2,2-dinitroethyl) formal, $C_5H_6N_4O_{10}F_2$	1.607	320.1	−742.7
FM-1	23% FEFO, 52% MF-1, 25% BDNPF, $C_6H_{9.1}N_4O_8F$	1.509	320.1	−667.8
MF-1	(2-Difluoro-2,2dinitroethyl-2,2-dinitropropyl)formal	1.905	316.1	−669.9
TMETN	Trimethylolethane trinitrate, $C_5H_9N_3O_9$	1.47	255.15	−443.0
TEGDN	Triethylene glycol dinitrate, $C_6H_{12}N_2O_8$	1.335	240.2	−608.8
TATB	1,3,5-Triamino-2,4,6-trinitrobenzene	1.938	258.2	−154.2
HMX	Octahydro-1,3,5,7-tetranitro-1,3,5,7-tetrazocine	1.905	296.17	74.8

explosive liquids with polymeric binders are not well understood. Because these paste and injection-moldable explosives will be injected into parts under low pressure (0.6–2.8 MPa) and at relatively low shear rates, parallel plate rheometry was chosen to characterize them.

Injection-moldable and paste explosives are usually processed in vertical, high-shear mixers. Current formulations are sufficiently viscous to require processing in a special deaerator loader developed at LLNL when the solids content exceeds ~70%. A series of TIE formulations evaluated in this work are given in Table 12.4.

Table 12.4. Formulations of RX-52-series transferable insensitive explosives.

Formulation: RX-52-	TATB WA[a]	TATB uf[a]	Solids (%)	FEFO	FM-1	PVF	PCL
AB	50.75	19.25	70	27.42		1.17	1.4
AE	47.12	17.88	70	31.99		1.37	1.63
AG	29.00	11.00	40	54.85		2.34	2.80
AH	29.00	11.00	40	53.00		2.30	2.74
AI	39.15	21.10	60	38.54		1.50	1.80
AJ	42.25	22.75	65	31.03		1.32	1.58
AK	45.5	19.5	65	31.03		1.32	1.58
AL	48.72	16.24	65	31.03		1.32	1.58
AM	47.12	17.87	65 200 μm	31.01		1.32	1.58
AV	28	12.5	59.5	32		1.36	1.64
AW	28	12.5	59.5	16	16	1.36	1.64

[a] TATB WA is wet aminated grade and TATB uf is ultrafine grade.

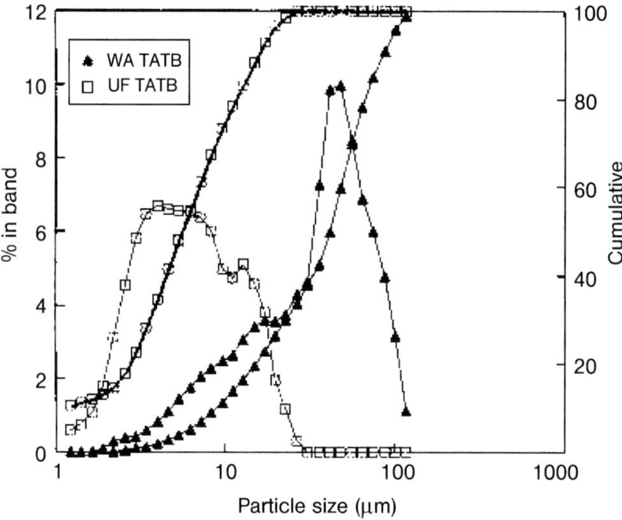

Figure 12.25. Particle size distributions for wet-aminated and ultrafine TATB gave mean particle sizes of 36.4 and 3.4 μm, respectively. A Malvern 3600 small-angle laser light-scattering apparatus was used to measure the particle size distribution.

Wet-aminated TATB is available in two particle size distributions, shown in Fig. 12.25. Samples were formulated with different solids loadings and different coarse to fine ratios. The steady shear viscosity for each sample was measured using 2.54-cm diameter parallel plates with a constant gap of 0.2 cm.

12.7.1 Effect of Solids Content

The effect of solids content on TIE formulations at 72.5 : 27.5 coarse to fine ratio was evaluated. Comparison of steady shear viscosities in Fig. 12.26 showed a large increase in viscosity above 65% solids loading. Note that above about 1 s^{-1} the higher solids loading viscosities drop off dramatically. This is because these compositions are thrown out of the parallel plate fixture at these shear rates. A discussion of edge failure in viscous fluids for parallel plate rheometers can be found elsewhere [12.4]. Viscosity versus shear rate for 40–70% bimodal TATB (27.5 : 72.5 uf : WA) showed large reductions in viscosity between 65 and 70% solids, smaller changes between 55 and 65% and settling after very short times when the solids content was below 40%. Clearly, a significant increase in viscosity occurred above 65% solids.

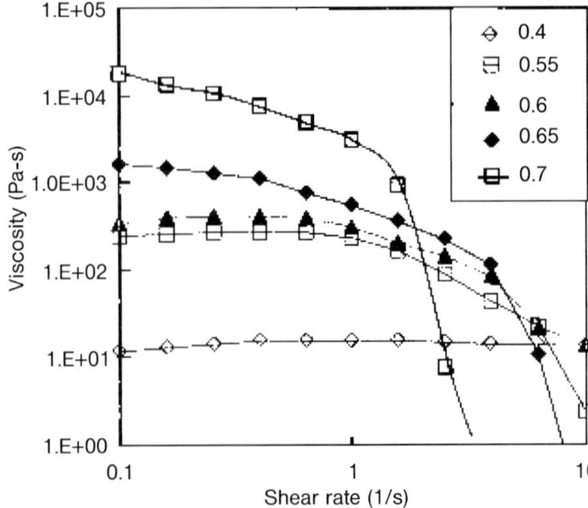

Figure 12.26. Viscosity versus shear rate for 40 to 70% bimodal TATB (27.5/72.5 : uf/WA) showed large reductions in viscosity between 65–70% solids; smaller changes between 55–65% and settling at 40% solids.

12.7.2
Effect of Particle Size Distribution

Comparison of different TATB coarse to fine ratios at 65% solids loading showed a minimum in apparent viscosity at a ratio of 72.5 : 27.5, in agreement with the theoretical predictions of Farris [12.39]. Figure 12.27 shows a fishnet plot of viscosity as a function of shear rate for four different TIE formulations with coarse to fine ratios as given in Table 12.4. There was a definite minimum in viscosity when the volume fraction of wet aminated TATB was 72.5%. Note that the viscosity axis is

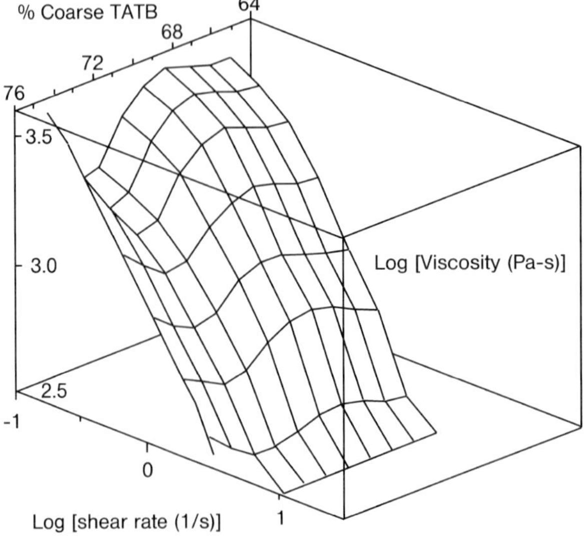

Figure 12.27. The surface plot of viscosity versus shear rate, with increasing wet-aminated TATB content of 65% bimodal solids mixture in TIE formulations, showed a distinct minimum at 72.5%, which corresponds to that predicted by Farris theory.

Table 12.5. Composition of RX-08-series formulations.

Composition	% by weight	% by volume
HMX	73.95	66.70
TMETN	19.33	22.59
Tone 260	5.04	8.10
Tone 6000	0.78	1.25
Desmondur N-100	0.91	1.36
Dabco T-131	0.007	
Total	100.00	100

logarithmic so small changes in coarse to fine ratio result in dramatic changes in viscosity. Still, the change in viscosity at 65% solids is not quite an order of magnitude. It has been tacitly assumed that this ratio does not shift substantially with solids content, although the Farris theory implies that it should.

Once an optimum coarse to fine ratio has been established, what is the most effective size differential between coarse and fine? A series of 74% solids formulations were prepared as indicated in Table 12.5 using various coarse and fine grades of HMX. These injection-moldable explosives were formulated from a solution of TMETN and the Tone polyester polyols, called RX-44-BJ, combined with the bimodal distribution of HMX in a sigma-blade mixer. The TMETN and Tone polyols were dissolved at 60 °C for several hours with stirring then cooled to ambient temperature. T-131 mercaptotin catalyst was added and allowed to coordinate for 10–20 min. The HMX solids were added and mixed remotely under vacuum to constant viscosity. This RX-08-series paste could be stored for several weeks to months without degradation of the catalyst. Prior to use, the injection-moldable paste and N-100 isocyanate were mixed and allowed to cure. Table 12.6 lists the test matrix of formulations evaluated in 50-g mixes and their steady shear viscosity at $0.1\ s^{-1}$ shear rate and dynamic viscosity at 0.1 rad/s oscillatory frequency, which were used to characterize their rheology. Table 12.6 lists nine RX-08- formulations

Table 12.6. Viscosity and HMX composition of various RX-08-series formulations.

RX-08-	η^* (Pa s/10^3)	η (Pa s/10^3)	Coarse	Size[a] (μm)	Fine	Size[a] (μm)
GX	5.30	3.89	Class 1	162	5 μm FEM	5
GY	8.70	7.05	> 43 μm C2	90	5 μm FEM	5
GZ	5.14	4.14	> 43 μm LX04	100	< 43 μm LX04	4.4
HA	10.20	8.70	> 43 μm C2	90	< 43 μm LX04	7.7
HB	6.38	3.54	> 43 μm LX04	100	< 43 μm C2	9.8
HC	180.00	10.30	> 43 μm C2	90	< 43 μm C2	9.8
HD	3.54	3.09	> 43 μm LX04	100	Class 5	4.0
HE	8.80	7.05	> 43 μm C2	90	Class 5	4.0
HF	26.70		Class 1 (m)	250	3 μm FEM	3

[a] Particle size distribution was measured on a Malvern 3600 laser light-scattering apparatus.

where the mean particle size of the HMX was varied. Coarse HMX was either class 1, the coarse fraction of LX-04 grade HMX or the coarse fraction of class 2 HMX collected on a 43-μm sieve. Fine HMX was either 3- or 5-μm fluid energy milled HMX or the HMX, which passed through a 43-μm sieve from LX-04 or class 2 HMX or class 5 HMX used directly. The mean particle size, given in Table 12.6, was obtained from the distribution measured with a Malvern 3600 light-scattering apparatus. This instrument has a size cut-off of 1 μm at the low end. Recently, a new Malvern apparatus showed that class 5 HMX obtained from Holston Army Ammunition Plant is trimodal with about 20% sub-micron material and perhaps as much as 70% about 10–12 μm.

Both dynamic and steady shear viscosities were measured on the HMX-based formulations in Table 12.6. The Cox-Merz rule [12.4]:

$$\eta^*(\omega) \approx \eta(\dot{\gamma}) \tag{12.46}$$

seems to apply up to about 1 s^{-1}.

Figure 12.28 shows a typical set of dynamic and steady shear viscosity measurements for RX-08-HD. Table 12.6 lists the dynamic and steady shear viscosities for each of the nine samples measured at 0.1 s^{-1}. The lowest viscosity was observed in this formulation where the mean particle size difference between coarse sieved LX-04 grade HMX and fine class 5 HMX is slightly more than an order of magnitude. Although the class 5 HMX (fines) average particle size was 4–5 μm, there are two peaks in the band distribution at 0.4 and 8–12 μm. This essentially makes

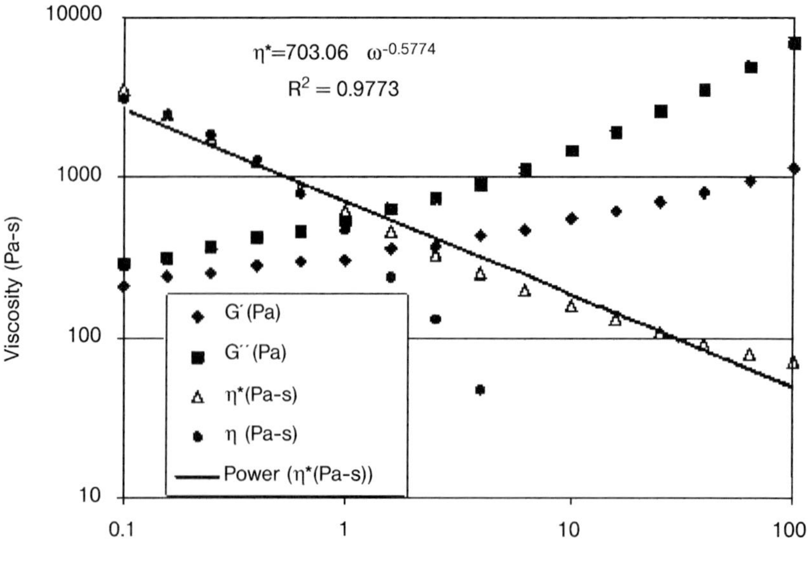

Figure 12.28. Dynamic viscosity and steady shear viscosity measurements of RX-08-HD can be fitted to a power law relationship over a short range of rates. Steady shear viscosity deviates from the fit near 1 s^{-1}.

RX-08-HD trimodal, which may account for the lowest viscosity. When the coarse to fine HMX ratio is bimodal and about an order of magnitude different (-GZ, -HA or -HB), viscosities only slightly higher than for the -HD formulation were obtained. When the separation between coarse and fines is larger than an order of magnitude (-GX or -GY), the viscosity increases by a factor of ~2 compared with -HD. When 3-μm fines were used, with a coarse to fine ratio of almost an order of magnitude (-HF), the viscosity increased by a factor of ~4. This is apparently due to the increasing surface area of the fines. When the coarse particles were slightly smaller than LX-04 coarse and mixed with the trimodal class 5 HMX fines (-HE), the viscosity also increased slightly. When the coarse to fine ratio was less than an order of magnitude (-HC), the highest viscosity occurred and the Cox-Merz rule was not followed.

As is well known, different HMX manufacturers make their classes in different ways. When HMX is precipitated to obtain fines, a much more uniform distribution results than when the explosive is ground. Recent improvements in scattering technology allow particles down to ~0.05 μm to be resolved in a distribution measurement and ground class 5 HMX samples A and D had a substantial concentration of these very fine particles (see Fig. 12.29, A and D) in their distribution. On the other hand, precipitated HMX samples B and C had a remarkably narrow particle size distribution (see B and C in Fig. 12.29). The largest mean particle size was from sample B at 45 ± 17 μm and then sample C at 30 ± 4 μm. Both ground class 5 HMXs had mean particle sizes similar to the original estimates from

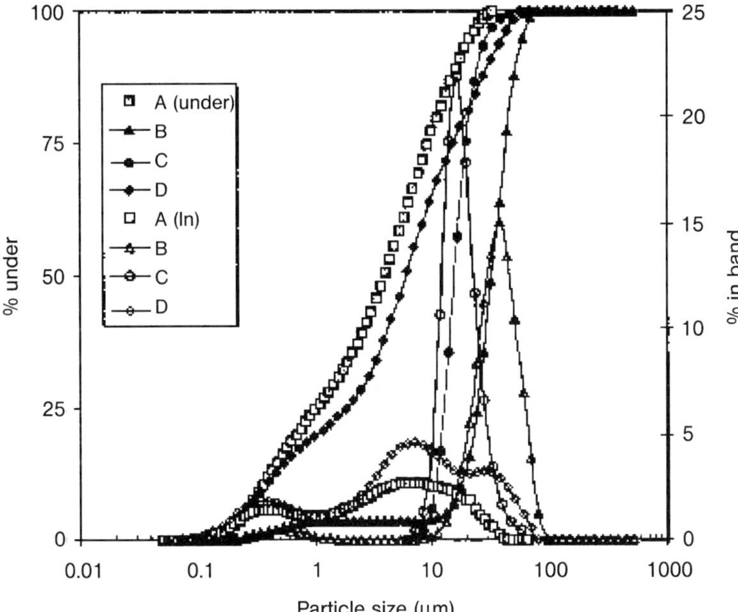

Figure 12.29. Four different class 5 HMXs (A and D were ground and C and B were precipitated) have very different particle size distributions.

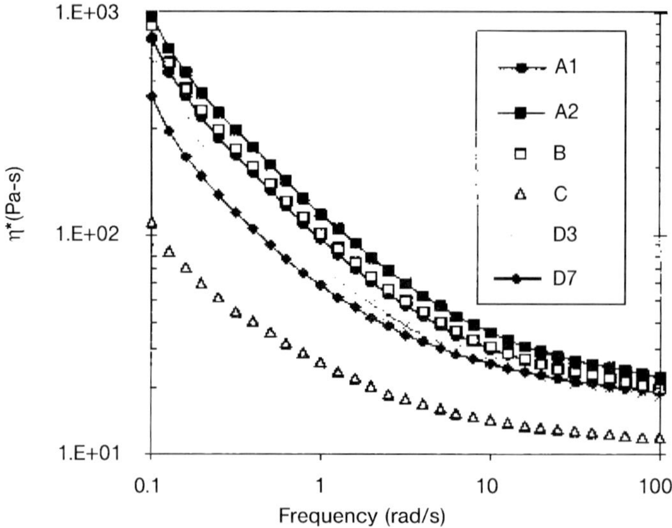

Figure 12.30. The dynamic viscosity of RX-08-HD formulated with different particle size distributions of fine HMX with coarse held constant showed that broad distributions with substantial quantities of sub-micron fines were more than the larger, nearly monomodal fines such as C.

conventional light-scattering measurements (A = 3.8 ± 1 μm and D = 6.4 ± 1 μm), but the sub-micron component was resolved.

When these four different class 5 HMXs were formulated into RX-08-HD and their viscosity was measured as described above, sample C had a much lower dynamic viscosity than the others. Figure 12.30 shows the dynamic viscosities of each formulation as a function of oscillating frequency in rad/s. At 0.1 rad/s the 30-μm class 5 HMX (C) had a dynamic viscosity of 10 P s compared with 400–900 Pa s for the other formulations. This result was observed in three separate tests. However, when steady shear viscosities were measured on these formulations, their viscosities were nearly identical as a function of shear rate.

Figure 12.31 shows steady shear viscosities of each variation of RX-08-HD. The values are very near the original viscosity of RX-08-HD given in Table 12.6. The variability of the Cox-Merz relationship with particle size distribution is at present not well understood. This should serve as a caveat to formulators and rheologists alike that the explosive should be tested by several methods and finally in a system or mock-up of a system rather than relying on one set of measurements alone.

12.7.3
Effect of Thixotropy

Another concern as the solids content increases is thixotropic behavior. One method for evaluating the changes in viscosity due to changes in structure in the paste or

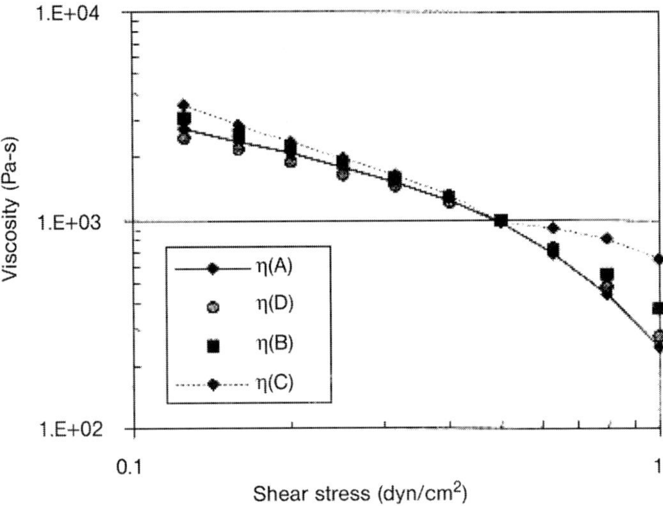

Figure 12.31. Steady shear viscosities of RX-08-HD formulated with different class 5 HMXs were nearly identical.

IMX is to follow the dynamic viscosity as a function of strain amplitude. The shear storage and loss moduli can be measured dynamically as a function of strain amplitude using Eq. (12.23). A simplistic attempt to estimate thixotropic behavior and yield stresses of these pastes was made using the relationship

$$\sigma_{12}(0) = G'(\omega)/\dot{\gamma}_{12}\cos(\delta) \qquad (12.47)$$

where $\sigma_{12}(0)$ is the shear stress; $\dot{\gamma}_{12}$ is the shear rate and δ is the phase angle. The viscosity of RX-08-FK, an unstabilized colloidal paste explosive, showed dramatic differences in transient characteristics on the first runs compared with subsequent runs, as shown in Fig. 12.32. The fumed silica in RX-08-FK tends to form weak aggregates over time that can be disrupted by shear. Without sufficient startup pressure this formulation will not flow. Dynamic viscosity and shear stress measurements as a function of amplitude (strain) showed yield and flow in first runs as the amplitude of oscillation increased. Second runs showed no evidence of yielding behavior.

Thixotropic behavior measured in this way is also dependent on the formulation's solids content and particle size distribution. When no colloidal particles were used in the TIE and IMX, formulations with solids contents > 65% showed thixotropic behavior. Figure 12.33 shows dynamic viscosity as a function of strain amplitude for TIE formulations at 65 and 70% solids with optimum particle size distribution with the second run of RX-08-FK from Fig. 12.30 for comparison. Both formulations with ≥ 70% solids show a reduction in viscosity as the amplitude of oscillation increases. The viscosity of RX-08-FK is about an order of magnitude higher with only 2.5% more solids because it does not have the 72.5 : 27.5 coarse to fine particle size ratio. The reason for the lack of thixotropic behavior at 65% solids is not known. It may be

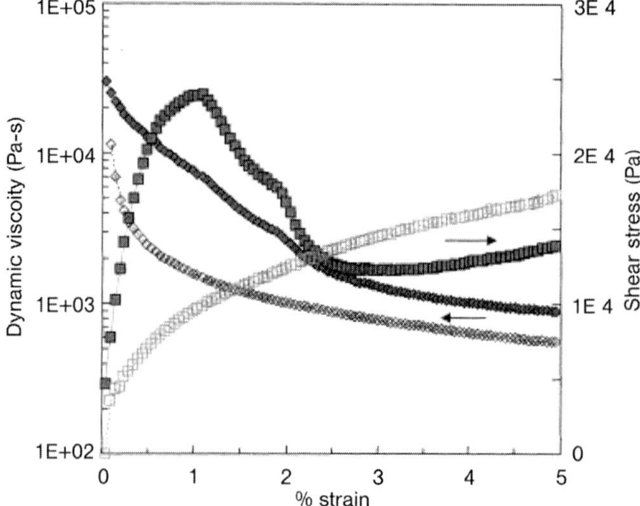

Figure 12.32. Thixotropic behavior in RX-08-FK, a fumed silica-based paste, is measured by strain sweeps of dynamic viscosity increasing the strain amplitude. The first run (solid symbols) is very different to the second because of the structure in the silica. This characteristic takes several days to develop.

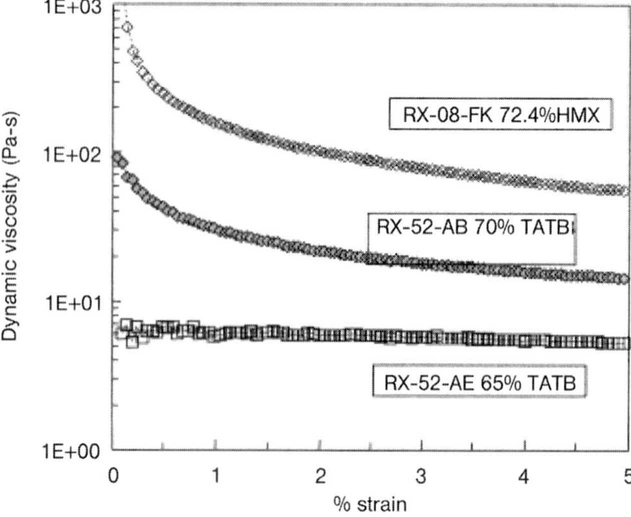

Figure 12.33. Comparison of strain sweeps at different solids loadings and with multimodal particle size distributions.

that above some critical solids concentration, packing is disrupted in a different way to initiate flow. Once particles begin to travel over each other, it will take some time for them to resettle into an optimum configuration above this critical concentration.

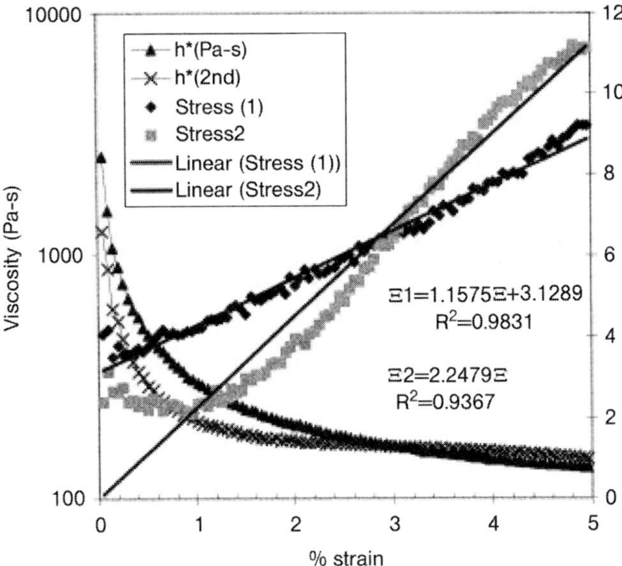

Figure 12.34. Dynamic viscosity and apparent stress as functions of strain amplitude show very mild thixotropic behavior in RX-08-HD.

If there is a volume change or hysteresis associated with settling, voids could develop in the loaded part. This could have undesirable consequences during functioning of a weapon.

In Fig. 12.34, first and second strain sweep measurements of dynamic viscosity and shear stress (from Eq. (12.47)) are shown as a function of strain amplitude at 1 Hz (6.28 rad/s) for RX-08-HD. The apparent dynamic viscosity of the second run is initially lower than the first at low amplitudes and approaches the first run viscosity above 3% strain. The shear stress approaches a limiting value of about 3 kPa, reminiscent of a Bingham fluid.

RX-08-series explosive filling characteristics were evaluated visually on a 50-g deaerator loader. This mock-up consisted of two pistons above and below an orifice plate. The paste explosive is loaded into the upper chamber and passed through the orifice plate under vacuum to remove any entrapped air. The lower piston is extended, driving the explosive out of a tube just below the orifice into the mold. A polycarbonate cylindrical mold (2.4 cm high × 2.3 cm i.d.) with a copper mandrel (1.27 cm high × 1.2 cm diameter offset from the center by 0.9 cm) was filled with each explosive in the series. Two configurations are possible, as shown in Fig. 12.35. Explosive can be injected either over the copper mandrel or away from it. Low-viscosity RX-08-series explosives (-GX, -GZ and -HD), when injected over the mandrel, fill in three phases. First the mandrel is covered and the explosive starts down the small side. Next it bridges and enters the vacuum line. At this point the vacuum must be closed and filling continues until the part is completely filled. The alternative configuration (fill side away from the mandrel) loaded better. The low-

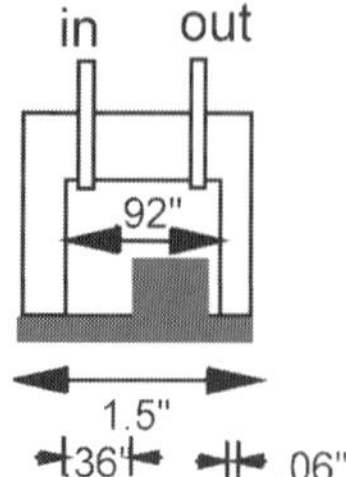

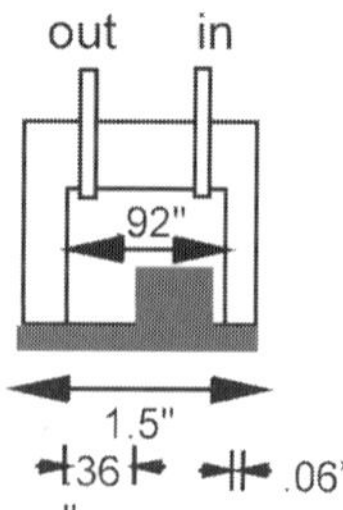

Figure 12.35. Alternative loading configurations of test fixture and the beginning of a typical load for RX-08-series explosives in the left-hand side configuration.

viscosity explosive flowed down the side, to the bottom and part way around the mandrel before filling the upper section over the mandrel and bridging. With good vacuum, low-viscosity formulations could fill the mold completely without voids. Intermediate viscosity RX-08-series explosives (-GY, -HB, -HF, -HA and HE) tend to bridge before completely filling the large gap (38 cm). Vacuum had to be turned off before the part was half full. This often resulted in small voids along the top of the mold. High-viscosity paste (-HC) did not bridge badly or flow into the vacuum line until very near the end of the fill. However, this formulation filled slowly, was difficult to deaerate and tended to relax after pressure release, forming small voids along the edge of the mold. This formulation showed the largest difference between first and second dynamic viscosity versus strain amplitude runs. Based on the rheological and small-scale safety results, RX-08-HD was down selected and successfully loaded into shaped charge munitions.

In IMXs and cast cure explosives, rheological measurements can be used to estimate pot life and time to handling strength. The cure characteristics of RX-08-HD were followed as a function of the latent cure catalyst T-131 concentration to determine how much pot life could be expected for filling large components. Figure 12.36 shows that changes in the catalyst concentration between 10 and 70 ppm level changed the pot life of the formulation from 10 to 2 h. Using 20 ppm of Dabco T-131 catalyst gave a 4–6-h pot life. With this information, 130 lb (~ 59 kg) of IMX were successfully loaded into a 500-mm shaped charge in 3 h with 1 h of equipment clean-up time. The fill was cured after 24 h at ambient temperature.

Dynamic viscosity at constant frequency is excellently suited for such measurements since instead of going to infinity as the explosive cures, the dynamic viscosity goes to a limiting value. Gelation or pot life can be evaluated in various ways. DOD explosive processors have used a viscosity of 4000 Pa s [12.36]. The crossover point in shear storage (G') and loss modulus (G'') was suggested by Tung and Dynes and others [12.51]. When a peak in the tangent delta has been used when it is observable. Explosives with a wide variety of solids loadings may be measured. Cure temperature above ambient can be followed when the rheometer has an oven.

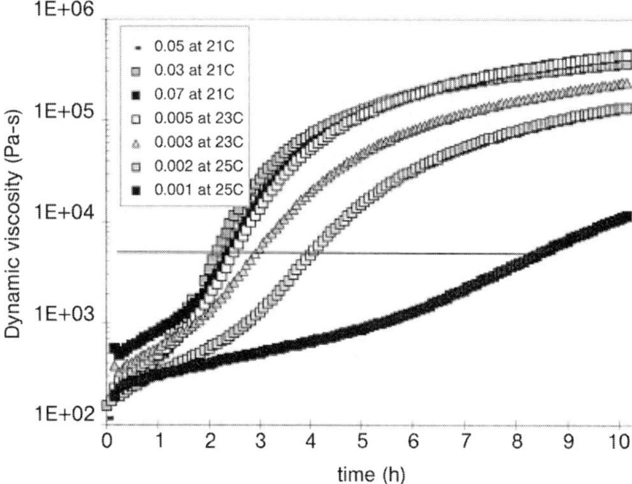

Figure 12.36. Cure characteristics of RX-08-HD as a function of T-131 catalyst content show that above 70 ppm a minimal increase in cure is observed. Pot life as measured by a critical viscosity of 4000 Pa s can be varied from 2 to 9 h depending on the level of catalyst. Above 70 ppm, additional catalyst seems to have no effect.

12.8 Computer Simulation of Rheological Behavior of Suspensions

The wide application of suspensions in industry makes it necessary to study and simulate these objects. The specific behavior of particles within a fluid determines their internal structure in suspension. The ability of particles in suspension to coagulate and form space continuous structures is known.

The purpose of this work was the computer simulation of suspensions. The essential feature of the model offered is the description of the behavior of the studied system on the basis of descriptions of the behavior of each particle. Such an approach allows us to describe investigated processes which take place in suspensions and to predict the influence of technological parameters on the behavior of suspensions.

To solve this task, a computer model consisting large number of uniform structural elements was created. There are available all necessary data on the behavior of each structural element. This approach has been widely used in physics and is called 'the method of particles' [12.52]. The term 'the method of particles' is a general term for a class of computing models to describe discretely different physical phenomena. The state of a multiphase system is determined by the attributes of the final ensemble of particles. The evolution of the system is defined by the laws of the interaction of these particles.

The application of 'the method of particles' is attractive to research systems from the computing point of view. It is connected with the particularity of the method which makes it possible to preserve the attributes of particles in the computing model with its development over time.

The computer model offered is based on research on the evolution of spherical particles in three-dimensional space [12.53, 12.54]. The evolutionary process is divided into discrete temporary intervals. The movement of the particle is defined over each temporary interval considering its current coordinates and the forces acting on the particle. The value of the temporary interval is chosen from the condition of constancy of the velocity movement of particles. The maximum displacement of particles does not exceed the previously chosen value.

On the basis of the created computer model, calculations for *N* particles originally located in the fluid medium in a casual way were carried out. Each particle has a set of attributes such as mass, size, position and velocity. The particles are located in a Newtonian fluid, which is characterized by viscosity, temperature and shear rate. The size of the particles and their concentration in the studied system is also set. The calculation of fulcrum reaction forces is applied in the model for the exception of particles penetrating into each other and a description of joint moving particles in the case of their contact [12.55]. In this algorithm, a total projection of the forces acting on particles on a line connecting the centers of particles is calculated. The fulcrum reaction forces acting on the line connecting the centers of particles is subtracted from the total force. The value of the difference between forces acting on each of the considered particles determines their further movement in space.

In the model, the velocity and coordinates of particles are determined with respect to the following. The evolution of a system is considered to be the result of the interaction of particles with each other and with the medium. In the model, Brownian motion of particles and van der Waals and hydrodynamic interactions of particles are taken into account. In our model for the account of Brownian motion of particles the classical Einstein-Smoluchowski equation is used.

It is known that the Einstein equation which determines the dependence of the relative viscosity of a medium on the concentration of particles, correctly reflects experimental data at low concentrations of disperse phase. The explanation is that in the equation interaction of particles with each other is discounted. For the description of the behavior of high-concentration suspensions it is necessary to take into consideration hydrodynamic interactions between particles.

In the developed model, processes of coagulation of particles are counted. The stability of spatial structures is determined by forces of interaction of particles. On these theoretical bases, the following algorithm of the program for the study of processes was developed. Input data for a computer model are number and size of particles of suspension, concentration filler, temperature, viscosity of medium, shear rate of system and time interval. Before the beginning of work on the model, an initial distribution of particles in three-dimensional space with the help of a generator of random numbers is produced (Fig. 12.37). The size of a cell determines the dependence on the concentration of suspension. It is accepted that at the first instant all particles have zero velocities. For many temporal pitches it does not influence the outcome of calculations. Subsequently the basic cyclical algorithm starts working. Forces operating on particles by the medium, forces defining

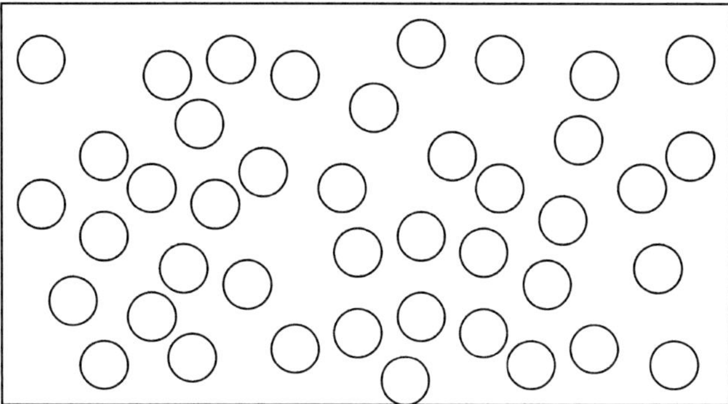

Figure 12.37. Initial distribution of particles of a suspension in a simulated system.

Brownian motion, forces of interaction of particles and formed forces in a shift flow are taken into account in the algorithm. In the model, forces of hydrodynamic interaction of particles are also taken into account. The first group of forces determine the velocities of movement of particles. The Brownian motion was calculated using the known equation with the use of three generators of random numbers which can create a normal distribution. Then, using earlier calculated distances between particles, we determined the forces of interaction of particles. In contrast to Brownian motions, which are summed in each step with the previous ones, in this model the forces of interaction equal zero at the beginning of each time step. This appears possible because they are short-range forces and their value is not proportional to time and is determined only by the disposition of particles of the suspension.

The algorithm for the calculation of forces operating on particles in a shift stream is also used. These forces are directed along the axis of flow and defined by the shear rate of flow. The direction of flow is determined from the condition that the center of the considered system is in the center of the cell and the gradient of shear rate is directed along the y-axis. Its value is proportional to the difference in coordinates on the y-axis and of the shear rate. Subsequently, forces of hydrodynamic interaction of particles are determined. After the calculation of forces and velocities, the trajectories of movement of particles in space are calculated. For this purpose, the most acceptable approach is selected. In the first stage, we consider movement of particles under the influence of Brownian forces. At the second stage, we consider the movement of particles under the action of forces of interaction and shift and hydrodynamic forces. Such an approach makes it possible to describe the process and the existence of the space configuration of particles.

The choice of axes for the movement of particles in each calculated step is produced by a method of random numbers. Thus, we use the algorithm taking into account the disposition of particles and the possibility of their coagulation. The sequence of choice of particles in the calculation of trajectories of their movement is determined by a method of random numbers. In the algorithm, two possible variants of the behavior of simulated suspension are taken into account:

1. In the absence of shift in the modeled system, particles move in closed space, not exiting the limits of the closed volume.
2. In the presence of shear rate in the considered system, permeability of walls of cell is assumed.

Thus, the exiting particles change the sign of the *x*-coordinate that the constancy of the degree of volumetric filling of the system ensures. After the realization of all calculations, the next picture at a given instant of the disposition of particles is deduced from the display on the computer (Fig. 12.38). At this stage, possible modifications of the technological parameters of the simulated suspension are temperature, viscosity, shear rate and the value of the time step that allows us to investigate their influence on the behavior of suspensions.

Calculations were carried out before a dynamic equilibrium of the simulated system. The sharp increase in the time of calculation with increase in the selected number of particles limits the number of particles of suspension in the simulated system. For each variant of data the calculations will be carried out no less than 3–5 times, taking the mean of the results into account.

The dispersion of carbon black in an oligomer medium was used for the experimental investigations. It is known that the processes of structuring of the carbon black particles in oligomerleads to the creation of continuous spatial structures of particles in suspension [12.56, 12.57]. The particles in a fluid medium are moved by Brownian motion and the collisions of particles can form agglomerates. During the agglomeration, endless structures appear. The existence over time of the structures formed is a result of the activity of intermolecular forces, particularly van der Waals forces. The formation of endless agglomerates leads to electrical conduction of the suspension.

Hence an observed increase in electrical conductivity of a studied composition indicates the existence of processes of agglomeration of particles. These agglomer-

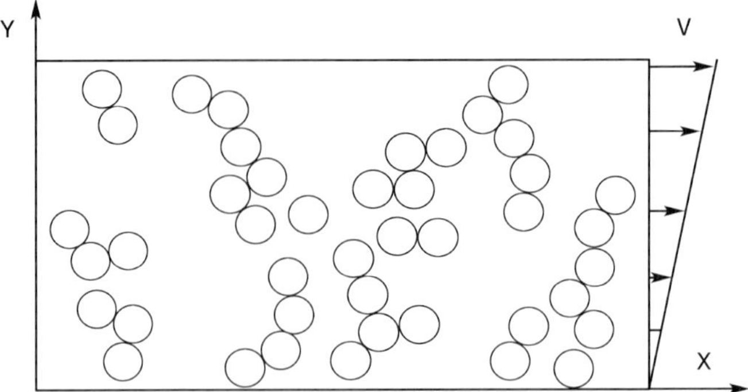

Figure 12.38. Flowing state of simulated suspension.

ates penetrate the whole volume of the measuring cell. The basic idea of applying our computer model to the suspension of carbon black is that the suspension becomes electrically conductive when the coordination number of particles in the suspensions becomes ≥ 2.

The coordination number 2 characterizes the structure of the suspension when each particle has two contacts with closely located particles, i.e. this particle is an element of the continuous chain of particles that penetrates the whole volume of the suspension. The existence in the suspension of these chains is the reason for the electrical conductivity of the studied system. The results of the computer calculations and experimental investigations are given in Figs 12.39–12.41. Figure 12.39 shows the typical calculation dependence.

The average calculation and experimental kinetic curves of processes of structuring of the carbon black particles in an oligomer are shown in Fig. 12.40. These results indicate adequacy of the designed computer model for real systems.

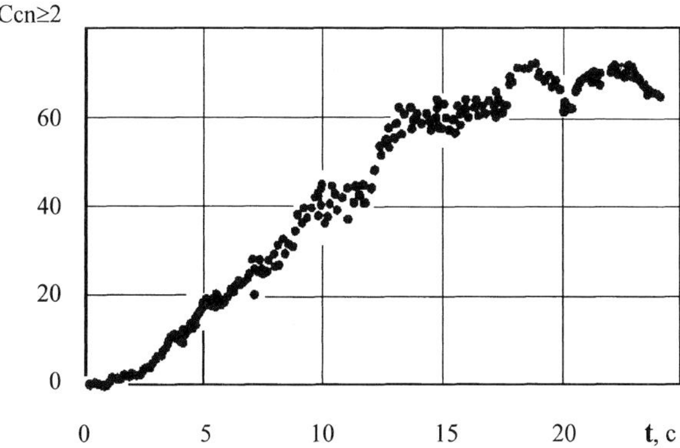

Figure 12.39. Results of calculation.

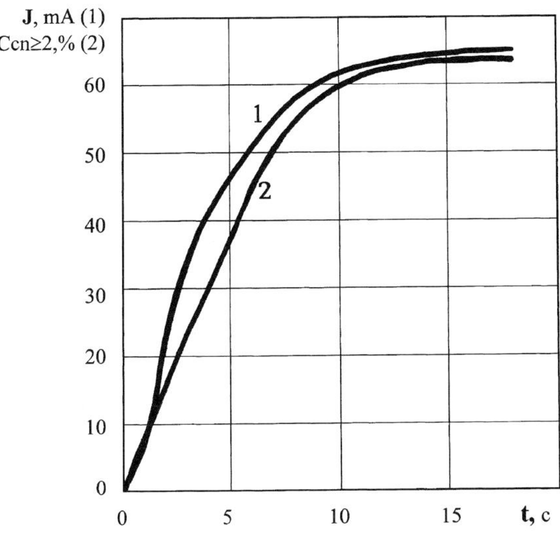

Figure 12.40. Kinetic dependence of agglomeration of carbon black in oligobutadiene. Temperature, 60 °C; concentration of carbon black, 4 %. Curves: 1, experiment; 2, calculation.

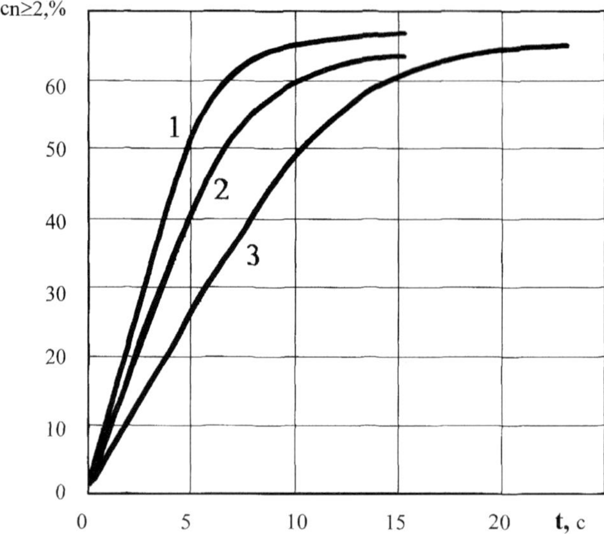

Figure 12.41. Kinetic dependence of agglomeration of carbon black in oligobutadiene. Temperature: 1, 80; 2, 60; 3, 40 °C.

In Fig. 12.41, the results of calculations for different temperatures are given. It is obvious that an increase in temperature leads to acceleration of the process of structuring of particles, which can be explained as follows. An increase in temperature leads to a decrease in viscosity of the oligomer and a reduction in the hydrodynamic friction forces acting on particles. Simultaneously with the increase in temperature, the influence of Brownian motion on the behavior of particles is increased, which in its turn accelerates the processes of structuring of suspensions.

The above results demonstrate the feasibility of the computer model allowing one to describe processes taking place in suspensions and to predict the influence of different technological parameters on their behavior.

12.9 Rheology of Solid Energetic Materials

Most modern energetic materials are of polymeric nature or certain polymers are used at least as one constituent. In most cases the polymers are used as the matrix, which contains rigid particles. Depending on the chemical structure of the polymer, plastic, elastic or viscoelastic behavior of the compound can be achieved. Therefore, energetic materials cannot be regarded simply as elastic bodies but also the concepts of rheology, describing the time-dependent behavior of materials, must be considered. The presence of rigid particles in the polymer and their characteristics (shape, size, size distribution, surface area) additionally influence the rheological peoperties of the compound.

The rheology of liquids, emulsions and suspensions is one of the most important fields of interest for the precursors of energetic materials during their manufacture owing to processing needs. There is a close interaction between the rheology of the energetic precursors and the mechanics of the solid energetic material. The better the processability, the better the mechanical properties will be in most cases.

The mechanical behavior of solid energetic materials is important with regard to

- quality control
- storage behavior
- lifetime prediction
- and of course for their use.

Furthermore, the mechanical behavior must be known for the use of finite element calculations in the design of systems containing energetic materials.

The mechanical properties can be used for quality control directly following the production of energetic materials. With their help, the completeness of the curing process and also certain aspects of the matrix-filler-bonding quality can be judged.

Large amounts of energetic materials are stored for long periods, sometimes for a few decades in the case of very expensive systems. In order to save money, one tends to keep these systems as long as possible, which requires a reliable lifetime prediction. During that storage period, a variety of stresses act on the energetic materials. The most important of these are temperature fluctuations, from which mechanical stresses can arise due to differences in the thermal expansion coefficients of the materials of which the system consists and mechanical loads such as shocks and vibrations during transport in maneuvers or flight practice.

In large rocket motors or rocket boosters, several tons of propellant are used. Therefore, the propellant is stressed by its own weight in addition to the stresses acting from outside. During long periods of time, no severe change in geometry of the propellant grains may occur and the propellant may not detach from the motor case.

Chemical processes in energetic materials such as additional cross-linking in the matrix can lead to severe changes in the mechanical properties and cause increasing brittleness with time.

The changes due to the aforementioned influences may not exceed certain limits of the mechanical properties of energetic materials for safe use.

The most critical point in the use of propellants is the time of their ignition. Here, the pressure which acts on the material rises into the range of milliseconds with a steep gradient. In large rocket motors, pressure waves pass through the material. Cracks which occur during ignition can cause catastrophic failure of the whole system. The response of the material to pressure shocks is therefore of crucial importance. Following the ignition, the propellant is stressed by high pressures under a high temperature gradient for a certain period of time (from milliseconds to a few minutes), which also may not cause cracks.

Finite element methods (FEMs) are a common tool for the design of technical systems and are also used for the design of systems containing energetic materials. The reliability of the results of the calculations is highly dependent on the constitutive equations describing the mechanical properties of the considered materials and the quality of the material data used. The material data which are determined can therefore be used not only for calculations during design but also for quality control needs and the judgement of aging for lifetime prediction.

Modern composite solid rocket propellants are often composite materials: a relatively soft matrix material (e.g. cross-linked polyurethanes) is used for embedding relatively hard particles which burn (e.g. aluminum or boron) or deliver the necessary oxygen for oxidation (e.g. ammonium perchlorate). Inherent to this combination is its weak point: the interface between the matrix and the filler surface. Here, large gradients in the material properties on a microscopic and mesoscopic scale exist. These gradients in the material properties lead to high local stresses in the case of acting loads which can cause detachment of the matrix material from the filler surface. The detachments, causing vacuoles, can act as starting points for cracks or can act as 'hot spots' which increase the sensitivity of the material to ignition by shock. An immediate compression can heat the gases captured in the vacuoles beyond the ignition temperature of the propellant.

12.9.1 Viscoelastic Behavior of Energetic Materials

A brief overview of viscoelasticity is given as far as it is useful for energetic materials. More detailed information can be found in the literature [12.58–12.61].

12.9.1.1 Stresses and Strains

The rheology of solid materials connects the effects of loads, which are measured and defined in terms of stresses (force per unit area with the units N/mm^2), to their effects, the deformations defined as the relation of differences in the distances between two material points to their distances in the initial unloaded condition (this relation has no units, but can be specified in terms of percent).

12.9.1.2 Material Laws and Constitutive Equations

The constitutive equations for the mechanical behavior are the mathematical formulations of the relation between stresses and strains. The huge number of equations can be divided up into different main categories regarding time dependence and the ability of the material to recover its initial state completely as shown in Table 12.7.

Within these categories, a further division into one-, two- and three-dimensional equations can be made. The dimensions indicate how many spatial directions are considered and described. The complexity of the formulations increases with the number of dimesions. In order to keep the more complex equations usable for

Table 12.7. The main categories of the mechanical behavior of materials.

Category	Time dependence	Complete recovery
Elastic	No	Yes
Viscoelastic	Yes	Yes
Elastic-plastic	No	No
Viscoelastic-(visco)plastic	Yes	Yes

calculations, simplifications due to symmetry of the problem at hand or other purposes are often possible.

One constitutive equation can be applied to several materials which behave in the same principal way in the range of interest, e.g. elastic. The distinction between the elastic behavior of different materials is then made according to material parameters. These parameters appear in the constitutive equation and have to be determined experimentally for each material with respect to the equation used.

Because a single constitutive equation gives a pure mathematical description of the behavior observed, the material parameters are not connected to the structure of the material. In recent years, there has been a tendency towards a molecular and micromechanical derivation of material parameters. This tendency is very strong in the area of composite materials, where attemts are made to achieve specific parameters with a purposeful variation of the constituents and their configuration (particles, fibers, fabrics, etc.).

A certain amount of knowledge about the viscoelasticity can be directly interchanged between energetic materials and filled elastomers which find wide applications (tires, seals, etc.) and where some research is undertaken.

12.9.1.3 Examples of Constitutive Equations in One Dimension

The simplest equation for the elastic behavior in one dimension is Hooke's law:

$$\sigma = E \cdot \varepsilon \tag{12.48}$$

where σ is the stress and ε is the strain, which are connected by Young's modulus E. When time dependence plays a role as in the case of solid rocket propellants and other polymeric materials at medium and higher deformations, this simple formulation has to be replaced, in the simplest case by an integral formulation:

$$\sigma(t) = \int_{-\infty}^{t} [E(t) \cdot \dot{\varepsilon}] \mathrm{d}t \tag{12.49}$$

where $E(t)$ is the time-dependent relaxation modulus. This equation considers the influence of the complete deformation history up to time t on the stress at time t due to the time dependence of the relaxation modulus. This case is called the linear viscoelastic case. In the non-linear viscoelastic case, the relaxation modulus is also dependent on the strain $E = E(t, \varepsilon)$, which is valid for most solid rocket propellants even at small strains. In this case a variety of equations exist and can be found in the literature.

12.9.1.4 Examples of Constitutive Equations in More Dimensions

The constitutive equations in more dimensions arise from theoretical considerations. A major problem is the performance of more-dimensional measurements in order to determine the necessary material parameters. Nevertheless, the well-known and established one-dimensional material laws must be derivable from the more-dimensional ones. In addition, several physical demands must be satisfied, e.g. objectivity and independence from the chosen coordinate system.

The three-dimensional formulation of Hooke's law exhibits 36 elastic constants. Owing to the unambiguity of the elastic potential and symmetry, the number of constants can be reduced. For an elastic isotropic material, the number of constants decreases to two. The equations for the single spatial components in this case, for small elongations, are

$$\sigma_{11} = \frac{E}{(1+\nu)\cdot(1-2\cdot\nu)}\cdot[(1-\nu)\cdot\varepsilon_{11}+\nu\cdot(\varepsilon_{22}+\varepsilon_{33})] \tag{12.50}$$

$$\sigma_{22} = \frac{E}{(1+\nu)\cdot(1-2\cdot\nu)}\cdot[(1-\nu)\cdot\varepsilon_{22}+\nu\cdot(\varepsilon_{33}+\varepsilon_{11})] \tag{12.51}$$

$$\sigma_{33} = \frac{E}{(1+\nu)\cdot(1-2\cdot\nu)}\cdot[(1-\nu)\cdot\varepsilon_{33}+\nu\cdot(\varepsilon_{11}+\varepsilon_{22})] \tag{12.52}$$

$$\sigma_{12} = \frac{E}{1+\nu}\cdot\varepsilon_{12} \tag{12.53}$$

$$\sigma_{13} = \frac{E}{1+\nu}\cdot\varepsilon_{13} \tag{12.54}$$

$$\sigma_{23} = \frac{E}{1+\nu}\cdot\varepsilon_{23} \tag{12.55}$$

Young's modulus E and Poisson's ratio ν describe the behavior of the material. In principle, it is possible to expand these equations to time-dependent behavior, which is rather complicated.

Some three-dimensional constitutive equations use the invariants, mathematical tensor functions, of the stress and strain tensors in order to achieve a coordinate system-independent description of the material behavior (e.g. Mooney-Rivlin). The invariants do not change in case of a change of the coordinate system.

The material parameters have no physical origin but arise from the mathematical formulation of the constitutive equation. In recent years, attempts were made to combine micromechanical deformation mechanisms with the measured material parameters. With the increasing performance of computer techniques it was possible to construct micromechanical models of the materials and to calculate the macroscopic observable behavior from micromechanical phenomena and the constitution of the material. This correlation is extremely useful for a purposeful development of energetic materials or for quality control [12.62–12.69].

12.9.1.5 Description of the Mechanical Behavior of Energetic Materials

The viscoelastic behavior of propellants can best be explained by viewing show-graphs of certain mechanical testing experiments. During a tensile test a sample of a given geometry (e.g. JANNAF) is clamped in a tensile test machine and strained with a certain velocity, and the resulting force is measured. In the following diagrams, the most common mechanical tests are shown, from which the material parameters can be derived. All tests can be performed both in tension and in compression.

In commonly used standard test methods, the samples are uniaxially loaded. Therefore, material parameters only for the one-dimensional constitutive equations can be measured. If the sample geometry is recorded in two dimensions (length and width) during the one-dimensional test, Poisson's ratio can be measured for isotropic materials.

Tensile test

In the normal tensile test of a viscoelastic material, the measured force is influenced by the deformation speed (ε) used, as shown in Fig. 12.42. The higher the deformation speed, the higher is the measured force. Figure 12.42(a) shows the excitation (strain over time), (b) shows the response of the material in the form of a stress-time

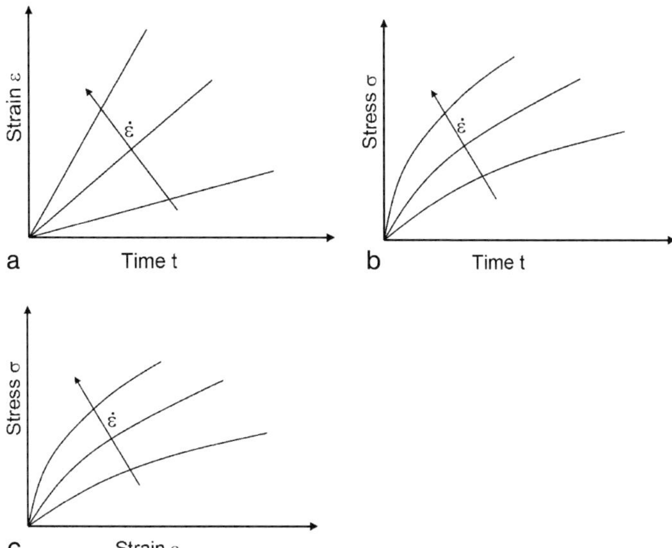

Figure 12.42. Diagrams for the tensile test with the deformation speed ε as parameter.

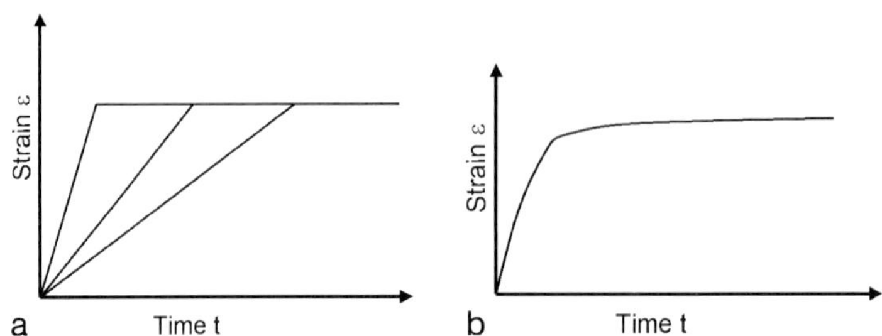

Figure 12.43. The relaxation experiment. (a) Excitation; (b) response of the material.

diagram and (c) shows the common representation of the behavior as a stress-strain diagram.

From the stress-strain diagram, Young's modulus *E*, the tangent to the curve at low strains, can be derived.

Relaxation test

In the relaxation test, the sample is strained and held at a certain strain while the force is recorded. The slower the velocity, the less maximum force and force decay can be observed. The decay of force at a constant strain means a decay of energy stored in the sample. The energy is dissipated in the internal relaxation processes (Fig. 12.43).

Retardation test

In the retardation test, the sample is loaded with a certain constant load and its deformation is recorded with time (Fig. 12.44). From this test, a distinction between liquid (no strain limit) and solid (strain limit for long times) behavior can be made.

Cyclic test

In the cyclic tensile test, one or several load-unload cycles are applied to the material. From the result, elastic, viscoelastic, plastic and viscoplastic behavior can be seen

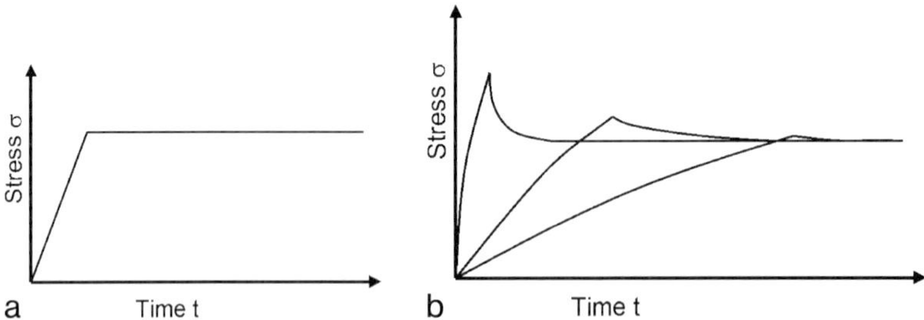

Figure 12.44. The retardation experiment. (a) Excitation; (b) response of the material.

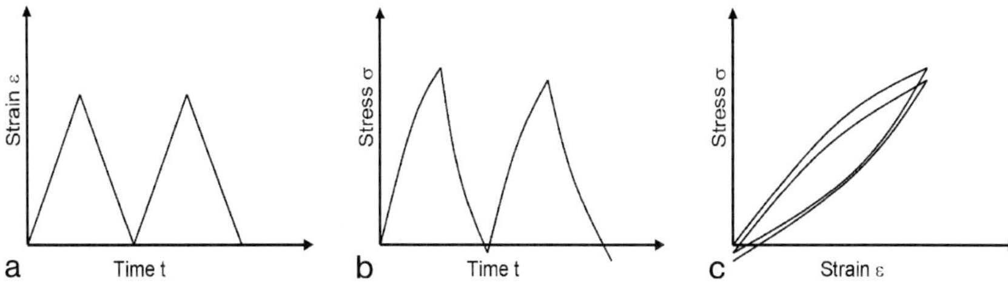

Figure 12.45. The cyclic experiment with constant peak strain.

(Fig. 12.45). From Fig. 12.45(c), the dissipated energy in one deformation cycle can be derived from the area between the load-unload stress-strain curves. Elastic behavior leads to no compressional stress between the cycles and no hysteresis in the stress-strain diagram. Plastic behavior causes a lasting elongation of the sample (lasting strain at zero stress).

Dynamic testing

Dynamic testing is a powerful tool for the determination of the fraction of deformation energy stored in the material and the fraction which is dissipated. The sample is excited with a sinusoidal strain signal (tension, compression or torsion) and the resulting stress or torque is recorded. Viscoelasticity results in a phase shift φ between the exciting strain and the measured stress. This method is strictly true only for linear viscoelastic behavior, because the resulting stress also needs to be sinusoidal. However, at small strains, the imperfections can often be neglected. The phase-shifted stress signal is divided into two parts: one part in-phase with the exciting strain (elastic part of the behavior) and one part 90° phase shifted to the exciting strain (viscous part of the behavior). The part that is in-phase is used for the determination of the stored part of the energy and the 90° out-of-phase part represents the dissipated part of the energy (Fig. 12.46).

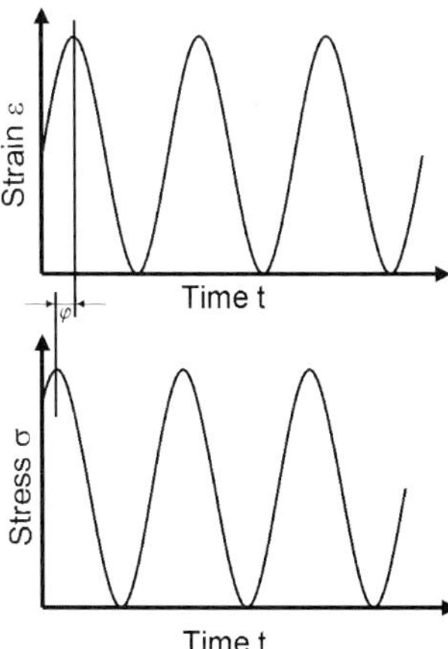

Figure 12.46. The phase shift φ in the dynamic experiment.

The results of all these tests are dependent on several parameters (temperature, pressure, humidity, velocity, time), of which the temperature is one of the most important. Solid rocket propellants are stiff at low temperatures and become softer at high temperatures. At the glass transition temperature, they change their mechanical properties by several orders of magnitude.

According to Williams, Landel and Ferry (WLF), there is a correlation between the characteristic time of an experiment (e.g. relaxation time, time to failure in retardation, excitation frequency in dynamic testing) and the temperature [12.70]. This correlation can be described by the WLF equation. In the case of the modulus it reads

$$E(t_0, T_0) = E(t_1, T_1) = E(t_2, T_2) \tag{12.56}$$

where

$$\log t_1 = \log t_0 + \log a_{T_1}^{T_0} \tag{12.57}$$

and

$$\log t_2 = \log t_0 + \log a_{T_2}^{T_0} \tag{12.58}$$

a_{T2}^{T0} is called the shift factor in relation to the reference temperature T_0, for which the glass transition temperature is often used. The shift factors can be derived from

$$\log a_T^{T_0} = \frac{-C_1(T - T_0)}{C_2 + T - T_0} \tag{12.59}$$

where C_1 and C are constants which are dependent on the material and have to be measured. With this knowledge, experiments at low temperatures can be performed to determine approximately the mechanical behavior at short excitation times at room temperature. High-speed properties can be measured only with an enormous experimental effort.

12.9.2 Measurement of Mechanical Properties of Solid Viscoelastic Energetic Materials

For the measurement of the mechanical properties of energetic materials, several kinds of test equipment are available. The normal quasi-static tensile test (< 2 m/min), the cyclic tensile test and compression tests can be performed with screw-driven tensile testing machines. The forces are measured with load cells. The deformations can be measured in the case of large deformations by recording the crosshead travel or, for better accuracy, with mechanical transducers which are directly clamped to the sample. These transducers influence the measurement in the case of very soft materials and can cause breaking of the sample by the formation of notches. Non-contact measurement devices work with the help of laser light or with video techniques and image processing [12.71–12.74]. With these techniques, a two-dimensional deformation measurement is possible.

High-speed tension or compression tests (> 1 m/s; < 50 m/s) are performed with (hydropulsing) test machines. Here the forces can be measured with the help of piezo force transducers. Contacting deformation measurement facilities cannot be used owing to the high inertia forces.

Ultra-high-speed tests can be made with powder-driven test devices or light gas canons where impact speeds up to 300 m/s can be realized. These facilities are not commercially available and are only useful for research needs owing to the high experimental effort and the consequent costs.

For retardation measurements, the samples are normally loaded with weights and the elongation is measured with optical devices over a long period of time.

Dynamic measurements can be performed with electrodynamic excitors in the range of low frequencies (< 100 Hz) and large deformations (< 5 mm) or with piezoelectric excitors (< 1000 Hz, < 0.1 mm). Ultrasonics can be used in the range of very low amplitudes and very high frequencies (< 25 kHz, < 0.01 mm). The dynamic measurements are performed either in torsion or in tension-compression.

The data acquisition and processing for the determination of the moduli are normally done automatically by computers.

12.9.3
Micromechanical Phenomena in Energetic Materials and their Influence on Macroscopic Mechanical Behavior

At the molecular level, mainly reorientations of parts of the polymeric chains are responsible for the macroscopic observable mechanical behavior of polymers. The most important phenomenon here is rotations of chain segments around the C-C bonds in the polymer backbone. These rotations can be hindered by voluminous side chains, entanglements, cross-links or the presence of filler particles. A second important phenomenon is the sliding of polymer chains along each other. These phenomena cause the elastic and time-dependent viscoelastic behavior of the polymeric binders.

Single filler particles are not considered at this point, because with respect to the matrix they can be regarded as rigid and therefore do not contribute to the viscoelastic behavior of the composite owing to their own mechanical properties. On the other hand, the filler regarded in its entirety is of high interest. In the filler alone, breakages of filler particles can take place, especially at high compressions, due to impacts. Agglomerates of single filler particles can break even more easily. Agglomerates are only formed in the case of small filler particles as they are used in energetics. The breakage of agglomerates leads to a decrease in stiffness of the energetic material and can cause the formation of vacuoles with a negative influence on the impact sensitivity.

Filler-filler interactions can be made responsible for unsymmetric mechanical behavior in the tension and compression ranges. In tension, the filler particles are separated from each other, whereas in compression they are pressed against each other. Therefore, the behavior of the matrix material becomes more important in tension and the behavior of the filler becomes more important in compression.

In energetic materials, the interface between the matrix material and the filler surface plays an important role. There are large differences in the mechanical properties of the single constituents which cause problems with regard to failures. The most important phenomenon which can cause severe problems during the service of energetic materials is the detachment of the matrix from the filler surface. This detachment leads not only to weakening of the material, a decrease in the modulus and an increased tendency to break, but also to vacuoles in the material. Breakages which occur when the propellant or gunpowders burn can cause an immediate pressure rise and therefore to a catastrophic failure of the system. The vacuoles which are formed by a detachment, even if no macroscopic observable breakage occurs, are filled with gases which emanate from the matrix or the filler (e.g. adsorbed humidity). In the event of an impact, these gases are very rapidly compressed and as a consequence heat up to temperatures which can ignite the energetic material and therefore increase its impact sensitivity. A superimposed hydrostatic pressure, as often occurs during the service of a propellant, decreases the tendency for the formation of matrix-filler detachments.

The detection of micromechanical phenomena, especially detachments of the matrix from the filler surface, are very important not only for the understanding of the mechanical properties of energetic materials and their service but also for the improvement of these materials with regard to sensitivity and lifetime. Suitable measurement methods can be used for quality control during the development and production of energetic materials and also as a tool for the determination of their aging. Owing to its great importance, the interface between the matrix and filler deserves special attention.

12.9.4 Special Measurement Techniques for the Detection of Micromechanical Phenomena

Normally, micromechanical phenomena are not directly observable and therefore must be detected by indirect methods. Some of the most important methods are listed below and are assessed with regard to their capability for the detection of the micromechanical phenomena which form the basis of the mechanical behavior of energetic materials.

12.9.4.1 Direct Methods

Ultrasonics

Ultasonics can be used to detect delaminations and cracks in composite materials. The length of the crack has to be greater than the characteristic length of the mictostructure of the material in order to be detectable. Therefore, this method is not suitable for energetic materials.

Microscopy

Microscopy is a widely used method for the detection of cracks in composite materials. A major disadvantage is that the materials must be prepared for the investigation, which can produce artefacts. Especially in elastomeric materials, cracks which were formed between a filler and matrix under load can close again when the material is unloaded again. Furthermore, in most cases, the investigation is only possible on the surface of the material.

X-Ray and Nuclear Magnetic Resonance

These methods allow a view through the surface into the material. If these methods are coupled with tomography, the achievement of a three-dimensional picture of the bulk material is possible in special cases. At present the resolution of these methods is not high enough to resolve single matrix-filler detachments in energetic materials. However, distributed microcracks in the material are detectable by a decreased overall density in the respective region.

12.9.4.2 Indirect Methods

Determination of Material Properties

Cracks can be detected in a standard tenslie test owing to their weakening effect on the material. Whether the weakaning is really due to cracks has to be determined by separate methods, which can be very costly. In these cases, the detection of the volume increase of the material during tension is a suitable solution. The microscopic vacuoles formed by cracks and delaminations lead to a macroscopic observable volume dilatation of the material. With this method, the detection of many cracks distributed in the material is possible, even in cases where the detection of single cracks is not possible with other methods owing to their small size.

Dilatometry

In gas and liquid dilatometers, the volume change of a sample in a tensile test is measured by the measurement of an amount of fluid (gas or liquid) displaced by the sample. The sample and parts of the deformation equipment are completely immersed in the fluid, which leads to an extremely high equipment effort, because the enclosing vessel must be absolutely tight. Furthermore, the temperature plays an important role and must be controlled very accurately. Possible infiltration of the liquid in the formed cracks can cause a false measurement.

Poisson's ratio

The measurement of Poisson's ratio can be used as a method for the determination of the volume dilatation of a sample in a normal tensile test. Poisson's ratio ν is defined as the negative ratio of lateral contraction and elongation:

$$\nu = \frac{\varepsilon_q}{\varepsilon_l} \tag{12.60}$$

where $\varepsilon_q = dq/q_0$ and $\varepsilon_l = dl/l_0$; q_0 and l_0 are the initial width and length of the sample, respectively, and dl and dq are their changes during deformation. In addition to its capability for the detection of vacuoles in the material, Poisson's ratio

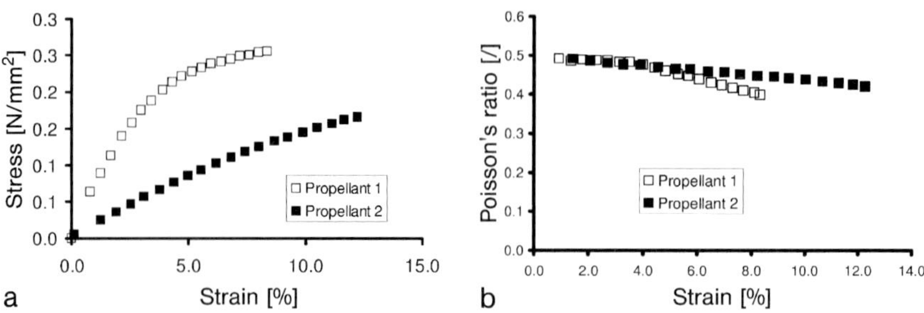

Figure 12.47. Comparison of well-bonded (propellant 1) and weakly bonded (propellant 2) propellants.

is a necessary parameter for the complete description of the material in the elastic region, as explained above [12.75].

Figure 12.47 shows (a) the stress-strain curves and (b) the Poisson's ratio strain curves for two different solid propellants. As can be seen for propellant 2, at the point where the weakening occurs, Poisson's ratio decreases. In this propellant, detachments of the matrix from the filler surface take place.

Direct methods for the determination of cracks have the advantage that they give direct information about the presence and the size of cracks or delaminations in the materials. They are not suitable from a statistical point of view, because they allow the observation of only small areas of the material. In order to obtain a comprehensive picture of the conditions in the material, tomography has to be used, which requires a high equipment effort and long measurement times.

The method for the determination of mechanical properties is much more cost-effective as it delivers not only information about cracks, but also the necessary data for FEM calculations. These methods allow an integrated view of the bulk material.

12.10 Injection Loading Technology

12.10.1 Introduction

Injection loading is an inter-disciplinary technology for transport operations being performed on highly filled suspensions through narrow flow channels. Similarly to injection-molding techniques in the plastics industry, a piston is used to transfer the viscous suspension from a reservoir into a mold cavity [12.76, 12.77]. However, unlike the traditional injection-molding techniques, the mold is a component of the product rather than a component of the machine. Similarly to the commercial bottle-filling machines used in the food and pharmaceutical industries, the mold is a container that approaches the dispensing device where it is filled and then taken

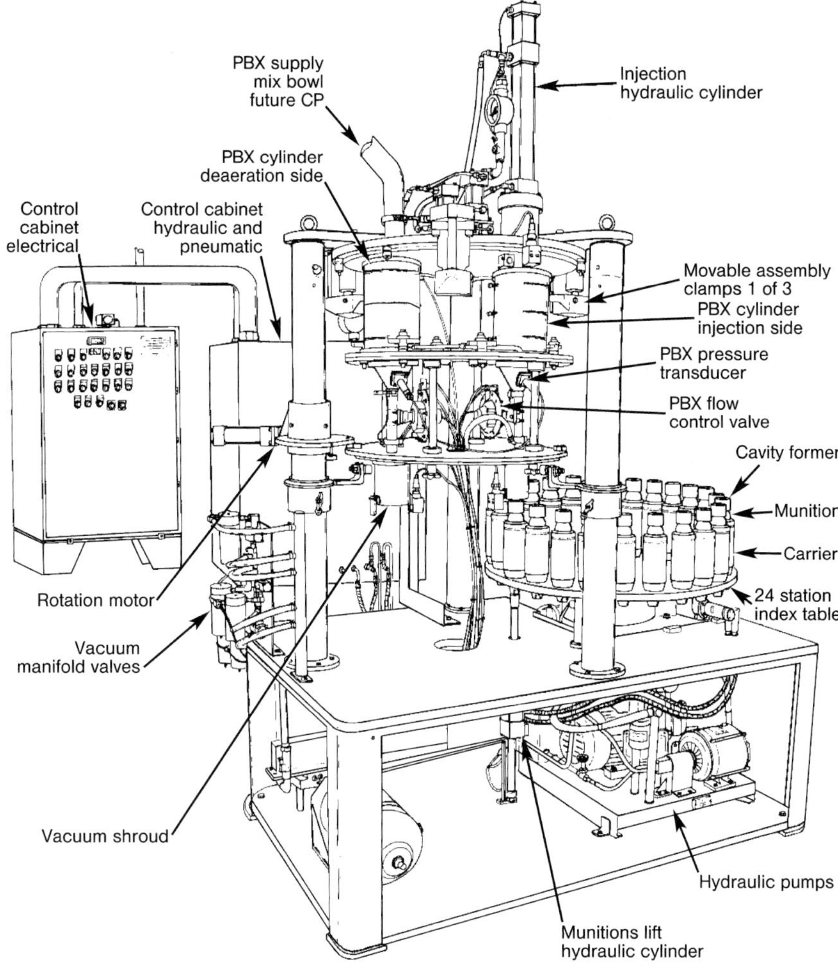

Figure 12.48. Injection loading apparatus.

away for final packaging [12.78]. However, injection loading features a unique dispensing device [12.79] (see Fig. 12.48). This device is designed simultaneously to degas and transport aliquots of a viscous suspension into containers. The injection loader can provide a substantial driving force for momentum transport and has a simple process geometry that minimizes mass transfer problems usually associated with flow through corner turns, abrupt contractions or multiple-port manifolds.

Even though injection loading technology can be relatively complex, it offers the potential to be an efficient automated unit operation. The product quality and the production rate are dependent on several factors. Some of these factors can be elucidated from the domains of energetic material formulation, process design and process control.

12.10.2 Energetic Material Formulation

The formulations for injection loading applications are usually multi-component, highly filled, viscoelastic thermosetting suspensions that undergo transition to a wet paste before final curing to take on the properties of an elastomer. These formulations are known as plastic bonded explosives (PBX). The particles are high-performance nitramines that have a high chemical energy value and a high crystalline density [12.80–12.82]. The desired concentration of nitramine particles in the formulation is close to the maximum packing fraction [12.83]. The thermosetting binder is plasticized and is usually either polyurethane or polyacrylate. The particles tend to be slightly negatively buoyant early in the polymerization process, but become neutrally buoyant in the binder as polymerization proceeds beyond a threshold molecular weight.

Processing PBX formulations that are designed for injection loading applications is more of an art than a science. The particle size distribution (PSD), binder selection and apparent rheological behavior are important characteristics to understand if processing is to be successful and reproducible.

12.10.2.1 Particle Size Distribution

The PSD must be optimized for injection loading formulations. It is well known that the maximum packing fraction of highly filled polymeric systems is dependent on several variables. Two principal variables are the number of grist modes and the aspect ratio of particles in each grist mode. McGeary [12.84] showed that a ternary mixture of hard spheres has a sufficiently high packing fraction to yield 90% of the theoretical density if the particle diameter of each of the three discrete component grist modes differs by at least a factor of seven. A ternary mixture of this type has the desirable property of being a free-flowing mixture of solid particles. Others [12.85, 12.86] have studied the packing of particles having grist size distributions within each mode of a multi-modal mixture and acknowledge the influence of infrastructure and microstructure upon packing. The packing of particles having a mixture size distribution may be very different from that with a discrete size distribution. There is some evidence that broader distributions in a grist mode may be helpful in reducing the resultant viscosity of a multi-modal PSD suspension [12.18]. Some of the packing efficiency appears to be gained by hexagonal alignment of non-spherical particles having favorable aspect ratios. However, there is a limit to the packing efficiency gained by the randomness in a broad PSD. Often the problem becomes unpredictable by computational methods, especially when the particle shape is insufficiently controlled to estimate using discrete aspect ratios. Hence the optimization of PSD for these energetic material formulations is performed experimentally. Usually a tri- or tetramodal mixture of relatively broad grist size distributions having approximately the same shape (axi-symmetric ellipsoids) can achieve the desired degree of fill for injection loading PBX formulations.

12.10.2.2 Binder Selection

The purpose of the binder is to insulate the nitramine particles and provide some structure to the final form of the energetic material. However, during processing, the binder plays at least two specific roles. First, the unreacted binder should contain components that reduce the interfacial tension at the surface of nitramine particles. These components are not necessarily surfactants. They can be liquid organic plasticizers of low molecular weight that have an affinity for the monomer and the resultant polymer. Plasticizers not only lower the glass transition temperature of the resultant polymer but also, during processing, they dilute the monomer and offer a free volume of fluid that can participate as a molecular lubricant. This latter phenomenon promotes wetting of the nitramine particles [12.87] and is also effective in desensitizing the nitramine particles to unplanned energy stimuli. Second, the binder should fluidize the nitramine particles so that there is a carrier fluid for transport operations. A common misconception is that the binder viscosity must be minimized to perform this role. However, the excessive use of dilution (or plasticization) can increase the negative buoyancy of larger particles in the PSD and contribute to flow problems. An emulsifier is sometimes added to the binder system to help maintain a suspension. Therefore, it becomes obvious that the binder system viscosity should be optimized to be sufficiently low to promote flow, but also sufficiently high to prevent particle settling.

The minimum acceptable viscosity of a PBX formulation (η_{min}) can be conceptualized by the term relative viscosity (η_{rel}). This is an empirical quantity that sums the unreacted and unfilled binder system viscosity (η_c) with the contribution from the optimized PSD and solids fraction used in the PBX formulation (c_v) as a function of maximum packing fraction ($c_{v,max}$). There are many expressions for the relative viscosity of a suspension [12.15, 12.39, 12.88–12.92]. The Dougherty-Krieger equation (see Table 12.1) is probably the most appropriate expression for large values of c_v, where $<\eta>$ is the intrinsic viscosity or the slope of the curve when c_v goes to zero (see Figure 12.49).

The only way to determine the minimum acceptable viscosity of PBX formulations and to optimize it is by experiment over the shear rate range of interest. The experimental methodology needs to have geometric similarity with the processing equipment. In this case, that is the injection loading process. The minimum acceptable viscosity for injection loading PBX formulations is higher than for traditional cast PBX formulations because the solids fraction (c_v) is higher.

The intrinsic viscosity ($<\eta>$) is the slope of the curve at $c_v = 0$. In practice, the maximum packing fraction ($c_{v,max}$) for this solids mixture is about 0.83. Therefore, the minimum acceptable viscosity for this formulation (η_{min}) is about a factor of 1000 greater than that for the unfilled and unreacted binder (η_c).

The maximum acceptable viscosity of a PBX formulation (η_{max}) can be conceptualized by the chemical reaction kinetics of the binder. If the thermosetting polymer is a step-type polymerization producing polyurethane, then the rate of reaction is first order and controlled by the catalyst concentration. If the thermosetting polymer is a free radical polymerization producing polyacrylate, then the rate of reaction is controlled stoichiometrically by the rate-limiting step [12.93]. Initiation of

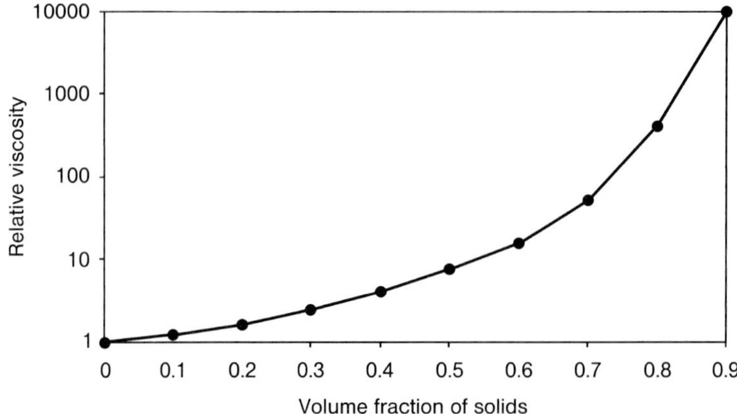

Figure 12.49. Relative viscosity as a function of solids packing. Fraction for a trimodal PSD of particles having broad grist size distribution.

the polyacrylate reaction is usually the rate-limiting step and, once initiated, the polymerization is sometimes difficult to control. Since the polyurethane reaction produces a binder of threshold molecular weight more quickly and in a more predictable fashion, polyurethane binders are preferred for injection loading. While the maximum acceptable viscosity of a PBX formulation is conceptualized by the reaction rate of the binder, in practice it is limited by the driving force provided by the processing equipment. The maximum acceptable viscosity for injection loading PBX formulations is higher than for traditional cast PBX formulations because the processing equipment can apply a greater driving force.

The PBX formulation is mixed in a unit operation that immediately precedes the injection loading unit operation. The mixing operation can be performed using high-shear equipment having clearances of at least twice the mean particle size of the largest grist mode in the PSD. This unit operation can be accomplished as a batch process or a continuous process; both are compatible with the injection loader design.

The time frame for mixing varies depending on the specific PBX formulation being processed. However, at the end-of-mix (EOM), the apparent viscosity of the PBX formulation is often very close to the minimum acceptable viscosity (η_{min}). The time frame represented by the binder polymerization, occurring from the EOM to the maximum acceptable viscosity (η_{max}), is referred to as the pot life (see Figure 12.50). This is also the practical time frame during which injection loading of containers must be done. The pot life for injection loading of PBX formulations is about twice that for traditional cast loading of PBX formulations because the driving force is an order of magnitude greater. In any case, the process engineer needs to have a system with a predictable and consistent pot life to ensure that processing parameters can be properly managed during the injection loading operation.

At time zero, the EOM viscosity corresponds to a typical minimum acceptable viscosity (η_{min}). The injection load PBX formulation has a higher η_{min} than the

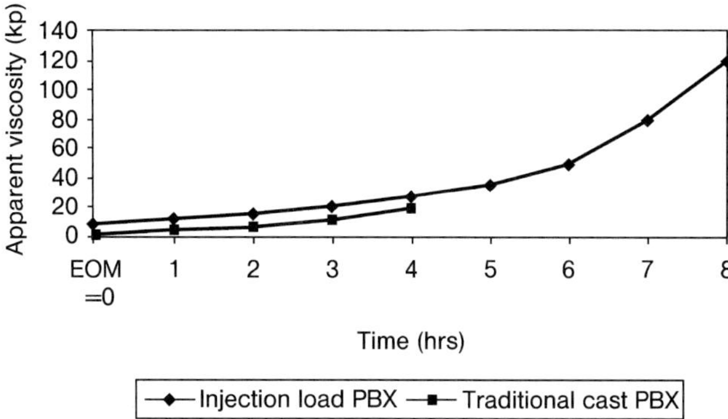

Figure 12.50. Pot life of PBX formulation.

traditional cast PBX formulation because of a higher solids content. The injection load PBX formulation also has a higher maximum acceptable viscosity (η_{max}) because the process equipment offers a greater driving force than the traditional cast approach. These advantages enable the pot life to be approximately doubled for the injection loading PBX formulation.

12.10.2.3 **Applied Rheology**

Rheology is important for any modeling efforts that support process design and process control. Shear stress, $\tau(t)$ and shear rate, $\dot{\gamma}(t)$, are dynamic phenomena that occur during transport operations, such as piston-driven dispensing of PBX formulations into containers. Characterization of the shear dependence of these highly filled thermosetting PBX formulations is elusive using commercial techniques. Rotational rheometers such as the Couette (concentric cylinders) device or a mechanical spectrometer that has parallel plates (or cone and plate) usually yield variable results because the clearances are too small. Rotating spindle-type rheometers can be useful if a helical path is swept through the material. Capillary rheometers can be more successful because they mimic the shear mechanisms of piston-driven flow that occur during injection loading. However, the diameter of the capillary should be widened significantly and appropriate correction procedures should be used. Typically, the Rabinowitsch correction is used to obtain the shear rate at the wall and the Bagley correction to eliminate the pressure drop due to the entrance region of the capillary [12.94] (see Sect. 12.4.2). When characterized, these PBX formulations exhibit four rheological features important to processing. First, the binders are shear thinning (or have pseudoplastic behavior). The apparent viscosity decreases as a function of increasing shear rate. Second, these viscous suspensions have a yield stress. This is a threshold stress (or driving force) that must to be applied before flow can be observed. A three-parameter model is necessary to describe this rheological behavior [12.95]. The preferred model is the Herschel-Bulkley equation (Eq. (12.13)), where τ is the yield stress, K is the experimentally determined apparent viscosity parameter and n is an experimentally determined pseudoplastic parameter that is specific for different PBXs. When c_v is large, the yield stress, τ, can be significant. During the processing pot life, the parameter K has a lower limit of η_{min} and an upper limit of η_{max}. Third, these PBX formulations display wall slip [12.96, 12.97]. This phenomenon indicates that a boundary layer of plasticizer-rich binder may be at the wall and transport may occur as a pseudo-plug flow. Fourth, at low transport velocity (or low Reynolds number), the bulk PBX flow appears to lose its pseudoplastic behavior and become similar to a Bingham plastic flow (see Sect. 12.2). This infers that the rheological behavior is a function Reynolds number and that phenomena observed at low transport velocities (or at low production rates) may intensify or change (from $n = 1$) to become more problematic at higher production rates (where $n < 1$). Therefore, it is important to characterize the rheology of PBX formations over three domains of interest. The first domain is the pot life and the rheology must be understood with respect to processing time. The second domain is the shear rate range of interest. The third domain is to characterize the transport phenomena over the desired range of Reynolds number.

An expression for volumetric flow-rate, $\dot{V}$, through a circular conduit for PBX formulations can be expressed in terms of rheological parameters n and K:

$$\dot{V} = [(\pi n R^3)/(3n + 1)][(R\Delta P^{1/n})/(2LK)] \tag{12.61}$$

The driving force applied for momentum transport is the pressure drop, ΔP, and cylindrical plumbing has radius R and length L. The flow of PBX formulations can appear to be predictable, but the rheological features that have been briefly discussed can change K and n values during processing and influence irregular flow with respect to production rate (the amount of applied shear) and PBX pot life. These changes can sometimes contribute to shear-induced phenomena that may result in demixing during transport through narrow flow channels. Potential problems such as these can be mitigated by using smart process design and adaptive process control techniques.

12.10.3
Process Design

There are many variables involved with processing thermosetting PBX formulations. Management of these variables becomes a challenge if the process design is not robust. Since the apparent viscosity is not constant during the pot life and these viscous suspensions (or pastes) have a tendency to change flow behavior, the process geometry design must incorporate features that can prevent (or minimize) the effects of undesired transport phenomena. If this is accomplished, injection loading has potential to offer pressed quality at a cast price.

12.10.3.1 Process Geometry

Many commercial filling machines force the fluid (or suspension) through corner turns and abrupt contractions before delivering it to containers. Many injection-molding machines force the fluid (or suspension) through a maze of extremely narrow flow channels and corner turns within a mold. However, unlike many commercial fluids (or suspensions), PBX formulations are very highly filled materials that are susceptible to particle jamming, binder filtration and shear-induced demixing if forced through these severe geometries. Therefore, manufacturing science and process engineering experience suggest that a simple geometry is always desirable for processing PBX formulations. There are at least three process design criteria that need consideration if the geometry is to be robust. These include elimination of 90° corner turns, elimination of abrupt contractions and minimization of the length of plumbing.

The injection loader design in Fig. 12.48 features two identical chambers on a movable assembly. These right cylinders have a smooth conical contraction into a diaphragm valve (that has no weir). A rotation motor rotates the movable assembly. The method of operation begins with filling one chamber with PBX from the mixing unit operation. As the PBX enters this chamber, vacuum is used to degas the material and remove any entrapped air. When that chamber is full, the movable assembly is rotated 180° to align the degassed aliquot of PBX with the inject piston. Injection loading of the degassed PBX into containers starts under piston-driven flow (see Fig. 12.51). Simultaneously, the second chamber has been positioned under the filling station and is ready to accept the next aliquot of PBX. This procedure is repeated until the manufacturing shift is completed. In this manner, the PBX never turns a corner. The straight plumbing design of the twin chambers in the injection loader apparatus eliminates any 90° corner turns and prevents processing problems associated with particle jamming and binder filtration.

Abrupt contractions include any reduction that contributes to flow instability. Two generalized examples include a step contraction and a severe taper. The step

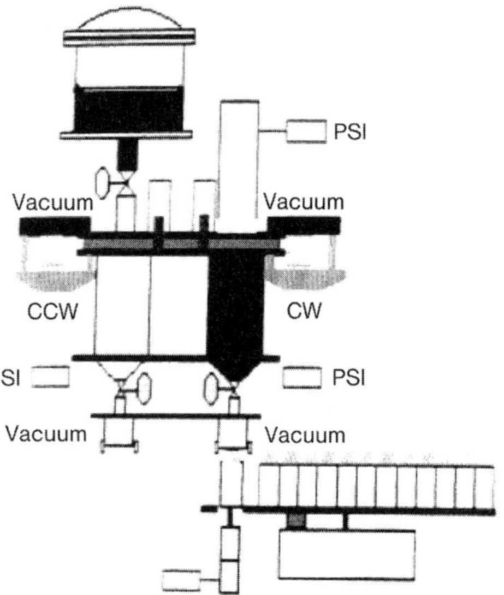

Figure 12.51. The injection loader with a batch mixed PBX reservoir as supply.

contraction is the joining of two circular conduits of different diameters. This design has an annular 'dead zone' where recirculation occurs. A severe taper is any conical reduction greater than a 4:1 ratio having a prohibitive angle of slide. Apparently well-mixed homogeneous suspensions can demix as a function of shear when forced through abrupt contractions. Altobelli *et al.* [12.98] have used nuclear magnetic resonance (NMR) imaging techniques (also referred to as MRI techniques) to observe and document demixing phenomena of piston-driven flow of suspensions through abrupt contractions. As the piston approaches the contraction, particles migrate toward the plumbing axis (or centerline) where the shear is minimal. The companion phenomenon is that binder can recirculate along the boundary layer and slowly accumulate at the piston head. This behavior can eventually produce a particle-rich volume element early in a piston stroke followed by a binder-rich volume element late in the same piston stroke. As a result, poor process designs that include abrupt contractions can produce injection loaded containers having density gradients.

Experiments indicate that it is important to minimize the length of plumbing used for transport of PBX. NMR imaging techniques and computational modeling have been used to examine phenomena in pressure-driven flow of suspensions through circular conduits [12.99]. Initially well-mixed suspensions can start to demix and develop microstructures relatively early after shear has been applied. First, a suspension velocity distribution starts to emerge. Then, a different solids fraction distribution begins to take shape. Beyond a threshold entrance length, the solids concentration profile can build to a sharp maximum at the plumbing axis (or centerline). Meanwhile, the suspension velocity profile becomes blunted, similar to Bingham plastic flow. One of the variables that contributes to the extent of this observed and undesired behavior is the ratio of particle radius (a) to circular conduit radius (R). If the suspension contains large particles (or the a/R ratio is large), especially at high solids concentration (c_v), the threshold entrance length of plumbing is short. As a result, poor process designs that include long lengths of plumbing can produce injection loaded containers having density gradients.

The two chambers are labeled clockwise (CW) and counter-clockwise (CCW). The injection loader rotates the first chamber filled (and all odd-numbered chambers) in the CW direction to position the degassed PBX under the injection piston. That is the position shown in Fig. 12.51. The injection loader rotates the second chamber filled (and all even-numbered chambers) in the CCW direction to position the degassed PBX under the injection piston.

12.10.3.2 **Transport Phenomena**

Numerical simulation of concentrated suspension flows has proven to be difficult. However, existing models can be useful in understanding observations and mitigating potential problems in transient flows at low Reynolds number. We know that particles tend to migrate from regions of high shear to regions of low shear owing to irreversible interactions. We also know that initially well-mixed concentrated suspensions can separate to develop non-uniform microstructures when subjected to inhomogeneous shearing motion. We also know that there are normal stress differences associated with these phenomena. The evolution of solids concentration profiles and suspension velocity profiles has been simulated with reasonable agreement between theory and experiment for some slow flow scenarios applicable to injection loading of PBX. The two approaches used to model observed shear-induced particle migration successfully in carefully controlled experiments are complicated and time consuming. The first approach is known as the diffusive flux model [12.100–12.103]. This model is a diffusive equation for the net particle flux derived through scaling arguments based on the consideration of a spatially varying particle interaction frequency and a concentration-dependent effective viscosity. The second approach is known as the suspension balance model [12.104–12.106]. This model is based on the conservation of mass and momentum for both the particle phase and the suspension phase. Both of these approaches involve challenging computations and require finite element Navier-Stokes solvers.

An over-simplified form of the diffusive flux model reveals fundamental relationships important to injection loading concentrated suspensions, such as PBX, into containers. The evolution of solids concentration profiles (dc_v/dt) is probably the most important concern and it can be generalized as a function of particle radius (x), shear rate ($\dot{\gamma}$) and apparent viscosity (η):

$$dc_v/dt = f(x^2, \dot{\gamma}, \eta^{-1}) \tag{12.62}$$

This means that there are some practical aspects to be considered that can minimize the transport phenomenon of shear-induced particle migration. First, small particles should be preferred to large particles in the PSD. Second, the viscosity of the binder should be maximized to maintain the suspension. Finally, the applied shear rate during transport should be minimized. Process design features can be employed to minimize and redistribute shear. In addition to the previous discussion, installing short lengths of in-line static mixers immediately downstream of a contraction can remix (or redirect) previously sheared material that may have become demixed. Process control techniques can also be used to limit the applied shear rate and eliminate the potential density gradient concern in the final product.

12.10.4 Process Control

There are two fundamental approaches to process control architecture. First, the traditional Ziegler-Nichols control theory provides for proportional, integral and derivative (PID) parameters that can be used to manipulate variables and correct predictable disturbances. The availability of inexpensive microprocessors and the familiarity of simple relationships make the PID approach attractive for processes at steady state. However, injection loading of PBX into containers never achieves steady state. Additionally, the shear-induced disturbances can be irregular. Therefore, the second approach of using adaptive process control techniques is more appropriate for injection loading of PBX. Adaptive techniques can include either fuzzy logic or neural networks. These are model-based and expert-based control strategies that use an integrated control scheme of multiple inputs and multiple outputs (MIMO). A design of experiments is typically used to determine the matrix of transfer functions that relate manipulated variables (or input variables) to the controlled variables (or output variables). The method includes a series of time steps in which one manipulated variable is altered at a time and the response of controlled variables is measured as a function of time. Selection of the appropriate adaptive control strategy usually follows an assessment of feasibility that considers stability and ease of implementation. It is not uncommon that different types of PBX can exhibit unique relationships or transfer functions, between manipulated and controlled variables.

The injection loading process for PBX uses a supervisory process control software that is run on a personal computer (PC) and monitors processing parameters necessary to track mass transfer and momentum transport phenomena. Some of these parameters are the vacuum level in the degassing chamber, vacuum level in the shroud where the container is evacuated, temperature of the PBX in the injection chamber, piston displacement, hydraulic pressure driving the piston, cavity pressure of the PBX entering the container and time. In addition, real-time calculations are performed to determine piston velocity, shear rate (as a function of $\dot{V}$), shear stress (as a function of ΔP) and apparent PBX viscosity (as a ratio of shear stress and shear rate). These parameters become a menu of inputs that can be used in the adaptive control strategy.

Standard back-propagation neural networks have been used successfully to recognize disturbances early in the injection loading cycle [12.107]. Figure 12.52 shows an example framework that has been used to recognize the onset of undesirable phenomena and invoke corrective action in time to avoid the manufacture of reject products. The input variables were cavity pressure (CP) of the PBX entering the container and the calculated apparent PBX viscosity (VIS). In limited testing and demonstration of an injection loading cycle having 10 time steps, this neural network output correctly predicted the result of either a good or reject product after only four time steps when compared with traditional post-mortem radiography. This means that a neural network framework provides sufficient time to take corrective action during processing. Once identified, the corrective action can be invoked within one time step to resolve the influence of the disturbance well before the injection loading cycle is complete. Early detection of CP and VIS disturbances allows recognition of shear-induced problems. Correction of these problems and avoidance of reject product are achieved by controlling the applied shear rate within acceptable limits. Therefore, this adaptive process control strategy has been useful for injection loading PBX into containers.

This network uses cavity pressure (CP) and viscosity (VIS) as input nodes. Using three hidden nodes, this framework was able to recognize patterns of difference between acceptable and unacceptable phenomena sufficiently early in the processing cycle to invoke corrective action that pre-emptively avoids the manufacture of reject products.

Injection loading technology for PBX materials is dependent on the proper

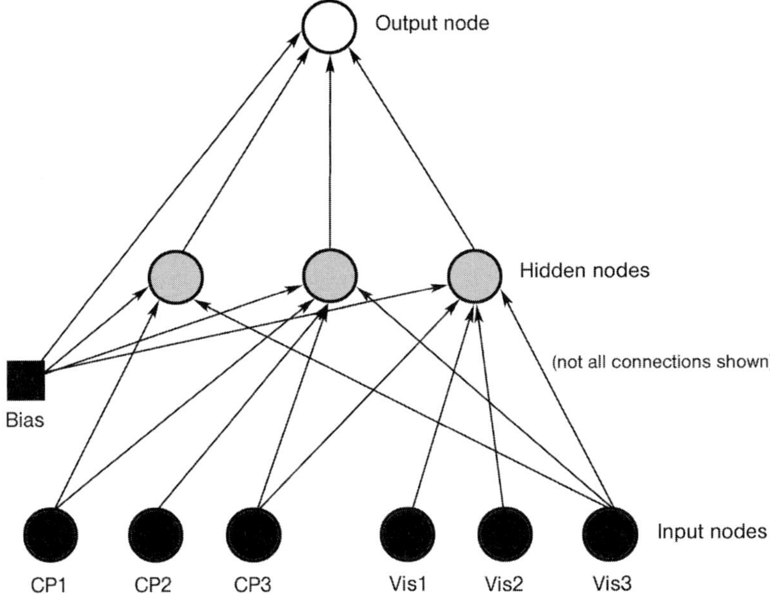

Figure 12.52. One of the neural network frameworks used to demonstrate how an adaptive process control strategy can recognize early onset of disturbances.

characterization of particles and understanding their behavior during processing. The PSD in PBX formulations is important for final product performance. Injection loading can be an efficient automated unit operation for many PBX applications if the PSD is maintained throughout the manufacturing process at the production rate desired. This can be achieved when the PBX formulation is appropriate, the process design is simple and robust and the applied shear can be controlled within acceptable limits.

Acknowledgments
The author of Sect. 12.10 would like to acknowledge the Office of Naval Research and the Insensitive Munitions Office of the Naval Sea Systems Command for sponsoring a significant portion of the work presented here.

12.11 References

12.1 Böhme G (1981) *Strömungsmechanik Nicht-Newtonscher Fluide*, B. G. Teubner Studienbücher: Mechanik, Stuttgart.

12.2 Giesekus H (1994) *Phänomenologische* Rheologie, Springer, Berlin.

12.3 Vinogradov GV, Malkin AYA (1980) *Rheology of Polymers*, Springer, Berlin,.

12.4 Macosko CW (1994) *Rheology: Principles, Measurements and Applications*, VCH, New York.

12.5 Weipert D, Tscheuschner H-D, Windhab E (1993) *Rheologie der Lebensmittel*, Behr's Verlag, Hamburg.

12.6 Chhabra R P (1993) *Bubbles, Drops and Particles in Non-Newtonian Fluids*, CRC Press, Boca Raton, FL.

12.7 Pahl M, Gleissle W, Laun H-M (1991) *Praktische Rheologie der Kunststoffe und Elastomere*, VDI Verlag, Düsseldorf.

12.8 Barnes HA, Hutton JF, Walters K (1989) *An Introduction to Rheology*, Rheology Series, Vol. 3, Elsevier, Amsterdam.

12.9 Teipel U (1999) Rheologisches Verhalten von Emulsionen und Tensid-Lösungen, *Dissertation*, Universität Bayreuth, Wissenschaftliche Schriften-Reihe des Fraunhofer ICT, Band 22.

12.10 Kamal MR, Mutel A (1985) Rheological properties of suspensions in newtonian and non-Newtonian fluids, *J. Polym. Eng.* **5**, 293–382.

12.11 Einstein A (1906, 1911) Eine neue Bestimmung der Moleküldimensionen, *Ann. Phys.* **19**, 289–306; **34**, 591–592.

12.12 Eilers H (1941) Die Viskosität von Emulsionen hochviskoser Stoffe als Funktion der Konzentration, *Kolloid-Z.* **97**, 313–321.

12.13 Mooney M (1951) The viscosity of a concentrated suspension of spherical particles, *J. Colloid Sci.* **6**, 162–170.

12.14 Maron SH, Pierce PE (1956) Application of Ree-Eyring generalized flow theory to suspensions of spherical particles, *J. Colloid Sci.* **11**, 80–95.

12.15 Krieger IM, Dougherty TJ (1959) A mechanism for non-Newtonian flow in suspensions of rigid spheres, *Trans. Soc. Rheol.* **3**, 137–152.

12.16 Thomas DG (1965) Transport characteristics of suspension: VIII. A note on the viscosity of Newtonian suspensions of uniform spherical particles, *J. Colloid Sci.* **20**, 267–277.

12.17 Frankel NA, Acrivos A (1967) On the viscosity of a concentrated suspension of solid particles, *Chem. Eng. Sci.* **22**, 847–853.

12.18 Chong JS, Christiansen EB, Baer AD (1971) Rheology of concentrated suspension, *J. Appl. Polym. Sci.* **15**, 2007–2021.

12.19 Quemeda D (1977) Rheology of concentrated disperse systems and minimum energy dissipation principle, I. Viscosity-concentration relationship, *Rheol. Acta* **16**, 82–94.

12.20 Batchelor GK (1977) The effect of Brownian motion on the bulk stress in a suspensions of spherical particles, *J. Fluid Mech.* **83**, 97–117.

12.21 Miller RR, Lee E, Powell RL (1991) Rheology of solid propellant dispersions, *J. Rheol.* **35**, 901–920.

12.22 Muthiah RM, Krishhnamurthy VN, Gupta BR (1992) Rheology of HTPB propellant. Effect of solid loading, oxidiser particle size and aluminum content, *J. Appl. Polym. Sci.* **44**, 2043–2052.

12.23 Hordijk AC, Sabel HWR, Schonewille E (1996) The application of rheological equipment for improved processing of HTPB based PBXs, In: *Proc. 27th Int. Annual Conference of ICT, Karlsruhe*, p. 3.

12.24 Yang K-X, Tao Z-M (1986) Viscosity prediction of composite propellant slurry, *Propell. Explos. Pyrotech.* **11**, 167–169.

12.25 Hordijk AC, Bouma RHB, Schonewille E (1998) Rheological characterization of castable and extrudable energetic compositions, In: *Proc. 29th Int. Annual Conference of ICT, Karlsruhe*, p. 21.

12.26 Ivanov GV, Tepper F (1997) Activated aluminum as a stored energy source for propellants, In: Kuo KK (ed.), *Challenges in Propellants and Combustion, 100 Years after Nobel*, Begell House, pp. 636–645, Stockholm.

12.27 Tepper F, Ivanov GV, Lerner M, Davidovich V (1998) Energetic formulations from nanosize metal powders, In: *Proc. 24th Int. Pyrotechnics Seminar, Monterey, CA*, pp. 519–530.

12.28 Glotov OG, Zarko VE, Beckstead MW (2000) Agglomerate and oxide particles generated in combustion of ALEX containing solid propellants, In: *Proc. 31st Int. Annual Conference of ICT, Karlsruhe*, p. 130.

12.29 Teipel U, Förter-Barth U (2001) Rheology of nano-scale aluminum suspensions, *Propell. Explos. Pyrotech.* **26**, 268–272.

12.30 Muthiah R, Krishnamurthy VN, Gupta BL (1996) Rheology of HTPB propellant: development of generalized correlation and evaluation of pot life, *Propell. Explos. Pyrotech.* **21**, 186–192.

12.31 Gupta BL, Varma M, Munjal NL (1986) Rheological studies on virgin and metallised unsymmetrical dimethylhydrazine gelled systems, *Propell. Explos. Pyrotech.* **11**, 45–52.

12.32 Rapp DC, Zurawski RL (1988) Characterization of aluminum/RP-1 gel propellant properties, *AIAA Paper 88–2821.*

12.33 Varghese TL, Gaindhar SC, David J, Jose J, Muthiah R, Rao SS, Ninan KN, Krishnamurthy VN (1995) Developmental studies on metallised UDMH and kerosene gels, *Def. Sci. J.* **45**, 25–30.

12.34 Varghese TL, Prabhakaran N, Thanki KP, Subramanian S, Rao SS, Ninan KN, Krishnamurthy VN (1999) Performance evaluation and experimental studies on metallised gel propellants, *Def. Sci. J.* **49**, 71–78.

12.35 Rahimi S, Hasan D, Peretz A, Benenson Y, Welner S (2001) Preparation of gel propellants and simulants, *AIAA Paper 2001–3264.*

12.36 Hoffman DM, Pruneda CO, Jessop ES, Walkup CM (1992) *RX-35-BX: A Low-Vulnerability, High-Performance Explosive for Main-Charge Applications*, Lawrence Livermore National Laboratory, Livermore, CA, UCRL-UR-110363.

12.37 von Holtz E, Scribner KJ, Whipple R, Carley JF (1990) *Paste-Extrudable Explosives and their Current Status*, Lawrence Livermore National Laboratory, Livermore, CA, UCRL-JC-103244.

12.38 Carley JF, von Holtz E (1997) Flow of RF-08-FK high-energy paste in a capillary rheometer, *J. Rheol.* 41, 473.

12.39 Farris RJ (1968) Prediction of the viscosity of the multimodal suspensions from unimodal viscosity data, *Trans. Soc. Rheol.* **2**, 281.

12.40 Hoffman DM, Walkup CM, Spellman L, Tao WC, Tarver CM (1995) *Transferable Insensitive Explosive (TIE)*, Lawrence Livermore National Laboratory, Livermore, CA, UCRL-JC-117245.

12.41 Fried LE (1994) *Cheetah 1.0, User's Manual*, Lawrence Livermore National Laboratory, Livermore, CA, UCRL-MA-117541.

12.42 Goods SH, Shepodd TJ, Mills BE, Foster P (1993) *A Materials Compatibility Study in FM-1, a Liquid Component of a Paste Extrudable Explosive, RX-08-FK*, Sandia National Laboratory, Livermore, CA, SAND93-8237 UC-704.

12.43 Chidester SK, Green LG, Lee CG (1993) In: *Proc. 10th International Detonation Symposium*, ONR 33395-12, Boston, MA, pp. 785–792.

12.44 Chidester SK, Tarver CM, Lee CG (1997) Impact ignition of new and age solid explosiv in: Proc. AIP Conference, Shock Compression of Condensed Molter, American Inst. of Physics, 707–710.

12.45 Schmidt SC, Dandekar DP, Forbes JW (1998) In: *AIP Conference Proceedings 429*, AIP Press, New York, pp. 707–710.

12.46 Chidester SK, Tarver CM, Garza (1998) *ONR 33395–12*, Aspen, CO, pp. 785–792.

12.47 Hoffman DM, Jessop ES, Swansiger RW (1997) In: *Insensitive Munitions and Energetic Materials Technology Symposium Proceedings*, Vol. 2, pp. 231–243.

12.48 Oberth AE (1987) *Principles of Solid Propellant Development*, CPIA Pub. 469, Johns Hopkins University, Laurel, MD.

12.49 Scribner KJ, Crawford P (1988) *PEX Energetic Carriers: Attempts to Crystallize*, Lawrence Livermore National Laboratory, Livermore, CA, UCID-20469.

12.50 von Holtz E, LeMay JD, Carley JF, Flowers GL (1991) *Dependence of the Specific Volume of RX-08-FK High Explosive Paste on Temperature and Pressure*, Lawrence Livermore National Laboratory, Livermore, CA, UCRL-JC-107078.

12.51 Hoffman DM, Walkup CM (1991) Following cure rheologically in ECX with standard and latent catalysts, presented at the JOWOG-9 Mechanical Properties Workshop, Lawrence Livermore National Laboratory, Livermore, CA.

12.52 Hockney RW, Eastwood JW (1981) *Computer Simulation Using Particles*, McGraw-Hill, New York.

12.53 Valtsifer VA (1991) Use of particle method for modelling the behavior of colloid systems, presented at the 7th ICSCS Symposium, France.

12.54 Valtsifer VA (1997) The influence of technology factors on rheological behavior of emulsions, In: *World Congress on Emulsion, Bordeaux*, Vol. 2, 3-1-021/03.

12.55 Valtsifer VA, Zvereva N (1999) Statistical packing of equal spheres, *Adv. Powder Technol.* **10**, 399–403.

12.56 Weymann HD, Chuang MC, Ross RA (1973) *Phys. Fluids* **16**, 775–783.

12.57 Kruif CG, Lersel EMF, Vrij A, Russel WB (1985), *J. Chem. Phys.* **83**, 4717–4725.

12.58 Ferry JD (1970) *Viscoelastic Properties of Polymers*, Wiley, New York.

12.59 Findley WN, Lai JS, Onaran K (1976) In: Lauwerier HA, Koiter WT (eds), *Creep and Relaxation of Nonlinear Viscoelastic Materials*, North-Holland Series in Applied Mathematics and Mechanics, Vol. 18, North-Holland, Amsterdam.

12.60 Christensen RM (1971) *Theory of Viscoelasticity*, Academic Press, New York.

12.61 Davenas A (1993) *Solid Rocket Propulsion Technology*, Pergamon Press, Oxford.

12.62 Hübner C, Geissler E, Elsner P, Eyerer P (1999) The importance of micromechanical phenomena in energetic materials, *Propell. Explos. Pyrotech.* 24, 119–125.

12.63 Schapery RA (1981) On viscoelastic deformation and failure behavior of composite materials with distributed flaws, In: Wang SS, Renton WJ (eds), *Advances in Aerospace Structures and Materials, Proceedings of the Winter Annual Meeting of the American Society of Mechanical Engineers, Washington DC.*

12.64 Schapery RA (1982) Models for damage growth and fracture in nonlinear viscoelastic particulate composites, In: *Proceedings of the 9th US National Congress of Applied Mechanics, New York.*

12.65 Schapery RA (1986) A micromechanical model for nonlinear viscoelastic behavior of particle reinforced rubber with distributed damage, *Eng. Fract. Mech.* **25**, 845–867.

12.66 Hübner C (1994) Ermittlung eines zweidimensionalen, zeit- und deformationsabhängigen Materialgesetzes für gefüllte Elastomere unter besonderer Berücksichtigung von Phasengrenzflächenphänomenen, *Dissertation*, University of Karlsruhe.

12.67 Oberth AE, Bruenner RS (1965) Tear phenomena around solid inclusions in castable elastomers, *Trans. Soc. Rheol.* **9**, 165–185.

12.68 Farris RJ (1968) The influence of vacuole formation on the response and failure of filled elastomers, *Trans. Soc. Rheol.* **12**, 315–334.

12.69 Stacer RG, Hübner C, Husband DM (1990) Binder/filler interaction and the nonlinear behavior of highly-filled elastomers, *Rubb. Chem. Technol.* **63**, 488–502.

12.70 Williams ML, Landel RF, Ferry JD (1955) The temperature dependence of relaxation mechanisms in amorphous polymers and other glass-forming liquids, *J. Am. Chem. Soc.* **77**, 701.

12.71 Eisenreich N, Fabry C, Geissler A, Kugler HP (1985) Messung der Querkontraktionszahl an Kunststoffen im Hinblick auf die Beurteilung der Füllstoffhaftung, In: *Vorträge der Tagung Werkstoffprüfung*, Deutscher Verband für Materialprüfung, Bad Nauheim, pp. 391–398.

12.72 Kugler HP, Stacer RG, Steimle C (1990) Direct measurement of Poisson's ratio in elastomers, *Rubb. Chem. Technol.* **63**, 473–487.

12.73 Kappel H, G'sell C, Hiver JM (1997) Détermination du comportement biaxial des élastomères par une méthode videométrique, In: G'sell C, Coupard A (eds), *Génie Méchanoque des Caoutchoucs et des Élastomères Thermoplastiques*, APOLLOR et INPL.

12.74 G'sell C, Hiver JM, Dahoun A, Souhai A (1992) Video-controlled tensile testing of polymers and metals beyond the necking point, J. Mater. Sci. 27, 5031–5039.

12.75 Freudenthal AM, Henry LA (1960) On 'Poisson's ratio' in linear visco-elastic propellants, In: Summerfield M (ed.), *Solid Propellant Rocket Research*, Academic Press, New York.

12.76 Tobin WJ (2000) *Fundamentals of Injection Molding*, 2nd edn, WJT Associates, Louisville, CO.

12.77 Rosato DV, Rosato GR (2001) *Injection Molding Handbook*, 3rd edn, Kluwer, Boston.

12.78 Soroka W (1998) *Fundamentals of Pack-*

aging Technology, 2nd edn, IoPP Press, Naperville, IL.

12.79 Mahoney TI, Newman KE, Gusack JA, Sallade GJ (1995) Apparatus for injection molding high-viscosity materials, *US Patent* 5 387 095.

12.80 Kamlet MJ, Jacobs SJ (1968) Chemistry of detonations, Part I: a simple method of calculation of detonation properties of C-H-N-O explosives, *J. Chem. Phys.* **48**, 23.

12.81 Baroody EE, Peters ST (1990) *Heat of Explosion, Heat of Detonation and Reaction Products: Their Estimation and Relation to the First Law of Thermodynamics*, IHTR 1340, NSWC, Indian Head, MD.

12.82 Fried LE, Murphy MJ, Souers PC, Wu BJ, Anderson SR, McGuire EM, Maiden DE (1998) detonation modeling with an in-line thermochemical equation of state, In: *Proc. of the 11th International Detonation Symposium, Snowmass Village, CO*, p. 889.

12.83 Ferguson J, Kemblowski Z (1991) *Applied Fluid Rheology*, Elsevier Applied Science, London.

12.84 McGeary RK (1961) Mechanical packing of spherical particles, *J. Am. Ceram. Soc.* **44**, 513.

12.85 Yu AB, Standish N (1993) A study of the packing of particles with a mixture size distribution, *Powder Technol.* **76**, 113.

12.86 Nolan GT, Kavanagh PE (1993) Computer simulation of random packings of spheres with log-normal distributions, *Powder Technol.* **76**, 309.

12.87 Adamson AW, Gast A (1997) *Physical Chemistry of Surfaces*, 6th edn, Wiley-Interscience, New York.

12.88 Kitano T, Karaoka T, Shirota T (1981) An empirical equation of the relative viscosity of polymer melts filled with various inorganic fillers, *Rheol. Acta* **20**, 207.

12.89 Sadler LY, Sim KG (1991) Minimize solid-liquid mixture viscosity by optimizing particle size distribution, *Chem. Eng. Prog.* **87**, 68.

12.90 Ferraris CF (1999) Measurement of the rheological properties of high performance, concrete: state of the art report, *J. Res. Natl. Inst. Stand. Technol.* 104, 461.

12.91 Lee JD, So JH, Yang SM (1999) Rheological behavior and stability of concentrated silica suspensions, *J. Rheol.* **45**, 1117.

12.92 Usui H, Li L, Kinoshita S, Suzuki H (2001) Viscosity prediction of dense slurries prepared by non-spherical solid particles, *J. Chem. Eng. Jpn.* **34**, 360.

12.93 Allcock H, Lampe F (1990) *Contemporary Polymer Chemistry*, 2nd edn, Prentice Hall, Englewood Cliffs, NJ.

12.94 Collyer AA, Clegg DW (1988) *Rheological Measurement*, Elsevier Applied Science, London.

12.95 Newman KE, Stephens TS (1995) *Application of Rheology to the Processing and Reprocessing of Plastic Bonded Explosives in Twin Screw Extruders*, IHTR 1790, NSWC, Indian Head, MD.

12.96 Yilmazer U, Kalyon DM (1989) Slip effects in capillary and parallel disk torsional flows of highly filled suspensions, *J. Rheol.* **33**, 1197.

12.97 Jana SC, Kapoor B, Acrivos A (1995) Apparent wall slip velocity coefficients in concentrated suspensions of noncolloidal particles, *J. Rheol.* **39**, 1123.

12.98 Altobelli SA, Fukushima E, Mondy LA (1997) Nuclear magnetic resonance imaging of particle migration in suspensions undergoing extrusion, *J. Rheol.* **41**, 1105.

12.99 Hampton RE, Mammoli AA, Graham AL, Altobelli SA (1997) Migration of particles undergoing pressure-driven flow in a circular conduit, *J. Rheol.* **41**, 621.

12.100 Leighton D, Acrivos A (1987) The shear-induced migration of particles in concentrated suspensions, *J. Fluid Mech.* **181**, 415.

12.101 Phillips RJ, Armstrong RC, Brown RA, Graham AL, Abbott JR (1992) A Constitutive equation for concentrated suspensions that accounts for shear-induced particle migration, *Phys. Fluids A* 4, 30.

12.102 Krishnan GP, Beimfohr S, Leighton DT (1996) Shear-induced radial segregation in bidisperse suspensions, *J. Fluid Mech.* 321, **371**.

12.103 Subia SR, Ingber MS, Mondy LA, Altobelli SA, Graham AL (1998) Modelling of concentrated suspensions using a continuum constitutive equation, *J. Fluid Mech.* **373**, 193.

12.104 Nott PR, Brady JF (1994) Pressure-driven flow of suspensions: simulation and theory, *J. Fluid Mech.* **275**, 157.

12.105 Morris JF, Brady JF (1998) Pressure-driven flow of a suspension: buoyancy effects, *Int. J. Multiphase Flow* 24, **105**.

12.106 Morris JF, Boulay F (1999) Curvilinear flows of noncolloidal suspensions: the role of normal stresses, *J. Rheol.* 43, 1213.

12.107 Smith RE, Parkinson WJ, Hinde RF, Newman KE, Wantuck PJ (1998) Neural network for quality control of submunitions produced by injection loading, In: *Proc 2nd International Conference on Engineering Design and Automation, Maui, HI.*

13
Performance of Energetic Materials

N. Eisenreich, L. Borne, R. S. Lee, J. W. Forbes, H. K. Ciezki

13.1
Influence of Particle Size on Reactions of Energetic Materials

13.1.1
Introduction

A substantial class of energetic materials consists of various solid organic and inorganic particles. These particles are formed by synthesis or treatment of the basic substances to allow handling and processing to the designed charge, which is mainly a pyrotechnic device, a gunpowder, a rocket propellant or a high explosive. The properties of the particles, mainly size and shape, quality and phase of crystallites and surface behavior, influence all steps of processing, especially rheology and mechanical properties. The reactive behavior defines the functionality of energetic materials, but it is strongly affected by the particles which compose the charge. The characteristics of the particles determine the desired conversion of the heterogeneous energetic material and the performance. Especially the particle size, mainly of the oxidizing component, is one important way to control burn rates or sensitivity. Therefore, a close interaction of processing with the resulting function acquisition is needed finally to fulfill all properties of production, chemical stability, mechanical resistance and performance. Whereas in the past particles of micrometer- to millimeter-scale were of interest, the recent success in nanotechnology has allowed the characterization and use of nanoparticles, but is practically restricted to research and development. Especially metals as fuels achieve high performance in rocket propellants. In pyrotechnic mixtures, particles at high temperature transfer energy to the material to be ignited. However, the utilization of nanoparticles leads to some serious problems and disadvantages. In gun propellants, solid particles could be erosive and produce a luminous muzzle plume and the high molecular weight of the reaction products reduces the effectiveness. The high molecular weight of the produced metal oxides, the exhaust plume signature and the impact on the environment have to be taken into account for rocket propulsion. In addition, a complete conversion of the metal particles is a precondition to make the most of

Energetic Materials. Edited by Ulrich Teipel
Copyright © 2005 WILEY-VCH Verlag GmbH & Co. KGaA, Weinheim
ISBN: 3-527-30240-9

the expected performance and depends on the phase transitions of the metals and of their oxides. The particle sizes of the metal powders used affect the conversion and can strongly influence the burning rate, as in the case of composite rocket propellants. In addition, the particle size of incorporated catalysts can also strongly define their activity in propellants.

This chapter summarizes some influences of particles on the ignition, combustion and detonation behavior of energetic materials, including nanoparticles. The effect of the particle size is the most important one and is therefore at the focus of this chapter. Theoretical descriptions are restricted to simplified approaches.

13.1.2
Principles of Reacting Particles

The conversion of solid energetic materials starts with the phase transitions and decompositions [13.1, 13.2]. In the case of some explosive materials such as AN and HMX, the particle size influences the appearance of the phases [13.3–13.6]: The sensitivity of HMX is highest for the δ-phase. For small particle sizes, thermally cycling of AN below the thermal decomposition leads to a preference for phase V in mixtures with phase IV at ambient temperature [13.4].

The phase behavior and the decomposition of substances are normally investigated by methods of thermal analysis, mainly TGA and DSC, to obtain simplified overall reaction mechanisms and kinetic data. The heating of a particle under various conditions, such as injection into a hot atmosphere, radiation, etc., and the resulting temperature profiles were well documented by Carslaw and Jaeger [13.7]. The slow thermal conversion of energetic materials can often be described by homogeneous kinetics:

$$\frac{d\alpha}{dt} = -k(T)f(\alpha) \tag{13.1}$$

where α = concentration of decomposing species, $k(T)$ = reaction constant and $f(\alpha)$ = a function describing the decomposition mechanism. The integral of Eq. (13.1) is given by

$$-k(T)t = \int \frac{d\alpha}{f(\alpha)} = g(\alpha) \tag{13.2}$$

At an early stage of the decomposition of energetic, inorganic, organic or polymeric substances the relation $f(\alpha) = \alpha^n$ often holds (with n order of reaction) [13.8, 13.9]. If the sample beingstudied is composed of crystallites, an ensemble of contracting three-dimensional particles often represented by spheres has to be considered. The order of the reaction n is often found to be a broken number. Theories such as those of Ginstling and Brounshtein (three-dimensional diffusion) and Avrami and Erofeev [13.10] (m-dimensional nucleation) take into account the various effects that define the order of this reaction. With functions $f(\alpha)$:

$$f(\alpha) = 1.5 \cdot [(1-\alpha)^{-1/3} - 1]^{-1} \qquad \text{3-D diffusion of Ginstling-Brounshtein} \tag{13.3}$$

$$f(\alpha) = 3 \cdot (1-\alpha)^{2/3} \quad \text{3-D phase boundary reaction} \tag{13.4}$$

$$f(\alpha) = m \cdot (1-\alpha) \cdot [-\ln(1-\alpha)]^{(m-1)/m} \quad m\text{-dimensional nucleation} \tag{13.5}$$

The sublimation of perchlorate, especially ammonium perchlorate (AP), occurs according to a boundary 2-D or 3-D phase reaction and its thermal decomposition according to the Prout-Tomkins or Avrami-Erofeev mechanisms [3.11].

Solid particles, especially metal particles, often form a solid oxide layer on the surface. Therefore, diffusion and reaction occur simultaneously. Fizer and Fritz described a combined model based on a quasi-steady-state approximation, which assumes a uniform temperature distribution of the particle [13.12]. The approach starts from the following equation for the profile of diffusing oxygen from the surface into a sphere to a reaction front with the metallic fuel under static conditions (solution of the 3-D diffusion equation):

$$c_{\mathrm{O}} = c_{\mathrm{O,K}} + \left(c_{\mathrm{O,S}} - c_{\mathrm{O,K}}\right)\frac{1 - R_{\mathrm{K}}/r}{1 - R_{\mathrm{K}}/R_{\mathrm{S}}} \tag{13.6}$$

where c = concentration, O = oxygen, S = surface, K = reaction front and R and r = radius variables.

The conversion of the diffusing oxygen occurs in a first-order reaction and consumes the oxygen flux completely:

$$\frac{\mathrm{d}n_{\mathrm{O}}}{\mathrm{d}t} = -4\pi R_{\mathrm{K}}^2 k(T) c_{\mathrm{O,K}} \qquad k(T) = Z_{\mathrm{K}} \mathrm{e}^{-E_{\mathrm{K}}/RT} \tag{13.7}$$

$$\frac{\mathrm{d}n_{\mathrm{O}}}{\mathrm{d}t} = -4\pi R_{\mathrm{K}}^2 D(T) \left(\frac{\mathrm{d}c_{\mathrm{O}}}{\mathrm{d}r}\right)_{r=R_{\mathrm{K}}} \qquad D(T) = Z_{\mathrm{D}} \mathrm{e}^{-E_{\mathrm{D}}/RT} \tag{13.8}$$

where n_{O} = number of moles of oxygen, Z_{K} and E_{K} = pre-exponential factor and activation energy of chemical reaction and Z_{D} and E_{D} = pre-exponential factor and activation energy of diffusion of oxygen.

When applied to the oxidation of boron [13.13] (Fig. 13.1), the reaction occurs according to

$$\frac{\mathrm{d}n_{\mathrm{O}}}{\mathrm{d}t} = \frac{3\mathrm{d}n_{\mathrm{B}}}{4\mathrm{d}t} \tag{13.9}$$

where n_{B} = number of moles of boron and M_{B} = molar mass of boron.

On combining Eqs (13.6–13.8) the equation for the reaction radius is then given by

$$\frac{\mathrm{d}R_{\mathrm{K}}}{\mathrm{d}t} = -\frac{\frac{4}{3} M_{\mathrm{B}} D(T)}{\rho_{\mathrm{B}}} \frac{c_{\mathrm{O,S}}}{\frac{D(T)}{k(T)} + R_{\mathrm{K}}\left(1 - \frac{R_{\mathrm{K}}}{R_{\mathrm{S}}}\right)} \tag{13.10}$$

Equation (13.10) can be integrated for isothermal conditions:

$$t = -\frac{3\rho_{\mathrm{B}}}{4 M_{\mathrm{B}} D(T) c_{\mathrm{O,S}}} \left[\frac{D(T)}{k(T)}\left(R_{\mathrm{K}} - R_{\mathrm{S}}\right) + \frac{1}{6}\left(3R_{\mathrm{K}}^{\ 2} - R_{\mathrm{S}}^{\ 2}\right) - \frac{R_{\mathrm{K}}^{\ 3}}{3R_{\mathrm{S}}}\right] \tag{13.11}$$

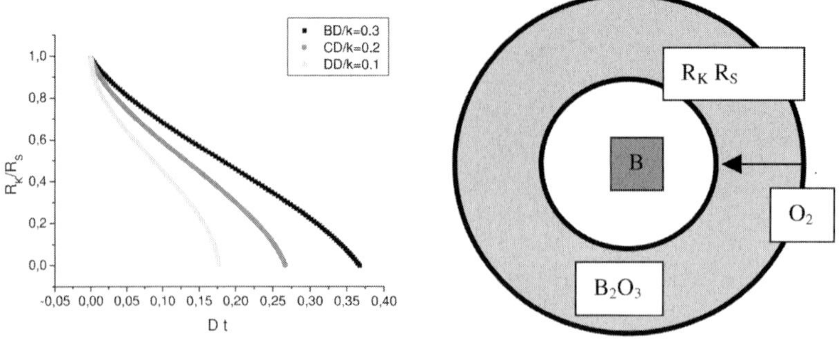

Figure 13.1. Progress of the reaction front into an oxidizing boron particle under isothermal conditions.

However, if it is applied to the evaluation of non-isothermal TGA curves then a numerical method has to be used [13.13].

The ignition and combustion of particles – fuel in the liquid/solid phase and the oxidizer in the gas phase or vice versa – has been widely investigated for liquid droplets and for metal and coal particles. The basic differential equations were summarized by Williams [13.14]. The solutions and the comparison with experiments and theoretical refinement of the description is still on-going research, mainly carried out under microgravity [13.15–13.19].

These diffusion-controlled heterogeneous reactions are often described using a quasi-steady-state approximation. The analytical procedure includes a chemical reaction between fuel and oxidizer, the heat transfer equation and the diffusion equations for a gaseous oxidizer (fuel) and an evaporating fuel (oxidizer) droplet, which can be represented by a Shvab-Zeldovich variable. The simplified steady-state theory is briefly outlined.

The theory proceeds from the assumption that a vaporizing or burning droplet generates a flow field quasi-independent of time in the gas phase. From the continuity equation,

$$\frac{\partial \rho_g}{\partial t} + \frac{1}{r^2}\frac{\partial}{\partial r}\left(r^2 \rho_g v_g\right) = 0 \tag{13.12}$$

follows for the quasi-steady state:

$$r^2 \rho_g v_g = \text{constant} = a \tag{13.13}$$

where ρ_g = gas density and v_g = gas velocity.

Profiles of the temperature and the reactants (mass fractions: Y_{OX} oxidizer, Y_F fuel and Y_P reaction product) in the gas are described by the equation

$$\frac{\mathrm{d}}{\mathrm{d}r}\left[\left(r^2 \rho_g v_g\right)\left(c_p T + Q_O Y_{OX}\right) - r^2 \rho_g D \frac{\mathrm{d}}{\mathrm{d}r}\left(c_p T + Q_O Y_{OX}\right)\right] = 0 \tag{13.14}$$

Similar equations arise where Y_{OX} is substituted by Y_F and Y_P. Boundary conditions can be set as follows:

$Y_F = 0,\ Y_{OX} = Y_{OX,\infty},\ Y_P = 0,\ T = T_\infty$ for $r = \infty$

$Y_{OX} = 0,\ Y_F = 1,\ Y_P = 0,\ T = T_S$ for $r = d/2$ (droplet surface)

$(\lambda dT/dr)_g = \rho_g v_g L + (\lambda dT/dr)_l + Q_r$ for $r = d/2$

where λ = heat conductivity, Q_O = reaction enthalpy related to the oxidizer, L = heat of vaporization and β = stoichiometric fuel to oxidizer mass ratio.

The solution of Eq. (13.14) is (C_1 and C_2 are integration constants):

$$c_p T + Q_O Y_{OX} = C_1 + C_2 e^{-\frac{a}{r\rho_g D}} \tag{13.15}$$

$$\rho_g v_g \frac{d}{2} = \rho_g D \ln(1+B) \tag{13.16}$$

$$B = \frac{\beta Q_O Y_{OX} + c_{p,g}(T - T_S)}{L + \lambda (dT/dr)_l + Q_R} \tag{13.17}$$

The case of evaporation only is obtained for $Q_O = 0$. Using the boundary conditions, an equation for the droplet radius can be derived:

$$\frac{d}{2}\frac{d}{dt}\left(\frac{d}{2}\right) = -\frac{K}{8} \qquad d^2 = d_0^2 - Kt \Rightarrow t_v = \frac{d_0^2}{K} \tag{13.18}$$

where

$$K = \frac{8\rho_g}{\rho_l} D \ln(1 + B)$$

d_0 = initial droplet diameter and t_v = time for complete evaporation.

The differential equations for the droplet diameter can be transferred to the mass conversion of the liquid to the gas phase:

$$\frac{dm}{dt} = -\frac{3}{2}\frac{m_0^{\frac{2}{3}}}{d_0^2} K \cdot m^{\frac{1}{3}} \qquad m(t) = m_0 \left(1 - \frac{t}{t_v}\right)^{\frac{3}{2}} \tag{13.19}$$

The flame is located:

$$\frac{d_f}{d} = \frac{\ln(1+B)}{\ln(1+\beta Y_{OX})} \qquad T_f = T_s + \frac{T_\infty - T_s + \frac{\beta Y_O (Q - L)}{c_{p,g}}}{(1+\beta Y_{OX})} \tag{13.20}$$

The quasi-static condition implies that a droplet acts a source of a potential flow $\Phi(r,r_i)$ with the droplet at the position r_i [13.20]. N droplets would act in a similar way. The source of the flow, the mass concentrations and the temperature are

$$\Phi(r,r_i) = -\sum_{i=1}^{N} \frac{q_i}{|r - r_i|} \qquad Y + c_p T = C_1 + C_2 e^{a\frac{\Phi(r,r_i)}{v_g D}} \tag{13.21}$$

given by Figure 13.2 shows the case of temperature profiles for two burning droplets at two droplet distances with a droplet temperature of 0 °C, a flame temperature of 1000 °C and an outside temperature of 300 °C. Using appropriate boundary conditions for the flame temperature and droplet temperature (see above), the concentration and temperature profiles can be obtained.

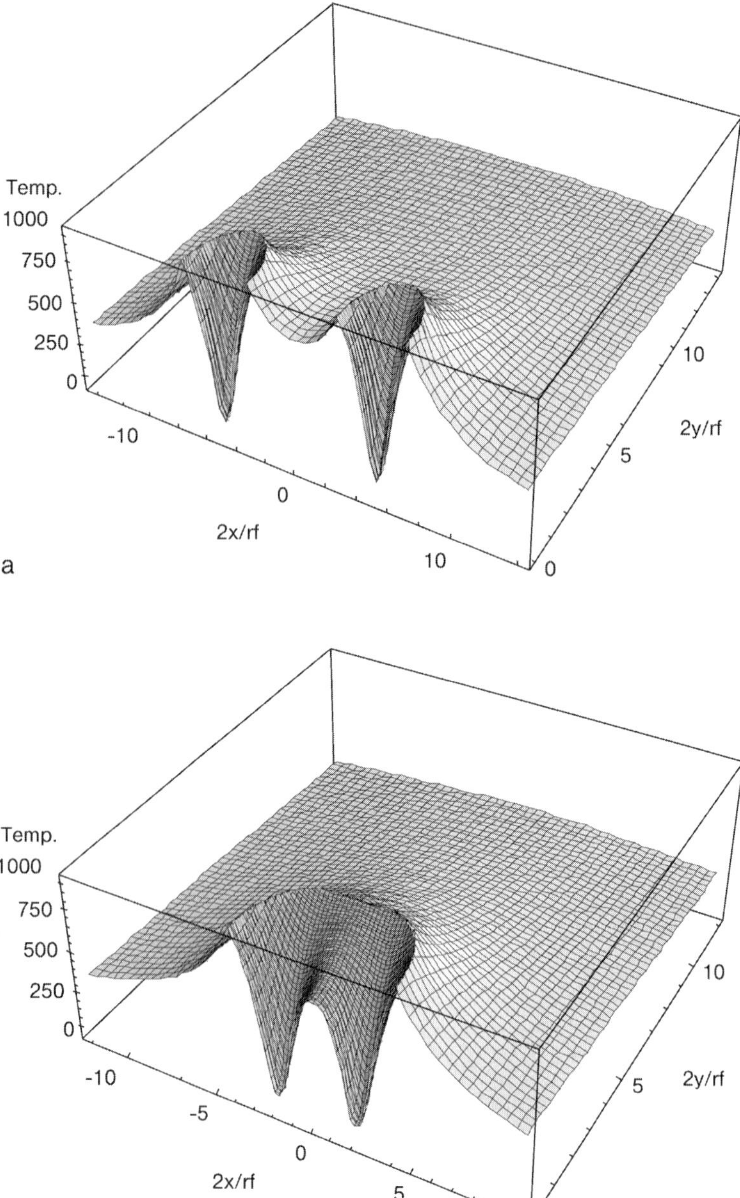

Figure 13.2. Temperature profiles of two neighbored burning droplets in the x–y plane (radius of the flame = r_f): (a) particles burn with low mutual influence; (b) particles evaporate independently but they are covered by one flame front.

Figure 13.2, illustrating interference of combustion of neighbored droplets in (a) and the overlap of two flame zones in (b), indicates that in a particle cloud the characteristics of combustion changes depending on the droplet distances if the flame radii of the individual combustion of particles interfere. Various regimes of burning exist and transitions occur during the progress of combustion: evaporation takes place only at the cloud boundary, overlapping of flame zones over two or more individually evaporating particles, individual burning of a single particle (see Fig. 13.3).

Moving particles or droplets with velocity v in a fluid of velocity u are subjected to drag:

$$v\frac{\mathrm{d}v}{\mathrm{d}x} = \frac{18\mu}{d^2\rho}(u - v)$$
$$v\frac{\mathrm{d}}{\mathrm{d}x}d^2 = -K \tag{13.22}$$

The heat-up is given by

$$\rho_1 c_p d\frac{\mathrm{d}T_p}{\mathrm{d}t} = 6\lambda\left[(T_\infty - T_p) - \frac{\mathrm{d}m}{\mathrm{d}t}L\right] \tag{13.23}$$

A spray injected into a fluid at rest will be consumed at a distance x_B (minimum length of a burning chamber):

$$x_B = \frac{1}{2}\frac{d_0^{\,2}v_0}{K} \tag{13.24}$$

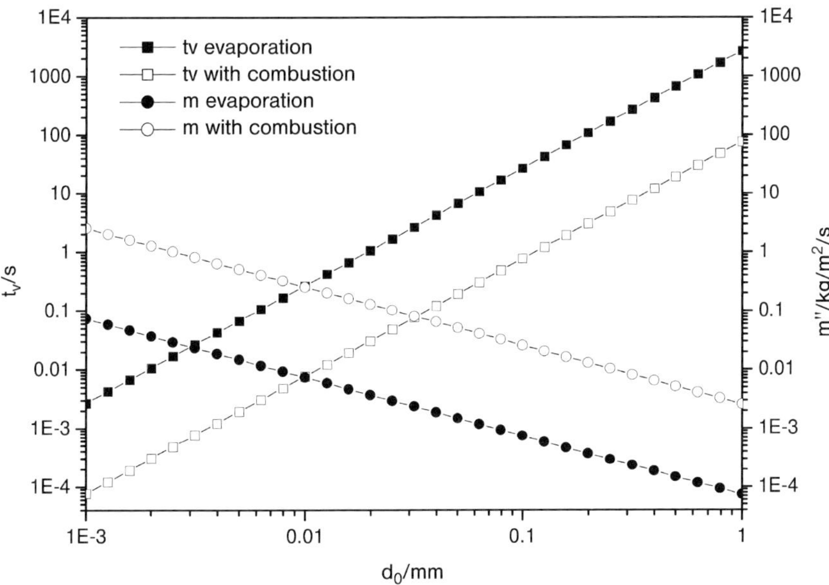

Figure 13.3. Comparison of the dependence of droplet life time and mass conversion of a vaporizing and a burning droplet.

Mono-propellant particles display a regression under quasi-steady conditions independently of the surrounding atmosphere [13.21]:

$$\frac{dm}{dt} = 8\pi d \frac{\lambda \left(\frac{dT}{dx}\right)_1}{c_p (T_s - T_\infty) + L} \tag{13.25}$$

Normally, the parameters of droplet evaporation and combustion can be found by observing the droplet diameter during reaction by cameras in micro-gravity experiments which allow the realization of spherical symmetry. Based on the d^2 law, the related evaporation parameters and, in addition, the kinetic parameters of thermal decomposition in the gas phase can be obtained by the measurement of gas evolution of an evaporating droplet in an IR cell [13.22, 13.23]. Normally, the droplet diameter is measured, and the accuracy is strongly reduced when the diameter approaches zero. In contrast, the error in measurement of the evolved vapor by IR absorption decreases. Assuming evaporation of a droplet of initial mass m_0 according to Eq. (13.19) and a first-order decomposition reaction of the vapor, the mass of the undecomposed vapor, m_v, is given by [13.22, 13.23]:

$$\frac{dm_v}{dt} = -\frac{dm}{dt} - k \cdot m_v \qquad \frac{dm_v}{dt} = \frac{3}{2}\frac{m_0}{t_v}\sqrt{1 - \frac{t}{t_v}} - k \cdot m_v \tag{13.26}$$

$$\frac{m_v}{m_0} = -\frac{3}{4k \cdot t_v}\left\{2e^{-k \cdot t} - 2\sqrt{1 - \frac{t}{t_v}} - \sqrt{\frac{\pi}{k \cdot t_v}}\left[\operatorname{erf}\sqrt{k \cdot t_v} - \operatorname{erf}\left(\sqrt{k \cdot t_v}\sqrt{1 - \frac{t}{t_v}}\right)\right]\right\} \quad t < t_v \tag{13.27}$$

$$\frac{m_v}{m_0} = -\frac{3}{4k \cdot t_v}\left(2e^{-k \cdot t} - \sqrt{\frac{\pi}{k \cdot t_v}} \operatorname{erf}\sqrt{k \cdot t_v}\right) \quad t > t_v \tag{13.28}$$

An illustration is shown in Fig. 13.4.

13.1.3 Composite Rocket Propellants

Typically, a composite solid rocket propellant consists of crystalline oxidizer particles embedded in a polymeric matrix [13.24]. In addition, metal particles increase the performance of the propellant. These composites are used in large-scale rocket motors. The most prominent propellant example of high thrust (up to 2600 N s) is the ammonium perchlorate (AP)-based propellant for space rocket boosters. It contains in addition hydroxy-terminated polybutadiene (HTBP) as a binder and Al as a metallic fuel component. The combustion products HCl and Al_2O_3 form a large rocket plume and contribute to environmental contamination. Therefore, substitute components have been investigated such as HMX, ammonium dinitramide (ADN) and ammonium nitrate (AN) to replace AP fully or partially.

The burning rate r of AP composites depends strongly on the particle size, especially the size of the oxidizer particles. The first successful approach, the granular diffusion flame model (GDF), described qualitatively the effects of pressure, the oxidizer particle size and fuel oxidizer ratio [13.25]. The GDF model

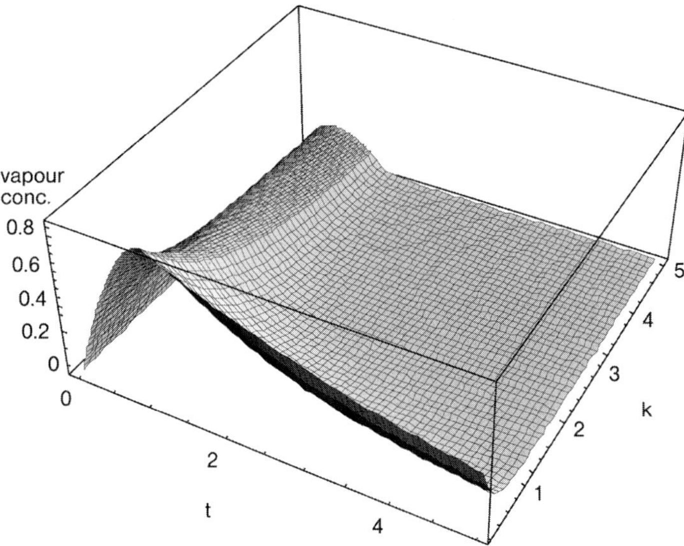

Figure 13.4. Vapor concentration of a droplet in an isothermal hot atmosphere including vapor thermal decomposition according to combined Eqs (13.27) and (13.28). $[k] = 1/t_v$.

assumed two parallel reactions of decomposition or evaporation products of the solid, a homogeneous chemical reaction and a diffusion-controlled reaction of a vapor or oxidizer 'pocket' with the surrounding gas. As shown above, the chemical reaction time is dominated by the Arrhenius rate constant and does not depend on the particle size. The diffusion-controlled reaction in the gas phase proceeds similarly to a droplet combustion and the conversion time is proportional to the square of the 'pocket' diameter (droplet size; see above). The regression rate of the solid is given by the heat transfer from the gas phase and the physical parameters of the solid:

$$r = \frac{\lambda_g \left(\frac{dT}{dx}\right)_s}{c_s (T_s - T_\infty) - L} \tag{13.29}$$

where T_S = temperature at the surface of the propellant, T_∞ = temperature of the cold propellant, c_S = specific heat in the solid. The resulting burning rate r was found to be

$$\frac{1}{r} = \frac{a}{P} + \frac{b}{P^{\frac{1}{3}}} \tag{13.30}$$

with

$$a = \left[\frac{\lambda_g (T_f - T_s)}{c_s (T_s - T_\infty) - L}\right]^{\frac{1}{2}} \frac{\rho_s R T_g}{\sqrt{Ze^{-\frac{E}{RT_g}}}} \tag{13.31}$$

$$b=\left[\frac{\lambda_g\left(T_f-T_s\right)}{c_s\left(Ts-T_\infty\right)-L}\right]^{\frac{1}{2}}\frac{\rho_s R^6 T_g^{\frac{5}{12}}\mu^{\frac{1}{3}}}{K_1^{\frac{1}{2}}} \qquad \mu=\rho_g d^3 \tag{13.32}$$

For the case of large particles (> 20–30 μm) the burning is controlled by diffusion, the pressure exponent n of the burning rate $r = p^n$ approaches 1/3. For smaller particles (< 5 μm), the chemical reaction rate dominates and the pressure exponents approaches 1. The burning rate varies in a hyperbolic way depending on the particle size, whereas it increases by a factor of ~ 10 on decreasing the AP particle size from 1000 to 1 μm. These relations are shown qualitatively in Figs 13.5 and 13.6. In addition, the temperature sensitivity is influenced by the particle size. The GDF model was modified to take into account the monopropellant combustion of AP and polydisperse particles as for high-performance bimodal oxidizer distribution are applied [13.26].

The heterogeneity of a burning surface induced by oxidizer particles was directly considered by Hermance [13.27, 13.28], including also endothermic and exothermic reactions on the conversion of the solid and in the gaseous zone. In a randomly packed bed of spherical particles, the average diameter d^* observed on the propellant surface, which forms a random plane of intersection to the particle size, is given by

$$d^*=\sqrt{\frac{2}{3}}d_0 \tag{13.33}$$

The regression of the surface is the sum of the decomposition/evaporation at the surface and forms a flux of the gaseous constituents which react controlled by diffusion. The model could fit the experimental data of composites at pressures > 20 MPa.

The more sophisticated model of Beckstead, Derr and Price [13.29] (BDP) predicts the dependence of the burning rate on the composition of the composite propellant

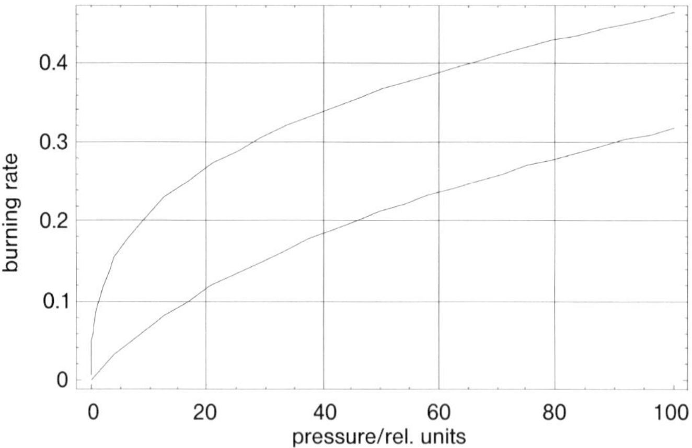

Figure 13.5. Upper curve, $b >> a$ – large particles; lower curve, $a >> b$ – small particles.

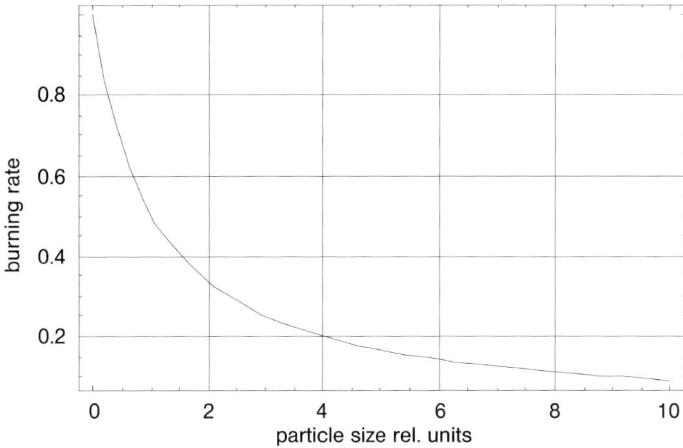

Figure 13.6. Dependence of the burning rate on the particle size.

even better. The components of the propellants can undergo solid-state reactions but are transferred to the gas phase by endothermic effects. The flame zone of a composite is composed of multiple flames, at least three:

- the AP monopropellant flame, which is a premixed flame dominated by chemical kinetics which determine the flame stand-off distance;
- the primary flame between evaporated or pyrolyzed fuel and AP decomposition products governed by chemical kinetics and diffusion;
- the final flame between fuel decomposition products and residual oxidizing products of the AP monopropellant flame, which is a diffusion flame.

The flame stand-off distances depend on the pressure and in relation to this dependence they control the pressure dependence of the burning rate. The mass flows from the burning surface are interrelated by the intersection average oxidizer diameters d^* (see above).

The implementation of statistical effects of the heterogeneity at the burning surface could additionally improve the BDP model [13.30]. The quantitative predictions of the BDP model concern the effects of oxidizer particle sizes, of the composition, the pressure dependence in a broader range and the temperature sensitivity and show better correlation withexperimental results, in contrast to the earlier theories. However, no simple unique equation is available to insert for propellant designers.

Extended models can also account for partial or full substitution of AP by, e.g., crystalline HMX (see the report of Blomshield [13.31]). For HMX crystallites used instead of AP, the variation of the burning rate is only a factor of 2. The main reason is that HMX is a 'better' monopropellant characterized by the oxygen balance, which is smaller than that of AP. Therefore, the diffusion characteristics of the granular flame do not strongly influence the combustion.

Some attempts have been made to model fully the combustion of heterogeneous energetic materials in the solid phase. The basic problem is that the scale of a static temperature profile in the solid propagating in the direction of the burning is of the order of magnitude of the particle size. Assuming such a steady-state thermal wave in a reacting solid [with reaction heats $\dot{q}_i$, specific heat of the solid $c(T)$ and a heat flux from the flame of $\dot{Q}$] it is given by [13.32])

$$\frac{\mathrm{d}}{\mathrm{d}x}\lambda(T)\frac{\mathrm{d}T}{\mathrm{d}x}+rc(T)\rho\frac{\mathrm{d}T}{\mathrm{d}x}=r\sum\dot{q}_i\frac{\mathrm{d}c_i}{\mathrm{d}x}+\dot{Q}\delta(x) \qquad \lambda(T)\left.\frac{\mathrm{d}T}{\mathrm{d}x}\right|_{T_s}=0 \tag{13.34}$$

The steady-state burning rate r does not depend on the values of physical parameters within the solid but only on those at the boundaries if all chemical reactions are completed in the solid and Eq. (13.34) can be integrated:

$$r=\frac{\dot{Q}}{\rho\int\limits_{T_\infty}^{T_s}c(T)\mathrm{d}T-\sum q_i} \tag{13.35}$$

The temperature profile of an inert solid is given by

$$x-x_0=\int\limits_{T_\infty}^{T}\frac{\lambda(\Theta)}{\rho\int\limits_{T_\infty}^{\Theta}c(\tau)\mathrm{d}\tau}\mathrm{d}\Theta \tag{13.36}$$

In an inert homogeneous solid (inert gas zone, substitute T_s by T_f) the profile is given by

$$T(x)=T_s\mathrm{e}^{-\frac{rc\rho}{\lambda}x} \tag{13.37}$$

For many solid propellants, $\lambda/c\rho$ is of the order of magnitude of 10^{-3} cm²/s.

The burning rates are of the order of magnitude of 1 cm/s at lower pressures, resulting in a scale thickness of the temperature profile of about 10 μm. Therefore, the temperature profile depends on time or even fluctuates on passing alternating crystallite and polymer areas. Only very small particle sizes at low pressures justify averaged steady-state profiles in the solid.

In the gas phase (physical parameters from the gas phase, v_g = gas velocity) the situation is similar [13.30, 13.31]:

$$\frac{\mathrm{d}T}{dx}=-\frac{v_g c\rho}{\lambda}T_f\mathrm{e}^{-\frac{v_g c\rho}{\lambda}x} \tag{13.38}$$

$$\dot{Q}=-\lambda\left.\frac{\mathrm{d}T}{\mathrm{d}x}\right|_{T_s}=v_g c\rho T_f\mathrm{e}^{-\frac{v_g c\rho}{\lambda}x_{\text{st-off}}} \tag{13.39}$$

$$x_{\text{st-off}}=-\frac{\lambda}{c\rho}\ln\left(\frac{\dot{Q}}{v_g c\rho T_f}\right) \tag{13.40}$$

Aluminum particles were originally used in composites for increasing the performance by the additional high heat of combustion. Situated beneath, Al or their oxides attenuate pressure oscillations when particles of a size in the range 20–30 μm burn in rocket motors. The dependence of the burning rate on the particle size was found to be low. This is different if ultra-fine particles (ALEX) are used, which have become available recently [13.33]. In order to investigate the burning behavior of composites containing ultra-fine Al particles, two test propellants were compared. Their compositions were both 75 % AP, 8 % Al, 12 % HTPB and 5 % DOA, but differed in the size of the Al particles: HTPB 1090: AL/Alex (nanoscale) and HTPB 1091: Al/15 μm

The burning rate increases by a factor of 1.5 or more [13.34] (Figs 13.6 and 13.7), which is in accordance with other results [13.35–13.41]. ALEX particles also lead to higher pressure exponents, which indicate a reduced dependence on the heterogeneous flamelet structure of the burning zone. Assuming that the droplet combustion model holds for this case also explains the fast oxidation of the ultra-fine Al-particles. As is obvious from the composite models described above, it is cumbersome to include also metal particle oxidation in detail, and estimations are therefore applied. These estimations reveal that particles of sizes > 10 μm would have completely reacted at stand-off distances orders of magnitude higher than those used for the different flamelet types of BDP theory if one applies a simplified droplet combustion theory (see Table 13.1) [13.42]. Particles of sizes 10^3 times smaller would react within distances 10^6 times smaller (see droplet combustion above) and allow Al oxidation very close to the burning surface within the standoff-distance of the flame.

Figure 13.7. Combustion of composites with (a) coarse Al particles and (b) ultra-fine Al particles (ALEX) at 4 MPa in an optical bomb.

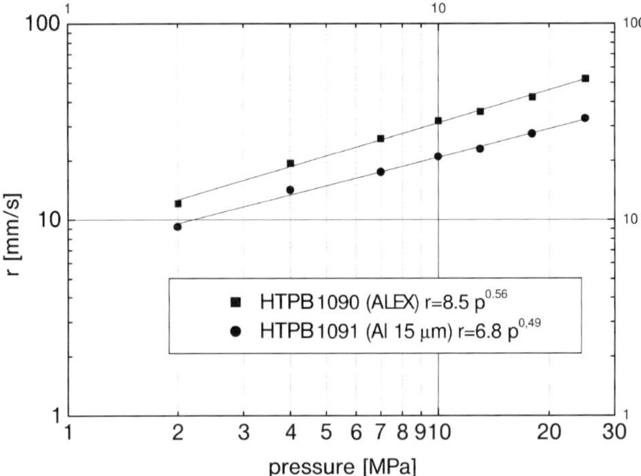

Figure 13.8. Burning rates of the test composites measured in a Crawford strand burner.

Experiments showed that a temperature maximum is found very close to the burning surface transferring the heat of ultra-fine Al combustion directly via the temperature gradient to the burning surface [13.43]. Using Eqs (13.29) and (13.40) for the analysis of the data in Fig. 13.7, it follows that at 10 MPa a heat flux $\dot{Q}$ of about 0.8 kW/cm^2 would lead to the measured burning rate ($T_s = 900$ K, $c_p\ \rho = 2$ J/(K cm^3), from [13.31]). This means that the ALEX combustion delivers additionally about 0.4 kW/cm^2. The stand-off distance is estimated to be in the region of 25 μm where the oxidation of the ALEX particles is completed in the gas phase. This is in accordance with the estimation of the completion of the Al-particle oxidation in an oxidizer stream of 10 m/s where particles with sizes < 1 μm are oxidized within < 50 μm and those < 100 nm within 500 nm (see Table 13.1; remember also the temperature profile from the solid surface to the flame temperature maximum).

Interesting phenomena are observed in the combustion of boron in the case of air-breathing rocket engines, which led to intensive investigations. Detailed descriptions can be found in the literature [13.44–13.50].

13.1.4 Pyrotechnic Mixtures

Practical experience with ignition of gunpowders and rocket propellants resulted in the requirement that igniters should produce hot particles beneath hot gases in time scales at least one order of magnitude smaller than the conversion of the energetic material itself [13.51–13.55]. The hot gas should generate a pre-pressure for combustion, allow conductive and convective heat transfer and assist stabilization of gas-phase reactions. The hot particles should transfer ignition energy effectively to the propellant by penetrating the surface. Theoretical approaches to describe ignition are predominantly based on the heat flow equation in a reactive solid phase. The energy input from an igniter is assumed to be a heat flow variable in time to the propellant surface, an approach which can simulate a hot gas contact, radiation or an averaged hot particle flow.

Pyrotechnic mixtures often consist of particles of inorganic oxidizers and metallic fuels of carefully controlled quality. Particle size is well known to be an important parameter to control and guarantee maintained quality [13.51–13.60].

Parametric investigations of the influence on the performance and the conversion rates were published only recently [13.59, 13.60]. They show that the particle size of both oxidizers and metallic fuels strongly influence the conversion rates, the pressure rises and the maximum pressures in closed volumes when varied in the range 1–100 μm [13.59]. To understand this relation is straightforward on a qualitative level: Smaller oxidizer particles decompose faster to set free an oxidizing atmosphere (Prout-Tompkins or Avrami-Erofeev mechanisms). According to the droplet (or particle) consumption in an oxidizing atmosphere, smaller metal particles are obviously consumed faster owing to the d^2 law and more complete because of the shorter particle combustion time t_v leading to the observed improvements. In detail, a rigorous mathematical modeling would be complex and the practical benefit of numerical solutions not guaranteed. Coarse particles of Ti with diameters > 20 μm would need about 1 ms to be oxidized in an atmosphere of KNO_3. Boron particles of 1 μm would need about 0.5 μs; Ti nanoparticles would need less time. Especially the simulation of gun propulsion would need an adequate input to estimate ignition delay time and pre-conditions to initiate the combustion of the main charge. The use of standardized pressure-time curves measured in small closed vessels (< 50 ml) for an igniter mixture of defined particle sizes and additives could be a useful approach.

The application of nanometal particles is of current interest also for igniters [13.61–13.64]. Igniting mixtures with various compositions including one with ultra-fine Ti particles were studied in a closed vessel with a volume of 50 ml [13.61–13.63]. Experiments included pressure measurements with the igniting mixture (350 mg) alone and together with gunpowder grains (3 g) were performed in a 50-ml mini-bomb. The ignition mixture of ultra-fine Ti particles reacted fastest (Fig. 13.8). Its reaction started earlier by about 0.8 ms, produced a steeper pressure rise of 7200 MPa/s and was mainly completed within 0.25 ms. This is in qualitative accordance with pyrometer results in open experiments. However, the ignition of the main charge by coarse particles works more effectively as the pressure-time curves of the propellant combustion show a (pressure maximum) delay for the ultra-fine particles. For four gun propellant grains, mainly experimental powders (JA2, a semi-nitramine powder, an NENA powder and a GAP powder), the pressure-time curves were compared. In all cases the ignition delay of the ultra-fine Ti in the ignition powder was longest. Evidently, the fine metal powder is oxidized and vaporized so fast that only a small fraction of the particles (the originally larger fraction) can reach the gunpowder to be ignited. A consequence of these results could be to use fine particles together with coarse ones to combine the effects, to accelerate the initial reaction and to have available the hot, large particles needed for effective ignition.

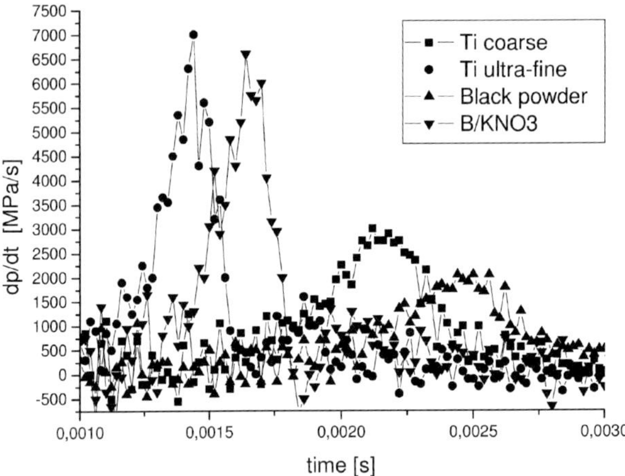

Figure 13.9. Conversion rates of booster mixtures with coarse and ultra-fine Ti particles compared with B/KNO_3 and Black Powder.

Flares consist mainly of inorganic oxidizers and metallic and organic fuels [13.65]. Currently used ones are composed of Mg particles embedded in Teflon/Viton (MTV flares) and are very effective in their brightness because of the hot reaction of Mg particles with F. Particle size plays an important role in burning characteristics and light emission [13.65, 13.66].

When the igniter reaction products should interact with the propellant, the effect of gaseous species with a well-defined temperature can be described by convective and conductive heat transfer to the grain surface. The ignition by multiple single hot particles produced and ejected by the igniter is more difficult to describe theoretically. To use an averaged energy transfer to the surface or near the surface could be a possibility but cannot distinguish between the individual particles and their characteristics. However, hotspot models can adequately take into account the size of particles, especially the size of the metal particles in booster mixtures and their energy input in random positions closely below the surface of the propellant where they have penetrated. It can be shown that many small particles need substantially less time to initiate a stable combustion front compared with a few large particles (the total energy input is the same) [13.62].

13.1.5 Detonation

Explosive charges are cast, molded or pressed energetic materials which consist of one or more components of different particle sizes or contain pores, voids or micro crazes [13.65–13.76]. The particles can vary in purity, internal defects and surface quality. These aspects do not substantially influence the performance, mainly detonation velocity and Gurney energy, of so-called ideal explosives such as cast

TNT, HMX and compound B, etc.. The effects are stronger for plastic bonded explosives or for underwater explosives and where high blasts have to be generated [13.75, 13.77, 13.78].

The imperfections interact with initiating temperature, shock and/or detonation waves, influencing the detonation characteristics. Initiation sensitivities, thresholds, run times of shock to detonations, detonation velocities and performance could be strongly affected by the particle characteristics of the constituents [13.72–13.76]. The recent use of sub-micron particles of energetic materials such as RDX and PETN has led to less sensitive charge formulations [13.79].

The progression of the detonation wave is fed by the chemical conversion of energetic materials. Simplified theories based on thermodynamics assume a reaction immediately completed behind the detonation front depending only on the pressure (Forest Fire model). If chemical reactions are taken into account by simplified mechanisms such as in the ignition and growth mechanism or the JTF model, the chemical reactions contribute by a dependence of the conversion on a fit parameter which is like a reaction order. Findings for an order of 2/3 indicate a dependence of the reactions on the 3-D crystallite structure (see particle decomposition above). The approaches to reactions described above show that particles of larger sizes would not follow this assumption. Deviations are found for heterogeneous high explosives (non-ideal explosives) when comparing measured values of detonation velocities and pressures with theoretically estimated values. Approaches to explain these phenomena are as follows [13.80–13.82]:

- Hotspot models, which assume the collapse of pores or bubbles (or other defects) on impact of the deformation wave and the use of the released energy for heating up the neighboring energetic material to stimulate fast chemical conversion. In a reactive medium these hotspots can enlarge at rates orders of magnitudes higher than normal burning rates, because of the high directly transferred temperature. Depending on the distribution of pores, the individual converting spots unite to form a more or less plane reaction front. Therefore, a thermal conversion wave can follow closely an initiating wave.
- The absorbed energy can be restored to the initial wave by the spherical acoustic waves emitted by the converting hotspots. The inhomogeneities scatter the waves and act as individual sources of spherically emitted pressure waves. These individual waves of various sources can merge into a plane wave depending on the geometric constitution of the medium and reinforce the initiating wave.
- Micro-structured models of wave propagation including all relevant effects could solve the problem. They account for the interaction (scattering, deformation, movements, etc.) with the inhomogeneities and the conversion of the initial solid to the reaction products in heterogeneous explosives. However, the numerical solutions of the equations require the use of super-computers. Results show that the scattering of propagating waves depends on the size and shape of the constituents of the explosives. The interference of the scattered waves forms patterns of high-temperature and -pressure spots, which could be understood as hotspots.

The dependence of the particle size was investigated in detail for HMX and RDX. Spherical particles show longer run distances to detonation and higher initiating shock pressures.

Fine particles decrease shock sensitivity; the shock sensitivity decreases with decreasing particle size. The fracture of particles and their elastic deformation play an important role. The yield stress τ on impact depends on the particle size [13.83]:

$$\tau = \tau_0 + \tau_p d^{-\frac{1}{2}} \tag{13.41}$$

An experimental relation of sensitivity against shock loading was found to be proportional to $d^{1/2}$. In contrast to RDX where the effect continues into the nanoscale region [13.79], for HMX the sensitivity rises when approaching particle sizes below 10–20 μm. This effect is not yet completely understood. An explanation could be that smaller HMX particles show a different phase transformation behavior to larger particles, thus influencing shock sensitivity and related temperature effects. In addition, smaller particles could be transformed to the more sensitive δ-phase more easily than larger particles on fast heating.

Aluminum in explosives is mainly used to increase the blast effect when after-reactions play an important role. This is assumed for detonations in air or underwater explosives. Particle sizes are used in the range 1–100 μm. As in the case of igniters and propellants, nanometal particles, especially ALEX, were of high interest, because it was expected that ultra-fine particles would contribute to the primary reactions and heat generation of the detonations. The results actually obtained do not fully confirm the expectations. Various groups have reported results on a large number of different compositions. Those compounds with Al nano-particles react at rates approaching those of high explosives [13.83–13.90]. For high explosives of the RDX, HMX and Cl-20 type, no measurable effect on the detonation velocity was found, although there were some smaller effects (e.g. maximum detonation velocity, depending on charge diameter) when Al was present in the reference samples. Experimental results compared with theoretical calculations (e.g. CHEETAH code) reveal that the particle size of Al strongly influences the detonation velocity of non-ideal explosives which consist, e.g., of ADN or AP [13.76, 13.77]. Especially ADN shows strongly enhanced detonation velocities and push-plate test velocities [13.83].

The reaction of Al particles is controlled by the effects of droplet combustion and should occur similarly in detonations. Values of K are estimated for Al using for the vapor diffusivity and ratio of Al densities $\approx$ 0.025 mm^2/s and $B \approx Q/L \approx$ 31 000/10 800 $\approx$ 3 to $K \approx$ 0.2–0.3 mm^2/s.

Related burning times are given in Table 13.1 and illustrate that 100-nm particles have completely reacted after 0.1 mm on following the product flow field of a detonation (> 2000 m/s). In contrast to the case of solid propellant combustion where flow velocities of the order of magnitude of 10 mm/ms are assumed, in detonations with high detonation velocities (> 5000 m/s) the Al reaction cannot contribute substantially to the detonation wave but for slower wave propagation a contribution seems possible.

Table 13.1. Relations between particle sizes and distances of completed Al combustions (estimated value for the estimated Al combustion constant $K \approx 0.2\ mm^2/s$).

Particle size (mm)	Burning time (ms)	Flow velocity (mm/ms)	Distances, Eq. (13.24) (mm)	Flow velocity (mm/ms)	Distances, Eq. (13.24) (mm)
1	5000			10	25000
0.01	0.5	2000	500	10	2.5
0.001	0.005	2000	5	10	0.025
0.0001	0.00005	2000	0.05	10	0.00025

In all cases, an increased sensitivity was found when investigating mixtures or charges of ultra-fine metal particles in energetic materials. Both thermal and shock stimuli initiate hazardous reactions at lower thresholds. Detailed results have been published for Al particles concerning thermal effects. Methods of thermal analysis [DSC, TGA, evolved gas analysis (EGA), etc.] showed an increased reactivity of the metal particle itself and of the mixture. Temperatures of onset of the reaction (oxidation with air) are lowered by 100 K or even more. In addition, the ESD (electrostatic discharge) sensitivity is strongly increased. Dusts of nanoaluminum explode easily and violently [13.91].

13.2 Defects of Explosive Particles and Sensitivity of Cast Formulations

In this section, we report accurate experimental correlations between the defects of explosive particles and the sensitivity of cast formulations. Two main experimental results are emphasized:

- The respective influence of internal and surface defects of explosive particles on the sensitivity of cast formulations is accurately demonstrated.
- The importance of the effects of defects of explosive particles on the sensitivity of cast explosive formulations is pointed out.

Two shock sensitivity tests were used to record the sensitivity of the cast formulations:

- The plane shock wave impact test is the most accurate. It is used to study the respective influence of internal and surface defects of the explosive particles. This test allows one to record tiny sensitivity variations.
- The small projectile impact test is the most demonstrative. This test provides a measurement of the detonation thresholds of the cast formulation. It will illustrate the importance of the effects of the defects of explosive particles on the sensitivity of cast formulations.

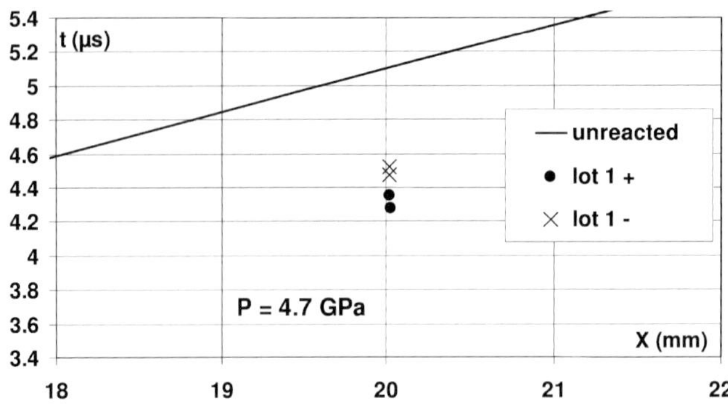

Figure 13.12. Effects of internal defects of RDX particles on the shock sensitivity of cast formulations (70% RDX 315/800 μm + 30% wax).

Internal defects are hotspots where chemical reactions can start. At low shock pressures (4.7 GPa), the shock-detonation transition is mainly controlled by the ignition of hotspots [13.92]. An increase in the amount of internal defects corresponds to an increase in the number of potential hotspots and this leads to an increase in the shock sensitivity of the formulation.

13.2.1.2 Effects of Surface Defects

A raw RDX lot was extracted by sieving from a commercial RDX lot. The crystal sizes were between 150 and 200 μm. A processed lot was manufactured from the raw lot by stirring in hot saturated acetone. This modified only the particle surface properties. Finally, the RDX processed particles were sieved to keep the narrow monomodal crystal size distribution of the raw lot.

Figure 13.13 shows the qualitative results of the surface processing. It removes twin crystals and provides smooth particle surfaces. Unfortunately, it was not possible to achieve an accurate quantification of the surface properties of the raw and processed particles. This point is discussed in detail in Sect. 9.2 [9.6].

The ISL sink-float experiment does not show any quantitative variation of the population of internal defects between the raw and processed lots (Fig. 13.14).

Two similar high-quality cast formulations (70% weight RDX + 30% wax) using the two RDX lots (raw and processed) were used to check the effects of surface defects of RDX particles on the shock sensitivity of cast formulations.

Figure 13.15 gives the results of the plane shock wave impact tests. At 4.7 GPa, the formulation based on the raw lot has the shortest transit times. This shows that the shock sensitivity of the formulation increases when the amount of surface defects on the RDX particles is increased.

As for internal defects, these accurate data show that surface defects are potential hotspots where chemical reactions can start. The shock transit time variation is now 150 ns. It is smaller than for internal defects but remains significant as the experimental accuracy is < 30 ns.

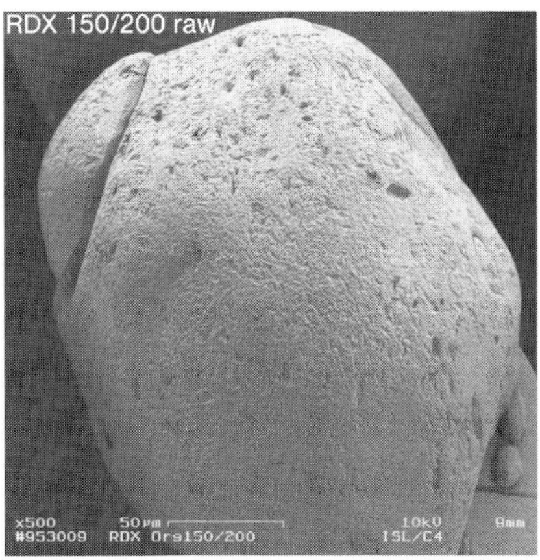

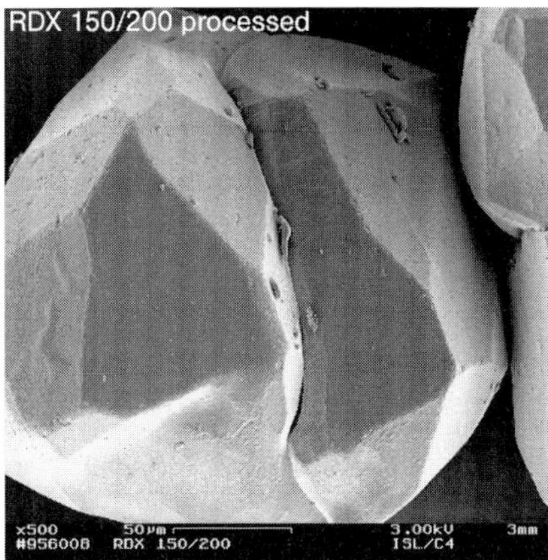

Figure 13.13. Scanning electron microscopy.

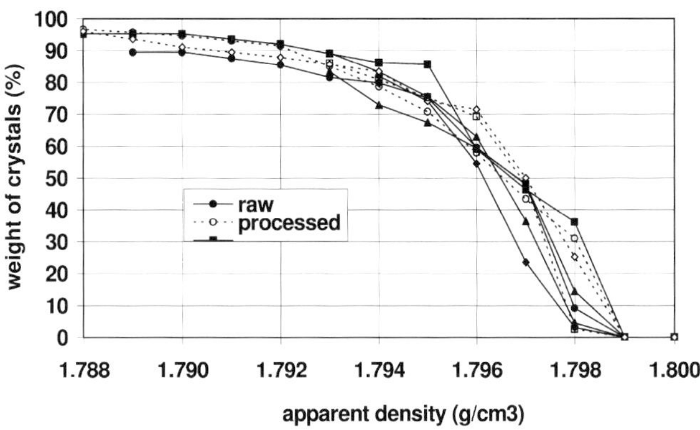

Figure 13.14. RDX 150/200 μm: ISL sink-float experiment.

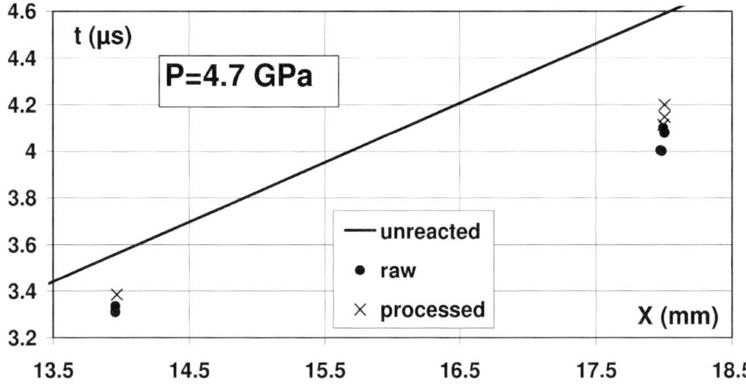

Figure 13.15. Effects of surface defects of RDX particles on the shock sensitivity of cast formulations (70% RDX 150/200 μm + 30% wax).

The dispersion of the experimental results recorded for the same formulation illustrates the dominant role of the hotspots at low shock pressures (4.7 GPa). At 4.7 GPa, the amount of ignited hotspots is limited because it is close to the initiation threshold of the formulations. The build-up of the detonation from a limited amount of localized sites gives small variations of the shock transit time for the same formulation [3.39].

A similar study based on larger RDX particle (400/500 μm) provides the same results. Sphere-like particles and a reduced amount of surface defects give less shock-sensitive cast formulations.

13.2.2
Magnitude of the Effects of the Defects of the Explosive Particles on the Sensitivity of Cast Formulations

The previous experimental data are very accurate and demonstrate the effects of internal and surface defects on the sensitivity of cast formulations. Nevertheless, these effects are small. The aim of this section is to check if an increase in the differences between the microstructures of the explosive particles provides an increase in the sensitivity variations [9.7, 9.9].

For this purpose, three HMX lots were selected among commercial HMX lots. The three lots have the same narrow monomodal particle size distribution: 200/300 μm. Qualitatively, lots 1 and 2 have similar particle shapes and surface properties (Fig. 13.16). Lot 3 exhibits particles with more spherical shapes.

The ISL sink-float experiment (Fig. 13.17) shows that the global amount of internal defects in the explosive particles is very different for the three HMX lots. The three apparent density curves are similar but are separated by a significant shift (0.002 g/cm^3). The particles of lot 1 have the smallest amount of internal defects and those of lot 3 the largest. These quantitative data are in agreement with the qualitative observations performed by optical microscopy with matching refractive index [9.7].

Three similar high-quality cast formulations (70% weight HMX + 30% wax) using the three HMX lots were used to check the shock sensitivity of cast formulations. These formulations are similar to the previous RDX-based formulations.

The small projectile impact test was used to check the sensitivity of the cast formulations. The experiment provides the impact velocity threshold which results

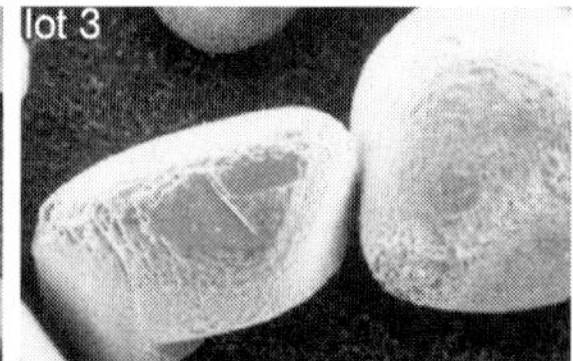

Figure 13.16. HMX 200/300 μm: scanning electron microscopy.

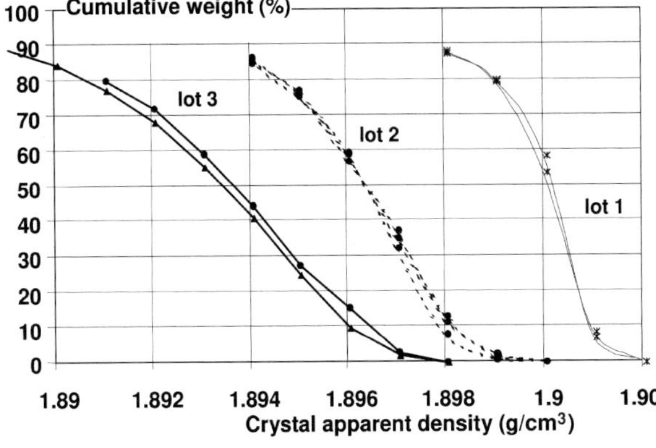

Figure 13.17. Results of the ISL sink-float experiment: HMX 200/300 μm.

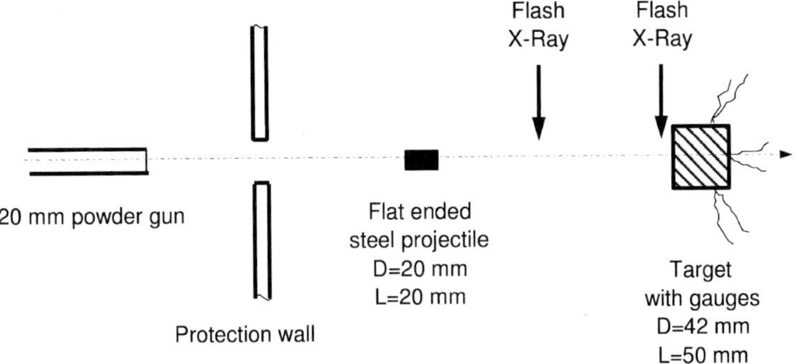

Figure 13.18. Small projectile impact experimental set-up.

in full detonation of the sample. A flat-ended steel projectile of diameter 20 mm and length 20 mm was launched against the formulation sample with a nylon sabot using a powder gun. A protection wall and a 2-m distance between the target and the gun muzzle prevented blast effects. Figure 13.18 gives a schematic view of the experimental set-up. A double-flash X-ray (Fig. 13.19) prior the impact gives the projectile impact velocity and the quality of the impact (stability of the projectile flight).

The detonation of the sample was checked by recording the shock wave transit time across the sample using simple short-circuit gauges on the front and rear sides of the formulation sample. The detonation diagnosis was confirmed with ionization gauges on the lateral sides of the sample.

Figure 13.20 gives the recorded shock wave transit time across the sample as a function of the incident projectile velocity. For the three HMX formulations, the curves exhibit sharp variations. In agreement with the ionization gauge responses and the absence of residual fragments, these transit time drops indicate the projectile velocity threshold to obtain the detonation of the sample.

The respective volume rate of internal defects in the explosive particles varies

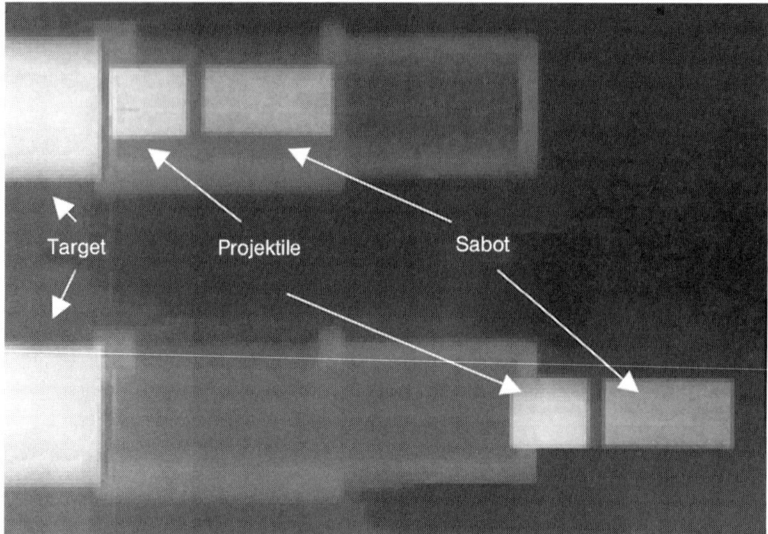

Figure 13.19. Flash X-ray images of the 20-mm projectile flight just before impact.

from 0.45 to 0.1 % between the HMX particles of lots 3 and 1. The velocity threshold to achieve detonation increases from 760 m/s for lot 3-based formulations to 1 100 m/s for lot 1-based formulation (~ 30 %). Reducing by a factor of 4 the very small amount of crystal internal defects leads to a large increase (30 %) in the impact velocity threshold to achieve detonation of the formulation.

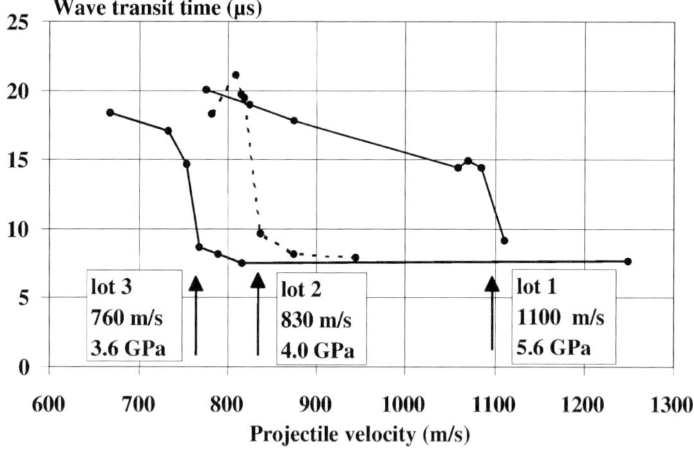

Figure 13.20. Projectile velocity threshold to achieve detonation of the formulation sample.

The cast formulations employed did not introduce any bias in the results. A comparative study between the previous formulations and an industrial PBX formulation with a higher solid load (78%) gave results with the same order of magnitude [9.9].

These last experimental data show that the magnitude of the effects of the defects of the explosive particle on the shock sensitivity of the cast formulations can be important and interesting for practical applications. These very accurate experimental correlations underline the important role of the crystal defects in the sensitivity of cast formulations.

It is difficult to have fine quantitative control of the various explosive crystal features (size, surface properties and internal defects).The main interest of the previous experimental data is to provide the simplest systems and data to relate accurately a specific crystal feature to shock sensitivity. Other studies confirm the global effects of the explosive crystal features on the sensitivity of cast formulations [3.37, 9.4, 9.5]. Efforts are now in progress to improve these experimental data:

- in using other tools to quantify accurately the microstructure of the explosive particles
- in trying to manufacture defect-free explosive particles and to record the performances of the resulting energetic materials.

13.3 A New, Small-Scale Test for Characterizing Explosive Performance

13.3.1 Motivation for Developing a New Test

It is expensive and time consuming to scale up the processing of a new explosive molecule or formulation to produce a sufficient quantity of material to characterize the detonation performance. The calculated performance does not always correspond to the measured performance, so there is also a concern that the time and resources expended in conducting performance tests on a new material may not be justified by the outcome. We have developed a small-scale test that permits a preliminary characterization of the performance of a new explosive using only a few grams of material. We have performed the test on several explosives in common use. These include LX-10 (95% HMX/5% Viton A binder), LX-16 (96% PETN/4% FPC 461 binder) and LX-17 (92.5% TATB/7.5% Kel-F 800 binder). The test results agree well with calculations which use equations of state that have been measured in cylinder tests [13.93]. We can also directly compare a new material with the test results on these well-known explosives. We have used the new test to characterize an explosive that has recently been synthesized, 2,6-diamino-3,5-dinitropyrazine-1-oxide (LLM-105), an insensitive energetic material with 25% greater power than TATB [13.94]. The energy content and thermal stability of this material make it very interesting for several applications, including insensitive boosters and detonators. Using this new test, we were able to measure the detonation pressure and detonation velocity using < 1 g of material.

13.3.2 Description of Test Fixture and Procedure

A schematic diagram of the test fixture is shown in Fig. 13.21. The test pellet is initiated by an initiation train consisting of an exploding foil initiator (EFI), which initiates a 6.35-mm diameter, 2-mm thick LX-16 pellet pressed to a density of 1.7 g/cm^3. The detonation of the LX-16 drives a 5.0-mm diameter, 0.127-mm thick aluminum flyer plate across a 1-mm gap to impact a 6.35-mm diameter, 5-mm thick test pellet. This initiation train provides a sufficient stimulus to initiate promptly ultra-fine TATB at a density of 1.8 g/cm^3. For explosives less sensitive than ultra-fine TATB, a booster pellet of LX-10 or other high-HMX formulation may be added. One must recognize, however, that for testing insensitive explosives there are critical diameter issues. The test is not appropriate for explosives with critical diameters much larger than the 6.35-mm diameter of the test pellet, although it may still be useful for screening purposes.

The test fixture is steel, consisting of reusable end plates that are bolted together to capture a stack of steel discs. Starting from the bottom of the stack, the first disc protects the bottom end plate and has a slot to contain the EFI and its flat cable. The next disc has a hole in the center, which contains a 6.35-mm diameter, 2.03-mm thick pellet of LX-16 pressed to a density of 1.7 g/cm^3. A 0.127-mm thick aluminum foil is then placed over the LX-16 pellet and is held in place by the next disc, which is 1 mm thick and has a 5.08-mm diameter hole in the center. Next, another containment disc holds the 6.35-mm diameter test pellet. This disc is carefully sized so that the test pellet protrudes ~25 μm above the surface of the disc.

A piece of 13-μm thick aluminum foil is cut to shape and flattened on a piece of glass with a steel roller. A tiny drop of mineral oil is placed on an LiF crystal and the

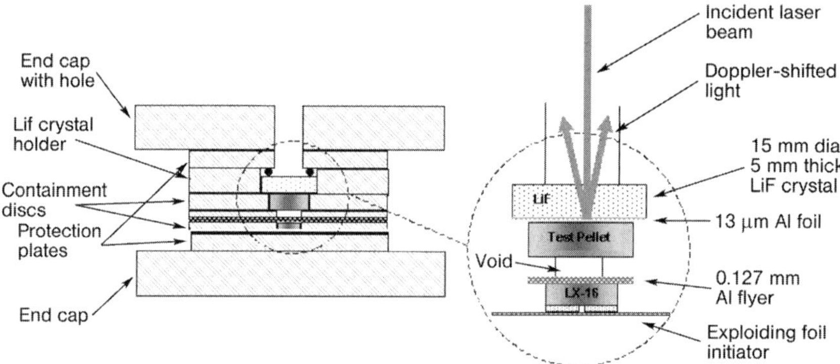

Figure 13.21. Cross-section of test fixture.

Al foil is placed on the crystal and flattened by pressing it between a glass cover-slip and the crystal. The cover-slip is removed and the LiF crystal is placed in the next containment disc and pressed against the slightly protruding test pellet with the foil against the pellet. The LiF crystal is held tightly against the test pellet by an O-ring, which is compressed by a final disc which protects the top end plate.

The motion of the 13-μm thick aluminum foil, which forms the interface between the aluminum foil and the LiF crystal, is measured with a Fabry-Perot laser velocimeter [13.95], which can be used to measure velocity to within ~1%. In this instrument, a laser beam is focused on to the 13-μm thick aluminum. The Doppler shift of the reflected light is analyzed by a Fabry-Perot étalon and recorded with a streak camera. The pressure wave transmitted into the LiF may be determined from the foil velocity and the Hugoniot of the LiF. The laser velocimeter was also used to measure the velocity-time history of the 0.051-mm thick Kapton flyer from the EFI and the 0.127-mm aluminum flyer accelerated by the detonating LX-16 pellet. The fill time of the Fabry-Perot étalon is estimated to be ~10 ns, but the time resolution of the system is estimated to be ~1 ns.

13.3.3 The CALE Hydrodynamics Code

The axial symmetry of the test fixture makes the test results amenable to simulation using a 2-D hydrocode. We compared test results with calculations made using CALE, a 2-D ALE hydrocode [13.96].

13.3.4 Experimental Results and Comparison with Calculations

Figure 13.22 shows the timing schematic for the experiments. Times of events are referenced to the bridge foil burst in the EFI. A common fiducial signal allows us to relate electrical records to the film records from the laser velocimeter. The laser velocimeter records the velocity-time histories of the EFI slapper and the aluminum flyer plate. Flyer arrival times are determined from the time at which a jump in interface velocity is observed when the flyer is allowed to strike an aluminized glass surface placed at the impact position. Arrival time of the detonation wave at the end of the test pellet is taken as the time at which the 13-μm thick aluminum foil starts to move.

The detonation velocity is of interest in evaluating a new material. In our test, we can determine the transit time of the detonation through the test pellet from the difference between the impact time of the aluminum flyer on the test pellet and the detonation breakout time at the center of the pellet. We assume that the time from bridge foil burst to the impact of the aluminum flyer on the test pellet is constant but, of course, there is some jitter due to variations in the EFI function time and to variations in the density of the LX-16 pellet. We estimate this jitter to be ~± 0.02 μs. For an explosive with a detonation velocity of 8 km/s, this uncertainty leads to a relative error in velocity of about ± 3%. With a 1-μs long velocimeter record, we can

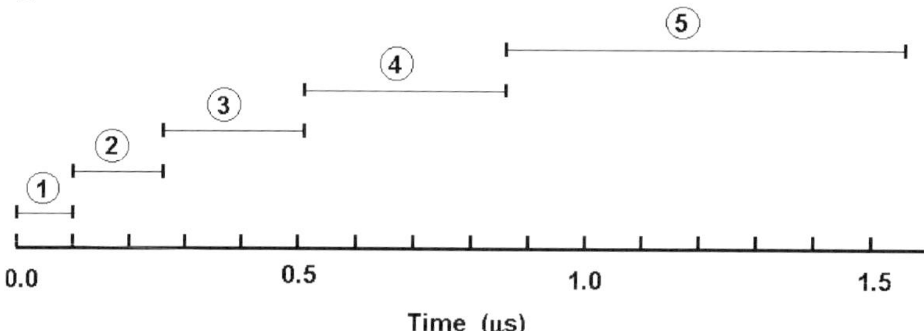

Figure 13.22. Timing schematic for the experiments.

easily read the time to within ~ ± 0.01 μs, so we estimate a total error of about ± 5% in determining the detonation velocity from the transit time through the test pellet.

Experimental results for the explosive LX-10 are shown in Fig. 13.23. The experimental interface velocity data, shown as gray lines, are from six different shots and the scatter in the curves shows the reproducibility of the measurement. The solid black line is a calculation using the CALE hydrocode [13.96] and a JWL equation of state [13.93], determined from cylinder tests. The agreement is good except near the peak at the shock arrival. The fact that the measured peak velocity is larger than the calculated peak velocity may be due to the Von Neumann spike, which has not been completely attenuated in its passage through the 0.13-mm Al foil. There is also more scatter in the experimental data near the shock jump, due to difficulties which will be discussed below.

Figure 13.24 shows an experimental streak record of fringes from the laser velocimeter record as the detonation wave collides with the interface. The interface velocity is determined from the spacing between the fringes and at the shock jump; the error is greater than the 1% error we estimate for the bulk of the record. This is because the fringe spacing decreases rapidly after the shock jump, which, coupled with the finite width of the fringes, makes the peak value very difficult to read. This difficulty can be overcome by streaking faster, so the slope of the fringes is reduced.

Figure 13.25 shows the six experimental records from Fig. 13.23 separated in time. The first two records were recorded at a sweep rate of 2 μs and the last four at 1 μs. There is a consistent trend for the faster sweeps to give higher peaks. At the time of the experiments we were limited to a 1-μs sweep because we did not have a comb generator that would generate time marks faster than every 100 ns. This meant that at sweeps faster than 1 μs, we could not generate an accurate time

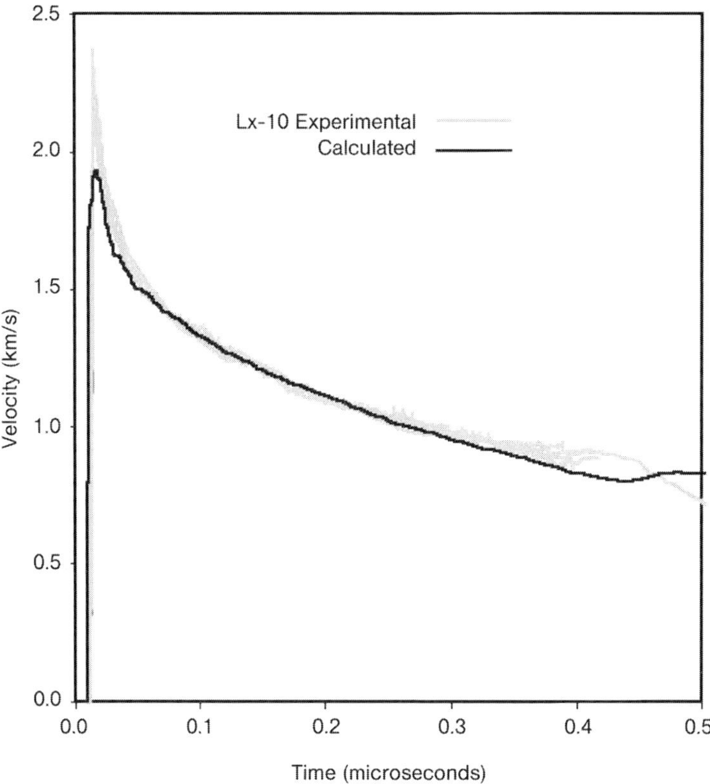

Figure 13.23. Velocity-time records for six different shots with LX-10 explosive, compared with a simulation using the CALE hydrocode [13.96].

fiducially to give us cross timing with the digitizer records of the electrical signals. In future work we will use a faster comb generator and, for some experiments, we plan to use a much thinner aluminum reflector in conjunction with a sub-

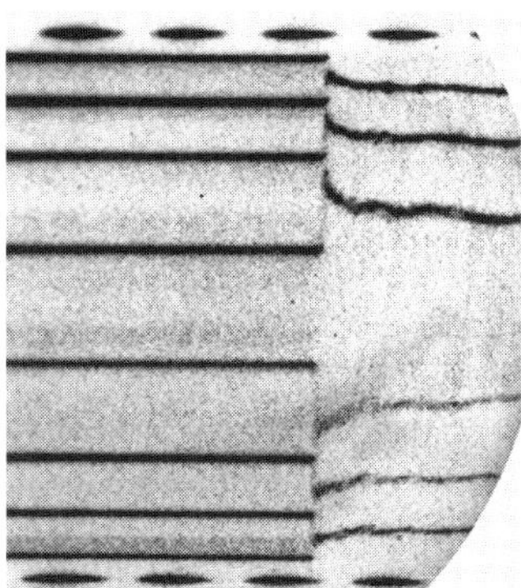

Figure 13.24. Streak record of velocimeter fringes. Interface velocity is determined from the spacing between the fringes. It is more difficult to determine the fringe spacing at the shock jump than for the rest of the record.

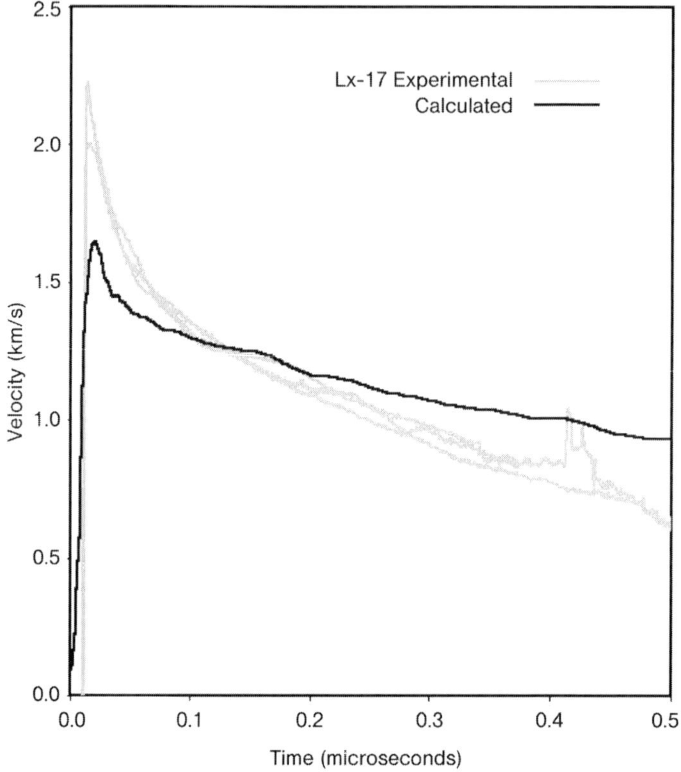

Figure 13.27. Three experimental velocity-time records for LX-17 explosive compared with a simulation using the CALE hydrocode [13.96].

13.3.5
Experimental Testing of LLM-105

LLM-105 (2,6-diamino-3,5-dinitropyrazine-1-oxide) was synthesized at LLNL as a possible insensitive explosive. The predicted power [13.94] is 125% of that of the extremely insensitive explosive TATB. The energy content, power and thermal stability of LLM-105 make it very promising for several applications, including insensitive boosters and detonators. Approximately 30 g of LLM-105 were produced for characterization by the new performance test. LLM-105 parts with adequate mechanical integrity could not be made by pressing neat material and, therefore, the pellets were formulated with 5 wt-% Viton A. The crystal morphology of LLM-105 was needle-like in these experiments. This morphology contributed to difficulties in obtaining high pressing densities. Only 92.4% of the theoretical maximum density was obtained.

Figure 13.28 shows the interface velocity record compared with a CALE hydrocode simulation for LLM-105 (neat) pressed to a density of 1.72 g/cm^3. The equation

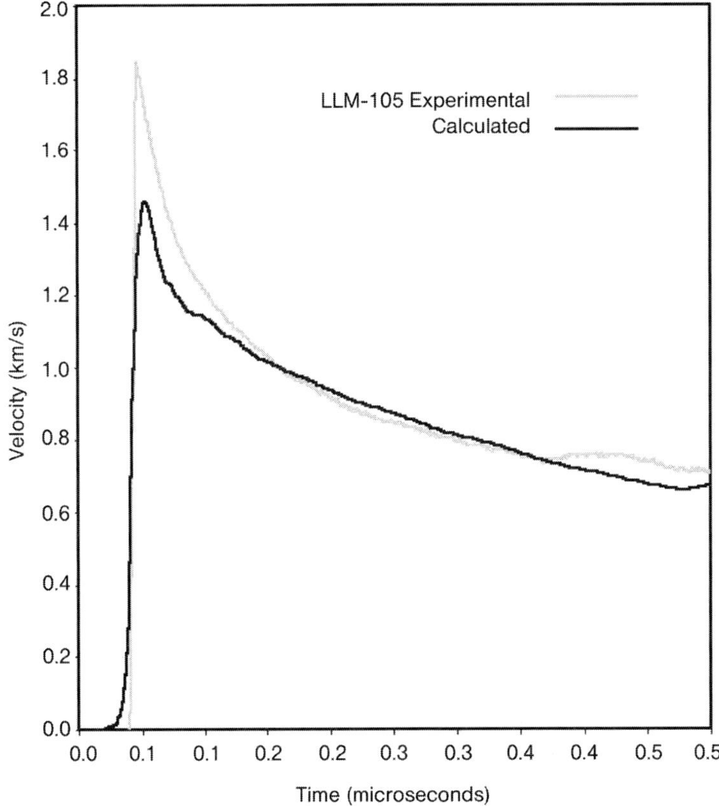

Figure 13.28. Experimental velocity-time record for LLM-105 explosive compared with a simulation using the CALE hydrocode [13.96].

of state used for the simulation was determined using the CHEETAH chemical equilibrium code [13.97] and the agreement is reasonable, indicating that expectations had been met.

13.4 Diagnostics of Shock Wave Processes in Energetic Materials

13.4.1 Introduction

With safety issues playing an important role in present-day energetic materials technology, concern is increasing about the relative safety of solid high explosives exposed to extreme conditions. Hazard scenarios can involve shock wave or dynamic loading of explosives. These scenarios can contain multiple stimuli such as explosives being heated to temperatures close to thermal explosion conditions followed by fragment impact, producing shock waves in hot explosives. In this section we present the approach used in our laboratory for experimentally exploring these conditions and with these data developing a computational model which allows the calculations of scenarios that would be difficult to explore experimentally. Only certain shock wave gauge techniques are presented since our intent is to review existing experimental techniques that we use in studies of shock loading of explosives. Selected references are cited for the reader to explore further these techniques and the data generated from them.

Experimental measurements of high-rate processes taking place in a shock wave dynamic environment require that the diagnostic systems have fast response and high resolution. This is not a trivial requirement because under shock loading one can expect not only sudden changes of state across the shock discontinuity but also subsequent changes in pressure, temperature and volume due to chemical reaction, phase change and other transformations, which may also take place in or behind the shock wave.

Among the various parameters that provide direct ties to theoretical studies of the equation of state and initiation of explosives and at the same time yield relatively accurate experimental measurements are shock velocity, particle velocity and pressure. Described here are various experimental techniques to measure these parameters. Although these techniques are not new, they have been continuously improved and regularly upgraded to yield greater accuracy, reliability and state-of-the-art performance. The emphasis in this section is on the operational features of the measuring techniques, but examples of experimental results are also included.

Most of our studies have one-dimensional shock waves that are unsteady, which require the solution of the conservation flow equations:

$$\left.\frac{\partial \rho}{\partial t}\right)_x + \left.\frac{\partial (\rho u)}{\partial x}\right)_t = 0 \qquad \text{Conservation of mass} \tag{13.42}$$

$$\rho \frac{\mathrm{d}u}{\mathrm{d}t} + \left.\frac{\partial P}{\partial x}\right)_t = 0 \qquad \text{Conservation of momentum} \tag{13.43}$$

$$\rho \frac{\mathrm{d}E}{\mathrm{d}t} + \left.P\frac{\partial u}{\partial x}\right)_t = 0 \qquad \text{Conservation of energy} \tag{13.44}$$

where ρ is density, u is flow velocity in the direction x, P is stress and E is internal energy.

Most experiments typically measure only two parameters such as pressure and shock velocity or particle velocity and shock velocity. The solution of these differential equations requires more information such as the equations of state for unreacted explosive and for the reacted explosive, which is usually in the JWL (Jones-Wilkins-Lee [13.93]) form and a reaction rate law. Experiments measure the shock wave equation of state for the unreacted explosive adiabat [13.98] (Hugoniot) and the reactive pressure profiles [13.99, 13.100] for calibrating the coefficients of an ignition and growth rate law. The reactive wave profiles also give the distance from the explosive impact surface to where it becomes a fully developed detonation for a specific input pressure. The reacted equation of state needs to be obtained by other experiments such as measuring the copper wall motion of a detonating explosive inside a cylindrical copper tube. The cylinder wall motion data are iteratively calculated with a hydrocode until a set of JWL constants are found which accurately reproduce the data. These constants vary over a limited range for all explosives. The JWL form is

$$P = A\exp(-R_1V) + B\exp(-R_2V) + \omega C_vT/V \tag{13.45}$$

where P is pressure in millibar, V is relative volume, T is temperature, ω is the Gruneisen coefficient, Cv is average constant volume heat capacity and A, B, R_1 and R_2 are constants.

The ignition and growth reactive flow of shock initiation and detonation of solid explosives has been incorporated into several hydrodynamic computer codes and used to solve many explosive and propellant safety and performance problems [13.101–13.107]. The model uses two JWL equations of state, one for the unreacted explosive and another for its reaction products, in the temperature-dependent form. The reaction rate law for the conversion of explosive to products is

$$\underset{(0<F<F_{\max})}{\mathrm{d}F/\mathrm{d}t = I(1-F)^b(\rho/\rho_0-1-a)^x} + \underset{(0<F<FG_{1\max})}{G_1(1-F)^cF^dP^y} + \underset{(FG_{2\min}<F<1)}{G_2(1-F)^eF^gP^z} \tag{13.46}$$

where F is the fraction reacted, t is time, ρ is the current density, ρ_0 is the initial density and I, G_1, G_2, a, b, c, d, e, g, x, y and z are constants. This three-term rate law models the three stages of reaction generally observed in shock initiation of heterogeneous solid explosives. The first term represents the ignition of the explosive as it is compressed by a shock wave creating heated areas (hotspots) as the voids in the material collapse. Generally, the amount of explosive ignited by a strong shock wave is approximately equal to the original void volume [13.101]. The second term in Eq. (13.46) represents the growth of reaction from the hotspots into the remaining solid. During shock initiation, this term models the relatively slow spread of reaction in a deflagration-type process of inward and/or outward grain burning. The exponents on the $(1-F)$ factors in the first two terms of Eq. (13.46) are generally set equal to 2/3 to represent the surface-to-volume ratio for spherical particles. The third term in Eq. (13.46) describes the rapid transition to detonation observed when the growing hotspots begin to coalesce and transfer large amounts of heat to the remaining unreacted explosive particles, causing them to react very rapidly.

13.4.2
Test Apparatus

The dynamic loading of material can be accomplished in many different ways. However, in order to describe properly the behavior of the material in such an unusual environment, one must have an accurate account of that environment and have good control of loading conditions such as the one-dimensional shock loading of gun experiments. One of the ways to control the shock strength or the characteristic features of impact loading is by the use of a gas or a powder breach gun to

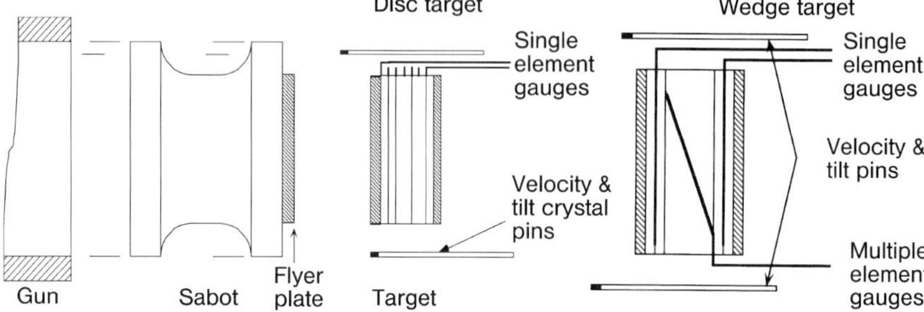

Figure 13.29. Schematic of one-dimensional strain gun experiment.

accelerate a projectile of known material with a known mass to provide a flat impact with the target. The gun in our facility at the Lawrence Livermore National Laboratory has a bore of 101 mm and is capable of reaching projectile velocities of 0.2–2.7 mm/μs. For explosive materials one can produce impact pressures in excess of 30 GPa. A schematic representation of such a gun experiment is shown in Fig. 13.29, where the projectile with the proper flyer plate is shown just after leaving the gun barrel but before impacting the target assembly.

The target can be of any form to accommodate any type of gauges and/or pins, which are best suited for the measurement of the desired parameter. Usually it consists of several layers of the investigated sample with gauges embedded between them. It can also consist of two wedges with a multiple element gauge between them. Each gauge element is then at a different depth in the flow and this depth can easily be controlled by the separation of the elements and by the angle of placement. This eliminates the multiple sources for flow disturbances and also enhances our measuring capability by giving us unlimited control of depth in positioning the gauge. At the same time, this technique increases the number of measurements in a single experiment. Included in the target are several crystal pins of different length, which are placed around the periphery of the target to measure the velocity of the impacting flyer plate and six flush-mounted crystal pins whose sole purpose is to measure the amount of projectile tilt relative to the sample surface during the impact.

13.4.3 Electromagnetic Particle Velocity Gauge

The electromagnetic particle velocity gauge is not new but, owing to the marginal quality of early results, remained dormant for a long time and had very limited application. Originally, Zavoiski developed the gauge in 1948 according to Dremin and co-workers at the Institute of Chemical Physics in Moscow. Many years later, Dremin *et al.* [13.108] reactivated this technique, which then found its way to other laboratories in the USA. Edward *et al.* [13.109] and Hayes and Fritz [13.110] further developed the system and improved its reproducibility and accuracy.

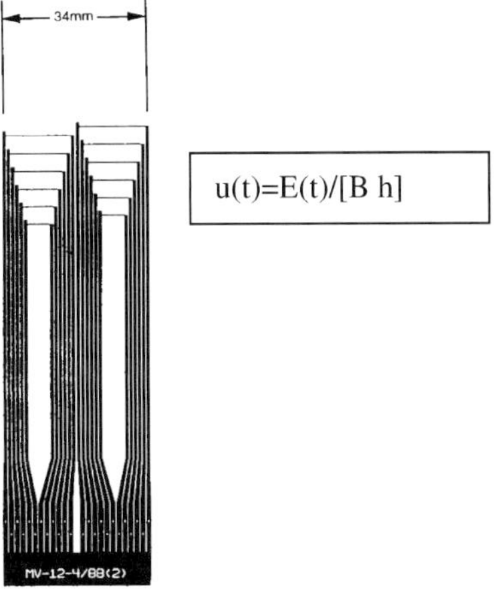

Figure 13.30. Single and multiple particle velocity emv gauges.

The thin foil particle velocity gauge was brought close to perfection and is now being used regularly at our laboratory [13.111] and at Los Alamos [13.100] as a standard diagnostic feature in one-dimensional strain gun impact experiments on explosive initiation and characterization. Note that one limitation of using this gauge is that no metal can be present in the target or impactor. This is because metal moving in the magnetic field will create electrical noise and distortions in the records.

At present, the gauge type shown in Fig. 13.30 is our preferred, one-dimensional, thin foil, hydrodynamic, particle-velocity sensor. It may be made of copper or aluminum, depending on the application. Note that copper, although not as good in terms of shock impedance match, survives better than aluminum in the severe environment of the detonation products. Also shown is the simple expression that was used to reduce the data. Here $E(t)$ is the electrical signal in V, B is the magnetic field in kG, h is the length of the gauge element in cm and $u(t)$ is the resulting particle velocity in mm/μs.

Particle velocity history data are very useful in validating theoretical models. Multiple gauge experiments reveal the information on the build-up to detonation and information on the release wave behind the detonation front. Particle velocity histories in LX-14 (HMX/Estane; 95.5/4.5 wt%) for a typical multiple gauge experiment is shown in Fig. 13.31. These profiles show the build-up to detonation as the reactive shock propagates through the LX-14 sample.

13.4.4 Manganin Pressure Transducer

Manganin was first used as a pressure transducer in a hydrostatic apparatus by Bridgeman [13.112]. More than a decade later manganin was used as a dynamic

additional voltage ΔV. The current in the gauge is also measured and the fractional resistance change is related to measured values of V_0, ΔV, I and R_0 as

$$\frac{\Delta R}{R_0} = \frac{V_0 + \Delta V}{IR_0} - 1 \tag{13.47}$$

where R_0 is the initial resistance of the gauge measured to ± 0.01% accuracy, V_0 is the ambient voltage, I is the current through the gauge and ΔV is the change in voltage due to the pressure.

The conversion to pressure from the voltage records is not straightforward. The manganin is permanently changed by the shock by creating defects during plastic deformation [13.118, 13.121–13.124]. Gupta [13.121] derived a model for the plastic flow causing the hysteresis in the gauge resistance that requires experimental determination of the coefficients in order to correct for the hysteresis. In addition, if the test material has a large shear strength then the stress in the gauge and stress in the test material do not correspond directly [13.125].

For explosives, the material strength is very small, and therefore longitudinal stress in the gauge corresponds closely to that in the explosive. This allowed Vantine *et al.* [13.124] to use an empirical approach for a reasonable correction of gauge hysteresis. They assumed that the irreversible changes in the gauge all occur in the first shock jump. The hysteresis correction was done by assuming that the irreversible change in relative resistance of the gauge can be represented as a second-order polynomial of the relative resistance change for the initial peak shock pressure. With this correction to the resistance, the steady value of pressure reached in double wave experiments (double shock or shock and release to a steady pressure) was used to fit the coefficients of a third-order polynomial of resistance change corrected for the hysteresis. This was done iteratively by guessing the coefficients of the irreversible resistance change and then fitting the third-order polynomial. This iterative process was repeated until the entire data set fitted the polynomial to within 2% standard error. This correction for hysteresis is not based on solid theoretical grounds such as Gupta's [13.121, 13.122] but pressures are determined with this approach to within ± 5%. This polynomial is used by us to determine the pressure for the measured change in relative resistance.

When placed in an explosive target such as PAX-2A (HMX/BDNPA-F/CAB; 85:79:6 wt%), the multiple manganin gauge is very valuable for measuring pressure along a Lagrangian space coordinate as the pressure builds up and at some point turns over into the detonation wave. Such a case is illustrated in Fig. 13.33, where the turnover happened within the 4.14 mm depth between the gauge elements 4 and 6 of the six-element gauge package. Note that gauge element 5 is missing. In addition to having an accurate account of the pressure build-up in the pre-detonation regime, this gauge also provides accurate information on the wave run distance to detonation.

To study explosives that are pre-heated and confined, one uses the configuration [13.99, 13.126] shown in Fig. 13.34. The profiles are used to calibrate the coefficients of the ignition and growth model for these pre-existing conditions. This new set of parameters can be used to calculate the shock sensitivity of the explosive

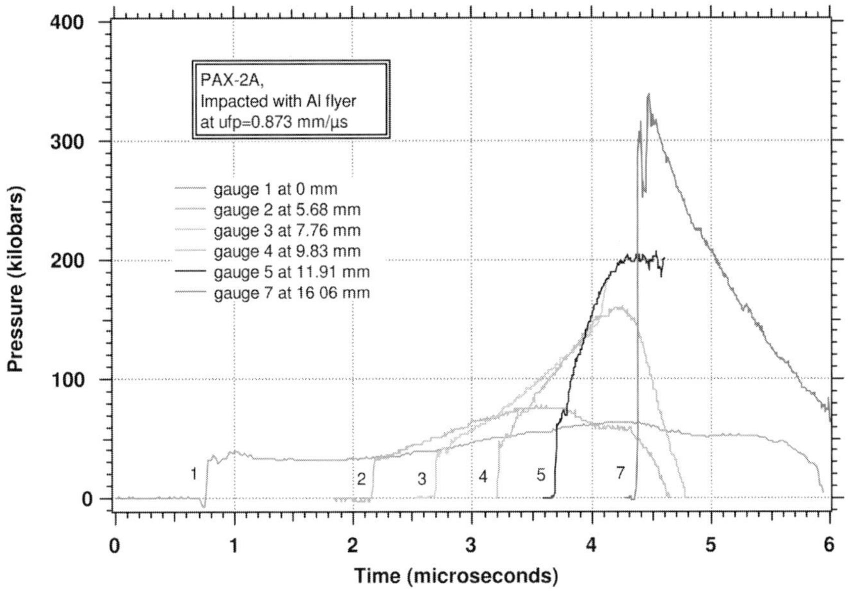

Figure 13.33. Reactive pressure profiles in shocked PAX-2A.

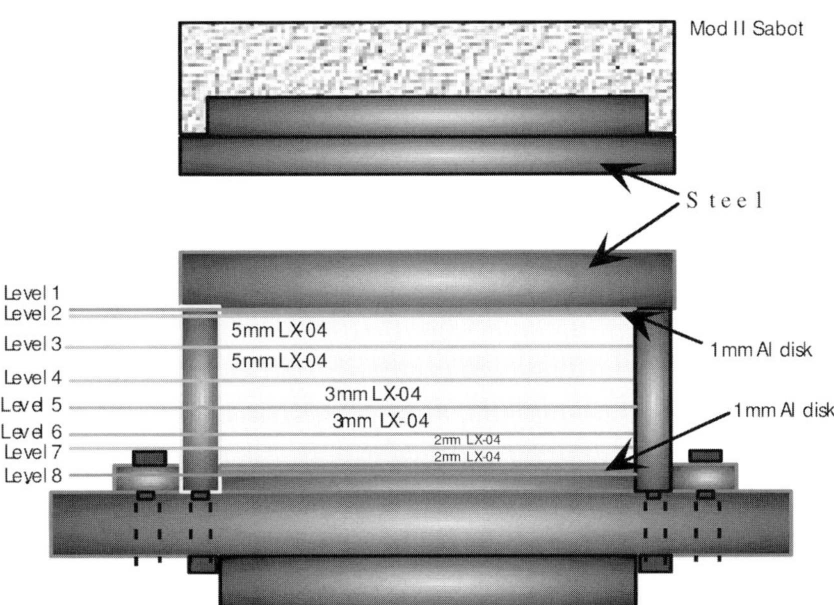

Figure 13.34. Geometry of the heated and 304 stainless-steel confined manganin gauge experiments.

for scenarios where pre-heating and confinement exist. The six manganin gauges were at various depths between the right cylindrical discs of the explosive for recording the pressure profile histories. This configuration gives an accurate location of the gauge elements with regard to the discs of explosives. Slanted gauges are not used here because they are known occasionally to slip while adhesives dry. Also, the linear thermal expansion of the explosives is not always accurately known. The front steel plate was fastened to the rear steel plate with several steel bolts. Each disc of explosive was radial contained by close fitting 10-mm thick steel rings of the same height as the explosive discs.

The gauge packages typically contain both thermocouples and manganin pressure gauges between two foils of 0.13-mm thick Teflon armor. The leads for the gauges and thermocouples were brought out of the sides between the steel rings. The flat spiral ribbon was used as the heater and it was placed on both sides of the target assembly between the steel plate and the aluminum plate (levels 1 and 8). The aluminum plate served the purpose of facilitating a faster and more uniform distribution of heat into the explosive from both sides of the assembly. PZT pins were placed flush with the face of the target to measure tilt and also nominally 15 mm in front of target face to measure projectile velocity. Flash X-rays are also used to give a second measure of projectile velocity. The explosive samples typically are heated at a rate of 1.5 °C/min. The gun was fired when the temperatures between layers of the target were within ± 5 °C of the desired value. Figure 13.35 gives the experimental and calculated wave profiles using reactive flow modeling with the ignition and growth reaction rate model for heated LX-04 (HMX/Viton A; 85:15 wt%). Each gauge record is labeled with its depth from the initial LX-04 impact surface. Ambient confined experiments were done using a similar gauge configuration but without the confinement, heaters and thermocouples.

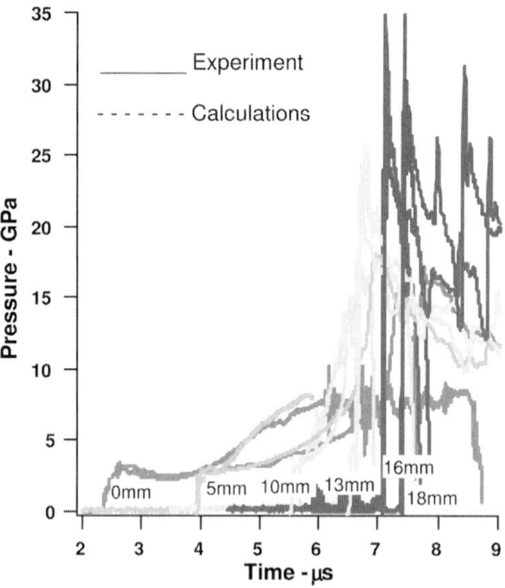

Figure 13.35. Pressure histories for steel confined 170 °C LX-04-01 impacted by a steel flyer at 0.535 km/s.

13.4.5
Other Gauges

A variety of gauges have been used in various experiments on shock loading of explosives [13.127–13.135]. Table 13.2 summarizes many of these and gives nominal values for the gauge characteristics. All the pressure coefficients for these gauges have some sensitivity to temperature. For the foil gauges the lower time resolution was determined by assuming a 25-μm thick foil and the upper number assumed the foil gauge package to have insulation of 50-μm layers on both sides of it (i. e. a total package thickness of 125 μm). The resistor gauge was assumed to have a 12.5-μm glue layer on both sides of it. To reach equilibrium it was assumed that the principal wave and its reflections transited the gauge element five times (roughly 4.5 times the package thickness) at a nominal velocity of 5 km/s.

Table 13.2. Selected additional gauges used in shock wave experiments on explosives in our laboratory.

Gauge name	Generally used for	Nominal range of applicability	Special features	Selected references
Ytterbium foil	Low-pressure 1-D experiments	0–20 kb; $P \pm$ 5–10%; typically 25–115 ns temporal resolution	High piezo-resistance coefficient giving high sensitivity; large hysteresis on pressure release	13.121
Carbon foil	Low-pressure 1-D experiments	0–50 kb; $P \pm$ 8–15%; typically 25–115 ns temporal resolution	Good for use in radiation environment; peak pressure measurement; large hysteresis on pressure release	13.127, 13.128
Carbon resistor	Low-pressure 1-D and 2-D experiments	0–50 kb; $P \pm$ 8–15%; typically 1.4–1.5 μs temporal resolution	Very durable and can last for milliseconds; used in difficult environments such as large grain materials; good peak pressure measurement; large hysteresis on pressure release	13.129–13.131
PVDF foil	Medium-pressure 1-D experiments	0–100 kb (some claim up to 300 kb); P ± 5–10%; typically 25–115 ns temporal resolution	Very fast response piezo-electric gauge, which requires no external power supply. Current mode recording gives pressure derivatives directly; very sensitive to lateral strain	13.132–13.135

13.5
Diagnostics for the Combustion of Particle-Containing Solid Fuels with Regard to Ramjet Relevant Conditions

13.5.1
Introduction

Reacting multiphase flows occur in many practical applications such as in power plants, propulsion systems, welding processes, production processes for carbon black, powders and metal suspensions and so forth. Especially the combustion of dispersed fine particles in an oxidizing atmosphere or other environments can mainly be found, for example, in pulverized coal combustion, pyrotechnic devices and propulsion systems for space and aerospace applications. In this connection, aluminum or boron particles are attractive as additives to the commonly used hydrocarbon binders [13.136–13.139] for propulsion systems with solid fuels or propellants such as solid rockets, solid rocket boosters, solid fuel ramjets, ducted rockets and hybrid rockets. Because of their high combustion energy per unit volume in comparison with the hydrocarbon binders, they offer the possibility of building more compact vehicles or vehicles with longer flight duration. Not only for solid fuels but also for gel and slurry fuels these particles are attractive as additives for the same reasons as mentioned earlier. For future applications on Mars-like sample return missions the combustion of Mg particles with CO_2 seems to be promising, because the Martian atmosphere consists mainly of CO_2 and therewith less fuel and oxidizer masses need to be carried with from Earth [13.140]. In addition to these and other desirable combustion processes not mentioned here, the knowledge of flame propagation in dusty media and the related ignition conditions are also important for explosion risk assessment in mills, mines, storage devices and so forth and the design of necessary safety devices [13.141].

Up to now, the level of understanding of combustion processes of dispersed energetic particles in a gaseous carrier phase is in many areas relatively low in comparison with pure gas-phase combustion processes. Dispersed systems are micro-heterogeneous in character and for the building of reliable models for the characterization of the combustion process both macroscopic and microscopic effects have to be taken into account and consequently have to be investigated. It is well known that the combustion process is an interplay of reaction, heat diffusion and radiation [13.142], whereas radiation in dense dispersed systems is much more important than in gaseous flames and is difficult to handle in numerical calculations.

Furthermore, many organic particles such as coal and starch produce volatiles, so that not pure solid combustion processes occur, but a combination of a gaseous flame (combustion of volatiles) and heterogeneous combustion processes at the surface of the particles. Here, the rate of flame propagation is controlled by the rate of streamwise molecular diffusion of oxygen and volatiles, together with heat conduction from hot gas to the particles [13.143]. It must be noted in this context that the heating rate influences the amount and the composition of the volatiles, as has been shown for the pyrolysis of coal particles in a shock tube in comparison with that in other devices [13.144].

In the case of propulsion-relevant materials such as aluminum or boron, liquid oxide layers or caps may occur on the surface of the particles, which hinder the further combustion of these particles. This effect led, for example, for Al particles to slag formation and deposition of up to 2 tons inside the ARIANE 5 boosters during their development phase [13.145]. Symmetry-breaking effects such as oxide cap formation on aluminum particles or oxide layer rupture and removal are processes of 'metal' combustion, which must be understood better in order to obtain high-efficiency engines [13.146]. In this connection it must be noted that 'metal' indicates here a group of elements which are important for propulsion applications and which have a strong tendency for the oxides towards the condensed phase. They include not only the species called metals in chemistry but also other related species such as boron and silicon.

In order to obtain a better understanding of these and further effects not mentioned here and their complex interplay in the combustion of dispersed systems, well-defined experiments and efficient diagnostic tools are necessary. Basic investigations are important for the building of well-grounded models and their evaluation. The different types of experimental set-ups such as flat flame and Bunsen-type burners, strand burners, basic experimental set-ups for the investigation of single particles, particle arrays or particle clouds and so forth have different aims of investigation. Hence the conditions for the use of diagnostic tools are different but access to the object of investigation can be realized in most cases relatively simply. In practical combustion systems, however, the processes that occur are more complex and further difficulties for the use of diagnostic tools exist caused by the hostile conditions in the test facilities and the test cells such as noise, vibration and pollution.

The literature on investigations of non-reacting multiphase flows, reacting single and multiphase phase flows, combustion of energetic particles and materials and also the use of diagnostic techniques for these tasks is very large. The books by Eckbreth [13.147], Taylor [13.148], Boggs and Zinn [13.149], Kuo and Parr [13.150] and Hetsroni [13.151] are cited here as examples of detailed descriptions of fundamentals of diagnostic techniques and applications to experimental set-ups. The determination of qualitative and especially quantitative results in real propulsion systems or model combustors, however, is often difficult because of internal problem areas such as optically dense conditions, soot, agglomerates, broad particle size distributions and condensed oxides, in addition to outer environmental conditions in the test cell such as high sound levels and long optical pathlengths. To the authors' knowledge, only a few publications so far exist on the use of (laser-)optical diagnostic tools for the investigation of combustor processes in solid fuel ramjets, etc., and can mainly be found in some of the above-cited books.

The aim of this section is to discuss how some (laser-)diagnostic tools and developments of some 'older' diagnostic techniques have been used for investigations under ramjet-relevant conditions at the DLR Space Propulsion Institute at Lampoldshausen. Special attention is paid to the accuracy of measurements and to the limits of these techniques. It should be mentioned that in addition to physical limitations of measuring techniques, limitations are also often given because parts or components of a measuring system cannot be realized with the actual state of the art of the technologies used. Maybe in future years these techniques may nevertheless be successfully used if there is the necessary progress in the relevant technology areas.

Additionally, a simple classification and a very short survey of diagnostic techniques, which had been used for investigations of reacting multiphase flows in the past by various investigators, will be given here, especially for newcomers to this field. A detailed description of the state of the art of the whole field would be beyond the scope of this survey.

13.5.2
Classification of Diagnostic Techniques and General Considerations

The enormous number of papers dealing with flow and combustion measurements has often been systematized by various classification schemes. Hewitt [13.152], for example, listed different forms for the classification of measurement tools for two-phase flows. This presentation serves as a basis for the classification of reacting multiphase flows in Table 13.3. From all these possible schemes, the classification by the type of measuring methods has been chosen for the presentation in Tables 13.4 and 13.5 and the following sections of this chapter, because common or similar basic principles of the measuring tools, advantages, problem areas and so forth are easier to present and to discuss. It must be mentioned in this context that the survey presented here does not cover all existing and possibly useful techniques, because the main emphasis is put on reacting multiphase flows with special regard to propulsion-relevant conditions. Furthermore, it must be remarked that some of the

Table 13.3. Classification scheme for measurement tools.

Classification by	Examples
Parameters, which will be measured	Pressure; density; temperature; velocity
'Type' of measuring methods that are used	Optical methods; electrical methods; tracer methods
Extent of interference with flow field and combustion process	Intrusive; non-intrusive
Conditions of flow and combustion field	Steady; non-steady; pulsating; premixed; non-premixed; compressible; incompressible; 1-D; 2-D; 3-D; single phase; two phase; multiphase
Localization of measurement	Local; overall, average; integrated on a line of sight; 2-D-plane (e.g. in a light sheet)
Temporal resolution	Fast response; temporal average
Temporal response	On-line analysis; off-line analysis; diagnostics of residues; diagnostic tools are separated from experiment by long tubes, etc.

Table 13.4. Classification by types of methods.

Types of methods	Examples
Optical methods	
• Methods based on transmitting radiation from – conventional sources – lasers	Shadowgraphy; (Color) Schlieren; interferometry Laser-Schlieren; holographic methods
• Scattering of incident radiation (from lasers)	LDV, PIV, droplet or particle tracking; holographic methods, CARS
• Laser-induced emission techniques (LIE)	LI(P)F
• Observation of self-emissions	Emission spectroscopy; pyrometry; planar imaging (photography; video imaging; high-speed photography, etc.); streak techniques
Flow separation methods	Suction probes; sampling probes
Tracer methods	LDV, PIV
Heat and mass transfer methods	Heat flux gauges
Methods using the Seebeck effect (thermocouple)	Wall temperature measurements; intrusive probes with thermocouples
Force, momentum and stress measurement	Wall shear stress; thrust
Pressure-measuring methods	Pneumatic probes: Pitot tubes, Prandtl tubes, 3- and 5-hole probes; wall pressure measurements

techniques listed in Table 13.4 could belong to more than one of these types of methods (for example, LDV: tracer methods and scattering).

Optical methods can generally be divided into methods that use external radiation sources and methods where the self-emissions of the combustion process are observed. Different types of radiation sources are available such as lasers, conventional UV, visible or IR light sources, microwave sources, sources for X-rays, γ-rays, β-rays or neutron rays. The last kinds of sources for high-energy radiation (X-rays, etc.) have often been used in the past for investigations of two-phase flows, propellant combustion, flows in guns and so forth. Further information can be found elsewhere [13.152–13.154] and in the contributions in the book on non-intrusive combustion diagnostics of Kuo and Parr [13.150]. In recent decades, considerable advances have been made in non-intrusive spectroscopic and laser-based diagnostic techniques. Especially the invention of the laser was fundamental to the development of novel non-intrusive diagnostic tools such as LDV, PIV, CARS, LI(P)F, LII and many others. Together with on-line data acquisition and analysis by fast computers, optical methods are nowadays commonly used techniques in research and partly even in development departments in industry. From the large variety of novel diagnostics only a few are listed in Table 13.4 as examples, whereas the laser-based techniques mainly can be summarized as scattering techniques and laser-induced emission (LIE) techniques, whereas the incident laser light beam or sheet interacts with the molecules or particles of the flow. They will partly be described later in connection with Table 13.5 and Sects 13.5.6 and 13.5.7.

Flow separation methods are used to separate part of the (reacting) fluid flow for subsequent or on-line analysis. Valve-operated probe units are well known in the process industry for monitoring of production processes. Suction methods through porous walls are applied for the separation of boundary layers or liquid film flows. Intrusive suction probes are used for sampling of both gaseous and condensed-phase species. Further details on intrusive sampling probes are given in Sect. 13.5.3.2.

In tracer methods, a tracer is added to one of the phases of the flow in order to obtain information on mixing, velocity distributions and so forth either by sampling or detection downstream of the tracer injection or production point. For the determination of velocity distribution in multiphase flows by LDV or PIV, the addition of tracer particles is in many cases not necessary, because the existing droplets or particles are used as the measuring object. Here it should be mentioned that only very small particles are able to follow the carrier gas phase flow so that care has to be taken with the interpretation of the results obtained. Further information is given in Sect. 13.5.6. Radioactive tracers and other techniques such as the neutron activation technique [13.155] are not considered here.

'Classical' diagnostic tools such as pressure gauges, thrust measurement devices, heat flux gauges and thermocouples are commonly used for the determination of performance characteristics of solid rocket motors, ramjet engines and so forth and are often described in the literature and are therefore also not treated here.

It is obvious that the influence of every measuring tool on the flow and combustion field and therewith the parameters to be measured should be as low as possible. Therefore, care must be taken as how a tool can be used and consequently how the results obtained should be interpreted. For example, intrusive probes for velocity vector determination have to be calibrated in separate devices with a well-known and homogeneous flow field in order to obtain reliable results. Also, the so-called 'non-intrusive' (laser-)optical methods may influence the flow field by the high energy density inside the measuring volume caused by focusing of the laser beams. This could lead, for example, to unwanted ignition, which may change or stabilize the combustion and flow process inside the combustor. Furthermore, windows inside the combustor walls, which provide access for (laser-)optical tools, are often polluted by combustion residues so that these tools are not able to work correctly. In many cases, especially in flows with a high particle loading, window devices with additional protection flows [13.156] are used so that care has to be taken that the influence of this additional flow on the governing processes is negligible. Therewith non-intrusive techniques could be interpreted as techniques where no solid part of the measuring tool is in or near the measuring volume.

Optical techniques, where external light sources are used, can be further sub-classified in terms of the modulation of the radiation which is introduced in the investigation area. Table 13.5 presents a classification by the light modulation process, where various methods and also the parameters that can be measured with them are listed. Further, some publications are cited in the last column that give detailed information on the basic principles of these techniques and on applications. Also in Table 13.5 some of the listed methods belong to more than one group. LDV,

Table 13.5. Classification of optical techniques (in the IR, visible and UV ranges) by the light modulation processes.

Light modulation process	Method (examples)	Parameters to be measured	Literature (examples)
Absorption	Absorption spectroscopy	Temperature (gas phase); species concentration	13.159
Elastic scattering			
• by particles, Mie, $d/\lambda \approx 1$	Fraunhofer diffraction	Particle size	13.148 (Ch. 5)
	LDV	Velocity of (tracer) particles	13.148 (Ch. 3)
	PIV	Velocity of (tracer) particles	13.160
• by molecules, $d/\lambda << 1$	Rayleigh	Total density (gas phase); temperature (gas phase) in constant-pressure situations	
Extinction (= absorption + scattering)	DQ method	Average particle diameter; particle number density	13.158
Refraction	Shadowgraphy	Refractive index gradient $\left(\frac{\partial n}{\partial x}\right)$; $\frac{\partial \rho_g}{\partial x}$	13.161
	Schlieren photography	$\left(\frac{\partial^2 n}{\partial x^2}\right)$; $\frac{\partial^2 \rho_g}{\partial x^2}$	13.161
	LDV	Velocity of (tracer) particles	See above
Interference	Holography	Particle size distribution	13.162
	LDV	Velocity of (tracer) particles	See above
Excitation (inelastic stattering)	Coherent anti-Stokes Raman spectroscopy (CARS)	Temperature (gas phase); species concentration	13.147, 13.148 (Ch. 4)
	Laser-induced fluorescence (LIF)	Species concentration	13.148 (Ch. 6)
	Laser-induced incandescence (LII)	Soot particle concentration	13.163
Complex	Photography; video imaging		

for example, can be grouped in both interference and scattering owing to the fringe pattern generated by scattered light of particles passing through the crossing volume of the two coherent laser beams. The extinction of a light beam transmitting a particle-laden flow is caused by both absorption and scattering and is presented in this table as a separate division. Based on the work of Teorell [13.157], Wittig and Lester [13.158] used this modulation process for the determination of average diameters of soot particles by the determination of the dispersion quotient (DQ method).

Excitation is subdivided into techniques in which molecules are excited (e.g. CARS and LIF) and particles (e.g. LII). CARS will be described later in Sect. 13.5.7. LIF is a commonly used tool for investigations on gas-phase synthesis of particles. A review by Wooldridge [13.164] gives a detailed list of investigations with LIF and other diagnostic tools. LII allows one to measure soot concentrations in flames.

Incandescence occurs when particles are heated near to its vaporization temperature by a laser beam.

Tools which are useful in gaseous reacting single-phase flows are confronted with additional problems in reacting multiphase flows. Thermal radiation, absorption, Mie scattering and so forth of the particle phase cause often low signal-to-noise ratios. Rayleigh scattering, for example, is very sensitive to small particles in the flow field so that it can be used only in very clean conditions because typical derivative cross-sections are ca 10^{-27} cm^2/sr for Rayleigh scattering in comparison with 10^{-7}–10^{-13} cm^2/sr for Mie scattering. Additional difficulties exist for the investigation of the processes in combustors, where the conditions are closer to real engines. In the case of optically thick or semi-thick conditions due to high particle loadings, only a few of the successful optical combustion diagnostic methods for gaseous flows can be applied. In addition to the necessity for access to the combustor for (laser-)optical diagnostic tools, problems can arise, for example, because of high sound pressure levels, which may lead to vibrations of components of the laser and of the optical equipment. Consequently, for some of these diagnostics, the expensive and sensitive lasers and parts of the optical equipment should be placed in separate rooms or protected boxes beside the test object. This causes often long optical pathlengths, which may lead to other problems.

In addition to the use of 'classical' diagnostic tools such as pressure gauges, thrust measurement devices, mass flow-rate gauges, thermocouples, etc., other tools are necessary for a better understanding of the governing multiphase combustion processes inside the combustion chambers of airbreathing and non-airbreathing propulsion systems. In the following sections the experience with the use of some of these techniques, namely sampling probes, Color Schlieren and the laser-based techniques LDV, CARS and PIV, at the DLR Space Propulsion Institute at Lampoldshausen is reported.

13.5.3 Intrusive Probes

13.5.3.1 General Remarks

Intrusive probes, which are also called physical probes, have been commonly used measurement tools during recent decades for investigations in many areas of reacting and non-reacting flows. An enormous number of publications exist describing their application to the measurement and determination of various parameters such as velocity (Pitot probes, five-hole probes, hot-wire anemometers, etc.), temperature (thermocouples, etc.), heat flux and many other parameters especially in reacting and non-reacting gas-phase flows. In addition to their use in research, they are particularly also applied in development testing in industry owing to their mostly simple handling characteristics and low costs in comparison with laser-based techniques. Nowadays, non-intrusive laser-optical diagnostics are more in vogue, because with their help new insights into the processes that are occurring can be obtained. Unfortunately, these tools are often difficult to handle for the evaluation of reliable results in combustor test facilities, where high particle loadings exist, owing to the problem areas mentioned above. Especially in optically dense systems laser-based techniques and diagnostic tools with conventional light sources have severe problems due to absorption, multiple scattering and so forth, as has been mentioned earlier. Nevertheless, often intrusive probes offer a possibility to obtain information from the processes inside a combustor under these severe conditions. However, it is obvious that care has to be taken whether the probe influences the combustion process and therewith the measurement results by acting either as a heat sink due to the use of a cooling fluid or in contrast as a flame holder. The additional visual observation of the investigation area with the probe, if possible, is sometimes very helpful for monitoring the correct operating conditions during the measurement.

In particle-laden reacting systems, additional care must be taken regarding interactions of the particle phase with the probe, caused, for example, by thermal radiation, deposition of particles on the sensor area, blockage of entrance orifices, separation effects and so forth. The use of intrusive probes in flows with soot, where the particle diameters are often clearly beyond 1 μm, is nowadays very common in both research and industrial applications as in the exhaust duct of piston engines or in power plants. In reacting flows with larger particles, however, the above-mentioned problems are much more pronounced. The chemical and physical properties of the object to be measured, such as the kind of particles, particle size and particle size distribution, particle number density, velocity, tendency to form condensed oxides and so forth, strongly affect the design and use of probes.

Often individual solutions have been developed for special applications. For example for the investigation of fluidized bed combustion processes, Becker and co-workers [13.165, 13.166] have built different kinds of sampling probes, which will be discussed later. Furthermore, they used a micro-thermocouple probe for the determination of the gas-phase temperature and a zirconia oxygen probe for the determination of oxygen concentration. Chedaille and Braud [13.167] described the use of Pitot probes and other probes for velocity measurements in particle-laden flows. Further examples are described in the literature, for example, see Becker [13.168]. Optical (intrusive) probes, which are introduced into a flow field, use light modulation processes. For example, Danel and Delhaye [13.169] developed a fiber-optic sensor for the determination of void fractions in liquid/gas two-phase flows. The difference in the refractive indices of gases and liquids, which flow around the glass-fiber, causes different reflection behavior of light at the fiber wall, which is led through the fiber. Further kinds of optical probes have been discussed [13.154]. Smeets [13.170] developed a probe for the determination of particle or droplet velocities. In this probe, the optics for the light guidance of a Michelson spectrometer are integrated, as can be seen in Fig. 13.36. Coppalle and Joyeux [13.171] used a probe system consisting of two water-cooled probes with internal glass-fibers, the tips being oriented in-line with a 5-mm distance. The absorption of a laser beam between the two probe tips allows one to determine the soot volume fraction and the soot (surface) temperature. It must be mentioned that there could be some misunderstanding with the definition of optical probes, because sometimes some of the laser-optical tools are also called optical probes.

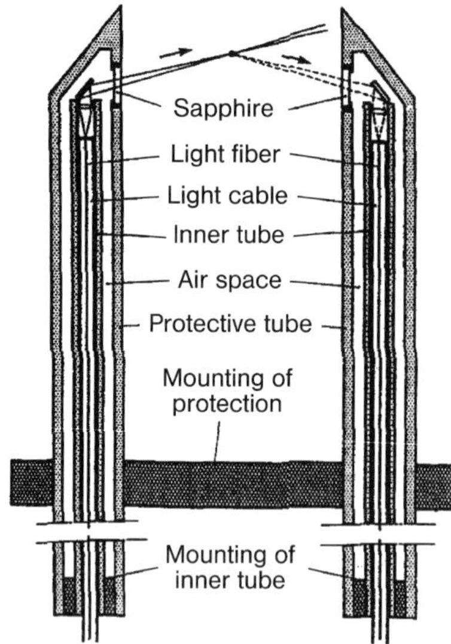

Figure 13.36. Probe of Smeets [13.170] with integrated light guidance for a Michelson spectrometer for measurements in spray detonations.

13.5.3.2 Sampling Probes

For the determination of concentration distributions of gaseous and solid species, many different types of probes have been developed in recent decades and have been presented and discussed in the literature, e.g. [13.167, 13.168, 13.172]. Gaseous species are typically sampled with intrusive probes by sucking gas at a definite point in the flow field through an orifice mainly at the probe tip and a following internal tube so that they can be transmitted to an on-line gas analyzer or a sampling bottle for subsequent analysis. Particles and condensed reaction products, however, have been sampled with different techniques. A large number of publications exist dealing with probe sampling of soot and particles in furnaces, fluidized bed combustion facilities and so forth. In addition to suction probes with internal or subsequent filtration devices, which are useful only for relatively low particle loadings, other solutions for special tasks and applications have been developed. Becker and co-workers, for example, reported a spring trap probe [13.166], which captures the particles of a definite volume from a fluidized bed combustor. They also developed different types of dust sampling probes, which are described in detail in the same report.

Not many publicly available publications, however, exist to the authors' knowledge about probe sampling in combustors, behind exit nozzles or in model combustors under conditions relevant for solid fuel ramjet, ducted rocket or solid rocket propulsion. For this task, for example, water-cooled rods [13.173] or plates [13.174] with cavities have been used for the sampling of condensed combustion products. However, the results, especially those obtained with flat plates, should be interpreted with caution, because on the one hand the sampled combustion products remain in the hot flow region and chemical reactions may be possible at the surface and in the inner region of the sample, and on the other hand smaller particles can more easily follow the flow field around these sampling devices so that larger particles from a particle phase with a broad size distribution may dominate the sample.

Several investigations in particle-laden combustion processes relevant for propulsion devices have been conducted with suction probes in the past. For example, Abbott *et al.* [13.175] used a water-cooled probe in particle-laden burning jets. Concentration profiles of gaseous species inside burning PE tubes, which simulate a solid fuel ramjet, were measured by Schulte and Pein [13.176]. Girata and McGregor [13.177] used water-cooled probes in high-altitude test cells for particle sampling in the exhaust of solid rocket motors. Sampling and analysis of residues of the fuel grain or (slag) deposition on the combustor walls after an experimental run, however, are more familiar but are not subject of this chapter.

An ideal sampling probe sucks a definite amount of the reacting particle-laden flow out of the combustor in such a way that the flow field and combustion process will not be disturbed and further reactions inside the extracted sample stream are terminated immediately. In reality, a distinct temperature decrease (for gas-phase flows) down to ~ 1000 K with cooling rates of about 1 K/μs should be realized, as noted by Becker [13.168] and Bilger [13.178], who summarized the experiences of other researchers. In order to achieve this goal, different ways to quench the sample stream are commonly used. They are presented in Table 13.6, partly making use of the presentation of Becker [13.168]. There the advantages and disadvantages of these three quenching methods in gas-phase flows are critically reviewed, giving attention to aerodynamics, heat transfer and cooling and so on, whereas special attention is given to radical removal, especially on the side walls.

In reacting multiphase flows, further difficulties have to be taken into account that influence the selection of a suitable quenching process and thereby also the design of useful sampling probes for application, for example, in model combustors under solid fuel ramjet conditions. This is discussed in detail below.

Table 13.6. Basic quench mechanisms commonly used in gas sampling probes.

Type	Mechanism	Example
Convective quench	Cooling of sample stream by heat transfer to coolant streams across the sampling passage wall	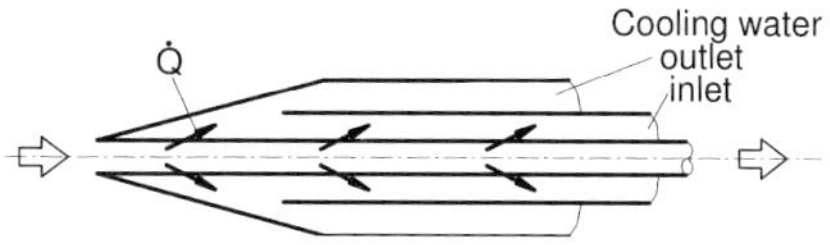
Expansion quench	Cooling of sample stream by a fast expansion	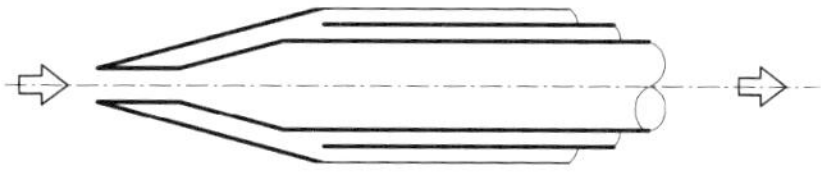
Mixing quench	Cooling and dilution of sample stream by mixing with a cold stream of inert gas	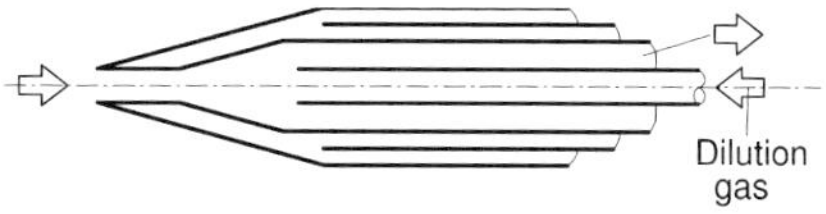

Theoretically, a sample should be taken from of a flow field without any disturbance of the flow field and the combustion process. In a non-reacting flow, minimum disturbance would be reached with a thin-walled probe oriented exactly in the upstream direction, whereas a sampling rate should be chosen for which the diameter of the extracted flow d_1 would be the same as the entrance diameter d_e of the probe, as can be seen in Fig. 13.37(a). In this case the entrance velocity u_e would be the same as the velocity u_1 in front of the probe. This behavior is called isokinetic sampling. In reality, the outer probe diameter is larger than the inner diameter d_e especially in reacting flows, where water cooling is often necessary to withstand the high temperatures and heat fluxes. Consequently, sampling procedures are often conducted in such a way that the Pitot pressure inside the probe mouth is the same as in the undisturbed flow field. This can be approached either with a sampling probe with an internal and an additional external Pitot tube in the flow field beside the sampling probe, as can be seen on Fig. 13.37(a), or with a Pitot probe, which is positioned at the same measuring position in a separate experiment without the sampling probe. More information about probes for isokinetic sampling in reacting and non-reacting flows can be found elsewhere [13.168, 13.178]. Furthermore, the outer diameter d_a of the probe and the tip angle should be as small as possible in order to decrease the influence of the probe on the flow and combustion field and thereby on the sampling procedure. This is shown, for example, by investigations of Lenze [13.179], who showed the influence of the shape of the probe head on concentration distributions in hydrocarbon flames.

In turbulent flows and flows with distinct vertical structures, care should be taken with how the sampling procedure is conducted and how the results obtained are interpreted. In theory, small vortices with diameters below the suction area diameter d_1 are completely sucked in. If the vortex diameters are significantly larger than d_1, the direction of the flow in comparison with the probe axis and also the concentration distributions, for example in mixing layers, change with time. Consequently, isokinetic sampling is nearly impossible in most cases. However, if the suction time

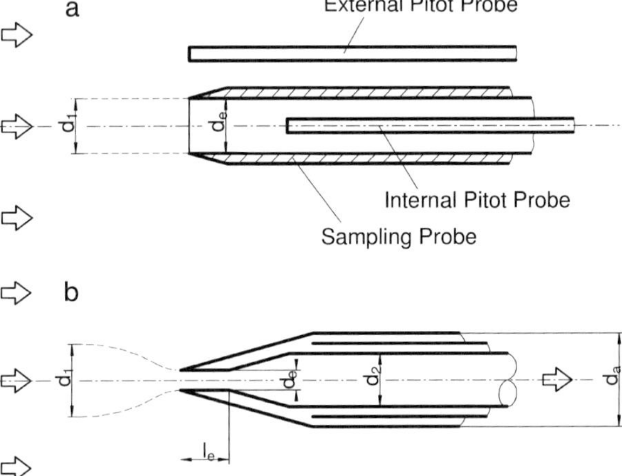

Figure 13.37. (a) Sampling probe for isokinetic operation; (b) sampling probe with expansion quench and cooling jacket.

is long in comparison with the moving time of the vortices and the demand for isokinetic sampling is neglected, the determined species distributions of the sample inside the sample bottle can be assumed to be average values at the measuring position. Further information about the influence of turbulent flow and instationary conditions on probe sampling is available [13.168].

In particle-laden flows, non-isokinetic sampling can lead to separation effects due to the inertia of particles to follow changes of direction and velocity of the gas-phase flow. The most important characteristic factor, which describes this behavior, is the particle Stokes number:

$$St_p = \frac{\tau_p}{\tau_g} \tag{13.48}$$

where τ_p is the particle relaxation time and τ_g the fluid time scale:

$$\tau_p = \frac{\rho_p d_p^{\,2}}{18\eta_g} \tag{13.49}$$

where τ_p is the time that a spherical particle initially at rest needs to acquire $1/e$ of the velocity of a suddenly applied air stream. Lieberman [13.180] defined τ_g as the ratio of the probe entrance diameter to the velocity u_1 in the flow tube of the extracted fluid far in front of the probe; u_1 can therefore be assumed to be the same velocity as in the outer undisturbed flow:

$$\tau_g = \frac{d_e}{u_1} \tag{13.50}$$

It is obvious that with decreasing entrance diameter d_e, increasing entrance velocities u_e occur if a constant suction rate and subsonic conditions in the probe mouth are assumed. This leads to increasing particle Stokes numbers and to increasing separation effects. Especially for broad particle size distributions this may lead to distinct losses and to possibly non-representative results. Figure 13.38, which is taken from Lieberman [13.180], shows the influence of the ratio of u_1, which he calls duct air speed, to u_e, which he calls sample tube speed, on the particle number loss. It can be seen that with increasing particle diameters these losses become more pronounced. For the expansion quench method with sonic conditions in the probe inlet, which seems to be useful for particle-laden reacting flows, as will be shown later, u_1/u_e will be of the order of 0.2 or below. This indicates that for this sampling method, particle diameters should be of the order of 1 μm or smaller. According to Lieberman [13.180], St_p should be < 10 so that errors due to anisokinetic sampling are negligible.

Additional care must be taken because particles tend to stick on the interior walls of the sampling passage. In order to reduce these losses, the length l_{ts} between the probe tip and the separation (or filter) element for the particle or condensed phase should be minimized. A not too small sample passage diameter should be chosen so that the ratio of the sampling passage wall area to the sample volume inside the sampling passage will be lowered. The cooling fluid, the whole sample passage and the separation unit should be heated in order to avoid water condensation from the

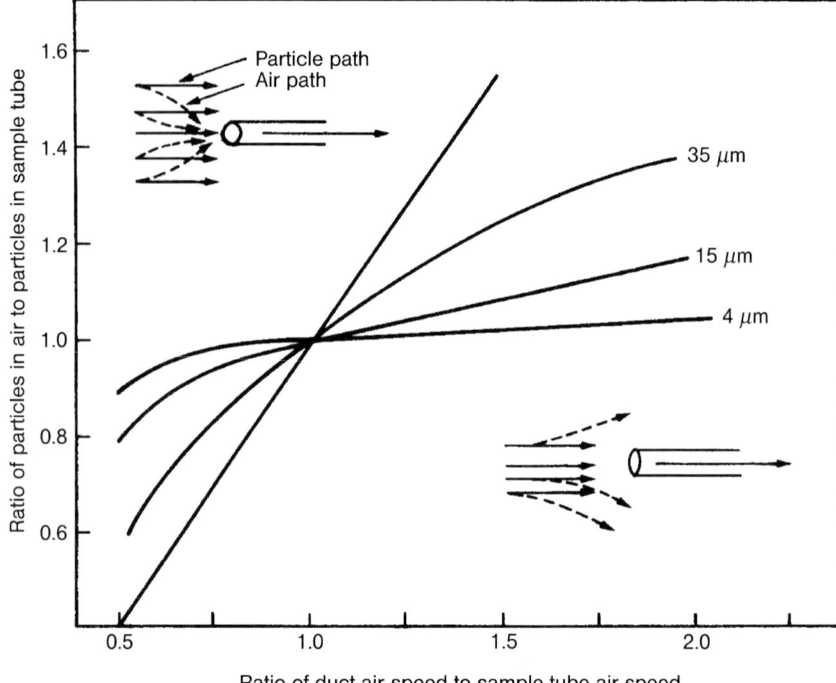

Figure 13.38. Anisokinetic sampling effects [13.180].

combustion products on the sampling passage walls, which can also increase the particle deposition. Furthermore, the above-proposed increase in the sampling passage cross-section needs longer pathlengths for the convective cooling procedure and consequently more time to achieve the temperature drop to the desired 1000 K.

In order to achieve a rapid and more homogeneous temperature drop over the whole probe passage cross-section (if possible immediately behind the probe mouth), the expansion quench method, which is also called aerodynamic quench, is more effective. Additionally, the expansion leads to a decrease in the particle number density so that agglomeration effects can be reduced. On the other hand, radical removal is more difficult owing to a reduced number of side wall collisions in comparison with the convective quench method. Additionally, the suction process leads to an acceleration of the sample flow upstream of the probe mouth, so that separation effects have to be taken into account, as mentioned above.

Measurement accuracy

Care has to be taken in flow and combustion processes, where the structures to be observed are of the size of the suction area diameter or distinctly larger. In order to estimate the fluid flow conditions during the suction period and the spatial resolution of measurement, typical experimental conditions in the planar RFSC at DLR, which can be seen on Fig. 13.39, will be used here as an example.

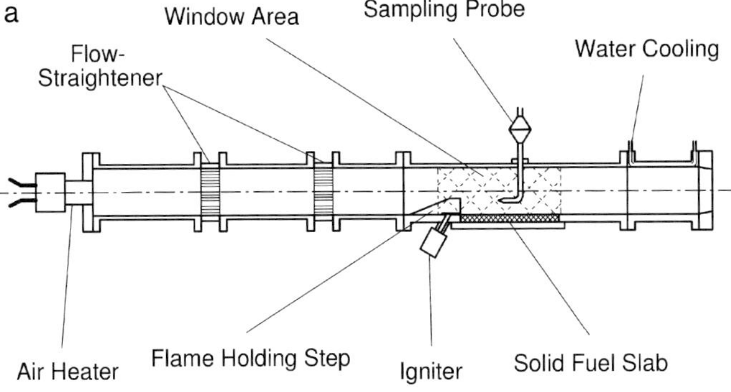

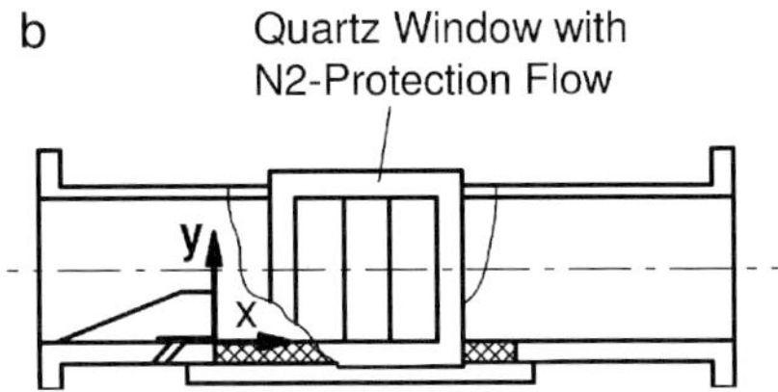

Figure 13.39. (a) Test facility with rearward-facing step combustor and sampling probe; (b) enlarged diagram of the combustor segment with window arrangement for laser-optical diagnostics application.

This facility has been used for basic investigations, because it allows easy access to the processes immediately above the solid fuel slabs and the test conditions concerning the temperature and the velocity of the air inlet flow are typical for solid fuel ramjets. Table 13.7 gives the dimensions of the test facility, typical test

Table 13.7. Typical test conditions, planar rearward facing step combustor.

Dimension size			***Solid fuel slab***		
Combustor height	H	45 mm	Fuel composition	HTPB	64.1 wt%
Combustor width	B	150 mm		IPDI	5.9 wt%
Step height	h	25 mm		Boron	30.0 wt%
Solid fuel slab length	L_s	200 mm			
Solid fuel slab width	B_s	100 mm	*Boron particles*		
			Sauter size	$d_{p3,2}$	0.96 μm
Inlet conditions			Purity		> 95 %
Air mass flow-rate	$\dot{m}_{air}$	0.15 kg/s	Density	ρ_p	2.34 g/cm^3
Air mass flux	$\dot{g}_{air}$	40 kg/m^{2}s			
Air velocity above step	u_{air}	93 m/s	*Fuel regression*		
Air mass flow temperature	T_{air}	800 K	Regression rate (800 K)		0.06 mm/s
Reynolds number	Re_h	22 300	Calculated particle number		2 ×
Combustor pressure	p	1 bar	density immediately above fuel slab ($\dot{r}$ = 0.06 mm/s)		10^4 mm^{-3}

conditions and the composition of the mostly used solid fuel slabs. The amorphous boron particles used were obtained from by H. C. Starck (Germany). Vitiated hot air, which is produced by heating the air flow with a H_2/O_2 burner, was made homogeneous by two flow straighteners and sieves. No nozzle was attached to the exit of the combustor, henve nearly ambient pressure was maintained inside. Further information about the test facility and the test conditions is available [13.181, 13.182].

A typical expansion quench probe, which had been used in various investigations, e. g. [13.181, 13.183], is of following dimensions in the tip region: $d_e = 1.5$ mm, $l_e = 5$ mm, $d_2 = 3$ mm and $d_a = 7$ mm [see Fig. 13.37(b)]. The tip passage and the widening area are made of copper, which allows an additional cooling and radical removal from the sample stream.

For the following very simple assumption of the measurement accuracy in the RFS-combustor flow field, dry air ($R = 287$ J/kg K) will be used and the particle phase and turbulent structures will be neglected. The probe will be oriented exactly in the upstream flow direction as shown in Fig. 13.37(b). Position 1 of the extracted fluid flow is located upstream of the probe mouth, where the disturbance of the flow field is negligible and the conditions can be assumed to be the same as in the outer flow. Temperature and velocity distributions were determined inside the combustor in earlier investigations with CARS [13.156] and LDV [13.184] so that the data for a typical point in the combustion zone above the solid fuel slab are assumed to be $p_1 = 10^5$ N/mm^2, $T_1 = 2000$ K and $u_1 = 50$ m/s. If the boundary layer thickness inside the short intake tube ($l_e/d_e \approx 3.5$) and the heat transfer to the wall are neglected, the mass flow that passes the probe mouth can be estimated in a first assumption as the mass flow through a sonic orifice if the pressure p_2 is low enough. With the equation given, for example, by Shapiro [13.185]:

$$\dot{m} = \frac{p_0 A_e}{\sqrt{T_0}} \sqrt{\frac{\kappa}{R}\left(\frac{2}{\kappa+1}\right)^{\frac{\kappa+1}{\kappa-1}}} \tag{13.51}$$

the sample mass flow-rate $\dot{m} = 0.16$ g/s was determined, using the average value $\kappa = 1.33$ (at 1800 K) for the temperature decrease from '1' to 'e'. Making use of the following equations for sonic conditions in the intake passage:

$$\frac{p_e}{p_0} = \left(\frac{2}{\kappa+1}\right)^{\frac{\kappa-1}{\kappa}} \tag{13.52}$$

$$\frac{T_e}{T_0} = \frac{2}{\kappa+1} \tag{13.53}$$

$p_e = 0.54$ bar and $T_e = 1710$ K were calculated, whereas the conditions at '1' can be approximately equated to the conditions at '0'. With the equations for the mass flow-rate:

$$\dot{m} = \rho_{g,e} u_e \frac{\pi}{4} d_e^2 = \rho_{g,1} u_1 \frac{\pi}{4} d_1^2 = \text{constant} \tag{13.54}$$

and the ideal thermal equation of state:

$$p = \rho_g RT \tag{13.55}$$

the diameter d_1 of the suction area A_1 was determined as $d_1 = 4.8$ mm. This shows that d_1 is smaller than the outer diameter $d_a = 7$ mm of the probe. Becker gives an approximation for the spatial resolution of gas sampling probes. He defines two gradient parameters:

$$G_{SS} = \frac{d_1}{B} \tag{13.56}$$

$$G_{Pr} = \frac{D_{out}}{B} \tag{13.57}$$

for the influence of the upstream sample stream tube diameter d_1 and the influence of the outer sample tube diameter D_{out} in the tip region. For a tapered probe head, D_{out} can be assumed to be equal to d_e. Both equations contain the characteristic spatial scale

$$B = \frac{|\Delta\bar{\phi}|}{|\nabla\bar{\phi}|} \tag{13.58}$$

where $\nabla\bar{\Phi}$ is the local gradient of a field $\bar{\Phi}$ and $\Delta\bar{\Phi}$ is a measure of the range of $\bar{\Phi}$ in the region about the measurement point (the height of the hill, as it were, whose slope at the measuring point is $\nabla\bar{\Phi}$). For flows of the boundary layer type, B is the boundary layer thickness [13.168].

In the RFSC the combustion process behind the recirculation zone can be characterized as a combustion process in a boundary layer. Consequently, the layer thickness is of the order of the step height h. This leads to gradient parameters $G_{SS} = 0.24$ and $G_{Pr} = 0.075$. Both gradient parameters should be as small as possible and $G_i < 0.1$ would be sufficient for an excellent resolution. Both determined values seem to be satisfactory for investigations under the environmental conditions in the RFSC. It should be noted again that the internal sampling passage from the probe mouth up to the end of the widening area is made of copper so that a convective heat transfer occurs and the mass flow rate decreases owing to a decrease in the velocity of sound at 'e'. Thereby the sample stream diameter d_1 decrease and also G_{SS}.

The estimation of St_p with Eq. (13.48) for boron particles of initial diameter $d_{p3,2} = 0.96$ µm gives $St_p = 0.076$, whereas the Sutherland equation [13.186] was used for the determination of the viscosity. SEM images of the collected solid phase indicate that most of the particles on the SEM sampling plate and hence also on the filter have diameters > 3 µm. It should be noted that agglomerated particles in the combustor cannot be identified on these images. Mie scattering images, however, which will be discussed later in detail in Sect. 13.5.6, show that the smaller particles dominate in the combustor. If we assume that most of the particles or agglomerates have diameters $d_p > 10$ µm, then St_p for these particles would be < 7.2, which is clearly below the above-mentioned value of 10. Hence the separation effects for these particles are not too pronounced under the conditions in the RFSC.

Furthermore, it should be mentioned that the orientation of the probe tip with respect to the local flow direction also influences the separation behavior. Small angles of attack should be chosen to reduce these losses. Velocity vector determinations with LDV or PIV should be made if possible for a better orientation of the probe or for the determination of regions with large angles of attack, where measurements should not be conducted. For the fixed and horizontally oriented probe tip in the RFSC, results are presented here only for angles < 15°.

Sampling procedure

In addition to the above-described basic conditions for the use of sampling probes in a reacting multiphase flow under ramjet-relevant conditions, the design and the operating procedure of the developed modular sampling probe system is also described here. A diagram of the sampling system that was used for investigations in various combustors under solid fuel ramjet-relevant conditions concerning the air inlet temperature and also under ducted rocket-relevant conditions is shown in Fig. 13.40. It consists of a water-cooled sampling probe, an exchangeable filter housing, a remote-controlled valve unit, a vacuum pump and exchangeable gas sampling cylinders for subsequent analysis. The sampling probe is mounted in a device on the upper combustor wall, which allows the horizontally oriented probe tip to be placed at any position $y \geq 4$ mm inside the combustor.

The sampling probe consists of three thin-walled inner stainless-steel tubes and an outer copper tube with an outer diameter $d_a = 7$ mm, as can be seen in Fig. 13.41. The three outer tubes serve for water cooling of the probe. Pressurized water cooling is necessary to withstand the high thermal radiation fluxes and the high temperatures. In order to suppress the condensation of the water vapor content of the combustion products inside the inner probe tube, an inlet temperature of the

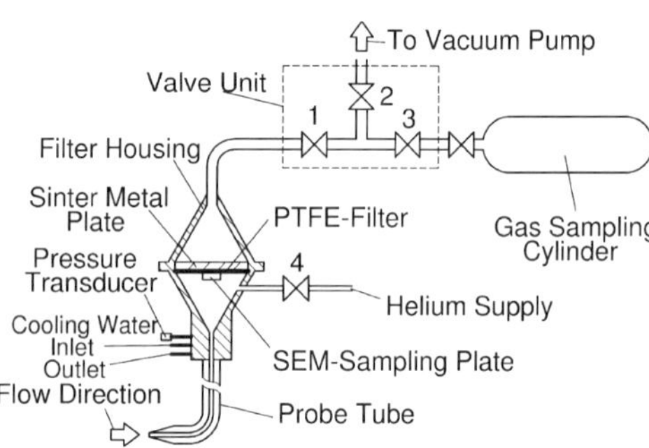

Figure 13.40. Gas and particle sampling probe system.

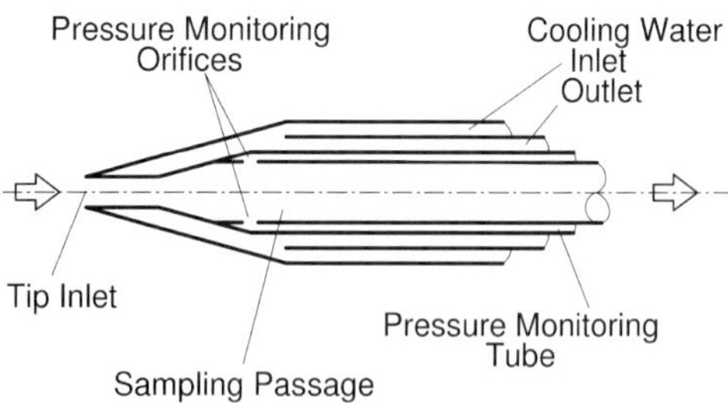

Figure 13.41. Tip region of the sampling probe.

cooling-water of 80 °C was chosen. In the filter housing a PTFE filter of 0.45 μm pore size separates at nearly ambient temperature the condensed phase from the gas phase. For newer investigations, which are not presented here, the filter housing is also heated in order to reduce water condensation on the filter, which diminishes the permeability and thereby the suction rate. In this connection it is noteworthy that care has to be taken, because boric acid is volatile with water vapor. For these newer experiments a heating temperature of 60 °C for the filter housing was chosen, because at this temperature level the losses during the sampling time are low, as has been shown in separate experiments.

A pressure transducer is connected to the annulus between the two inner tubes and allows the observation of the pressure immediately after the widening area in the tip through small holes, which are drilled into the inner probe tube as shown in Fig. 13.41. The continuous monitoring of the pressure in the sampling passage during the experiments allows one to draw conclusions about the sufficient expansion and the necessary temperature decrease in order to suppress further reactions. Particles and condensed reaction products can be deposited inside the inner probe tube or the sampling passage in the tip and may lead to a partial or total blockage, which can be seen on the pressure traces as a distinct pressure drop. On the other hand, partial blockage of the filter element can be recognized on the pressure trace as a too high pressure level inside the sampling passage. This condition may lead to insufficient expansion and cooling of the sample fluid flow, which may lead to incorrect results.

In flows with very high particle loads, and especially with particles of which the oxides have a strong tendency towards the condensed phase, the blockage of the probe mouth by combustion residues can occur very quickly. In order to protect the probe tip during the start procedure of the test facility, an additional helium protection-flow out of the probe mouth before starting the sampling procedure is very helpful.

The experimental procedure with the sampling system at the RFSC is described as follows. Before starting a test run the valve (4) of the helium protection-flow is opened. Then the air heater is started for heating up the facility for 2 min. Subsequently the solid fuel slab is ignited by a further H_2/O_2 burner, which is positioned beyond the rearward-facing step. A few seconds after ignition of the solid fuel slab, the valve of the helium line is closed and valves (1) and (2) to the vacuum pump are opened simultaneously to draw the combustion products up through the probe and the filtering assembly. In the filter housing a PTFE filter of 0.45 μm pore size separates the ambient temperature-condensed phase from the gas phase. A few seconds later, valve (2) to the vacuum pump is closed and valve (3) to the sampling cylinder, which had been evacuated before the experimental run, is opened. Now the pressure in the probe cylinder rises until valves (1) and (3) are closed.

Figures 13.42 and 13.43 show results obtained with an earlier version of the sampling system [13.181, 13.183], which was not heated in total and where the pressure in the sampling cylinder rose to the pressure level inside the combustor. Figure 13.42 presents concentration profiles as volume ratios of the ambient temperature-stable gaseous intermediate reaction products, which were collected in

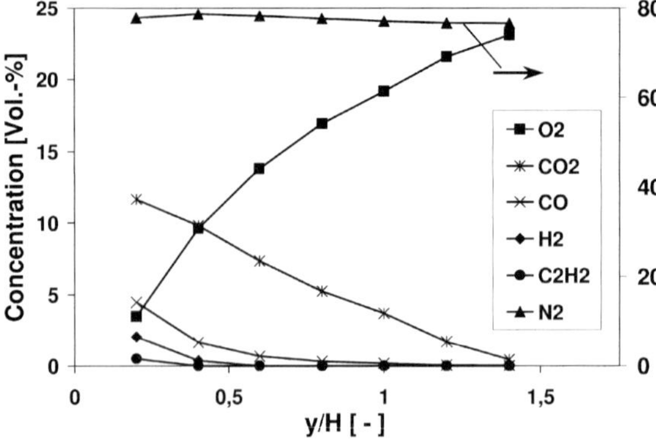

Figure 13.42. Concentrations of gaseous stable (intermediate) reaction products versus the dimensionless vertical distance y/h from the combustion chamber bottom at the measuring cross-section at $x = 171$ mm.

the sampling cylinders and analyzed by gas chromatography after the experimental runs. The measuring cross-section was located at $x = 171$ mm behind the rearward-facing step. The diagram shows profiles with decreasing values for CO_2, CO, H_2 and C_2H_2 with increasing distance y from the combustion chamber bottom with the inserted solid fuel slab. In contrast to this, the oxygen concentration increases as a function of the same parameter y up to the value of the outer air flow. These profiles show the characteristic features of a diffusion flame, which is embedded in a boundary layer.

Analysis of collected condensed phase

The collected condensed phase is analyzed after the experiments by wet chemical analysis using the well-known mannitol method. The probes were ground and treated for 15 min with warm water (40 °C). This temperature was chosen to avoid losses of boric acid, which is volatile with water vapor. Under these conditions, the boric acid and the boron oxide were dissolved. Then mannitol was added and the mixture was titrated with dilute NaOH with phenolphthalein as indicator. This method is not able to distinguish between different boron oxides and acids. Hence the sum of the components is measured and the results obtained are calculated and presented as B_2O_3. Further details are described elsewhere [13.187–13.189]. Owing to the small suction area and the short experimental run time, only a low probe mass could be obtained during the sampling time and only the boron oxide content could be determined.

Figure 13.43 shows the boron oxide content of the collected condensed (solid) phase for three different measuring cross-sections behind the rearward-facing step at $x = 60$, 97 and 171 mm. With increasing distance y from the combustor bottom the B_2O_3 contents at all measuring cross-sections show an upward tendency. This indicates a higher conversion of the boron particles to boron oxide with increasing distance y from the combustor bottom and increasing distance x from the rearward-facing step. At $x = 60$ mm a slight maximum of the B_2O_3 content occur inside the recirculation zone, which is located immediately behind the rearward facing step.

Scanning electron microscopy (SEM) allows a magnified view of the sample and

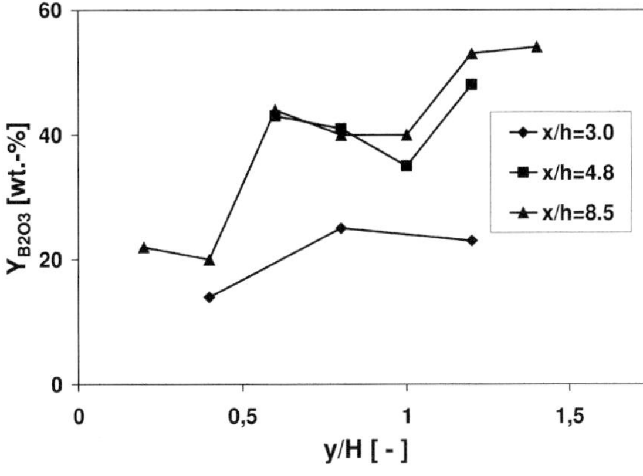

Figure 13.43. Boron oxide content of the collected solid phase versus the dimensionless vertical distance y/h from the combustion chamber bottom for three measuring cross-sections.

gives information about the shape and size of the collected particles. Ciezki and Schwein [13.183], for example, showed SEM images obtained with the above-described water-cooled sampling probe at distinct heights above a burning solid fuel slab. Figure 13.44 presents two SEM images of samples taken at two heights above the combustor bottom. Figure 13.44(a), obtained in the outer zone of the reacting boundary layer flow at x = 171 mm and y = 28 mm (see also Fig. 13.47), shows mainly particles with average diameters up to 3 μm and also some larger particles with diameters of the order of 10 μm. Immediately above the combustion chamber bottom at y = 8 mm bulged areas on the surface of the boron particles can be seen. These can be interpreted as the residues of the liquid boron oxide layer which cover every particle immediately after the first ignition stage and hinder the further combustion process. Some of the collected particles show porous structures, which

Figure 13.44. SEM images of collected solid phase of HTPB/IPDI with 30 wt% boron. (a) x = 171 mm, y = 28 mm; (b) x = 171 mm, y = 8 mm.

may be caused by the second ignition stage of these particles when the boron oxide layer will be ruptured at distinct points, as Meinköhn and Sprengel [13.146] have shown theoretically.

Obtaining additional information about the internal structure of single particles and their internal species distribution is partly possible with instrumental physical methods such as Auger electron spectroscopy (AES), X-ray photoelectron spectroscopy (XPS) and X-ray diffraction (XRD). Information about atomic concentrations of O, C and B and also the boron oxide content depending on the distance from the particle surface have been obtained from samples of particles that had been collected in the afterburner chamber of a solid fuel ramjet model combustor with a water-cooled rod with cavities [13.173]. To obtain reliable and statistically safe results, a large number of particle samples have to be investigated owing to the individual history of each particle. Further information about this and other diagnostic tools for the analysis of the collected condensed phase can be found elsewhere [13.173].

13.5.4 Self-Emissions

The observation of self-emissions of combustion processes with cameras, spectrometers, pyrometers and so forth is widely used in research and industrial applications. Both normal and high-speed camera systems are very common for obtaining information on flame structures and locations, fuel regression rates [13.190], particle and flake movement [13.191], etc. (Interference) filters are often added to select the spontaneous band emissions of intermediate species such as OH and CH in order to obtain information about their distributions in the reaction zones. In multiphase combustion processes. the self-emissions in the visible and IR regions are often superimposed by strong thermal emissions. Hence care must always be taken with the interpretation of the results obtained. For example, Hensel [13.192] observed spontaneous BO_2 emission of the combustion process of boron particles containing solid fuel slabs of GAP/N100 using two video cameras with interference filters simultaneously. The first filter ($\lambda_{m,1}$ = 5 477 Å, half-width FWHM = 27 Å) allows the band emissions of BO_2 to be registered and the second filter ($\lambda_{m,2}$ = 5 309 Å, FWHM = 35 Å) the thermal emissions in a nearby region. The subtraction of the two images (after the calibration of the whole measuring device) gives a more realistic image of the spontaneous BO_2 emission. Emission spectroscopic methods and pyrometers for the determination surface temperatures of the particle phase have also been used at the step combustor test facility [13.193, 13.194]. The use of high-speed cameras together with the image subtraction technique can give further detailed information about the processes in staged ignition processes, as Klimov *et al.* [13.195] have shown for the multiphase diffusion flame of an aluminum particle-laden hydrogen jet in a hot air environment.

13.5.5 Color Schlieren

Schlieren methods are well known for the visualization of structures in flow fields. Light is deflected due to a refractive index gradient (caused, for example, by a density, temperature or concentration gradient), which is oriented normal to the direction of propagation of the light beam [13.161, 13.196]. In a homogeneous gaseous medium the refractive index n is proportional to the density ρ according to the Gladstone-Dale law:

$$n-1=K(\lambda)\rho \tag{13.59}$$

where K is the wavelength-dependent Gladstone-Dale constant.

The deflection angles $\varepsilon_{y'}$ and $\varepsilon_{x'}$ of a light beam transmitting the area under investigation at the location x'_0, y'_0, as can be seen in Fig. 13.45, can be expressed with $\tan\varepsilon_i << 1$ according, for example, to Schardin [13.161] as

$$\left(\varepsilon_{x'}\right)_{x'_0,y'_0}=\int_{z'_1}^{z'_2}\frac{1}{n}\left(\frac{\partial n}{\partial x'}\right)_{x'_0,y'_0}\mathrm{d}z' \tag{13.60}$$

$$\left(\varepsilon_{y'}\right)_{x'_0,y'_0}=\int_{z'_1}^{z'_2}\frac{1}{n}\left(\frac{\partial n}{\partial y'}\right)_{x'_0,y'_0}\mathrm{d}z' \tag{13.61}$$

Assuming a two-dimensional flow field in the x'–y' plane, the equations can be written as

$$\left(\frac{\partial\rho}{\partial x'}\right)_{x'_0,y'_0}\approx\frac{1+K(\lambda)\rho}{L'\cdot K(\lambda)}\varepsilon_{x'} \tag{13.62}$$

$$\left(\frac{\partial\rho}{\partial y'}\right)_{x'_0,y'_0}\approx\frac{1+K(\lambda)\rho}{L'\cdot K(\lambda)}\varepsilon_{y'} \tag{13.63}$$

where $L = z'_2 - z'_1$ represents the length of the light pass through the investigation area.

In a Schlieren set-up, the refractive index gradients can be visualized by the separation of the deflected light from the undisturbed light by means of a Schlieren

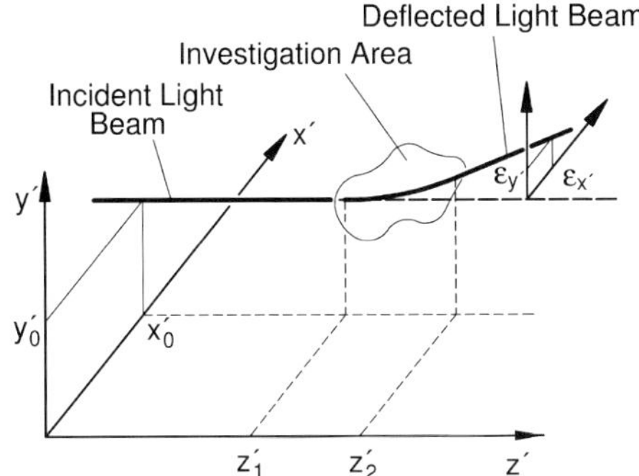

Figure 13.45. Deflection of a light beam due to a refractive index gradient in the investigation area.

stop, which is positioned in the cutoff plane. This Schlieren stop can be realized in a black/white Schlieren (B/W-Schlieren) or monochrome Schlieren set-up, for example with a knife edge, or in a Color Schlieren set-up, for example with a slit aperture. In a Töppler Z set-up, as can be seen in Fig. 13.46, the cutoff plane is located in the focal plane of the second concave mirror. With this kind of Schlieren stop, only the gradient perpendicular to the edge of the knife or to the slit can be visualized. The image of the investigation object shows, in B/W-Schlieren, these gradients as different shades of gray. In strongly particle-laden flows additionally a reduction of the light intensity on the image is caused by scattering, absorption, etc., by the particle phase itself. Consequently, the interpretation of the B/W-Schlieren images of the flow and combustion processes is more difficult.

Color Schlieren methods, however, based on the dissection technique developed by Cords [13.197], offer the possibility of working under these conditions. For further information about other Color Schlieren techniques, see for example the review by Settles [13.198], who summarizes publications up to 1980. Using, for example, a set-up similar to that developed by Ciezki [13.199, 13.200], a color filter source mask is positioned in the focal plane of the first Schlieren mirror as can be seen in Fig. 13.46. Illuminated by a light source, the color mask produces separated light beams with different colors, which represent different deflection angles and thereby different refractive index gradients. While these light beams transmit the particle-containing investigation area, their light intensity is lowered but the different colors are not changed. Hence the Schlieren image gives information about the (macroscopic-averaged) refractive index gradients in the gas phase in a first approximation.

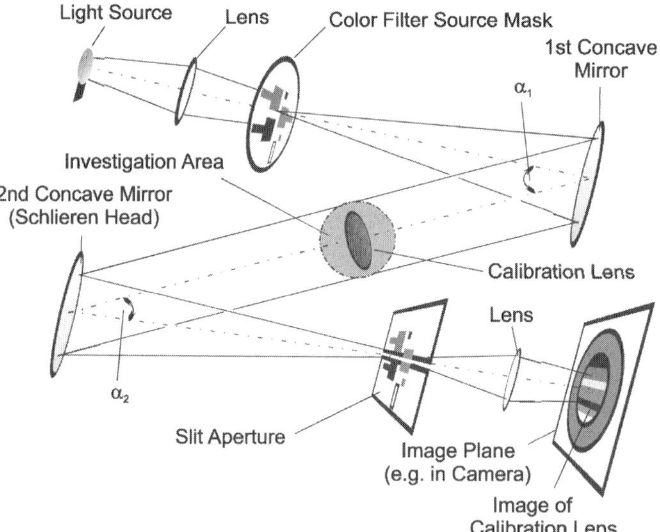

Figure 13.46. Color Schlieren set-up based on the dissection technique with a lens of very long focal length as calibration object.

The limitation of this technique is due on the one hand to the sensitivity of the film material or the video or CCD camera system used in relation to the intensity of the light from the Color Schlieren source mask and the properties of the transmitted media. On the other hand, the fractional intensity of scattered light may lead to color changes in the direction to white color on the Schlieren images owing to symmetric scattering in the direction of the original light beam, especially for multiple scattering.

Besides the above-mentioned advantages, especially for particle-laden (combustible) flows, the addition of color makes the interpretation of the processes that occur easier, because different colors are easier to distinguish by the human eye than the different shades of gray. Furthermore, contamination of windows, lenses, etc., and disturbances caused by film processing and image production or the CCD chip may lead to wrong colors on false color images of B/W-Schlieren exposures. On Color Schlieren images, however, the intensities of the different colors are only lowered, and not changed, and still give the same information about the refractive index gradients as described above. Consequently, the Color Schlieren method often produces more informative images.

In order to work with an efficient Color Schlieren set-up, further demands have to be met:

- The color coding should be unambiguous in order to avoid misinterpretation. An origin of the colors as in a rainbow fulfills this task. The Schlieren apparatus is adjusted in such a way that the green color represents refractive index gradients near to zero. This kind of color coding has been described in detail [13.199].
- In order to obtain good color quality on the Schlieren images, the light source, the color filters and the film material or the spectral sensitivity of the CCD camera should be optimized. The easiest way to achieve this goal is to change the width of the filter segments on the color filter source mask as has been described by Ciezki [13.199]. A calibration lens with a long focal length, which is positioned at the investigation area instead of the object to be investigated, as can be seen in Fig. 13.46, is very helpful for this task.
- The planar Z set-up, where parallel light transmits the object under investigation, should be chosen if possible. In this arrangement, aberration coma can be eliminated by the symmetrical layout, i.e. equal angles $\alpha_1 = \alpha_2$ and equal focal lengths of the concave mirrors. Astigmatism can be reduced by choosing small $\alpha_i < 10°$ [13.161].
- Achromatic lenses must be used to reduce chromatic differences of focal lengths.
- Windows should be of Schlieren quality.

Netzer and Andrews successfully used a Color Schlieren technique that is different from the technique described above for the investigation of burning ammonium perchlorate [13.201]. Difficulties, however, were observed in combustion processes of other propellants, where significant self-emissions occur. These are mainly caused by thermal emissions of soot particles, flakes and other hot (condensed or solid) materials inside the combustion chambers. Netzer and Andrews recognized that their flash light source was not powerful enough to overpower this self-luminosity, so they opted for the laser-Schlieren method. For the investigation of the combustion behavior of boron particles containing solid fuel slabs, Ciezki [13.200] used an 18-J xenon flash light with a flash duration of 13 μs. Additionally, a shutter unit was positioned in front of the camera in order to reduce the strong thermal emissions from the combustion process on the Schlieren images. The unit with variable exposure times down to 300 μs consisted of a conventional iris diaphragm with a minimum exposure time of 1 ms and a liquid crystal shutter with a minimum exposure time of 300 μs. The iris diaphragm must additionally be used because the thermal emissions are so intense that the closed liquid crystal shutter cannot block them totally.

Figure 13.47(a) shows a Color Schlieren image taken with an exposure time of 1 ms during the combustion process of HTPB/IPDI without boron particle addition under the conditions described in Table 13.7. Quartz slices without any protection-gas flow on both side walls of the combustor allow one to observe the combustion process over the whole length of the fuel slab. In Fig. 13.47(b) a diagram of the characteristic features of the combustion process is shown. Included also are some technical terms derived from a study by Natan and Gany [13.202] of the combustion

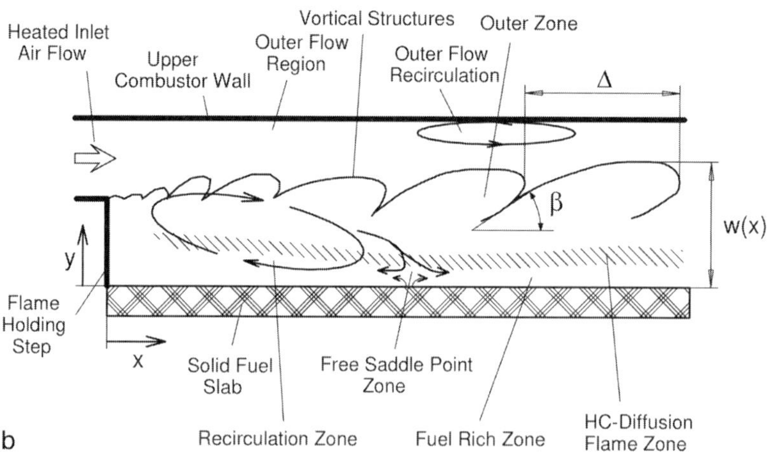

Figure 13.47. (a) Color Schlieren image of HTPB/IPDI without boron particle addition [13.200]; (b) flow field and combustion characteristics visible on the image in (a) with regard to LDV measurements.

processes in the boundary layer of a solid fuel ramjet. Large-scale coherent vortex-like structures can be seen on the Schlieren image in the outer zone above the diffusion flame in the developing boundary layer downstream of the recirculation zone. In the lower region of the image a yellow glowing band can be seen, which is not visible on the Color Schlieren images, where much shorter exposure times than the 1 ms used were chosen. This indicates that this band originates from thermal emissions from the condensed phase (boron, B_2O_3, soot, etc.), which is superimposed on the Schlieren image. Because the thermal emission of soot dominates under the presented experimental conditions [13.193, 13.194], the glowing band can be related to the hydrocarbon diffusion flame zone.

At the end of the recirculation zone, which was identified with LDV measurements [13.184], the distance between the yellow glowing band and the combustor bottom is smaller. This is caused by a downward movement of the flow in this region. On the left side of the Schlieren image, i. e. the area immediately behind the rearward-facing step, no information about the processes occurring can be obtained, because of the contamination of the windows by soot and combustion residues.

13.5.6 Velocity Measurements with Scattering Techniques

Information about velocity distributions in flow and combustion fields can be obtained by methods such as laser-Doppler velocimetry (LDV), particle image velocimetry (PIV) and phase Doppler anemometry (PDA), which use the scattered light from illuminated small particles following the flow. In many applications, gaseous or liquid single-phase flows are seeded with very small tracer particles to allow Mie scattering by these particles. In multiphase flows, however, the use of the existing particles or droplets as scattering sources seems to be useful, but care has to be taken in the interpretation of the results obtained for large particles and broad size distributions owing to the inertia of the particles to follow the flow. Furthermore, boron and aluminum particles, which are used in propulsion systems, have an irregular shape so that techniques which are limited to spherical particles, such as PDA, cannot be used. The literature about non-intrusive methods using scattered light for velocity determination is very large so that only a very short and simple description of the basic principles of LDV and PIV will be given here. LDV was developed by Yeh and Cummings in 1964 [13.203]. Two focused coherent laser beams intersect and form a measuring volume. Particles that pass through this volume cause the the light of these intersecting beams to stray and produce an oscillating intensity signal on the detector, as can be seen on right upper diagram in Fig. 13.48.

For a very simple explanation of the processes occurring, the fringe pattern model will be used here, where the two coherent laser beams produce a pattern of bright and dark stripes (or planes because the measuring volume is three-dimensional). When a particle traverses this fringe pattern, the scattered light fluctuates in intensity. The velocity u of the particle, which has to be determined, is equal to the fringe spacing d_f divided by the time interval Δt_f which is registered by a detector:

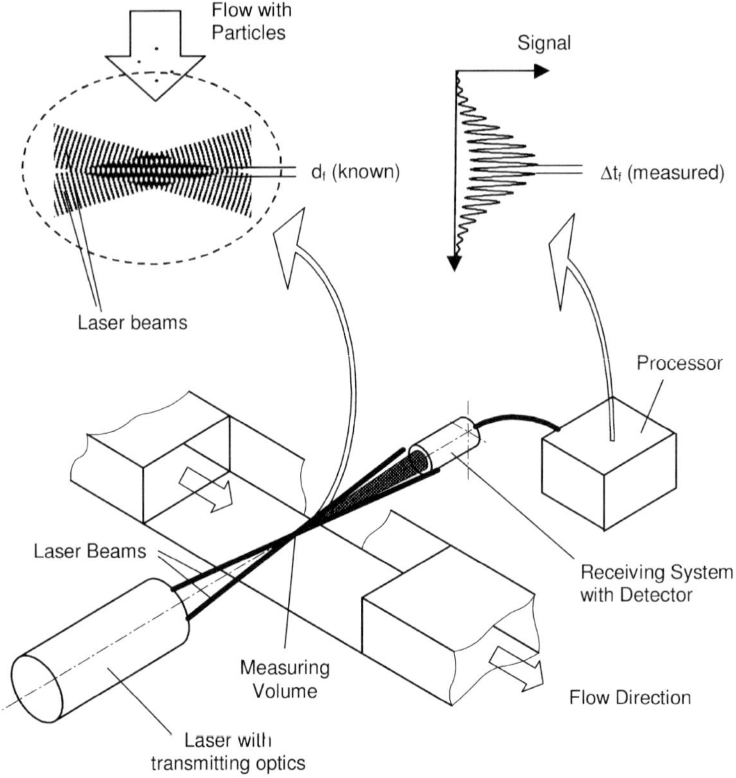

Figure 13.48. Experimental arragement for 1-D LDV (forward scattering).

$$u = \frac{d_f}{\Delta t_f} \tag{13.64}$$

$$d_f = \frac{\lambda}{2\sin(\theta/2)} \tag{13.65}$$

whereas d_f is a function of the angle θ between the two laser beams and the laser wavelength λ. For further detailed information about the basic principle, various arrangements and applications see [13.204–13.206].

The LDV system used for investigations [13.184] at the DLR Space Propulsion Institute at Lampoldshausen consists of a 5-W argon ion laser and two burst spectrum analyzers, which were used to measure the velocities using forward scatter sampling. The beam pairs were rotated at 45° along the z-axis to improve the resolution for both the x and y components. A total of 6100 samples were taken at each measuring point, scanning the incoming LDV signal at 5 kHz. The LDV was mounted on a 3-D traversing unit controlled by a PC. Data reduction was performed using standard software.

Forward scattering has the advantage that the signal intensity is higher than for backward scattering for the particle diameter range 0.5–10 μm. In optically dense systems, however, backscatter might be better if the signal intensities for forward scatter are too low. Unfortunately, only regions near the windows of the walls of the test chamber can be examined with the backscatter method. Pre-tests using the RFSC have shown that forward scattering can be used under the prevailing conditions. Furthermore, interference filters are necessary to reduce the superimposed thermal emissions on the LDV signal in order to obtain better signal-to-noise ratios. Generally, the major concern in LDV measurements should be to achieve a homogeneously distributed seeding if the flow is seeded, proper selection of the optical set-up, data reduction settings and statistical parameters.

Particle image velocimetry (PIV) is a non-intrusive, two-dimensional flow visualization technique [13.207–13.209], where the scattered light from double-illuminated particles in the flow is observed and collected by, for example, a CCD camera, as shown in Fig. 13.49. The double illumination of the particles inside a thin light sheet can be achieved by means of a laser system with an adjustable time delay between two laser pulses. Two different ways to perform the measurements have been used for the experiments using the step combustor [13.210]. The first method uses one pulse per image and tracks the particles between two consecutive images. In the second method, the two pulses are recorded on one image so that the particles are tracked within the same frame. In the first case, with two individual

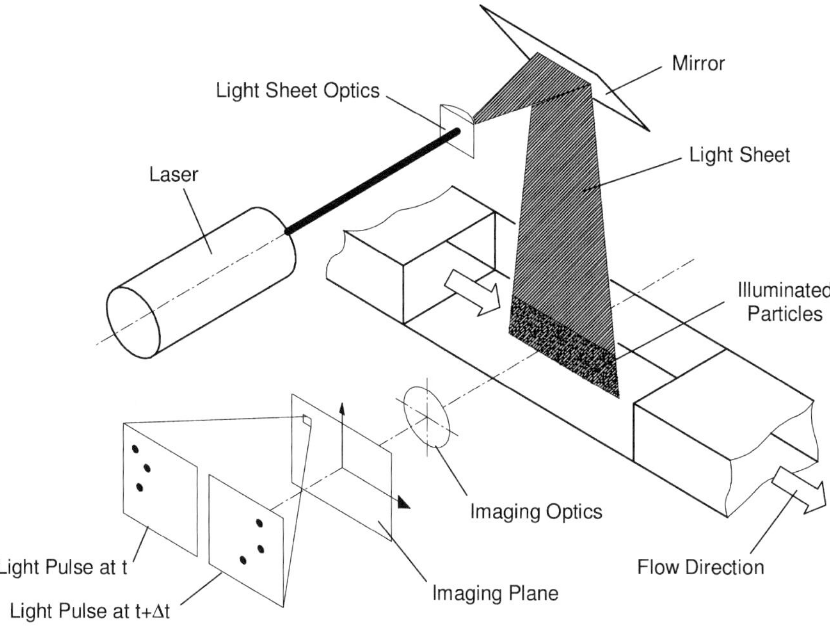

Figure 13.49. Experimental arrangement for particle image velocimetry (PIV).

images, the cross-correlation technique is utilized. In the case of the double-exposure experiment, the measured frame is analyzed by an auto-correlation technique. More details about data analysis can be found elsewhere [13.160, 13.209, 13.211, 13.212]. The result of this evaluation is in both cases a displacement of individual particles or particle assemblies. By dividing the measured displacement by the known time delay between the two lasers pulses, the velocity of these particles can be obtained. The results from the step combustor test facility [13.210] were mainly achieved with the cross-correlation technique.

Comparing LDV and PIV, it can be said that LDV is a single-point measuring method for the determination of average structures of the flow field. It is unable to capture the instantaneous spatial structure of the flow, but parameters for turbulence statistics such as mean velocity, turbulence intensity, Reynolds stress and higher order momentums can be obtained. The actual flow structure in a 2-D plane of the flow field, however, can be determined with PIV. The average structures can then be determined from a series of instantaneous velocity fields that have been determined from PIV images.

In particle- or droplet-laden flows, their scattered light can be used for velocity determination if the intensity of the scattered light is high enough to deliver sufficient signal-to-noise ratios and data rates. Furthermore, care must be taken with the interpretation of the results obtained owing to the limited ability of large particles to follow accelerations of the gas-phase flow. Ruck [13.206], for example, reported that too short recirculation zone lengths have been determined in non-reacting RFS flows with large tracer particles. Furthermore, in broad particle size distributions the amount of large particles may change the average velocity values and also the turbulence statistics parameters. Consequently, the results obtained with an LDV or PIV system show only the movement of the particle phase and further estimations have to be made if the results obtained allow conclusions on the movement of the gas phase.

A simple way to assume the particle-following behavior can be done using the particle Stokes number St_p as described earlier in Sect. 13.5.3. The particle relaxation time τ_p was defined in Eq. (13.49). The flow field can be assumed to be a reacting boundary layer flow and the fluid time scale τ_g can now be defined as the ratio of the step height h to the air inlet velocity:

$$\tau_g = \frac{h}{u_{air}} \tag{13.66}$$

For the presented estimation of St_p, a point at the border of the reacting boundary layer with $T = T_{air} = 800$ K was chosen. At this point the highest values occur, because in the hotter regions of the combustor lower densities and higher viscosities occur, leading to lower St_p values.

The Stokes number $St_p = 0.016$ can be calculated for the Sauter size $d_{3,2}$ of the original boron particles. The size distribution of the original boron particles shows that only 10% of the particles have diameters > 5 μm and 1% have diameters > 10.7 μm. The corresponding Stokes numbers are $St_p(5\ \mu m) = 0.42$ and $St_p(10.7\ \mu m) = 1.76$. This indicates that most of the original particles have St_p

clearly > 1.0 and might follow the gas-phase flow fairly well. It should be mentioned that the requirements on the particle-following behavior for LDV and PIV are more strict than for sampling probes and therefore smaller St_p are necessary. A more detailed analysis of the particle-following behavior has been made by various groups. For example, Durst *et al.* [13.213] discussed various theories for the movement of small particles in a turbulent flow with regard to LDV.

A quantitative calculation of the ability of particles to follow a flow has been done for an oscillating flow. The movement of a spherical particle in a turbulent flow field can be described by the Basset-Boussinesq-Oseen (BBO) equation. Low particle Reynolds numbers, i.e. a low relative velocity between the particle and the fluid, a locally homogeneous and stationary turbulence and a low particle concentration have to be assumed. If the fluctuating velocity is described by a Fourier integral, the BBO equation can be solved. For further information see Ref. [13.206]. Ruck and co-workers [13.206, 13.214] described the limit of the frequency which, for example, water droplets in air can follow with a ratio of 99% of the amplitudes. They found that for a frequency of 10 kHz, which is for most cases the maximum occurring frequency in turbulent flows, the droplet diameter should be < 1 µm.

For an assumption of the following behavior of boron particles inside the hot RFSC environment, simplified conditions (1 bar, 1800 K, dry air) for the solution of the BBO equation were used. Figure 13.50 presents the frequency f with which the particles follow the oscillating hot air flow against the particle diameter d_p. The ratio ε is an additional parameter, which is the amplitude of the oscillating particle with respect to the amplitude of the oscillating flow. The results presented in Fig. 13.50 show that boron particles of 1 µm diameter show the same qualitative behavior as the water droplets in ambient air aforementioned. This is mainly due to the

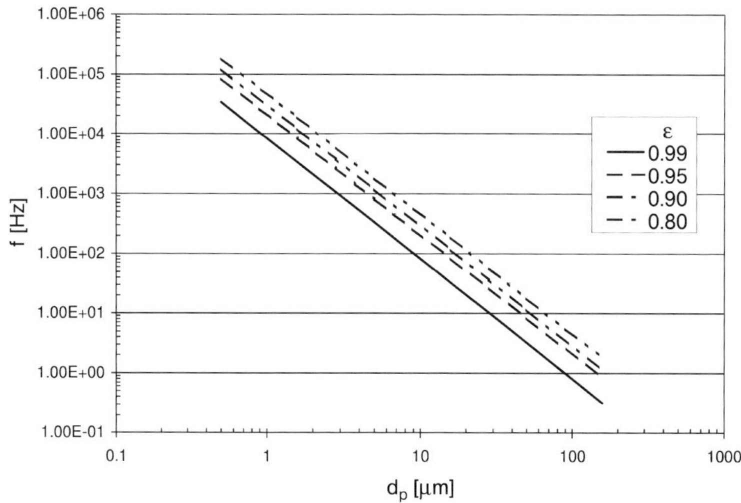

Figure 13.50. Particle-following frequency f versus particle diameter d_p for different amplitude ratios ε for boron particles in dry air at 1700 K.

increased temperature, which strongly affects the viscosity of the hot air. Taking into account an overall error of 1 % for LDV measurements, no further contributions due to particle inertia are assumed.

The aim of the investigations at the step combustor test facility with LDV and PIV was to visualize the movement of the reacting particle phase that is ejected from the solid fuel slabs. The PIV images are dominated by small particles but also larger particles or agglomerates up to 2 mm in size have been observed, as can be seen in the image in Fig. 13.51 of the area immediately behind the rearward-facing step. For the velocity determination from the PIV images, only the smaller particles were used. Furthermore, boron particles might have proper motion as has been found in earlier investigations such as those by Clauss [13.215] anf Ulas *et al.* [13.216].

An examination of this phenomenon using LDV found no significant difference between the motions of inert TiO_2 particles (with an average diameter of 0.2 μm of the original particles) and boron particles, as can be seen in Fig. 13.52. For these experiments, solid fuel slabs with boron particles and with TiO_2 particles instead of boron were burnt. Additionally, measurements with and without seeding of the outer flow with TiO_2 were made to distinguish separate flow regions, which are presented in Fig. 13.47. In summary, for small particles not larger than 30 μm no significant proper motion of boron particles was found. The larger particles and flakes, however, show a proper motion due to their inertia as has been found in the PIV measurements.

13.5.7 Coherent Anti-Stokes Raman Spectroscopy (CARS)

Coherent anti-Stokes Raman spectroscopy (CARS) is a well known laser-spectroscopic tool for measuring temperature and concentration in gas-phase combustion processes. Especially N_2-vibrational CARS is often used for investigations of com-

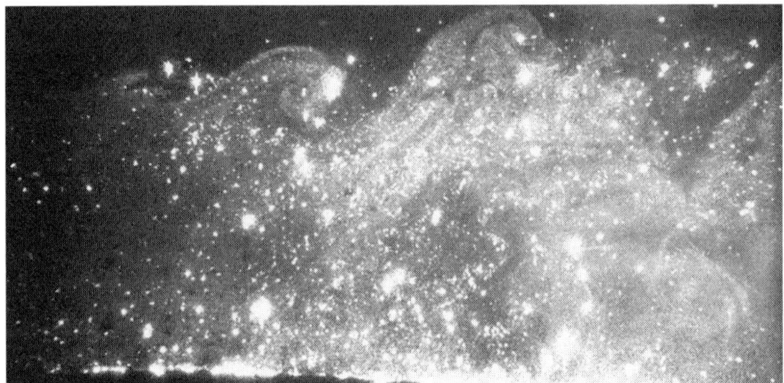

Figure 13.51. PIV-image of the area immediately behind the rearward-facing step [13.210].

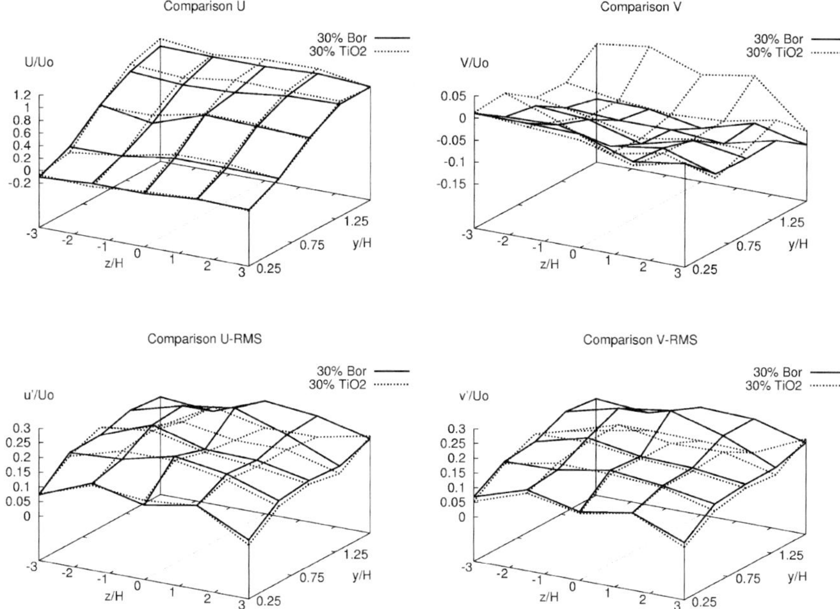

Figure 13.52. Comparison of velocities and RMS for test fuels with 30 wt% boron and 30 wt% TiO_2 additives at the measuring cross-section at $x = 96$ mm.

bustion processes with air because of its high N_2 content. CARS is a three-wave mixing process where, in its degenerate form, two photons of frequency ω_1 (pump laser) and one photon of frequency ω_2 (Stokes laser) interact with the third-order non-linear susceptibility of matter $\chi^{(3)}$ to generate an oscillating polarization and a resulting coherent radiation according to the equations

$$I_3(\omega_3) \propto I_1^2(\omega_1) I_2(\omega_2) \left|\chi^{(3)}\right|^2 \tag{13.67}$$

$$\chi^{(3)} = \frac{2c^4}{\hbar\omega_2^{\,4}} \left(\frac{\mathrm{d}\sigma}{\mathrm{d}\Omega}\right) \frac{\omega_j \cdot \Delta N}{\omega_j^2 - (\omega_1 - \omega_2)^2 - i(\omega_1 - \omega_2)\Gamma_j} + \chi_{\mathrm{NR}} \tag{13.68}$$

where I_1 and I_2 are the intensities of the Stokes laser (ω_1) and the pump laser (ω_2) and I_3 the intensity of the CARS signal [13.147, 13.217, 13.218]. $\chi^{(3)}$ can be determined for a single isolated Raman line of frequency ω_j with Eq. (13.68), where $\mathrm{d}\sigma/\mathrm{d}\Omega$ is the Raman scattering cross-section, ΔN the population difference between the states involved in the Raman transition, Γ_j the Raman linewidth of the jth transition, c the speed of light, $\hbar$ Planck's constant and χ_{NR} the dispersionless real non-resonant susceptibility.

Figure 13.53 shows for the folded BOXCARS scheme (as an example of a possible experimental realization) the beam geometry and the related aspects of the process. If the frequency difference $\Delta\omega = \omega_1 - \omega_2$ is tuned to a Raman-active transition ω_R of the probed species, the coherent laser-like signal at $\omega_3 = 2\omega_1 - \omega_2$ of the molecules is

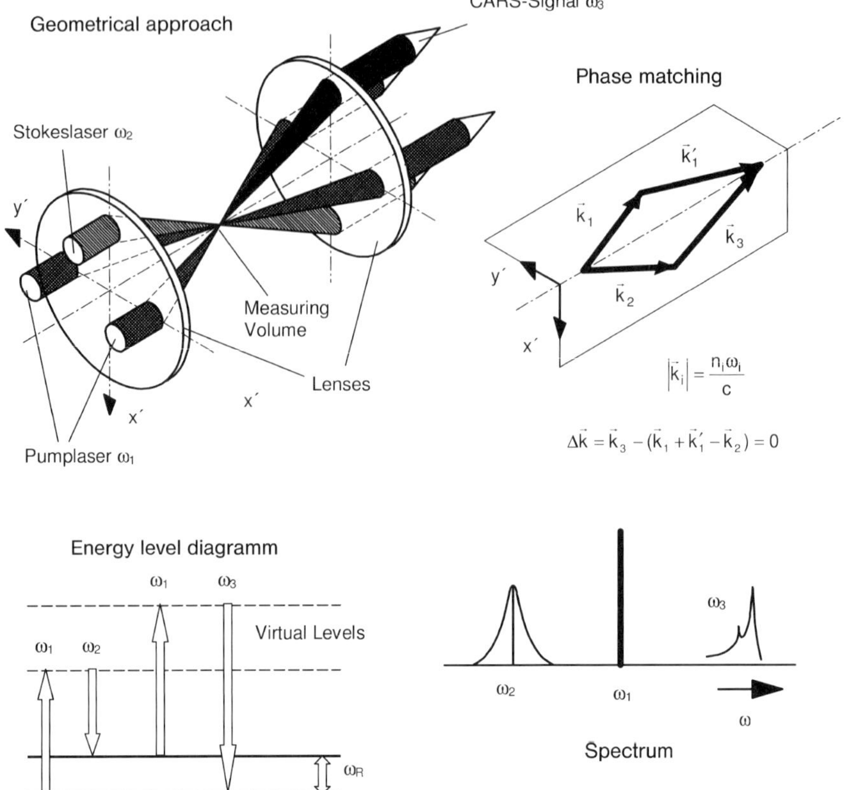

Figure 13.53. Folded BOXCARS.

resonantly enhanced and forms the CARS signal. The spectral distribution of the resonant molecule's CARS spectrum is dependent on the square of the absolute value of the molecule's complex third-order non-linear susceptibility $\chi^{(3)}$, as can be seen in Eq. (13.67). After recording the spectra, the evaluation of the related temperature is performed via a best-fit library method by comparing the experimentally measured spectra with theoretical calculated spectra, for example by a least-squares fit. Further information about the basic principles of the measurement technique are given, for example, in Refs [13.147, 13.219, 13.220].

A large number of papers, reviews and contributions to books have been presented on the use of the CARS technique in laboratory-scale set-ups. This technique has also been used successfully in practical applications, such as in piston engines [13.221, 13.222], jet engines [13.223], coal gasifiers, coal-fired MHD generators and furnaces; Eckbreth [13.147] gives a detailed list of publications up to 1988. In applications which are nearer to solid fuel ramjet applications, CARS spectroscopic measurements have been performed, for example, in the exhaust plume of a liquid-fueled rocket engine by Williams *et al.* [13.224], in a liquid hydrogen-fueled model combustor [13.225] and in propellant flames by Aron and Harris [13.226], Kurtz *et al.* [13.227] and Stufflebeam and Eckbreth [13.228].

The CARS system available at the DLR Space Propulsion Institute at Lampoldshausen, which has been used for measurements using the RFSC [13.156, 13.229], consists of a frequency-doubled Nd:YAG laser operating at 10 Hz repetition rate with a pulse width of 8 ns, where 70 mJ from the 130-mJ pulse energy are split off and taken to pump the Stokes laser, which is a conventional broadband dye laser with a longitudinally pumped amplifier stage. With Sulforhodamine 101 dissolved in methanol, an emission maximum at 607 nm was achieved. The second YAG beam with the remaining 60 mJ and the dye laser beam form a USED-CARS (Unstable-resonator Spatially Enhanced Detection) beam geometry with a dye laser beam with a ring-shaped cross-section, which can be seen in Fig. 13.54. This geometry was chosen, because losses in beam overlap by beam-steering effects can be minimized.

The CARS system is transportable in modular units and was situated in a reinforced room near the test cell. Only beam-guiding optics, the focusing lens, which was mounted on a movable table for traversing the probe volume to the measuring locations, and the optics for fiber coupling were located close to the combustion chamber. Thereby the laser beams were sent over a distance of about 10 m to the combustion chamber and focused into the probe volume, which was located at the central plane of the combustor. The length of the probe volume was 4 mm with a pulse energy of 14 mJ from the YAG laser and 4 mJ from the dye laser measured at the focusing lens. The CARS signal was split off the laser beams by a series of mirrors and filters and finally focused into a 600-μm single fiber and guided to a spectrograph. The fiber output was matched to the spectrograph's aperture. For the set-up described, minor changes were made, losing resolution in favor of intensity. The CARS spectra were sequentially recorded using an intensified 1024 diode array operating in gated mode to suppress continuum radiation from the

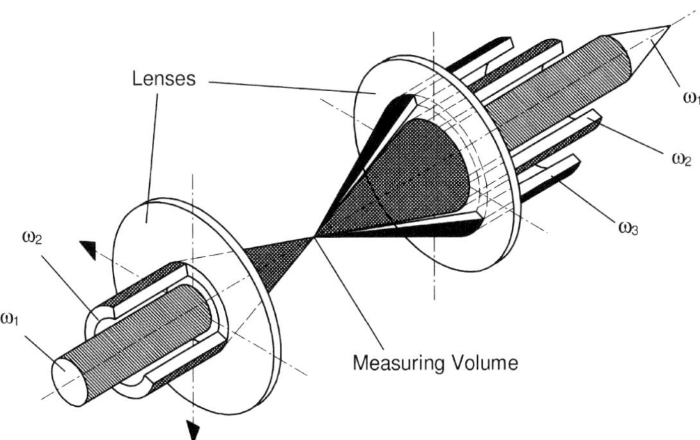

Figure 13.54. Beam geometry of USED-CARS.

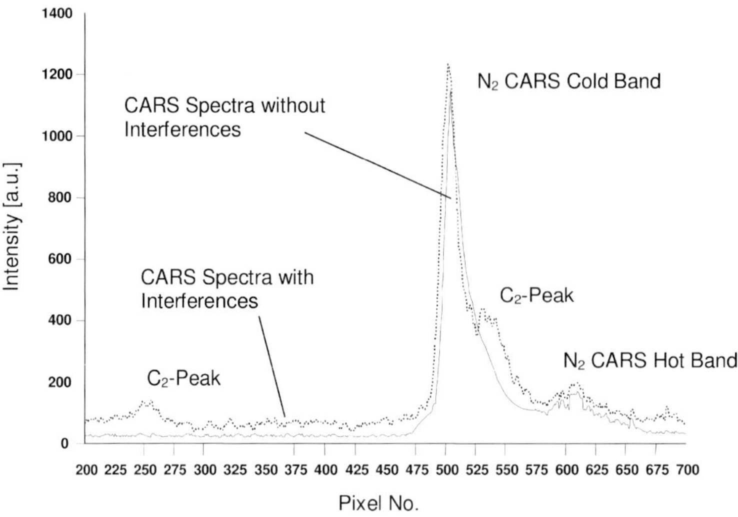

Figure 13.55. CARS spectra at different heights above the combustor bottom [13.229].

soot and other particles. Further information about the CARS apparatus, the measuring procedure, etc., can be found elsewhere [13.156, 13.229].

Figure 13.55 presents two nitrogen CARS spectra, which were obtained in the region downstream of the recirculation zone. The abscissa shows the pixel numbers of the diode array of the detector, which is proportional to the Raman shift. The full line displays a normal spectrum from a measuring position high above the combustion chamber bottom. In the fuel-rich zone beyond the band of high thermal emission, which can be seen on the Color Schlieren image in Fig. 13.47, a large number of disturbed spectra were obtained. The dotted line in Fig. 13.55 represents a spectrum at the measuring position at x = 170 mm behind the rearward-facing step and at y = 4 mm above the chamber bottom. Here the N_2 CARS spectra are superimposed by interferences from soot and soot precursors, which were present in this area. As has been reported [13.229], these interferences consist of incoherent anti-Stokes fluorescences excited by the broadband Stokes laser and coherent emission from laser-produced C_2 which degrades to the UV region. The coherent emission originates in an electronically enhanced wave-mixing process from the C_2. The incoherent part due to the Stokes laser is in a spectral region far away from the signal, but the coherent part interferes with the cold band of N_2 CARS from the 532-nm CARS laser. These emission interferences occur mainly in and beyond the band of high thermal emission and in the area of the recirculation zone. Since the hot band of the CARS signal was not involved, temperatures could be calculated from this hot band only with low accuracy. C_2 emission interferences have often been reported, for example by Aldén *et al.* [13.230] in sooting flames and by Eckbreth and Hall [13.220] and by Williams *et al.* [13.224] in the exhaust plume of a rocket engine.

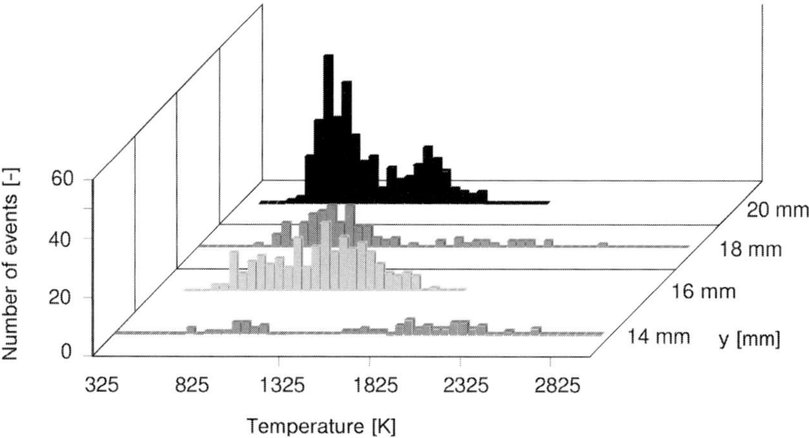

Figure 13.56. Temperature histograms at the measuring cross section at $x = 170$ mm [13.229].

Figure 13.56 presents temperature histograms obtained at the measuring cross-section $x = 170$ mm. Bimodal temperature distributions can occur owing to the turbulent mixing processes, which can be seen as large-scale coherent vortex structures in Fig. 13.47, which lead 'colder' air to lower regions. Special software has been adapted to calculate temperatures from such spectra [13.229]. Tolerating a loss of accuracy, this software is based on the assumption that only two temperatures, one very high and the other rather low, form the spectral shape of the CARS signal. The open symbols in Fig. 13.57 represent the upper and lower average temperatures in a histogram and the full symbols the temperature which is averaged over the whole histogram. It can be seen that even at the lowest measuring position at $y =$ 4 mm temperatures of ~800 K occur, which is similar to the temperature of the outer air flow. A detailed presentation of the results obtained has been published [13.229].

Owing to the above-mentioned C_2 interferences and the low signal-to-noise ratios caused by the high particle number density (boron particles, soot, etc.), especially near the combustor bottom and in the recirculation zone, only a small number of the acquired spectra could be evaluated. In order to obtain reliable results with low errors, a large number of spectra have to be acquired, which may lead to further difficulties owing to short run times of the experiments, regression of the fuel surface, ignition by the high energy density in the measuring volume and so forth. It must be mentioned that the use of the CARS technique under these difficult conditions needs a lot of time and experience to obtain reliable results. Nevertheless, the results obtained provide interesting insights that are very useful for the understanding of the combustion process.

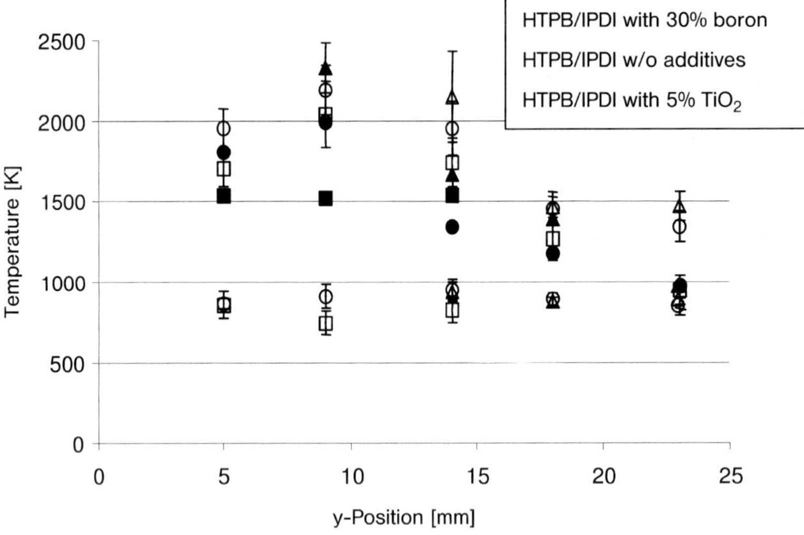

Figure 13.57. Gas-phase temperatures of HTPB/IPDI with different particle additives [13.229].

13.6 References

13.1 Zarko VE, Gusachenko LK, Rychkov AD (1996) Simulation of combustion of melting energetic materials, *Def. Sci. J.* **46**, 425–433.

13.2 Gusachenko LK, Zarko VE, Rychkov AD (1997) Instability of a combustion model with evaporation on a surface and overheat in the condensed phase, *Combust. Explos. Shock Waves* **33**, 320.

13.3 Hermann M (1997) Temperaturverhalten von Ammoniumnitrat, *PhD Thesis*, Wissenschaftliche Schriftenreihe des ICT, p. 15.

13.4 Herrmann M, Engel W, Eisenreich N (1992) Thermal expansion, transitions, sensitivity and burning rates of HMX, *Propell. Explos. Pyrotech.* **17**, 190–195.

13.5 Saw CK (2002) Kinetics of HMX and phase transitions: effects of grain size at elevated temperature, In: *Proc 12th Int. Detonation Symp., San Diego, CA.*

13.6 Smilowitz L, Henson B, Asay B, Dickson P (2002) A model of the β–δ phase transition in PBX9501, In: *Proc 12th Int. Detonation Symp., San Diego, CA.*

13.7 Carslaw HS, Jaeger JC (1973) *Conduction of Heat in Solids*, 2nd edn, Clarendon Press, Oxford.

13.8 Schmalzried H (1981) *Solid State Reactions*, Verlag Chemie, Weinheim.

13.9 Opfermann J, Hädrich W (1995) Prediction of the thermal response of hazardous materials during storage using an improved technique, *Thermochim. Acta* **263**, 29–50.

13.10 Sestak J, Berggren G (1971) Study of kinetics of the mechanism of the solid-state reactions at increasing temperature, *Thermochim. Acta* **3**, 1–12.

13.11 Jacobs PWM, Whitehead HM (1969) Decomposition and combustion of ammonium perchlorate, *Chem. Rev.* **69**, 551–590.

13.12 Fizer E, Fritz W (1982) *Technische Chemie*, Springer, Berlin.

13.13 Eisenreich N, Schmid H (1989) Thermoanalytische Untersuchung der Oxidation von Bor, *FhG-Ber.* **2**, 37–40.

13.14 Williams FA (1985) *Combustion Theory*,

2nd edn, Benjamin/Cummings, Menlo Park, CA.

13.15 Kuo KK (1996) *Recent Advances in Spray Combustion, Vol. I: Spray Atomization and Drop Burning Phenomena*, Progress in Astronautics and Aeronautics, Vol. 166, AIAA, New York.

13.16 Kuo KK (1996) *Recent Advances in Spray Combustion, Vol. II: Spray Combustion Measurements and Model Simulation*, Progress in Astronautics and Aeronautics, Volume 171, AIAA, New York.

13.17 Peters N (1999) *Turbulent Combustion*, Cambridge University Press, Cambridge.

13.18 Moriue O, Eigenbrod C, Rath HJ, Sato J, Okai K, Tsue M, Kono M (2000) Effects of dilution by aromatic hydrocarbons on staged ignition behavior of n-decane droplets, *Proc. Combust. Inst.* **28**, 969–975.

13.19 Schnaubelt S, Moriue O, Coordes T, Eigenbrod C, Rath HJ (2000) Detailed numerical simulations of the multistage self-ignition process of n-heptane isolated droplets and their verification by comparison with microgravity experiments, *Proc. Combust. Inst.* **28**, 953–960.

13.20 Marberry M, Ray AK, Leuny K (1981) *Combust. Flame* **57**, 237.

13.21 Barrere M, Moutet H (1956) Etude experimental de la combustion de gouttes de monergol, *Tech. Aero.* **9**, 31.

13.22 Eisenreich N, Krause H (1989) Verbrennung von Flüssigkeiten: Bestimmung der Verdampfungs- und Zeretzungskinetik durch schnelle FTIR-Spektroskopie, *FhG-Ber.* **2**, 32–36.

13.23 Krause H, Eisenreich N, Pfeil A (1989) Kinetics of evaporation and decomposition of isopropyl nitrate by rapid scan IR spectroscopy, *Thermochim. Acta* **149**, 349–356.

13.24 Kubota N (2000) *Propellants and Explosives*, Wiley-VCH, Weinheim.

13.25 Summerfield M, Sutherland GS, Webb WJ, Taback HJ, Hall KP (1960) The burning mechanism of ammonium perchlorate propellants, In: *ARS Progress in Astronautics and Rocketry, Vol. 1, Solid Propellant Rocket Research*, Academic Press, New York, pp. 141–182.

13.26 Miller RR, Hartmann KO, Myers RB (1970) *Prediction of Ammonium Perchlorate Particle Size Effects of Composite Propellant Burning Rate*, CPIA Pub. 196, pp. 567–591.

13.27 Hermance CE (1966) A model of composite propellant combustion including surface heterogeneity and heat generation, *AIAA J.* **4**, 1629–1637.

13.28 Hermance CE (1967) A detailed model of the combustion composite solid propellants, In: *Proc 2nd Solid Propulsion Conf., Anaheim, CA.*

13.29 Beckstead MW, Derr RL, Price CF (1970) A model of composite solid propellant combustion based on multiple flames, *AIAA J.* **8**, 2200–2207.

13.30 Glick RL (1974) On statistical analysis of composite solid propellant combustion, *AIAA J.* **12**, 384–385.

13.31 Blomshield FS (1989) *Nitramine Composite Solid Propellant Modelling*, Naval Weapon Center, China Lake, CA.

13.32 Eisenreich N, Fischer TS, Langer G, Kelzenberg S, Weiser V (2002) Burn rate models for gun propellants, *Propell. Explos. Pyrotech.* **27**, 142–149.

13.33 Ivanov GV, Tepper F (1996) Activated aluminum as a stored energy source for propellants, In: *Proc 4th Int. Symp. Special Topics in Chemical Propulsion, Challenges in Propellants and Combustion 100 Years after Nobel, Stockholm.*

13.34 Weiser V, Eisenreich N, Kelzenberg S (2001) Einfluß der Größe von Metallpartikeln auf die Anzündung und Verbrennung von Energetischen Materialien, In: *Proc. 31st Int. Annual Conf. ICT, Karlsruhe*, p. 34.

13.35 Mench MM, Yeh CL, Kuo KK (1998) Propellant burning rate enhancement and thermal behavior of ultra-fine aluminum powders (ALEX), In: *Proc. 29th Int. Annual Conf. ICT, Karlsruhe*, p. 30.

13.36 Simonenko VN, Zarko VE (1999) Comparative studying the combustion behavior of fine aluminum, In: *Proc. 30th Int. Annual Conf. ICT, Karlsruhe*, p. 21.

13.37 Arkhipov VA, Ivanov GV, Korotkikh AG, Medvedev VV, Surkov VG (2000) Features of ignition and burning of composite propellants with nanosized aluminum powder, In: *Proc. Burning and Gas Dynamics of Dispersion System*,

Mat. 3rd Int. School Inter Chamber Processes, St Petersburg, pp. 80–81.

13.38 Dokhan G, Price EW, Sigman RK, Seitzman JM (2001) The effect of Al particle size on the burning rate and residual oxide in aluminized propellants, In: *AIAA/SAE/ASME/ASEE 37th Joint Propulsion Conf.*, AIAA Paper 2001–3581,

13.39 Bashung B, Grune D, Licht HH, Samirant M (2000) Combustion phenomena of a solid propellant based on aluminum powder, In: *Proc. 5th Int. Symp. Special Topics in Chemical Propulsion (5-ISICP), Stresa.*

13.40 Lessard P, Beaupré F, Brousseau B (2001) Burn rate studies of composite propellants containing ultra-fine metals, In: *Proc. 32nd Int. Annual Conf. ICT, Karlsruhe*, p. 88.

13.41 Vorozhtsov A, Arkhipov V, Bondarchuk S, Kuznetsov V, Korotkikh A, Surkov V (2002) Ignition and combustion of solid propellants containing ultrafine aluminum, In: *Proc. Rocket Propulsion: Present and Future, Pozzuoli, Naples*, pp. 78–79.

13.42 Law CK (1973) Surface reaction model for metal particle combustion, *Combust. Sci. Technol.* **7**, 197–212.

13.43 Weiser V, Roth E, Plitzko Y, Poller S (2002) Experimentelle Untersuchung aluminisierter Kompositetreibstoffe mit nanoPartikel, In: *Proc. 32nd Int. Annual Conf. ICT, Karlsruhe*, p. 122.

13.44 King MK (1974) Boron particle ignition in hot gas streams, *Combust. Sci. Technol.* **8**, 255–273.

13.45 Natan B, Gany A (1987) Ignition and combustion characteristics of individual boron particles in the flow field of a solid fuel ramjet, In: *AIAA/SAE/ASME/ASEE 23rd Joint Propulsion Conf., San Diego, CA*, AIAA Paper 87–2034.

13.46 Yeh CL, Kuo KK (1996) Ignition and combustion of boron particles, *Prog. Energy Combust. Sci.* **22**, 511–541.

13.47 Eisenreich N, Liehmann W (1987) Emission spectroscopy of boron ignition and combustion in the range of 0.2 μm to 5.5 μm, *Propell. Explos. Pyrotech.* **12**, 88–91.

13.48 Eisenreich N, Krause HH, Pfeil A, Menke K (1992) Burning behaviour of gas generators with high boron content, *Propell. Explos. Pyrotech.* **17**, 161–163.

13.49 King MK (1974) Boron particle ignition in hot gas streams, *Combust. Sci. Technol.* **8**, 255–273.

13.50 Meinköhn D (2000) Metal combustion modelling, In: *Proc. 2nd ONERA-DLR Aerospace Symp.*, ODAS, Berlin.

13.51 Kuznetsov VT, Marusinl VP, Skorik AI (1974) On a mechanism of ignition in heterogeneous systems, *Combust. Explos. Shock Waves* **10**, 526–529.

13.52 McSpadden HJ (1985) Improvement in particle size and moisture analysis methods for lead nitrate, In: *Proc. 16th Int. Annual Conf. ICT, Karlsruhe*, p. 74.

13.53 Köhler H (1986) Qualitätskontrolle von Formkörpern aus Anzündmischungen auf Basis Bor/Kaliumnitrat, In: *Proc. 17th Int. Annual Conf. ICT, Karlsruhe*, p. 74.

13.54 Baier A, Weiser V, Eisenreich N, Halbrock A (1996) IR-Emissionsspektroskopie bei Verbrennungsvorgängen von Treibstoffen und Anzündmitteln, In: *Proc. 27th Int. Annual Conf. ICT, Karlsruhe*, p. 84.

13.55 Eckl W, Kelzenberg S, Weiser V, Eisenreich N (1998) Einfache Modelle der Anzündung von Festreibstoffen, In: *Proc. 29th Int. Annual Conf. ICT, Karlsruhe*, p. 154.

13.56 Rochat E, Berger B (1999) Unempfindliche Anzündmittel für moderne Treibladungspulver, In: *Proc. 30th Int. Annual Conf. ICT, Karlsruhe*, p. 24.

13.57 Eisenreich N, Ehrhard W, Kelzenberg S, Koleczko A, Schmid H (2000) Strahlungsbeeinflussung der Anzündung und Verbrennung von festen Treibstoffen, In: *Proc. 31st Int. Annual Conf. ICT, Karlsruhe*, p. 139.

13.58 Kosanke KL, Kosanke BJ, Dujay RC (2000) Pyrotechnic particle morphologies – metal fuels, *J. Pyrotech.* 11.

13.59 Berger B, Haas B, Reinhard G (1996) Einfluß der Korngröße des Reduktionsmittels auf die Reaktionsparameter pyrotechnischer Systeme, In: *Proc. 30th Int. Annual Conf. ICT, Karlsruhe*, p. 13.

13.60 Weiser V, Kuhn D, Ludwig R, Poth H (1998) Einfluß der Partikelgröße auf das Abbrandverhalten von B/KNO_3-Anzündmischungen, In: *Proc. 29th Int. Annual Conf. ICT, Karlsruhe*, p. 75.

13.61 Weiser V, Koleczko A, Kelzenberg S,

Eisenreich N, Müller D (2000) Ti-Nanopartikel zur Anzündung von Rohrwaffentreibmitteln, In: *Proc. 31st Int. Annual Conf. ICT, Karlsruhe*, p. 146.

13.62 Weiser V, Kelzenberg S, Eisenreich N (2001) Influence of metal particle size on the ignition of energetic materials, *Propell. Explos. Pyrotech.* 26, 284–289.

13.63 Weiser V, Eisenreich N, Kelzenberg S (2001) Einfluß der Größe von Metallpartikeln auf die Anzündung und Verbrennung von Energetischen Materialien, In: *Proc. 32nd Int. Annual Conf. ICT, Karlsruhe*, p. 34.

13.64 Simonenko VN, Zarko VE, Kiskin AB, Sedoi VS, Birukov YA (2001) Effect of ALEX and boron additives on ignition and combustion of Al-KNO_3 mixture, In: *Proc. 32nd Int. Annual Conf. ICT, Karlsruhe*, p. 122.

13.65 Koch EC (2002) Metal-fluorocarbon-pyrolants: III. Development and Application of magnesium/Teflon/Viton (MTV), *Propell. Explos. Pyrotech.* **27**, 262–266.

13.66 Kuwahara T, Ochiachi T (1992) Burning rate of Mg/TF-pyrolants, In: *Proc. 18th Int. Pyrotech Seminar, Breckenridge, CO*, pp. 539–549.

13.67 Campell AW, Davis WC, Ramsay JB, Tarver JR (1961) Shock initiation of solid explosives, *Phys. Fluids* **4**, 511–521.

13.68 Campell AW, Davis WC, Tarver JR (1961) Shock initiation of detonation in liquid explosives, *Phys. Fluids* **4**, 498–510.

13.69 Lindstrom LE (1970) Planar shock initiation of porous tetryl, *J. Appl. Phys.* **41**, 337.

13.70 Boyle V, Howe P, Ervin L (1972) Effect of heterogeneity on the sensitivity of high explosives to shock loading, In: *Proc. 14th Annual Explosives Safety Seminar, New Orleans, LA*, p. 841.

13.71 Taylor B, Ervin L (1976) Separation of ignition and buildup to detonation in pressed TNT, In: *Proc. 6th Int. Symp. Detonation, Coronada, CA*, pp. 3–10.

13.72 van der Steen A, Meulenbrugge JJ (1986) Effect of RDX particle shape and size on the shock sensitivity and mechanical properties of PBXs, In. *Proc. 21st Int. Annual Conf. ICT, Karlsruhe*, p. 11.

13.73 Simpson RL, Helm FH, Crawford PC, Kury JW (1989) Particle size effects in the initiation of explosives containing reactive and non-reactive phases, In: *Proc. 9th Int. Symp. Detonation, Portland, OR*, pp. 25–38.

13.74 Bernecker RR, Simpson RL (1997) Further observations on HMX particle size and buildup to detonation, In: *Proc. APS Shock Topical Conf., Boston, MA*.

13.75 Tulis AJ, Austing JL, Dihu RJ, Joyce RP (1999) Phenomenological aspects of detonation in non-ideal heterogeneous explosives, In: *Proc. 30th Int. Annual Conf. ICT, Karlsruhe*, p. 2.

13.76 Moulard H (1989) Particular aspects of the explosives particle size effects on the sensitivity of cast PBX formulations, In: *Proc. 9th Int. Symp. Detonation, Portland, OR*, pp. 18–24.

13.77 Tulis AJ (1986) The influence of particle size on energetic formulations, In: *Proc. 17th Int. Annual Conf. ICT, Karlsruhe*, p. 40.

13.78 Tulis AJ, Sumida WK, Dillon J, Comeyne W, Heberlein DC (1998) Submicron aluminum particle size influence on detonation of dispersed fuel-oxidizer powders, *Arch. Combust.* **18**, 157–164.

13.79 Gifford MJ, Chakravarty A, Greenaway M, Watson S, Proud W, Field J (2001) Unconventional properties of ultrafine energetic materials, In: *Proc. 32nd Int. Annual Conf. ICT, Karlsruhe*, p. 100.

13.80 Baer MR, Nunziato JW (1989) Compressive combustion of granular materials induced by low velocity impact, In: *Proc. 9th Int. Symp. Detonation, Portland, OR*, pp. 293–305.

13.81 Leiber CO (2001) Physical model for explosion phenomena – physical substantiation of Kamlet's complaint, *Propell. Explos. Pyrotech.* **26**, 302–310.

13.82 Langer G, Eisenreich N (1999) Hot spots in energetic materials, *Propell. Explos. Pyrotech.* 24, 113–118.

13.83 Coffee CS (1997) Initiation by shock or impact, the effect of particle size, In: *Physics of Explosives, Technical Exchange Meeting, Berchtesgaden*, pp. 45–59.

13.84 Miller PJ, Bedford CD, Davis JJ (1998) Effect of metal particle size on the detonation properties of various metallized explosives, In: *Proc. 11th Int. Symp. Detonation, Snowmass, CO.*

13.85 Lefrançois A, Le Gallic C (2001) Expertise of nanometric aluminum powder on the detonation efficiency of explosives, In: *Proc. 32nd Int. Annual Conf. ICT, Karlsruhe,* p. 36.

13.86 Brousseau P, Cliff MD (2001) The effect of ultrafine aluminum powder on the detonation properties of various explosives, In: *Proc. 32nd Int. Annual Conf. ICT, Karlsruhe,* p. 37.

13.87 Ritter H, Braun S (2001) High explosives containing ultrafine aluminum ALEX, *Propell. Explos. Pyrotech.* **26**, 311–314.

13.88 Danen WC, Jorgensen BS, Busse JR, Ferris MJ, Smith BL (2001) Los Alamos Nanoenergetic Metastable Intermolecular Composite (Super Thermite) Program, In: *Proc. 221st ACS National Meeting, San Diego, CA.*

13.89 Ilyin AP, Proskurovskaya LT (1990) Two stage combustion of ultradispersed aluminum powder in air, *Combust. Explos. Shock Waves* **26**, 71–72.

13.90 Beaudin G, Lefrançois A, Bergues D, Bigot J, Champion Y (1998) Combustion of nanophase aluminum in the detonation products of nitromethane, In: *Proc. 11th Int. Symp. Detonation, Snowmass, CO.*

13.91 Jones DEG, Turcotte R, Fouchard RC, Kwok QSM, Turcotte A-M, Abdel-Qader Z (2003) Hazard characterization of aluminum nanopowder compositions, *Propell. Explos. Pyrotech.* **28**, 120–131.

13.92 Moulard H (1989) Particular aspects of the explosive particle size effect on shock sensitivity of cast PBX formulations, In: *Proc. 9th Int. Symp. Detonation, Portland, OR,* pp. 18–24.

13.93 Kury JW, Hornig HC, Lee EL, McDonnel JL, Ornellas DL, Finger M, Strange FM, Wilkins ML (1965) Metal acceleration by chemical explosives, In: *Proc. 4th Int. Symp. Detonation,* Naval Ordnance Laboratory, ACR-126, pp. 3–13.

13.94 Pagoria PF Synthesis and characterization of 2,6-diamino-3,5-dinitropyrozine-1-oxide, *Propell. Explos. Pyrotech.*

13.95 Goosman DR (1979) *Measuring Velocities by Laser Doppler Interferometry,* Lawrence Livermore National Laboratory Report UCRL-5200–7-3, Lawrence Livermore National Laboratory, Livermore, CA, pp. 17–24.

13.96 Tipton R (1988) *Modeling Flux Compression Generators with a 2-D ALE Code,* Lawrence Livermore National Laboratory Report UCR-99900, Lawrence Livermore National Laboratory, Livermore, CA.

13.97 Fried L (1997) *Improved Detonation Modeling with CHEETAH,* Lawrence Livermore National Laboratory Report UCR-5200-9-11, Lawrence Livermore National Laboratory, Livermore, CA, pp. 21–23.

13.98 Sutherland GT, Forbes JW, Lemar ER, Ashwell KD, Baker RN (1994) Multiple stress-time profiles in a RDX/AP//Al/HTPB plastic bonded explosive, *High-Pressure Science and Technology,* AIP Conf. Proc. 309, New York, pp. 1381–1384.

13.99 Forbes JW, Tarver CM, Urtiew PA, Garcia F (2001) The effects of confinement and temperature on the shock sensitivity of solid explosives, In: *Proc. 11th Int. Symp. Detonation, Snowmass, CO,* pp. 145–152.

13.100 Sheffield SA, Gustavsen RL, Hill LG, Alcon RR (2001) Electromagnetic gauge measurements of shock initiating PBX9501 and PBX9502 explosives, In: *Proc. 11th Int. Symp. Detonation, Snowmass, CO,* pp. 451–458.

13.101 Tarver CM, Hallquist JO, Erickson LM (1985) Modeling short pulse duration shock initiation of solid explosives, In: *Proc. 8th Int. Symp. Detonation, Albuquerque, NM,* Naval Surface Weapons Center, NSWC MP 86–194, pp. 951–959.

13.102 Urtiew PA, Erickson LM, Aldis DF, Tarver CM (1989) Shock inititation of LX-17 as a function of its initial temperature, In: *Proc. 9th Int. Symp. Detonation, Portland, OR,* Office of the Chief of Naval Research, OCNR, pp. 112–122.

13.103 Bahl K, Bloom G, Erickson L, Lee R, Tarver C, von Holle W, Weingart R (1985) Initiation studies on LX-17 explosive, In: *Proc. 8th Int. Symp. Detonation, Albuquerque, NM,* Naval Surface Weapons Center, NSWC MP 86–194, pp. 1045–1056.

13.104 Urtiew PA, Cook TM, Maienschein JL, Tarver CM (1993) Shock sensitivity of IHE at elevated temperatures, In: *Proc. 10th Int. Symp. Detonation, Boston, MA*, Office of Naval Research, ONR 33 395–12, pp. 139–147.

13.105 Urtiew PA, Tarver CM, Maienschein JL, Tao WC (1996) Effect of confinement and thermal cycling on the shock initiation of LX-17, *Combust. Flame* **105**, 43–53.

13.106 Tarver CM (1990) Modeling shock initiation and detonation divergence tests on TATB-based explosives, *Propell. Explos. Pyrotech.* **15**, 132–142.

13.107 Urtiew PA, Tarver CM, Forbes JW, Garcia F (1998) Shock sensitivity of LX-04 at elevated temperatures, In: *Shock Compression of Condensed Matter*, AIP Conf. Proc. 429, Woodbury, NY, pp. 727–730.

13.108 Dremin AN, Pershin SV, Pogorelov VF (1965) Structure of shock waves in KCl and KBr under dynamic compression to 200,000 atmospheres, *Combust. Explos. Shock Waves* **1**, 1–4.

13.109 Edward DJ, Erkman JO, Jacobs SJ (1970) *Electromagnetic Velocity Gauge and Applications to Measure Particle Velocity in PMMA*, Naval Ordnance Laboratory Report, NOLTR-70-79.

13.110 Hayes B, Fritz JN (1970) Measurement of mass motion in detonation products by an axially-symmetric electromagnetic technique, In: *Proc. 5th Int. Symp. Detonation*, Office of Naval Research, pp. 447–464.

13.111 Urtiew PA, Erickson LM, Hayes B, Parker NL (1986) Pressure and particle velocity measurements in solids subjected to dynamic loading, *Combust. Explos. Shock Waves* **22**, 597–614.

13.112 Bridgeman PW (1950) Bakerian Lecture. Physics above 20,000 kg/cm^2, *Proc. R. Soc. London, Ser. A* **203**, 1–16.

13.113 Bernstein D, Keough DD (1964) Piezoresistivity of manganin, *J. Appl. Phys.* **35**, 1471–1474.

13.114 Wackerle J, Johnson JO, Halleck PM (1976) Shock initiation of high density PETN, In: *Proc. 6th Int. Symp. Detonation*, Office of Naval Research, pp. 20–28.

13.115 Kanel GI, Dremin AN (1977) Decomposition of cast TNT in shock waves, *Fiz. Goreniya Vzryva* **13**, 85–92.

13.116 Burrows K, Chivers DK, Gyton R, Lambourn BD, Wallace AA (1976) Determination of detonation pressure using a manganin wire technique, In: *Proc. 6th Int. Symp. Detonation*, Office of Naval Research, pp. 625–636.

13.117 Weingart RC, Barlett R, Cochran S, Erickson LM, Chan J, Janzen J, Lee R, Logan D, Rosenberg JT (1978) Manganin stress gauges in reacting high explosive enviromnent, In: *Proc. Symp. High Dynamic Pressures, Paris*, pp. 451–461.

13.118 Erickson L, Weingart R, Barlett R, Chan J, Elliott G, Janzen J, Vantine H, Lee R, Rosenberg JT (1979) Fabrication of manganin stress gauges for use in detonating high explosives, In: *Proc. 10th Symp. Explosives and Pyrotechnics, San Francisco, CA*, pp. 1–7.

13.119 Vantine H, Chan J, Erickson LM, Janzen J, Lee R, Weingart RC (1980) Precision stress measurements in severe shock-wave enviromnents with low impedance manganin gauges, *Rev. Sci. Instrum.* **51**, 116–122.

13.120 Urtiew PA,. Forbes JW, Tarver CM, Garcia F (2000) Calibration of manganin pressure gauges at 250 °C, In: *Shock Compression of Condensed Matter*, AIP Conf. Proc. 505, pp. 1019–1022.

13.121 Gupta YM (1983) Analysis of manganin and ytterbium gauge data under shock loading, *J. Appl. Phys.* **54**, 6094–6098.

13.122 Gupta S, Gupta YM (1987) Experimental measurements and analysis of the loading and unloading response of longitudinal and lateral gauges shocked to 90 kbar, *J. Appl. Phys.* **62**, 2603–2609.

13.123 Rosenberg Z, Partom Y (1985) Longitudinal dynamic stress measurements with in-material piezoresistive gauges, *J. Appl. Phys.* **58**, 1814–1818.

13.124 Vantine HC, Erickson LM, Janzen J (1980) Hysteresis-corrected calibration of manganin under shock loading, *J. Appl. Phys.* **51**, 1957–1962.

13.125 Gupta YM (1983) Stress measurements using piezoresistance gauges: modeling the gauge as an elastic inclusion, *J. Appl. Phys.* **54**, 6256–6266.

13.126 Tarver CM, Forbes JW, Urtiew PA, Garcia F (2000) Shock sensitivity of LX-04 at 150 °C, In: *Shock Compression of Condensed Matter*, AIP Conf. Proc. 505, pp. 891–894.

13.127 Charest JA, Keller DB, Rice DA (1972) *Carbon Gauge Calibration*, AFWL TR-7-207.

13.128 Lynch CS (1995) Strain compensated thin film stress gauges for stress measurements in the presence of lateral strain, *Rev. Sci. Instrum.* **66**, 5582–5589.

13.129 Ginsberg MJ, Asay B (1991) Commercial carbon composition resistors as dynamic stress gauges in difficult environments, *Rev. Sci. Instrum.* **62**, 2218–2227.

13.130 Wilson WH (1992) Experimental study of reaction and stress growth in projectile-impacted explosives, In: *Shock Compression of Condensed Matter*, North-Holland, Amsterdam, pp. 671–674.

13.131 Austing JL, Tulis AJ, Hrdina DJ, Baker DE (1991) Carbon resistor gauges for measuring shock and detonation pressures I. Principles of functioning and calibration, *Propell. Explos. Pyrotech.* **16**, 205–215.

13.132 Bauer F (2000) Advances in PVDF shock sensors: applications to polar materials and high explosives, In: *Shock Compression of Condensed Matter*, AIP Conf. Proc. 505, pp. 1023–1028.

13.133 Anderson MU, Graham RA (1996) The new simultaneous PVDF/VISAR measurement technique: applications to highly porous HMX, In: *Shock Compression of Condensed Matter*, AIP Conf. Proc. 370, pp. 1101–1104.

13.134 Bauer F, Moulard H, Graham RA (1996) Piezoelectric response of ferroelectric polymers under shock loading: nanosecond piezoelectric PVDF gauge, In: *Shock Compression of Condensed Matter*, AIP Conf. Proc. 370, pp. 1073–1076.

13.135 Charest JA, Lynch CS (1992) A simple approach to piezofilm stress gauges, In: *Shock Compression of Condensed Matter*, North-Holland, Amsterdam, pp. 897–900.

13.136 Kuo KK, Summerfield M (1984) Fundamentals of solid propellants combustion, In: *Progress in Astronautics and Aeronautics 90*, AIAA, New York, pp. 479–513.

13.137 Timnat YM (1990) Recent developments in ramjets, ducted rockets and scramjets, *Prog. Aerospace Sci.* 27, 201–235.

13.138 Gany A (1993) Combustion of boron-containing fuels in solid fuel ramjets, In: Kuo KK, Pein R (eds), *Combustion of Boron-based Solid Propellants and Solid Fuels*, CRC Press, Boca Raton, FL, pp. 91–112.

13.139 King MK (1993) A review of studies of boron ignition and combustion phenomena at Atlantic Research Corporation over the past decade, In: Kuo KK, Pein R (eds), *Combustion of Boron-based Solid Propellants and Solid Fuels*, CRC Press, Boca Raton, FL, pp. 1–80.

13.140 Shafirovich EYa, Shiryaev AA, Goldshleger UI. (1993) Magnesium and carbon dioxide: a rocket propellant for Mars missions, *J Propulsion Power* 9, 197–203.

13.141 Eckhoff RK (1997) *Dust Explosions in the Process Industries*, 2nd edn, Butterworth-Heinemann, Oxford.

13.142 Sarofim AF (1986) Radiative heat transfer in combustion: friend or foe, In: *Proc. 21st Int. Symp. Combustion*, pp. 1–23.

13.143 Mazurkiewicz J, Jarosinski J, Wolanski P (1993) Investigations of burning properties of cornstarch dust-air flame, *Arch. Combust.* **13**, No. 3–4.

13.144 Frieske H-J, Adomeit G, Ciezki H (1985) Das Chemische Stoßwellenrohr als untersuchendes Instrument auf dem Gebiet der Pyrolyse bei hohen Aufheizgeschwindigkeiten, In: *12. Deutscher Flammentag, Karlsruhe.*

13.145 Fabignon F, Marion P (1998) *3ème Colloque R&T Ecoulements Internes en Propulsion Solide, Châtillon*, CNES – Information Scientifique et Publications, Paris.

13.146 Meinköhn D, Sprengel H (1997) Thermo-hydrodynamics of thin surface films in heterogeneous combustion, *J. Eng. Math.* **31**, 235–257.

13.147 Eckbreth AC (1988) *Laser Diagnostics for Combustion Temperature and Species*, Abacus Press, Cambridge, MA.

13.148 Taylor AMKP (1993) *Instrumentation for Flows with Combustion*, Academic Press, London.

13.149 Boggs TL, Zinn BT (1978) Experimental diagnostics, in combustion of solids, In: *Progress in Astronautics and Aeronautics 63*, AIAA, New York.

13.150 Kuo KK, Parr TP (eds) (1994) *Non-Intrusive Combustion Diagnostics*, Begell House, New York.

13.151 Hetsroni G (ed.) (1981) Handbook of Multiphase Systems, Hemisphere, New York.

13.152 Hewitt GF (1981) Measurement techniques, In: Hetsroni G (ed.), *Handbook of Multiphase Systems*, Hemisphere, New York, pp. 10-3–10-8.

13.153 Hewitt GF (1978) *Measurement of Two Phase Flow Parameters*, Academic Press, New York.

13.154 Jones OC (1983) Two-phase flow measurement techniques in gas-liquid systems, In: Goldstein RJ (ed.), *Fluid Mechanics Measurements*, Hemisphere, New York.

13.155 Kehler P (1978) *Two Phase Flow Measurement by Pulsed Neutron Activation Technique*, Report ANL-NUREG-CT-78-17, Argonne National Laboratory, Argonne, IL.

13.156 Clauß W, Vereschagin K, Ciezki HK (1998) Determination of temperature distributions by CARS-thermometry in a planar solid fuel ramjet combustion chamber, In: *Proc. 36th Aerospace Science Meeting, Reno, NV*, AIAA-98-0160.

13.157 Teorell T (1931) Photometrische Messung der Konzentration und Dispersität in kolloidalen Lösungen III, *Kolloid-Z.* **54**, 150–156.

13.158 Lester TW, Wittig SLK (1975) Particle growth and concentration measurements in sooting homogeneous hydrocarbon combustion systems, In: *Proc. 10th Int. Symp. Shock Tubes and Waves*, pp. 632–647.

13.159 Lu YC, Freyman TM, Hernandez G, Kuo KK (1994) Measurement of temperatures and OH concentrations of solid propellant flames using absorption spectroscopy, In: *Proc. 30th AIAA Joint Propulsion Conf., Indianapolis, IN*, AIAA 94–3040.

13.160 Raffel M, Willert C, Kompenhans J (1998) *Particle Image Velocimetry*, Springer, Berlin.

13.161 Schardin H (1942) Die Schlierenverfahren und ihre Anwendungen, *Erg. Exakt. Naturwiss.* **20**, 303–439.

13.162 Briones RA, Wuerker RF (1978) Holography of solid propellant combustion, In: Boggs Th, Zinn BT (eds), *Experimental Diagnostics in Combustion of Solids*, Progress in Astronautics and Aeronautics 63, AIAA, New York, pp. 251–276.

13.163 Vander Wal RL (1998) Soot Precursor carbonization: visualization using LIF and LII and comparison using bright and dark field TEM, *Combust. Flame* **112**, 607–616.

13.164 Wooldridge MS (1998) Gas-phase synthesis of particles, *Prog. Energy Combust. Sci.* **24**, 63–87.

13.165 Becker HA, Code RK, Gogolek PEG, Poirier DJ, Antony EJ (1991) Detailed gas and solids measurements in a pilot scale AFBC with results on gas mixing and nitrous oxide formation, In: *Proc. Int. Conf. Fluidized Bed Combustion*, ASME, Miami, FL, pp. 91–98.

13.166 Becker HA, Code RK, Gogolek PEG, Poirier DJ (1991) *A Study of Fluidized Dynamics in Bubbling Fluidized Bed Combustion – Part of the Federal Panel of Energy R&D (PERD) Program, Final Report*, Technical Report QFBC.TR. 91.2, Queen's Fluidized Bed Combustion Laboratory, Queen's University of Kingston, Kingston, ON.

13.167 Chedaille J, Braud Y (1972) Industrial flames, 1. In: Beér JM, Thring MW (eds), *Measurements in Flames*, Edward Arnold, London.

13.168 Becker HA (1983) Physical probes, In: Taylor AMKP (ed.), *Instrumentation for Flows with Combustion*, Academic Press, London, pp. 53–112.

13.169 Danel F, Delhaye JM (1971) Sonde optique pour mesure du taux de présence local en ecoulement diphasique, *Mes. Regul. Autom.* 99–101.

13.170 Smeets G (1994) Doppler velocimetry

measurements using a phase stabilized Michelson spectrometer, In: Kuo KK, Parr TP (eds), *Non-Intrusive Combustion Diagnostics*, Begell House, New York, pp. 518–531.

13.171 Coppalle A, Joyeux D (1994) Temperature and soot volume fraction in turbulent diffusion flames: measurements on mean and fluctuating values, *Combust. Flame* **96**, 275–285.

13.172 Zinn BT (1977) Experimental diagnostics in gas phase combustion systems, In: *Progress in Astronautics and Aeronautics 53*, AIAA, New York.

13.173 Pein R, Ciezki HK, Eicke A (1995) Instrumental diagnostics of solid fuel ramjet combustor reaction products containing boron, In: *Proc. 31st AIAA Joint Propulsion Conf.*, AIAA-95-3107, San Diego, CA.

13.174 Besser HL, Strecker R (1993) Overview of boron ducted rocket development during the last two decades, In: Kuo KK, Pein R (eds), *Combustion of Boron-Based Solid Propellants and Solid Fuels*, CRC Press, Boca Raton, FL, pp. 133–178.

13.175 Abbott SW, Smoot LD, Schadow K (1972) *Direct Mixing and Combustion Measurements in Ducted Particle-Laden Jets*, AIAA-72-1177.

13.176 Schulte G, Pein R, Högl A (1987) Temperature and concentration measurements in a solid fuel ramjet combustion chamber, *J. Propulsion Power* 3, 114–120.

13.177 Girata PT, McGregor WK (1984) Particle sampling of solid rocket motor (SRM) exhausts in high altitude test cells, In: Roux JA, McCay TD (eds), *Spacecraft Contamination: Sources and Prevention*, Progress in Astronautics and Aeronautics 91, AIAA, New York, pp. 293–311.

13.178 Bilger RW (1977) Probe measurements in turbulent combustion, In: Zinn BT (ed.), *Experimental Diagnostics in Gas Phase Combustion Systems*, Progress in Astronautics and Aeronautics 53, AIAA, New York, pp. 49–69.

13.179 Lenze B (1979) Probeentnahme und Analyse von Flammengasen, *Chem. Ing. Tech.* **42**, 287–292.

13.180 Lieberman A (1981) Aerosol measurement and analysis, In: Hetsroni G (ed.), *Handbook of Multiphase Systems*, Hemisphere, New York, pp. 10–119–10–165.

13.181 Ciezki HK (1999) Determination of concentration distributions of gaseous and solid intermediate reaction products in a combustion chamber by a water cooled sampling probe, In: *Proc. Sensor 99*, Vol. I, Nürnberg, pp. 327–332.

13.182 Ciezki HK, Sender J, Clauß W, Feinauer A, Thumann A (in press) Combustion of solid fuel slabs containing boron particles in a step combustor, *J. Propulsion Power.*

13.183 Ciezki HK, Schwein B (1996) Investigation of gaseous and solid reaction products in a step combustor using a water-cooled sampling-probe, In: *Proc. 32nd Joint Propulsion Conf., Lake Buena Vista, FL*, AIAA-96-2768.

13.184 Sender J, Ciezki HK (1998) Velocities of reacting boron particles within a solid fuel ramjet combustion chamber, *Def. Sci. J.* **48**, 343–349.

13.185 Shapiro A (1953) *The Dynamics and Thermodynamics of Compressible Fluid Flow*, Vol. 1, Wiley, New York.

13.186 Truckenbrodt E (1989) *Fluidmechanik*, Vol. 1, 3rd edn, Springer, Heidelberg.

13.187 Pein R, Vinnemeier F (1992) Swirl and fuel composition effects on boron combustion in solid-fuel ramjets, *J. Propulsion Power* **8**, 609–614.

13.188 Nemodruk AA, Karalova ZK (1965) *Analytical Chemistry of Boron*, Academy of Sciences of the USSR, Series Analytical Chemistry of Elements, Israel Program for Scientific Translations, Jerusalem, distributed by Oldbourne Press, London, pp. 184–185.

13.189 Berens T (1992) Stoffumsetzung und -transportvorgänge bei der Verbrennung fester Brennstoffe für Staustrahlantriebe, *PhD Thesis*, VDI Fortschrittsberichte, Series 6, No. 270, VDI-Verlag, Düsseldorf.

13.190 Zarko VE, Kuo KK (1994) Critical review of methods for regression rate measurements of condensed phase systems, In: Kuo KK, Parr TP (eds), *Non-Intrusive Combustion Diagnostics*, Begell House, New York, pp. 600–623.

13.191 Gany A, Netzer DW (1985) Combustion studies of metallized fuels for solid fuel ramjets, In: *Proc. 21st Joint Propulsion Conf., Monterey, CA*, AIAA-85–1177.
13.192 Hensel C (1995) Spektroskopische Untersuchung des Abbrandverhaltens von Festbrennstoffplatten mit Metallzusatz in einer ebenen Stufenbrennkammer, *Diploma Thesis*, DLR Internal Report IB 645-95/5, DLR, Cologne.
13.193 Blanc A, Ciezki HK, Feinauer A, Liehmann W (1997) Investigation of the combustion behaviour of solid fuels with various contents of metal particles in a planar step combustor by IR-spectroscopic methods, In: *Proc. 33rd Joint Propulsion Conf., Seattle, WA*, AIAA-97-3233.
13.194 Ciezki HK, Hensel C, Liehmann W (1997) Spectroscopic investigation of the combustion behaviour of boron containing solid fuels in a planar step combustor, In: *Proc. 13th Int. Symp. Airbreathing Engines, ISABE, Chattanooga, TX*, pp. 582–590.
13.195 Klimov, V, Ciezki, HK, Gemelli, E (2003) Spectroscopic investigation of the multiphase diffusion flame of an aluminum particle laden hydrogen jet in a hot air environment, In: *Proc. 34th Int. Annual Conf. ICT, Karlsruhe*, p. 152.
13.196 Merzkirch W (1981) *Density Sensitive Flow Visualization*, Methods of Experimental Physics, Vol 18A, Academic Press, London.
13.197 Cords PH (1968) A high resolution, high sensitivity Colour Schlieren method, *SPJE J.* **6**, 85–88.
13.198 Settles GS (1980) Color Schlieren optics – a review of techniques and applications, In: Merzkirch W (ed.), *Flow Visualization*, Hemisphere, New York.
13.199 Ciezki H (1985) Entwicklung eines Farbschlierenverfahrens unter besonderer Berücksichtigung des Einsatzes an einem Stoßwellenrohr, *Diploma Thesis*, Technical University Aachen.
13.200 Ciezki HK (1999) Investigation of the combustion behaviour of solid fuel slabs in a planar step combustor with a Colour Schlieren technique, In: *Proc. 35th AIAA Joint Propulsion Conf., Los Angeles, CA*, AIAA-99-2813.
13.201 Netzer DW, Andrews JR (1978) Schlieren studies of solid-propellant combustion, In: Boggs Th, Zinn BT (eds), *Experimental Diagnostics in Combustion of Solids*, Progress in Astronautics and Aeronautics 63, AIAA, New York, pp. 235–250.
13.202 Natan B, Gany A (1991) Ignition and combustion of boron particles in the flowfield of a solid fuel ramjet, *J. Propulsion Power* 7, 37–43.
13.203 Yeh Y, Cummings HH (1964) Localized fluid flow measurements with a HeNe laser spectrometer, *Appl. Phys. Lett.* **4**, 174–178.
13.204 Heitor MV, Starner SH, Taylor AMKP, Whitelaw JH (1993) Velocity, size and turbulent flux measurements by laser-Doppler velocimetry, In: Taylor AMKP (ed.), *Instrumentation for Flows with Combustion*, Academic Press, London, pp. 113–250.
13.205 Leder A (1992) *Abgelöste Strömungen: Physikalische Grundlagen*, Vieweg, Braunschweig.
13.206 Ruck B (1990) *Lasermethoden in der Strömungsmesstechnik*, AT-Fachverlag, Stuttgart.
13.207 Adrian RJ (1988) Statistical properties of particle image velocimetry measurements in turbulent flow, In: *Laser Anemometry in Fluid Mechanics*, Vol. III, Ladoan-Instituto Tecnico, Lisbon, pp. 115–129.
13.208 Grant I, Smith GH (1988) Modern developments in particle image velocimetry, *Opt. Lasers Eng.* **9**, 245–264.
13.209 Adrian RJ (1991) Particle-imaging techniques for experimental fluid mechanics, *Annu. Rev. Fluid Mech.* **23**, 261–304.
13.210 Thumann A, Ciezki HK (2000) Comparison of PIV and Colour-Schlieren measurements of the combustion process of boron particle containing solid fuel slabs in a rearward facing step combustor, In: *Proc. 5th Int. Symp. Special Topics in Chem Propulsion (5-ISICP), Stresa.*
13.211 Yamamoto F, Uemura T, Koukawa M, Itoh M, Teranishi A (1988) Application of flow visualization and digital image

processing techniques to unsteady viscous diffusing free doublet flow, In: *Proc. 2nd Int. Symp. Fluid Control, Measurement, Mechanics and Flow Visualization*, pp. 184–188.

13.212 Jambunathan K, Ju XY, Dobbins BN, Ashforth-Frost S (1995) An improved cross correlation technique for particle image velocimetry, *Meas. Sci. Technol.* **6**, 754–768.

13.213 Durst F, Melling A, Whitelaw JH (1981) *Principles and Practice of Laser-Doppler Anemometry*, 2nd edn, Academic Press, New York.

13.214 Schmitt F, Ruck B (1986) Laserlichtschnittverfahren zur quantitativen Strömungsanalyse, *Laser Optoelektronik* **18**, 107–118.

13.215 Clauss W (1988) *Verbrennung von Borstaubwolken in wasserdampfhaltigem Heißgas*, DLR Internal Report, IB 643–88/6, DLR, Cologne.

13.216 Ulas A, Kuo KK, Gotzmer C (2000) Ignition and combustion of boron particles in fluorinated environments: experiment and theory, In: *Proc. 5th Int. Symp. Special Topics in Chemical Propulsion (5-ISICP), Stresa.*

13.217 Antcliff RR, Jarrett O (1984) Comparison of CARS combustion temperatures with standard techniques, In: McCay TD, Roux JA (eds), *Combustion Diagnostics by Nonintrusive Methods*, Progress in Astronautics and Aeronautics 92, AIAA, New York, pp. 45–57.

13.218 Thumann A (1997) Temperaturbestimmung mittels der Kohärenten-Anti-Stokes-Raman-Streuung (CARS) unter Berücksichtigung des Druckeinflusses und nichteinheitlicher Temperaturverhältnisse im Meßvolumen, *PhD Thesis*, Berichte zur Energie- und Verfahrenstechnik 97.4, ESYTEC, Erlangen.

13.219 Druet SAJ, Taran, J-PE (1981) CARS spectroscopy, *Prog. Quantum Electron.* **7**, 1–72.

13.220 Eckbreth AC, Hall RJ (1979) CARS thermometry in a sooting flame, *Combust. Flame* **36**, 87–98.

13.221 Kajiyama K, Sajiki K, Kataoka H, Maeda S, Hirose C (1982) *N_2 CARS Thermometry in Diesel Engine*, SAE Paper 82–1038, 3243–3251.

13.222 Lucht RP (1989) Temperature measurements by coherent anti-Stokes Raman scattering in internal combustion engines, In: Durao DFG (ed.), *Instrumentation of Combustion and Flow in Engines*, Kluwer, Dordrecht, pp. 341–353.

13.223 Switzer GL, Goss LP, Trump DD, Reeves CM, Stutrud JS, Bradley RP, Roquemore WM (1985) *CARS Measurements in the Near-Wake Region of an Axisymmetric Bluff-Body Combustor*, AIAA-85-1106.

13.224 Williams DR, McKeown D, Porter FM, Baker CA, Astill AG, Rawley KM (1993) Coherent anti-Stokes Raman spectroscopy (CARS) and laser-induced fluorescence (LIF) measurements in a rocket engine plume, *Combust. Flame* **94**, 77–90.

13.225 Smirnov VV, Clauß W, Oschwald M, Grisch F, Bouchardy P (2000) Theoretical and practical issues of CARS application to cryogenic spray combustion, In: *Proc. 4th Int. Symp. Liquid Space Propulsion, Heilbronn.*

13.226 Aron K, Harris LE (1984) CARS probe of RDX decomposition, *Chem. Phys. Lett.* **103**, 413–417.

13.227 Kurtz A, Brüggemann D, Giesen U, Heshe S (1994) Quantitative nitric oxide CARS spectroscopy in propellant flames, In: Kuo KK, Parr TP (eds), *Non-Intrusive Combustion Diagnostics*, Begell House, New York, pp. 160–166.

13.228 Stufflebeam JH, Eckbreth AC (1989) CARS diagnostics of solid propellant combustion at elevated pressure, *Combust. Sci. Technol.* **66**, 163–179.

13.229 Ciezki HK, Clauß W, Thumann A, Vereschagin K (1999) Untersuchung zur Temperaturverteilung mittels CARS-Thermometrie in einer Modellstaubrennkammer beim Abbrand von Festbrennstoffplatten, In: *Proc. 30th Int. Annual Conf. ICT, Karlsruhe*, p. 31.

13.230 Bengtsson P-E, Aldén M (1991) C_2 Production and excitation in sooting flames using visible laser radiation: implications for diagnostics in sooting flames, *Combust. Sci. Technol.* **77**, 307–318.

Index

t